ASPARTIC PROTEINASES

Retroviral and Cellular Enzymes

ADVANCES IN EXPERIMENTAL MEDICINE AND BIOLOGY

Recent Volumes in this Series

Volume 427
DIETARY FIBER IN HEALTH AND DISEASE
Edited by David Kritchevsky and Charles Bonfield

Volume 428
OXYGEN TRANSPORT TO TISSUE XIX
Edited by David K. Harrison and David T. Delpy

Volume 429
BRAIN PLASTICITY: Development and Aging
Edited by Guido Filogamo, Antonia Vernadakis, Fulvia Gremo, Alain M. Privat, and Paola S. Timiras

Volume 430
ANALYTICAL AND QUANTITATIVE CARDIOLOGY
Edited by Samuel Sideman and Rafael Beyar

Volume 431
PURINE AND PYRIMIDINE METABOLISM IN MAN IX
Edited by Andrea Griesmacher, Peter Chiba, and Mathias M. Müller

Volume 432
HYPERTENSION AND THE HEART
Edited by Alberto Zanchetti, Richard B. Devereux, Lennart Hansson, and Sergio Gorini

Volume 433
RECENT ADVANCES IN PROSTAGLANDIN, THROMBOXANE, AND LEUKOTRIENE RESEARCH
Edited by Helmut Sinzinger, Bengt Samuelsson, John R. Vane, Rodolfo Paoletti, Peter Ramwell, and Patrick Y-K Wong

Volume 434
PROCESS-INDUCED CHEMICAL CHANGES IN FOOD
Edited by Fereidoon Shahidi, Chi-Tang Ho, and Chuyen Van Nguyen

Volume 435
GLYCOIMMUNOLOGY 2
Edited by John S. Axford

Volume 436
ASPARTIC PROTEINASES: Retroviral and Cellular Enzymes
Edited by Michael N. G. James

ASPARTIC PROTEINASES

Retroviral and Cellular Enzymes

Edited by

Michael N. G. James

University of Alberta
Edmonton, Alberta, Canada

PLENUM PRESS • NEW YORK AND LONDON

Library of Congress Cataloging in Publication Data

Aspartic proteinases: retroviral and cellular enzymes / edited by Michael N. G. James.
p. cm.—(Advances in experimental medicine and biology; v. 436)
"Proceedings of the Seventh International Conference on Aspartic Proteinases, held October 22–27, 1996, in Banff, Alberta, Canada"—T.p. verso.
Includes bibliographical references and index.
ISBN 0-306-45809-8
1. Aspartic proteinases—Congresses. I. James, Michael N. G. II. International Conference on Aspartic Proteinases (7th: 1996: Banff, Alta.) III. Series.
[DNLM: 1. Aspartic Proteinases—metabolism—congresses. 2. Aspartic Proteinases—chemistry—congresses. 3. Retroviridae—enzymology—congresses. 4. Protease Inhibitors—congresses. W1 AD559 v. 436 1998 / QU 136 S926 1998]
QP609.A86S77 1998
572′.76—dc21
DNLM/DLC 97-52057
for Library of Congress CIP

Proceedings of the Seventh International Conference on Aspartic Proteinases, held October 22–27, 1996, in Banff, Alberta, Canada

ISBN 0-306-45809-8

A Division of Plenum Publishing Corporation
233 Spring Street, New York, N.Y. 10013

http://www.plenum.com

10 9 8 7 6 5 4 3 2 1

Printed in the United States of America

PREFACE

The VIIth International Conference on Aspartic Proteinases was held in Banff, Alberta, Canada, from October 22 to 27, 1996. The venue was the Banff Centre in the Canadian Rockies, a setting well known worldwide for the scenic beauty and mountain grandeur. It was perhaps presumptuous of the organizers to call this the seventh Aspartic Proteinase Conference but it was felt that the meeting in 1982, organized by Tom Blundell and John Kay, was of an international stature and covered topics sufficiently broad to constitute a conference. Thus, there is a discontinuity in that the Gifu Conference organized by Prof. Kenji Takahashi was the fifth International Conference on Aspartic Proteinases. Officially, there has not been a sixth Conference and if there is confusion, it is the result of my desire to recognize the importance of the London meeting.

Banff hosted 106 scientists from 14 different countries. There were 26 invited speakers among the 44 oral presentations of the 7 main sessions. In addition, there were 53 contributed poster presentations that spanned the whole range of interest in aspartic proteinases.

In planning this conference, we departed slightly from the traditional order of the topics. The retroviral aspartic proteinases constituted the first day of presentations and it was necessary to have two sessions to cover all of the areas of interest. Cathepsin D and cathepsin E dominated the presentations in the Mammalian Aspartic Proteinases session, an area that has traditionally been dominated by papers on the stomach enzymes: pepsin, chymosin and gastricsin. Session 4 was on aspartic proteinase zymogens and the process of conversion of these molecules to active enzymes. Although this session was relatively small, the ensuing discussion was heated and lively. The aspartic proteinases from *Candida*, *Aspergillus niger* and *Rhizomucor miehei* replaced the fungal enzymes, endothiapepsin, rhizopuspepsin and penicillopepsin in the Microbial Aspartic Proteinases session. The aspartic proteinases and inhibitors from parasites were discussed in a new session at this meeting. The most widely discussed member of this branch of the aspartic proteinase family was malarial plasmepsin from *Plasmodium falciparum*. The seventh and final session was devoted to plant aspartic proteinases, in particular cardosin from *Cynara cardunculus*.

John Kay and Ben Dunn were extremely helpful in the planning stages of this conference. Their wise counsel in selecting speakers was most appreciated. The members of my laboratory devoted a great deal of time and effort to the successful organization and running of the conference. Masao Fujinaga was ever vigilant in scanning bibliographic databases for new papers on aspartic proteinases. In addition to their scientific contributions, Marie Fraser, Katherine Bateman, Nina Khazanovich and Amir Khan made sure that the conference paraphernalia, tote bags, T-shirts, coffee mugs, pens, paper etc. were available and ready at the time of the conference.

I want to express a very special note of thanks to Mae Wylie for her dedication and untiring hard work in making this conference a success. She did so many things that made the conference a pleasure for all of the attendees. I am sure that everyone joins me in thanking Mae for her devotion to the meeting.

The VIIth International Conference on Aspartic Proteinases would not have happened without the very generous financial support from the many public and private organizations listed on the accompanying page of Acknowledgements. The Government of Alberta through the offices of Alberta Economic Development and Trade and the Alberta Heritage Foundation for Medical Research was joined in their donations by grants from the Medical Research Council of Canada, the University of Alberta Conference Fund, and from the Dean of the Faculty of Medicine. I am very grateful to the corporate sponsors for their generosity in the light of ever decreasing funding for such meetings.

The concluding session of the conference was a Panel Discussion that looked to the future and to where the important questions in aspartic proteinase research were directed. As in the past, there was an interesting mix of structural and functional questions that were raised. Whereas one of the main themes of previous conferences dealt with mechanism, the discussion on this topic seems to have reached a consensus. However, the structural details of the reasons for the different pH optima exhibited by various members of the family have not been settled.

Members of the panel and conference delegates felt that we are at the brink of discovery regarding the widespread occurrence of aspartic proteinases in nature. In the future genomics will certainly bring to light how important these molecules are. Can the sequence of an aspartic proteinase reveal its membership in the family and its specificity for substrates? In other words, have our predictive methods been sufficiently refined to recognize these features for the authentic aspartic proteinases as well as those that might have adopted other functions (e.g. the pregnancy associated glycoproteins)? A very important question concerns identifying the naturally occurring substrates for the highly specific aspartic proteinases. This, as well as defining substrate specificity, will occupy many and challenge the phage biologists and combinatorial chemists for some time to come. Post-translational modifications clearly play a role in intracellular targetting. Some of these details have been revealed for cathepsin D and renin. There are many other effects of insertions or deletions to the canonical sequences that remain to be elucidated. What are the multitude of other roles for which nature has adapted aspartic proteinases? Can we deduce these functions from gene knock-outs? The most recent and exciting was the regulation of enzymatic activity by dimerization in the retroviral aspartic proteinases.

Clearly, from the wide ranging discussion in that concluding session the aspartic proteinase field will be active for years to come. I look forward to the next international conference wherever and whenever it will be held.

Finally, the generosity of the public and private sponsors of our conference allowed us to present four travel awards to the authors of the four best posters (as judged by an impartial panel), two for postdoctoral fellows and two for graduate students. The winners were: Deborah Dauber (San Francisco, USA), Carlos Frazao (Oeiras, Portugal), Karen Girdwood (Cardiff, Wales) and Ilya Kashparov (Moscow, Russia). My sincere congratulations to these most deserving young scientists.

The Aspartic Proteinase Conferences now have a 20-year history, since the first one was organized by Jordan Tang and held in Oklahoma in November 1976. To me the highlight of that meeting was the first public announcements of the tertiary structures of the three fungal aspartic proteinases endothiapepsin, rhizopus-pepsin and penicillopepsin, as well as that of porcine pepsin. In the pages of this volume, we are guided through the fascinating history of the aspartic proteinases in the text of the Plenary Lecture by Tom Blundell. I thank Tom for his insightful views. The success of this volume is due to all of the contributors who attended the conference and presented their results and interpretations.

Michael James
Edmonton, Alberta

BOOKS OF RELATED INTEREST

1. Tang, J. (1997) *Acid Proteases: Structure, Function, and Biology*, Advances in Experimental Medicine and Biology, vol. 95, New York: Plenum Press.
2. Kostka, V. (1985) *Aspartic Proteinases and Their Inhibitors*, Proceedings of the FEBS Advanced Course no. 84/07, Berlin: Walter de Gruyter.
3. Dunn, B. (1991) *Structure and Function of the Aspartic Proteinases: Genetics, Structures, and Mechanisms*, Advances in Experimental Medicine and Biology, vol. 306, New York: Plenum Press.
4. Tahahashi, K. (1993) *Aspartic Proteinases: Structure, Function, Biology, and Biomedical Implications*, Advances in Experimental Medicine and Biology, vol. 362, New York: Plenum Press.

ACKNOWLEDGMENTS

The financial support of the following institutions and companies is gratefully acknowledged:

PUBLIC INSTITUTIONS

Canadian Society for Biochemistry and Molecular Biology
Alberta Heritage Foundation for Medical Research
Alberta Economic Development and Trade
Faculty of Medicine, University of Alberta, Edmonton
University of Alberta Conference Fund, Edmonton
Medical Research Council of Canada

CORPORATE SPONSORS

Abbott Laboratories, Abbott Park, IL, USA
Advanced ChemTech, Inc., Louisville, KY, USA
Agouron Pharmaceuticals, Inc., San Diego, CA, USA
Bayer Corporation, West Haven, CT, USA
Beatrice Foods, Inc., Calgary, Alberta
Beckman Instruments (Canada) Inc., Mississauga, Ontario
Bio-Méga/Boehringer Ingelheim Research Inc., Laval, Quebec
Ciba-Geigy Canada Ltd., Calgary, Alberta
Chiron Corporation, Emeryville, CA, USA
Dupont Merck, Wilmington, DE, USA
Enraf-Nonius Co., New York, NY, USA
Genencor International, South San Francisco, CA, USA
Genentech, Inc., San Francisco, CA, USA
Glaxo Wellcome Inc., Mississauga, Ontario
Glaxo Wellcome Inc., Research Triangle Park, NC, USA
Glaxo Wellcome Research & Development, Stevenage, UK
Hoffmann-La Roche, Basel, Switzerland
Merck & Co., Inc., Whitehouse Station, NJ, USA

Molecular Simulations, Inc., La Jolla, CA, USA
Pharmacia & Upjohn, Kalamazoo, MI, USA
Protein Engineering Network Centres of Excellence, Edmonton, Alberta
Schering-Plough Research Institute, Kenilworth, NJ, USA
Scios Inc., Palo Alto, CA, USA
Selectide Corporation, Tucson, AZ, USA
Silicon Graphics Inc., Calgary, Alberta
Vertex Pharmaceuticals Inc., Cambridge, MA, USA

CONTENTS

Aspartic Proenzymes and Activation

Microbial Aspartic Proteinases

Parasite Aspartic Proteinases

Plant Aspartic Proteinases

1

THE ASPARTIC PROTEINASES

An Historical Overview

Tom L. Blundell, Kunchur Guruprasad, Armando Albert, Mark Williams, Bancinyane L. Sibanda, and Venugopal Dhanaraj

Department of Biochemistry
University of Cambridge
Tennis Court Road
Cambridge CB2 1QW, United Kingdom

1. INTRODUCTION

Research into the aspartic proteinases has had a rich and diverse history. It began with studies of the digestive juices of man and nepenthes using somewhat primitive technologies. It had a renaissance of biochemical characterisation, followed by a classical period when sequences were defined and catalytic activity identified. It is now very much under the influence of the structural school. Thus, in many ways its history parallels that of painting. And like painting it is now thoroughly commercialised, a mixed blessing which nevertheless keeps many of its practioners well above the breadline.

2. PRIMITIVE ORIGINS: PEPSIN, NEPENTHESIN, AND DARWIN

The existence of pepsin in digestive juices had been noted in the early part of the 19th century, but it was not until 1875 that similar digestive activity was identified elsewhere, surprisingly in the pitcher plant, nepenthes (Figure 1). Darwin (1) had detected the homology of nepenthesin with pepsin by the smell produced when it was heated with sulphuric acid. To quote from the passage in which he describes the result of his experiment *"Professor Frankland, as well as his assistant, observed (and this is an important fact) that the fluid when acidified with sulphuric acid, emitted a powerful odour like that of pepsin."* Although this comparative analysis of aspartic proteinases was based on somewhat dubious evidence, the hypothesis has proved correct (2). Today we have firm evidence for the roles of aspartic proteinases, not only as digestive enzymes but also in the regulation of many cellular processes.

Aspartic Proteinases, edited by James.
Plenum Press, New York, 1998.

Figure 1. Nepenthesin: A source of aspartic proteinase.

3. RENAISSANCE OF INTEREST: BIOCHEMICAL CHARACTERISATION

The pepsin family first attracted the attention of biochemists as a group of acid proteinases that were active in the pH range 1–5 (3). Crystals obtained in Svedberg's laboratory, led to the demonstration by Bernal and Crowfoot (4) that the crystals diffracted X-rays. They commented that *"Now that a crystalline protein has been made to give X-ray photographs, it is clear that we have the means of checking them and, by examining the structure of all crystalline proteins, arriving at far more detailed conclusions about protein structure than previous physical or chemical methods have been able to give."* These observations set the scene for modern protein crystallography and enzymology. Figure 2 shows hexagonal pepsin crystals of the type used in these early studies.

4. CLASSICAL STUDIES ON SEQUENCES

The aspartic proteinases really came of age when Tang and coworkers (5) defined the sequence of pepsin. Regions of this sequence were shown to be conserved in frag-

Figure 2. Hexagonal crystals of pepsin first studied by Bernal and Crowfoot (4).

ments previously defined for other enzymes such as chymosin (6) and penicillopepsin (7), indicating the existence of an homologous family. Tang and coworkers (8) later noted the remarkable similarity between the residues 30–42 and 213–225 of pepsin, regions identified with the active site aspartates sensitive to epoxides (Asp32) and diazo compounds (Asp 215). Figure 3 shows the sequences around these active site sequences.

```
                             20           30              80
hivm       - - - - - - g q l K e A L L D T G A  d d T V L e e - -     - p t - - - - - - - - - - - p v N I I G - -
rsvm       h p V k q r s V y I t A L L D S G A  D I T I I S e e d     - m V - - - - - - - - - - - r g S I L G - -

                                 30          40                      110               120
eprn       - - - - t p a q t L n L D F D T G S  S D L W V F S s e     k k V s - s s F t e d s t I D G L L G L A
appn       - - - - - - g t t L n L N F D T G s  A D L W V F S t e     q q I s - a q F q q d t n N D G L L G L A
rhn        - - - - t p G k k F n L D F D T G S  S D L W I A S t l     k r E a - a s F a s g - P N D G L L G L G
hexc148n   - - - - t p a q d F t V I F D T G S  S N L W V P S v y     e t E p g s f L y y A - p F D G I L G L A
chybxan    - - - - t p p q e F T V L F D T G s  S D F W V P S I y     t q E p g d v F T y A - e F D G I L G M A
eprc       s g t F k - s t s I d G I A D T G t  t L L y L p - - -     a g i - - - - - - - - - - - g i N I F G - -
appc       s q s - g - - d G f s G I A D T G t  t L L l L d - - -     s g i - - - - - - - - - - - g f S I F G - -
rhc        t s t V A - - s s F d G I L D T G t  t L L i L P - - -     n - - - - - - - - - - - - w g F A I I G - -
hexc148c   g e t I A C s g g C q A I V D T G t  s l L T G P - - -     d v p t s - - - - - - - s g e L W I L G - -
chybxac    g v v v A C e g G c q A I L D T G t  s k L V G p - - -     n h S - - - - - - - - - - - q k W i L G - -
                       β β β β β β β          β β β                                   β β β
```

Figure 3. The sequence alignment of aspartic proteinases in the region of the active sites achieved by comparing the three-dimensional structures. epr: endothiapepsin; app: penicillopepsin; rh: rhizopuspepsin; hexc148: hexagonal porcine pepsin; chybxa: calf chymosin; rsv: Rous sarcoma virus proteinase; hiv: human immunodeficiency virus proteinase. The last letter refers to the amino (n) or carboxy (c) terminal domains of the pepsins. The numbering above hivn corresponds to sequence numbering in HIV proteinase and that above eprn corresponds to that in pepsin N-terminal lobe. The coordinates of the three-dimensional structures were obtained from the PDB databank. The amino acid code is the standard one-letter code formatted using the following convention of Overington *et al.* (71). Italic for positive phi; upper case for solvent-inaccessible residues; lower case for solvent-accessible residues; bold type for hydrogen bonds to mainchain amide nitrogen; underline for hydrogen bonds to mainchain carbonyl oxygen; tilde~ for sidechain-sidechain hydrogen bonds.

5. IMPRESSIONS OF THE ACTIVE SITE

In a remarkable series of experiments Fruton (9) showed that pepsin cleaved preferentially between amino acid residues with aromatic sidechains and that longer substrates could make a relatively resistant bond more susceptible to cleavage. He also showed that pepsin appeared to have an extended active site that may accommodate a hexapeptide. Furthermore, secondary interactions of the peptide substrate, distant from the scissile bond, could markedly change k_{cat} without comparable change in K_m. Insights from these studies gave a remarkably accurate impression of the active site. They also strongly implied that the mechanism does not involve a covalent amino-acyl or anhydride intermediate. Later, the studies were extended to penicillopepsin, rhizopuspepsin and other aspartic proteinases (10,11). The consensus view now is that catalysis occurs through general acid base hydrolysis, with a non-covalent intermediate such as ($-C[OH]O^{-}-NH_2^{+}-$) or possibly ($-C[OH]_2-NH_2^{+}-$).

6. THREE-DIMENSIONAL STRUCTURES

The first aspartic proteinases to be studied by X-ray analysis were digestive enzymes of broad specificity such as pepsin (12–14) and chymosin (15,16) from mammalian sources, and endothiapepsin (17), penicillopepsin (18,19), rhizopuspepsin (20), mucorpepsins (21,22) and candidapepsins (23,24) from fungi. These bilobal enzymes have rather open active site clefts, consistent with binding a wide range of amino acid sequences.

More recently the three-dimensional structures of several other enzymes, some of which have more specific roles, have been reported. These include human renin, which is involved in the formation of angiotensin II, several crystal structures of which have been reported at medium resolutions (25–27). A higher resolution structure of a mouse renin complex with a decapeptide inhibitor has provided further detailed information about this important family of enzymes (27,28). Crystal structures have been reported for cathepsin D (29,30), a lysosomal proteinase implicated in the invasiveness of breast cancer and also the target for drug design. A related structure is found in the yeast vacuolar proteinase A (31), involved in proteolysis during nitrogen starvation and proteolytic maturation of other vacuolar enzymes. These structures have a similar general topology, but in addition a further flap covers the active site in a way that might contribute to specific interactions with the substrate. More recently three-dimensional structures have been defined for cyprosin from a plant (32) and plasmepsin from the malarial parasite (33). Some typical aspartic proteinases are shown in Figure 4.

7. PRECURSORS

Mammalian aspartic proteinases have inactive precursors and sophisticated mechanisms for activation. The three-dimensional structure of pepsinogen (34,35) demonstrates a beautifully organised precursor region that inhibits the catalytic activity using a conserved lysine and two tyrosine sidechains that form hydrogen bonds to the active site aspartic acids. Carboxylates involved in ion pairs that stabilise the precursor structure are protonated at acid pH, so leading to conformational changes and a two step proteolysis to

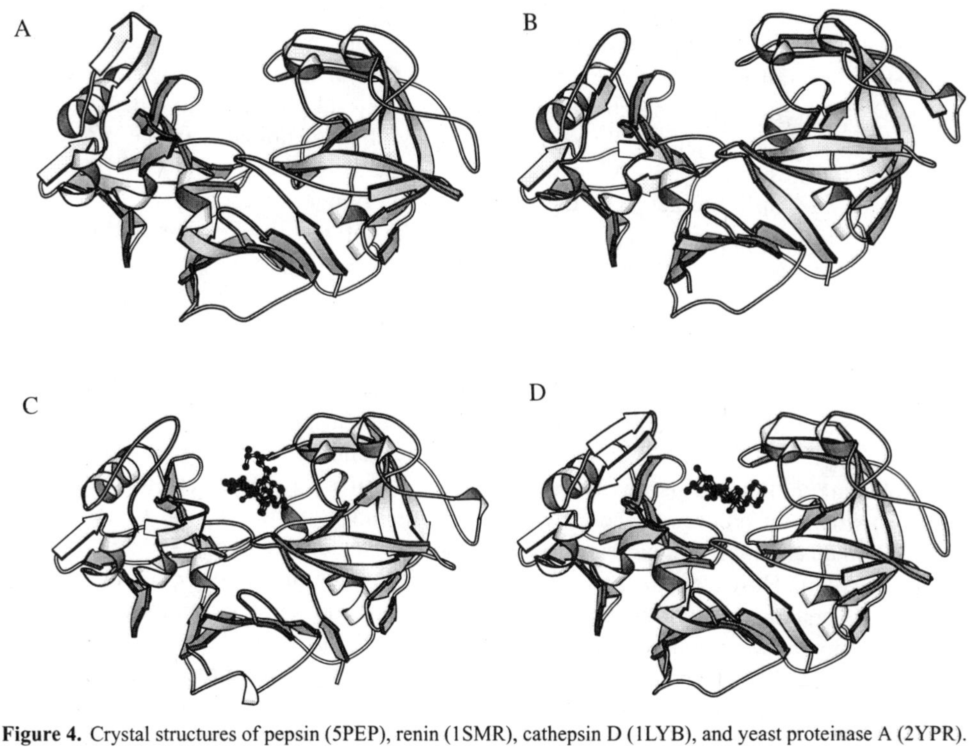

Figure 4. Crystal structures of pepsin (5PEP), renin (1SMR), cathepsin D (1LYB), and yeast proteinase A (2YPR).

give the mature enzyme (36). Aspartic proteinases, active at less acidic pH, appear to have specific activating enzymes.

8. NEW INHIBITORS

The study of aspartic proteinase inhibition began with pepstatin, the inhibitory character of which was correctly attributed to the statine hydroxyl group. This group was subsequently exploited in sequences otherwise similar to angiotensinogen (37). Substitution of a shorter hydroxyethylene (–CHOH–CH_2–) resulted in a very highly potent inhibitor (H261) with an inhibition constant of 0.7 nM (38). In this compound the –CHOH– function is assumed to mimic the tetrahedral gem diol and the –CH_2– function mimics the –NH_2^+ of the transition state. Statone analogues (–CO–CH_2–) are also highly potent (38) and were further modified to difluorostatone compounds (–CO–CF_2–) with even higher potency (39); these become partially hydrated to give –$C(OH)_2$–CF_2– even in the absence of complexation and are useful transition state analogues.

The first inhibitor complexes to be studied by X-ray analysis were pepstatin with rhizopuspepsin (40,41) and a pepstatin fragment with penicillopepsin (18); these confirmed that the hydroxyl group of the statine mimicks a tetrahedral intermediate, being hydrogen bonded to both catalytic aspartates and replacing the water bound at this position in the uncomplexed enzymes. Later crystallographic studies on statine-containing inhibitors showed that statine acts as a dipeptide analogue in the binding cleft (42). Further details of inhibitor enzyme interactions and their dependence on the nature of the transition state isostere have come from X-ray analyses of rhizopuspepsin (43), penicillopepsin (44) and endothiapepsin (45).

These studies showed that the active site cleft is formed between the two lobes, each of which provides one of the essential aspartates. The inhibitors have an extended β-strand conformation and make hydrogen bonds through their main chains to each lobe of the enzyme (Figure 5). The side chains alternate on either side of the main chain and make hydrogen bonds and van der Waals interactions with both domains in the molecule so defining subsites, S6 to S3′. An antiparallel β-sheet, comprising residues 71 to 82 and known as the 'flap', interacts with the central region of the inhibitor and shields the active site from the solvent region. The carboxylates of the active site aspartates hydrogen bond to the hydroxyl oxygen of the transition state isostere (Figure 5). Comparisons of the many inhibitor complexes show that the general mode of binding is conserved over very great evolutionary distances.

Subtle conformational changes allow even identical moieties in different inhibitors to interact with the enzyme in quite different ways. For example in modelling exercises it was often assumed that specificity derives from differences in the sizes of the residues in the specificity pockets (Sn) which complement the corresponding side chains at positions Pn in the substrate/inhibitor. This was correct for the S1′ subsites for the human and mouse renins (46,47) where differences in the enzymes are complementary to the valine and leucine at P1′ in human and mouse angiotensinogens; residue 213 is leucine in human renin and valine in mouse renin. However, the three-dimensional structure of human renin, complexed with a P4 to P1′ norstatine-containing inhibitor-CP-85,339, and mouse renin, complexed with a decapeptide hydroxyethylene isostere inhibitor CH-66 (27,28) show that the situation can be more complex. For example, the S1′ pockets of chymosin, pepsin and endothiapepsin have an aromatic side-chain at residue 189 whilst the renins have amino acids with smaller side-chains (valine in human and serine in mouse renins). This would be expected to make the

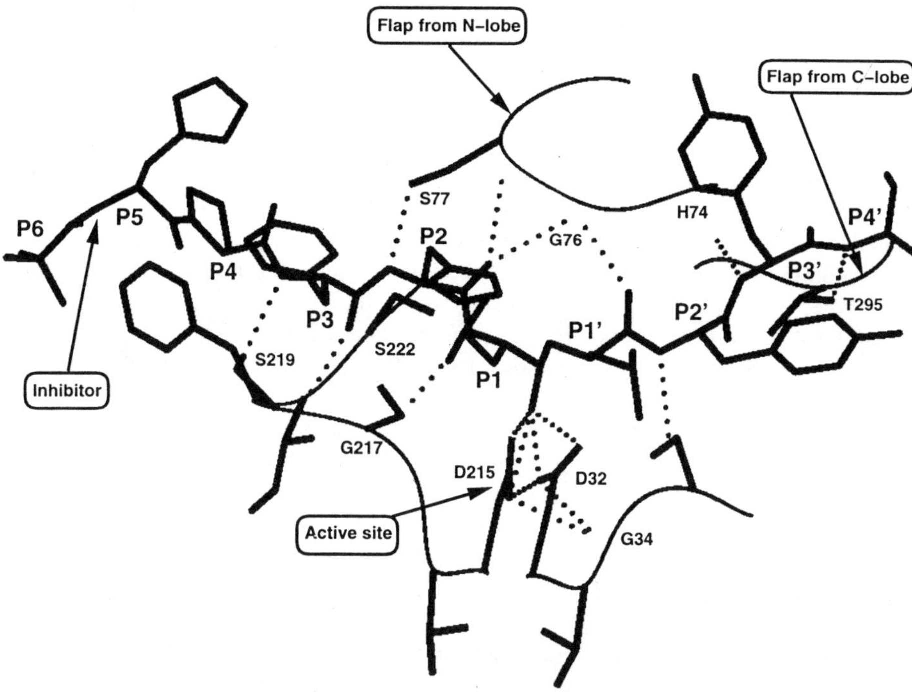

Figure 5. The active site of the mouse renin inhibitor complex (28).

pocket larger in renins, but the structure of the mouse renin complex shows that the substrate moves closer to the enzyme in renins as a result of the smaller residue at 189.

The specificity pockets are also influenced by the elaboration of loops on the periphery of the binding cleft in renins. For instance, in endothiapepsin, different conformations are adopted by inhibitors P3′ and the residue at P4′ is generally disordered. In contrast these residues are clearly defined in mouse renin. This is mainly a consequence of the poly-proline loop, involving residues 294 to 297. The X-ray analysis of the mouse renin complex shows that the S3′ and S4′ subsites are formed by the poly-proline loop together with residues of the 'flap'. The well-defined interactions of P3′ described in the mouse renin complex explain the significant affinity when inhibitors have phenylalanine or tyrosine at P3′ as well as the importance of a P3′ residue for catalytic cleavage of a substrate by renin. In cathepsin D and yeast proteinase, this proline-rich loop has a similar conformation but occupies a position further away from the inhibitors. This probably accounts for their decreased specificity for substrates at S3′ and S4′.

9. TRANSITION STATES AND INTERMEDIATES

Although there is no general agreement about the details of the catalytic mechanism, it is generally accepted that the nucleophilic attack by a water molecule leads to tetrahedral intermediates and/or transition states involving groups such as ($–C(OH)O^-–NH_2^+–$). There have been several attempts to simulate such tetrahedral intermediates. These have included phosphostatine ($–PO(OH)–CH_2–$) analogue of the tetrahedral transition state, crystal structures of complexes of which demonstrate that the second oxygen binds close to the outer oxygen of Asp32. However, this is probably not a very good model as there is likely to be one less proton in the complex than in the real intermediates/transition states. A better model is provided by the hydrated difluorostatone containing analogues where the pro-(R) hydroxyl oxygen of the inhibitor is within 3.4Å of all four carboxyl oxygens of the two active site aspartates, Asp32 and Asp215 (48). The distances to the two inner carboxylate oxygens are consistent with hydrogen bonds but the shortest distance (2.57Å) is to the outer oxygen of Asp 215. The second hydroxyl oxygen of the hydrate is located at an equivalent distance (2.6Å) from the outer carboxyl oxygen of Asp 32.

Most mechanisms envisage a nucleophilic attack on the substrate carbonyl carbon by a water molecule, via general acid-base catalysis by the active residues (49–52). Most now agree that the nucleophilic oxygen comes from the water bound to the active site. The strain mechanism of Pearl (53) suggests a distortion of the scissile peptide bond, facilitating the formation of the tetrahedral intermediate which is supported by T-torsion angle analysis of P1–P1′ peptide bond in the endothiapepsin inhibitor complexes (54). Veerapandian et al. (48) have suggested that the scissile bond carbonyl is protonated by Asp32, and concomitantly attacked by water polarized into a nucleophilic state by Asp215. This gives a tetrahedral intermediate with a proton arrangement consistent with that of the bound hydrate. Thus, the first intermediate is probably a gem-diol stabilized by hydrogen bonds to a negatively charged Asp32.

Jaskolski et al. (55) have suggested a catalytic mechanism for the HIV-1 aspartic proteinase reaction in which an electrophilic attack (of the proton from Asp215) weakens the scissile bond and facilitates a tighter binding between the proteinase and the substrate. Upon completion of the reaction cycle, the protonation states of the two aspartic acids would be reversed (56). Lee et al. (52) proposed that a second water molecule is involved in the product release. Silva et al. (57) based on semi-empirical quantum mechanical methods proposed that rotation around the C–N bond would readily occur after hydration of the scissile bond and would be driven by optimization of the interactions of peptide side chains with the enzyme to assume a *gauche* conformation.

10. PRIMITIVE ORIGINS REDISCOVERED: RETROVIRAL PROTEINASES

The discovery of pseudo two-fold symmetry in the 3-D structures of the pepsin-like aspartic proteinases by Tang et al. (58) gave rise to a productive series of hypotheses, concerning a dimeric ancestor. The two-fold symmetry suggested models of the retroviral proteinases as dimeric homologues of the pepsins (59,60), which were confirmed later by the crystal structures of the RSV (61) and HIV proteinases (62,63).

Figure 6 shows the 3-D structures of monomeric pepsin and dimeric HIV proteinase, viewed from equivalent directions perpendicular to the 2-fold axis. The monomeric pepsin is roughly 60% larger than the retroviral dimer. Each subunit of the retroviral proteinase and each lobe of a pepsin-like proteinase comprises two similar motifs formed from anti-parallel β-strands: *a,b,c* and *d* for the first and *a′,b′,c′* and *d′* for the second (17,62,63), organised together in a distorted sheet (sheet 1). Strands *c* and *d′* and strands *c′* and *d* form two pairs of parallel strands. Strands *b* and *c* and strands *b′* and *c′* form anti-parallel β-hairpins that are folded over sheet 1 and hydrogen-bonded together around intra-domain two-fold axis to give a second sheet (sheet 2), which is orthogonal to the first. In the retroviral proteinases the motifs are more symmetrical and less distorted than in the pepsin-like enzymes, where both sheets are fragmented.

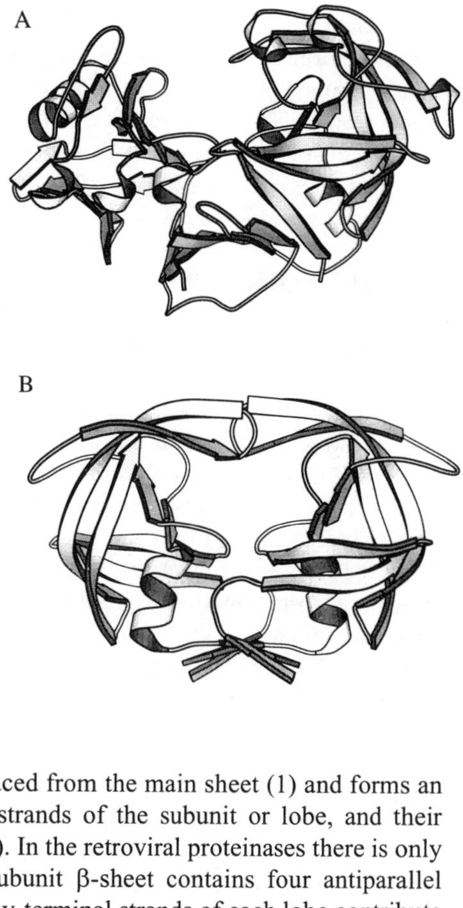

Figure 6. A schematic view perpendicular to the two-fold axis and along the active site cleft of (a) pepsin and (b) HIV-1 proteinase.

Much of strand *a* of the first motif is displaced from the main sheet (1) and forms an anti-parallel β-sheet with the carboxy-terminal strands of the subunit or lobe, and their equivalents in the second subunit or lobe (sheet 3). In the retroviral proteinases there is only one carboxy-terminal strand and so the inter-subunit β-sheet contains four antiparallel strands. In the pepsin-like proteinases two carboxy-terminal strands of each lobe contribute to a six-stranded antiparallel β-sheet. Figure 6 shows that the strands of sheet 3 form the base of the enzymes below the well defined cleft. In the retroviral and pepsin-like enzymes the equivalent sheets 3 occupy the same volume but have different orientations.

Although only the sequences around the active site aspartates are obviously equivalent between the lobes of the pepsins and the subunits of the retroviral proteinases, the topological similarities between the 3-D structures allows alignment of sequences, as shown in Figure 3.

One of the significant conservations in the retroviral and pepsin-like aspartic proteinases is the sequence hydrophobic-hydrophobic-glycine, for example Ile84–Ile85–Gly86 found in HIV-1 proteinase. Ile85 and its equivalents in other retroviral proteinases and in pepsin-like proteinases (Figure 6) is conserved as hydrophobic because it contributes to the core. Gly86 is packed close to the active site residues in a way that does not allow a sidechain. Ile84 is conserved for purposes of binding substrate (see below). This pattern is characteristic of both monomeric and dimeric aspartic proteinases.

The conserved active site residues, Asp25–Thr26–Gly27, of the HIV proteinase form a symmetrical and highly hydrogen-bonded arrangement, virtually identical to that in pepsin-like aspartic proteinases (residues Asp32–Thr33–Gly34 and Asp215–Thr216–Gly217 of pepsin). This includes the two threonines, which are inaccessible to solvent and hydrogen-bonded so that the γ-O forms hydrogen bonds with the mainchain NH and CO functions of the other subunit or lobe in a fireman's grip (50). They provide a good example of high conservation of buried, hydrogen-bonded polar residues that are important for maintenance of 3-D structure. The two aspartates lie approximately planar with their inner carboxylate oxygens hydrogen bonded to the NH functions of Gly27 and within hydrogen-bonding distance of each other. The conservation of the glycines appears to be a consequence of the fact that the existence of sidechains at this position would disrupt the structure of the aspartic acid sidechains.

11. NEW DIRECTIONS

Aspartic proteinases have been detected, extracted and characterised from seeds, leaves and flowers of a broad variety of plant species, where they probably have specific roles in processing protein precursors (64). As we have noted above, they are also found in the digestive pitcher fluid of an insectivorous plants such as nepenthes (65). They share typical features with their mammalian counterparts such as acidic pH-optimum, narrow hydrolytic specificity and sensitivity to the specific pentapeptide inhibitor pepstatin A.

The sequences of plant aspartic proteinases such as that from barley, HvAP (66) and from the flowers of cyanara, cyprosin, are similar to those of other aspartic proteinases, especially to mammalian lysosomal cathepsin D and yeast proteinase A. However, they also contain an inserted sequence of 99–104 amino acid residues bearing no similarity to animal or microbial aspartic proteinases, but sharing significant sequence similarity to the saposins. It has been shown that the association of procathepsin D with prosaposin (67) plays a role in targeting in the Golgi. There is probably a similar situation in plants, where the saposin-like domain is coded as part of the aspartic proteinase and plays a role in the transport of the enzymes to plant vacuoles.

Recently, the three-dimensional structure of the first plant proteinase has been defined by X-ray analysis (32). From this and earlier modelling studies (68) it appears that the overall three-dimensional structures of HvAP and cyprosin (excluding the plant-specific insert; 104 residues in HvAP) are closer to human cathepsin D than other aspartic proteinases of known three-dimensional structure. No specific similarities have been identified in the active sites of cyprosin and chymosin that could be correlated with their common milk clotting activity, although it was observed that the type of residues at P3, P2 and P2′ in the κ-casein sequences may be accommodated in the active site of both these enzymes.

The pregnancy-associated glycoproteins (PAGs) are secretory products synthesized by the outer epithelial cell layer (chorion) of the placentas of various ungulate species. All PAG molecules appear to have evolved from a pepsin-like progenitor molecule with the equine PAG most closely related to the pepsins. The presence of substitutions at the S_1 and other subsites relative to pepsin make it unlikely that either bovine, ovine or the porcine PAG-1 have catalytic activity. Only two of the eight PAGs examined (porcine PAG-2 and equine PAG-1) retain features of active aspartic proteinases with pepsin-like activity. Modelling studies indicate that the peptide binding specificities of PAGs probably differ significantly from each other and from pepsin, despite their high sequence identities (69). Analysis of the various peptide binding subsites demonstrates why both bovine and ovine PAG-1 are capable of binding pepstatin. The strong negative charge in the binding cleft of boPAG-1 and ovPAG-1 indicates a preference for lysine- or arginine-rich peptides. Thus, PAGs represent a family where the possible peptide binding function may be retained through their binding specificities, but where the catalytic activity may be lost in some cases (Figure 7).

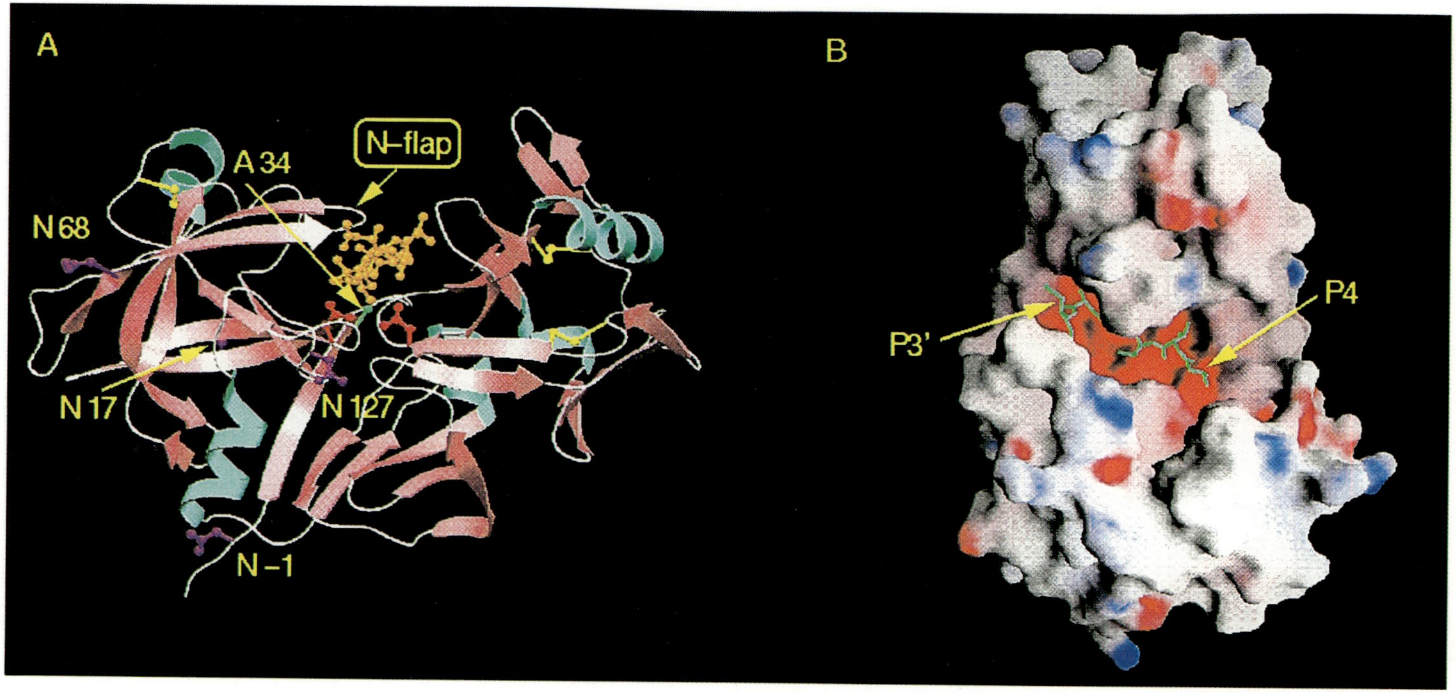

Figure 7. A model of a bovine pregnancy associated glycoprotein (boPAG-1) (69): (A) a ribbon representation viewed along the cleft, with pepstatin modelled in active site and (B) GRASP representation viewed perpendicular to the cleft, with red indicating negative charge.

12. THE ART COMMERCIALISED

Aspartic proteinases have been traditionally used in cheese, soya and cocoa processing and recombinant enzymes are in commercial use in the cheese industry. However, many fascinating new applications, especially in drug discovery, have emerged in the past twenty years.

Renin is the most specific aspartic proteinase yet discovered and is involved in the formation of angiotensin II, a potent pressor and aldosterogenic substance controlling vascular tone, fluid volume and sodium excretion. Because of its unique specificity, its inhibition was widely expected to provide selective therapy for hypertension, congestive heart failure and associated degenerative disorders linked to angiotensin II. Useful renin inhibitors have been developed by several companies in the 1980s. Structure-based approaches were first based on models (46) and much later on crystal structures (25–27) of mouse and human renins. The detailed analyses of inhibitor complexes (see above) indicated the general structural features that may contribute to specificity, especially of renin, but demonstrated the need for careful, high resolution X-ray analyses for more confidence in drug design. In particular, they show that even minor alterations in the positions of secondary structural elements can lead to major changes in the disposition of the subsites. Although considerable progress was made exploiting structure-based approaches in developing highly specific and orally active molecules, most research into in the synthesis of renin inhibitors was stopped when it became apparent that the sophisticated chemistry would demand a high price that would not be competitive with that of generic antihypertensives, based on the inhibition of angiotensin converting enzymes, as patent protection of these falls. However, the same technology offers possibilites for anticancer agents against cathepsin D and E, antimalarials against plasmepsin (33) and antifungal agents against candidapepsins (23,24), the latter of which can be used to treat infections associated with AIDS.

Most exciting of all has been the discovery of HIV proteinase and the production of antivirals, which are now proving their value in the clinic. This has been one of the first cases where structure-based approaches have convincingly played a role in discovery of drugs in the clinic (70). Structures of several hundred inhibitor complexes have been experimentally defined, providing a previously unparalleled structural database for design. These have exploited a range of different structural features, including two-fold symmetry in the ligand, replacement of a bound water molecule, cyclization and replacement of scissile peptide bonds. Several studies have involved the use of programs like DOCK or fragment searching to identify non-peptidic structures. Useful molecules are now exploited as cocktails in the treatment of AIDS, with encouraging results, although it is evident that mutation in HIV allows the virus to quickly escape if challenged with a single antiviral.

ACKNOWLEDGMENTS

We thank Dr Alan Barrett for the references and material relating to Darwin's identification of pepsin-like activity in nepenthes. We thank our colleagues at Cambridge, Birkbeck and elsewhere for stimulating discussions. We acknowledge that this is just one view of the history of research of this area, and others may describe the story differently.

REFERENCES

1. Darwin, C. (1875) Insectivorous Plants, p88. John Murray, London.
2. Takahashi, K. (1997) Personal communication

3. Northrop, J. H., Kunitz, M and Herriott, R.M. (1948) Crystalline enzymes, 2nd ed., Columbia University Press, New York.
4. Bernal, J. D. & Crowfoot, D. (1934) Nature, 133, 794–795.
5. Tang, J., Sepulvelda, P., Marciniszyn, J., Chen, K.C.S., Huang, W.Y., Tao, N., Liu, D. and Lainier, J.P. (1973) Proc. Natl. Acad. Sci. U.S.A. 70, 3437–3439.
6. Pedersen, V.B. and Foltmann, B. (1973) FEBS Lett., 35, 250–258
7. Sodek, J. and Hofmann, T. (1970) Can. J. Biochem. 48, 1014–1022
8. Sepulvelda, P., Marciniszyn, J., Liu, D and Tang, J., (1975) J. Biol. Chem. 250, 5082–5088
9. Fruton, J (1976) Adv. Enzymol. 11, 469–475.
10. Hofmann, T., Hodges, R.S. and James, M.N.G. (1984) Biochemistry 23, 635–643.
11. Rao-Naik, C., Guruprasad, K., Batley, B., Rapundalo, S., Hill, J., Blundell, T.L., Kay, J. & Dunn, B.M. (1995) Proteins 22:168–181
12. Andreeva, N. S., Zdanov, A. S., Gustchina, A. E. & Fedorov, A. A. (1984). J. Mol. Chem., 259, 11353–11365.
13. Sielecki, A. R., Fedorov, A. A., Boodhoo, A., Andreeva, N. S., & James, M. N. G. (1990). J. Mol. Biol., 214, 143–170.
14. Cooper, J. B., Khan, G., Taylor, G., Tickle, I. J. & Blundell, T. L. (1990). J. Mol. Biol., 214, 199–222.
15. Gilliland, G. L., Winbourne, E. L., Nachman, J. & Wlodawer, A. (1990). Proteins, 8, 82–101.
16. Newman, M. P., Safro, M., Frazao, C., Khan, G., Zdanov, A., Tickle, I. J., Blundell, T. L. & Andreeva, N. (1991). J. Mol. Biol., 221, 1295–1309.
17. Blundell, T. L., Jenkins, J. A., Sewell, B. T., Pearl, L. H., Cooper, J.B., Tickle, I. J., Veerapandian, B. & Wood, S. P. (1990). J. Mol. Biol., 214, 199–222.
18. James, M. N. G. & Sielecki, A. R. (1983). J. Mol. Biol., 163, 299–301.
19. James, M. N. G., Sielecki, A. R., Salituro, F., Rich, D. H. & Hofmann, T. (1982). Proc. Natl. Acad. Sci., U.S.A., 79, 6137–6142.
20. Suguna, K., Bott, R. R., Padlan, E. A., Subramanian, E., Sheriff, S., Cohen, G. H. & Davies, D. R. (1987). J. Mol. Biol., 196, 877–900.
21. Newman, M. P., Watson, F., Roychowdhury, P., Jones, H., Badasso, M., Cleasby, A., Wood, S.P., Tickle, I.J. & Blundell T. L. (1993). J. Mol. Biol., 230, 260–283.
22. Wilson, Q. (1997) Personal communication
23. Abad-Zapatero, C., Goldman, R., Muchmore, S.W., Hutchins, C., Stewart, K., Navaza, J., Payne, C.D. & Ray, T.L. (1996) Prot. Sci. 5, 640–652.
24. Cutfield, S. M., Dodson, E. J., Anderson, B. F., Moody, P. C. E., Marshall, C. J., Sullivan, P. A. & Cutfield, J. F. (1995). Structure, 3, 1261–1271.
25. Sielecki, A. R., Hayakawa, K., Fujinaga, M., Murphy, M. E. P., Fraser, M., Muir, A. K., Carilli, C. T., Lewicki, J. A., Baxter, J. D. & James, M. N. G. (1989). Science, 243, 1346–1351.
26. Rahuel, J., Priestle, J. P. & Grutter, M. G. (1991). J. Struct. Biol., 107, 227–236.
27. Dhanaraj, V., Dealwis, C. G., Frazao, C., Badasso, M., Sibanda, B. L., Tickle, I. J., Cooper, J. B., Driessen, H. P. C., Newman, M. , Aguilar, C., Wood, S. P., Blundell, T. L., Hobart, P. M. , Geoghegan, K. F., Ammirati, M. J., Danley, D. E., O'Connor, B. A. & Hoover, D. J. (1992). Nature, 357, 466–472.
28. Dealwis C., Frazao, C., Badasso, M., Cooper, J.B., Tickle, I.J., Driessen, H., Blundell, T.L., Murakami, K., Miyazaki, H., Sueiras-Diaz, J., Jones, D. M. & Szelke, M. (1994). J. Mol. Biol., 236, 342–360.
29. Metcalf, P. & Fusek, M. (1993). EMBO J., 12, 1293–1302.
30. Baldwin, E. T., Bhat, T. N., Gulnik, S., Hosur, M. V., Sowder, R. C., Cachau, R.E., Collins, J., Silva, A. M. & Erickson, J. W. (1993). Proc. Natl. Acad. Sci. USA., 90, 6796–6800.
31. Aguilar, C., Badasso, M., Cooper, J., Dreyer, T., Wood, S. P. & Blundell, T.L. (1997). J. Mol. Biol. (in press).
32. Frazao, C. (1996) Personal communication
33. Silva, M., Lee, A.Y., Gulnik, S.V., Majer, P., Collins, J., Bhat, T.N., Collins, P.J., Cachau, R.E., Luker, K.E., Gluzman, I.Y., Francis, S.E., Oksman, A., Goldberg, D.E. & Erickson, J.W. (1996) Proc. Natl. Acad. Sci. USA, 93, 10034–10039.
34. James, M.N.G and Sielecki, A.R (1986) Nature 319, 33–38
35. Hartsuck, J.A., Koelsch, G. & Remington, S.J. (1992) Proteins: Str. Fun. and Gen., 13, 1–25.
36. Kay, J. (1980) In the Enzymology of Post-Translational Modification of Proteins (R.B. Freedman and H. Hawkins, eds) pp424–460, Academic Press, London.
37. Rich, D.H., Bernatowiwicz, M.S., Agarwal, N.S., Kawai, M., Salituro, F.G. & Scmidt, P.G. (1985) Biochemistry, 24, 3165–3173.
38. Szelke, M., Tree, M., Leckie, B. J., Jones, D. M. Atrash, B., Beattie, S., Donovan, B., Hallett, A., Hughes, M., Lever, A. F., Morton, J. J. & Sueiras-Diaz, J. (1985). J. of Hypertension, 3, 13–18.

39. Gelb, M.H., Svaren, J.P. & Abeles, R.H. (1985) Biochemistry, 24, 1813–1817.
40. Bott, R., Subramanian, E. & Davies, D. R. (1982). Biochemistry, 21, 6956–6962.
41. Bott, R. & Davies, D. R. (1983). In *Proceedings of the Eighth American Peptide Symposium* (Hruby, V. & Rich, D., ed.), pp. 531–540, Pierce Chemical, Rockford, Illinois.
42. Cooper, J. B., Foundling, S. I., Boger, J., Jupp, R. A., Kay, J. & Blundell, T. L. (1989). Biochemistry, 28, 8597–8603.
43. Suguna, K., Padlan, E. A., Bott, R., Boger, J., Parris, K. D. & Davies, D. R. (1992). 13, 195–205.
44. Fujinaga, M., Chernaia, M. M., Tarasova, N. I., Mosimann, S. C. & James, M. N. G. (1995). Protein Sci., 4, 960–972.
45. Foundling, S. I., Cooper, J. B., Watson, F. E., Pearl, L. H., Hemmings, A., Wood, S. P., Blundell, T. L., Hallett, A., Jones, D. M., Sueiras, J., Atarsh, B. & Szelke, M. (1987). J. Cardiovas. Pharmacol, 10, S59-S68.
46. Blundell, T. L., Sibanda, B. L. & Pearl, L. H. (1983). Nature, (London), 304, 273–275.
47. Sibanda, B. L., Blundell, T. L., Hobart, P. M., Fogliand, M., Bindra, J. S., Dominy, B. W. & Chirgwin, J . M. (1984). FEBS Lett., 174, 102–111.
48. Veerapandian, B., Cooper, J. B., Sali, A., Dominy, B. W., Hoover, D. J. & Blundell, T. L. (1992). Protein Sci., 1, 322–328.
49. James, M. N. G. & Sielecki, A. R. (1985). Biochemistry, 24, 3701–3713.
50. Pearl, L. H. & Blundell, T. L. (1984). FEBS Lett., 174, 96–101.
51. Veerapandian, B., Cooper, J. B., Sali, A. & Blundell, T. L. (1990). J. Mol. Biol., 216, 1017–1029.
52. Lee, H., Darden, T.A. & Pederson, L.G. (1996) J. Amer. Chem. Soc. 118, 3946–3950.
53. Pearl, L.H. (1985) in "Aspartic Proteinases and Their Inhibitors" (Kostka, V., ed) pp 189–195, de Gruyter, Berlin.
54. Sali, A., Veerapandian, B., Cooper, J. B., Foundling, S. I., Hoover, D. J. & Blundell, T. L. (1989). EMBO J. 8, 2179–2188.
55. Jakolski, M., Tomasselli, A.G., Sawyer, T.K., Staples, D.G., Heinrikson, R.L., Schneider, J., Kent, S.B.H. & Wlodawer, A. (1991) Biochemistry 30, 1600–1609.
56. Hyland, L.J., Tomaszek, T.A. & Roberts, G.D. (1991) Biochemistry 30, 8441–8453.
57. Silva, A.M., Cachau, R.E., Sham, H.L. & Erickson, J.W. (1996) J. Mol. Biol. 255, 321–340.
58. Tang, J., James, J., Hsu, N., Jenkins, J.A. & Blundell, T. L. (1978). Nature 271, 618–621.
59. Pearl, L. H. & Taylor, W. R. (1987). Nature, 329, 351–354.
60. Blundell, T.L., Carney, D., Gardner, S., Hayes, F., Howlin, B., Hubbard, T., Overington, J., Singh, D. A., Sibanda, B. L. & Sutcliffe, M. (1988). Eur. J. Biochem. 173, 513–520.
61. Miller, M., Schneider, J., Sathyanarayana, B. K. Toth, M. V., Marshall, G. R., Clawson, L., Selk, L., Kent, S. B. H. & Wlodawer, A. (1989). Science 246, 1149–1152.
62. Wlodawer, A., Miller, M., Jaskolski, M. Sathyanarayana, B. K., Baldwin, E., Weber, I. T., Selk, L. M., Clawson, Schneider, J. & Kent, S. B. H. (1989). Science, 245, 616–621.
63. Lapatto, R., Blundell, T.L., Hemmings, A., Overington, J., Wilderspin, A., Wood, S., Merson, J. R., Whittle, P. J., Danley, D. E., Geoghegan, K. F., Hawrylik, S. J., Lee, S. E., Scheld, K. G. & Hobart, P. M. (1989). Nature 342, 299–302.
64. Kervinen, J., Sarkkinen, P., Kalkkinen, N., Mikola, L. & Saarma, M. (1993). Phytochemistry, 32, 799–803.
65. Tokes, Z. A., Woon, W. C. & Chambers, S. M. (1974). Planta, 119, 39–46.
66. Runeberg-Roos, P., Tormakangas, K., Ostman, A. (1991). Eur. J. Biochem., 202, 1021–1027.
67. Zhu, Y. & Conner, G. E. (1994). J. Biol. Chem., 269, 3846–3851.
68. Guruprasad, K., Tormakangas, K., Kervinen, J. & Blundell, T. L. (1994). FEBS Lett., 352, 131–136.
69. Guruprasad, K., Blundell, T. L., Xie, S. E., Green, J., Szafranska, B., Nagel, R. J., McDowell, K., Baker, C. B. & Roberts, R. M. (1996). Protein Engineering, 9, 849–856.
70. Wlodawer A. & Erickson, J. (1993). Ann. Rev. Biochem., 62, 543–585.
71. Overington, J., Johnson, M. J., Sali, A. & Blundell, T. L. (1990). Proc. Roy. Soc. Lond., B241, 132–145.

THE REGULATION OF SEQUENTIAL PROCESSING OF HIV-1 GAG BY THE VIRAL PROTEASE

Steve C. Pettit,[1] Nijing Sheng,[2] Radonna Tritch,[2] Susan Erickson-Viitanen,[2] and Ronald Swanstrom[1,3*]

[1]Lineberger Comprehensive Cancer Center
[2]Molecular Biology Department
DuPont Merck Pharmaceutical Co.
Wilmington, Delaware 19880
[3]Department of Biochemistry and Biophysics
University of North Carolina at Chapel Hill
Chapel Hill, North Carolina 27599

INTRODUCTION

A full understanding of the structure-function relationship of the retroviral aspartic proteinase (PR) requires a consideration of the role of PR in the biology of the virus. Very little is known about the mechanisms involved in the maturation of the retroviral particle. Viral maturation undoubtedly is a complex process since several thousand copies of the individual components that comprise the virion are generated by the action of PR on the polyprotein precursors, and these components must be sorted and properly assembled to form the mature, infectious virion. Recent studies suggest that cleavage of HIV-1 Gag at its processing sites occurs in an ordered or sequential manner, and that this sequential processing is a regulated event. Taken together, these results support a model in which ordered proteolytic processing of the Gag precursor is an integral part of the regulation and control of virion maturation.

RESULTS AND DISCUSSION

PR plays a role late in the virus life cycle during virion assembly and maturation. Without PR activity, virions are released but are not infectious and retain an immature,

* Author to whom correspondence should be addressed. (919)-966–5710. Fax (919) 966–3015

Aspartic Proteinases, edited by James.
Plenum Press, New York, 1998.

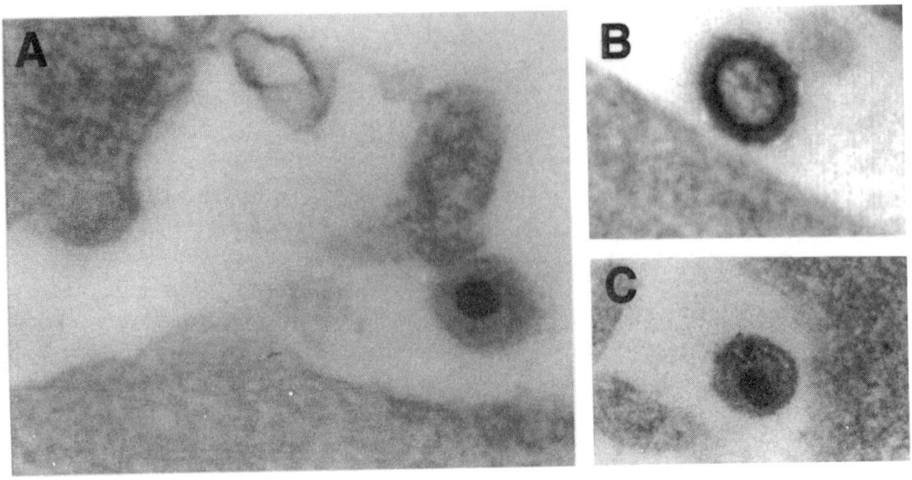

Figure 1. The morphology of budding, immature, and mature wild type HIV-1. HeLa cells transfected with an infectious molecular clone (a derivative of NL4–3) were examined by thin section electron microscopy. A) Left: an early "bud" showing viral proteins multimerized at the plasma membrane. Right: An HIV-1 virion with a condensed core. B) An immature virion. C) A mature virion showing evidence of a cone-shaped core.

ring-shaped morphology (Crawford et al., 1985; Katoh et al., 1985; Kohl et al., 1988; Sommerfelt et al., 1992) (Figure 1). In HIV-1, the protease is located in the Pro portion of the Gag–Pro–Pol precursor, which is formed by a ribosomal frameshift that occurs at a frequency of 5–10% during Gag translation (Jacks et al., 1988). Each virion is thought to contain approximately 2000 molecules of Gag and therefore approximately 100 molecules of Gag–Pro–Pol. The Gag precursor contains the viral structural proteins: matrix (MA), capsid (CA), nucleocapsid (NC), in addition to p6. The Pro–Pol portion of Gag–Pro–Pol encodes the viral enzymes: protease (PR), reverse transcriptase with its associated RNAse H activity (RT/RH), and integrase (IN) (Figure 2). During assembly, the Gag and Gag/Pol

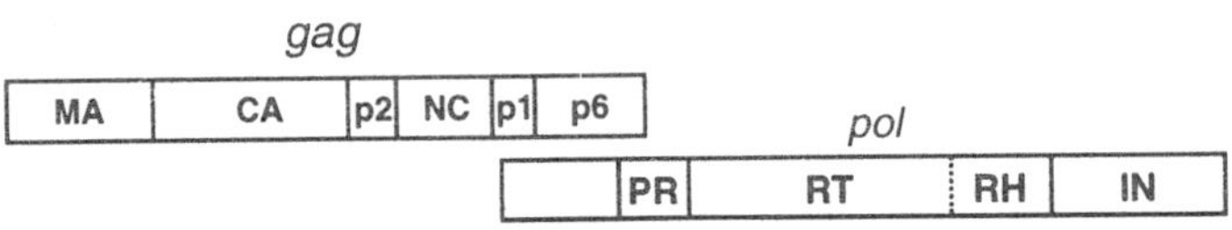

HIV-1 Processing Site Sequences

Gag		Pol	
MA/CA	SQNY/PIV	TF/PR	SFNF/PQI
CA/p2	ARVL/AEA	PR/RT	TLNF/PIS
p2/NC	ATIM/MQR	RT/RH	AETF/YVD
NC/p1	RQAN/FLG	RH/IN	RKVL/FLD
p1/p6	PGNF/LQS		

Figure 2. The location and sequence of the nine major HIV-1 Gag and Gag/Pol processing sites. Above: a schematic of the *gag* and *pol* open reading frames. The locations of protease cleavage sites are indicated by the vertical lines. Protein products are indicated in capital letters: MA, matrix; CA, capsid; NC, nucleocapsid; PR, protease; RT, reverse transcriptase; RH, RNAse H; IN, integrase. The Gag/Pol precursor is formed at a frequency of 5–10 percent by translational frameshift. The nucleotides responsible for the frameshift overlap the coding region the NC/p1 processing site (Jacks et al., 1988). Below: The sequence of the seven residues of each site that interact directly with the protease dimer. The sequence extends from the upstream P4 position to the downstream P3′ position. The position of the scissile bond is denoted with a slash (/). The locations and sequence of the processing sites in HIV-1 were determined by protein sequencing of HIV-1 virion proteins (Veronese et al., 1987; Henderson et al., 1988; Henderson et al., 1992). The sequences shown are of the HXB-2D isolate of HIV-1 (Ratner et al., 1987).

precursors are targeted to the plasma membrane of the infected cell (Rhee and Hunter, 1987; Gottlinger et al., 1989; Bryant and Ratner, 1990; Hansen et al., 1990; Wagner et al., 1992; Black et al., 1993). Aggregation and multimerization of the precursors at the membrane drives budding of the virus particle from the cell. The exact mechanism of PR activation within the virion is unknown. Likely, the concentration of Gag/Pol precursors during multimerization and budding promotes dimerization and activation of PR. Active PR dimer, once freed from the Gag–Pro–Pol precursor, would then initiate a processing cascade by trans-cleavage of the processing sites. Upon cleavage of the precursors, the virion undergoes the process of maturation, where the characteristic dark-staining core is formed (Ross et al., 1991; Schatzl et al., 1991; Kaplan et al., 1993). Maturation, or morphogenesis is thought to be necessary for the formation of an infectious particle. Very little is known about the mechanism of control of virion morphogenesis, but it is likely to be a controlled event given the complexity of virion structure and the importance of maturation in producing infectious virions.

Characterization of Retroviral Processing Site Diversity

Within the HIV-1 PR dimer a substrate binding cleft is formed that is capable of interacting with roughly seven consecutive amino acids of the substrate from positions P4 through P3′ (Darke et al., 1988; Moore et al., 1989; Swain et al., 1991; Tozser et al., 1991). Specificity is determined by the interaction of substrate side chains within a subsite pocket of the dimer. Bound substrate adopts an extended β-sheet conformation such that alternating residues are oriented to opposite sides of the binding cleft (Miller et al., 1989; Swain et al. 1991).

Retroviral processing sites are very diverse with no two sites within a virus having the identical sequence. To characterize the diversity of processing site sequences, we examined a database of 46 different retroviral processing sequences from ten different retroviruses (Pettit et al., 1991). We found that the sequences are hydrophobic in nature with residues becoming both less diverse and more hydrophobic closer to the scissile bond (Figure 3). The patterns of residues within processing sites were not necessarily random and definite patterns of preference and exclusion of residues were apparent. This indicated that the subsites of PR are not necessarily independent in recognition of substrate. Based on this collection, we classified processing site sequences into three groups according to the P1′ amino acid and the linked preference of the P1 residue. Type 1 sites contain Pro in the P1′ position with a preference for Phe or Tyr in P1. Type 2 sites contain Ala, Leu, or Val in the P1′ position with a preference for Leu over Phe or Tyr in the P1 position. Sites that fail to meet the requirements of these other two types with respect to the P1′ amino acid are designated type 3 and tend to have a sequence composition similar to type 2 sites. The fact that some amino acids are universally excluded from some positions (Figure 3) suggests certain residues may be incompatible with a mechanistic property common to all retroviral proteases. For example, residues with β-branched side chains are never found in the P1 position (Pettit et al., 1991).

Others have categorized the patterns of cleavage site sequences using a different collection of sequences (Pearl and Taylor, 1987). Henderson et al. grouped twenty cleavage site sequences from the primate lentiviruses HIV-1, HIV-2, and SIV into three classes (Henderson et al., 1988). In this classification, class 1 sites contain Phe or Tyr in the P1 position and show a preference for Gly or polar amino acids in the P3 position, Asn at P2, and hydrophobic residues at P2′. Class 2 sites have the sequence Phe-Leu in positions P1′–P2′ and a P4 Arg. Class 3 sites have a preference for hydrophobic residues in posi-

A

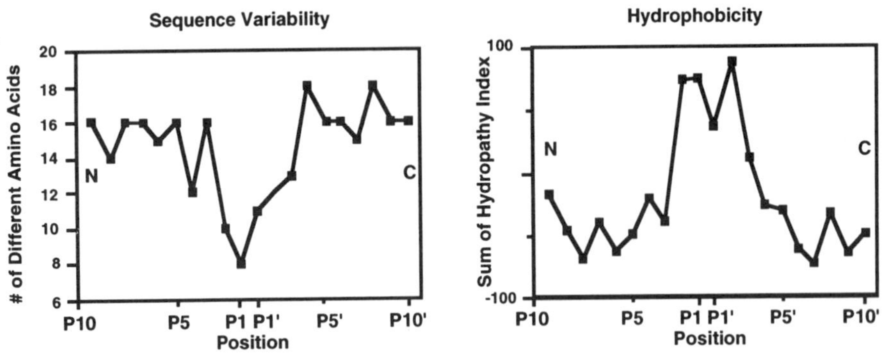

B

	P5	P4	P3	P2	P1	/	P1'	P2'	P3'	P4'	P5'
Cys	1	-	-	3	-		-	1	-	-	-
Ser	2	7	2	-	1		2	3	2	1	3
Thr	4	6	2	2	-		1	3	4	3	2
Pro	9	12	1	-	-		16	-	-	5	7
Ala	2	5	4	11	1		9	6	9	3	5
Gly	1	3	2	1	3		-	2	3	2	4
Asn	1	-	-	5	-		-	-	1	2	2
Asp	2	1	1	-	-		-	-	2	1	1
Glu	1	1	2	1	-		-	2	-	4	2
Gln	6	-	9	-	-		1	3	6	6	2
His	-	-	1	-	1		1	1	-	2	-
Arg	4	2	3	-	-		-	-	3	2	3
Lys	5	3	6	-	-		-	-	2	1	5
Met	1	-	3	1	7		1	1	3	3	2
Ile	3	-	-	6	-		-	6	2	4	4
Leu	2	2	5	3			7	7	3	3	2
Val	2	2	2	13	-		5	11	6	1	1
Phe	-	2	2	-	10		2	-	-	2	-
Tyr	-	-	1	-	7		-	-	-	1	-
Trp	-	-	-	-	-		1	-	-	-	1

Figure 3. The variability and hydrophobicity profiles of an alignment of processing site sequences collected from 46 different retroviruses. A) Left: Sequence variability shown as the number of different amino acid found in each position from P10 to P10′. Right: The average hydropathy index of the residues found in each position. The derived value is weighted for the frequency of appearance of each residue. B) The inclusion and exclusion of individual residues for the positions P5 through P5′. The slash (/) denotes the position of the scissile bond. Adapted from Pettit et al. (Pettit et al., 1991). Used with the permission of the publisher.

tions P2 through P1′ and either Gln or Glu at position p2′. Poorman et al. developed an algorithm in an attempt to defined the specificity of the HIV-1 and HIV-2 proteases from a database of known cleavage site sequences (Poorman et al., 1991). This algorithm assumed each PR subsite was independent and additive in selection of substrate. The predications of this algorithm correlated well with the results of a mutagenesis study of two HIV-1 processing sites, and it predicted some known processing sites in native protein precursors, but erroneously predicted the location of other sites (Partin et al., 1990). Nevertheless, linked amino acid preference in processing site sequences have been observed (Henderson et al., 1988; Pettit et al., 1991), rationalized structurally, and demonstrated experimentally (Griffiths et al., 1992). Thus, there are inherent constraints on the diversity of processing site sequences. These constraints may result from mechanistic limitations of PR activity or other viral functions that may lie within processing site sequences. We further examined the possibility that processing site sequences are functionally constrained by the rate in which they are cleaved. In this case, variability in processing site sequence would allow the sites to be cleaved at different rates. The variability in cleavage rate would, in turn, function in the control of virion morphogenesis.

Regulation of Gag Cleavage: A Role for PR

It has been recognized for some time that the processing sites in the HIV-1 Gag precursor are cleaved at different rates in infected cells, resulting in a sequential order of cleavage (Mervis et al., 1988; Gowda et al., 1989). During ordered processing of Gag in infected cells, defined intermediates appear transiently before the final products. Among the processing intermediates commonly observed are three species of approximately 39–41 kDa that result from single cleavage events of the Pr55 Gag precursor at either the MA/CA, CA/p2, or p2/NC processing sites (Veronese et al., 1987; Mervis et al., 1988; Veronese et al., 1988; Gowda et al., 1989). The p25–CA intermediate, which is the immediate precursor to p24–CA, contains a fourteen amino acid carboxy-terminal extension designated p2 (Veronese et al., 1987; Gowda et al., 1989). Another transient intermediate is p15, a 15 kDa species consisting of NC–p1–p6 (Veronese et al., 1987; Gowda et al., 1989) which is later cleaved in the formation of the final products (Mervis et al., 1988; Henderson et al., 1992). While the composition of Gag intermediates is well defined, less is known about the exact order of cleavage in cells, or the significance of sequential processing in virion maturation.

We utilized an in vitro Gag processing assay to clarify events associated with the cleavage of the HIV-1 Gag precursor by the PR and to circumvent some of the limitations associated with using infected cells (Erickson-Viitanen et al., 1989; Krausslich et al., 1989; Partin et al., 1990; Tritch et al., 1991; Sheng and Erickson-Viitanen, 1994). In this assay Gag is produced by translation of synthetic mRNA in a rabbit reticulocyte lysate and is mixed with recombinant viral protease. During the subsequent processing of Gag, transient intermediates and final products are generated that are similar to those observed in infected cells (Figures 4 and 5). The relative rate of cleavage of each individual

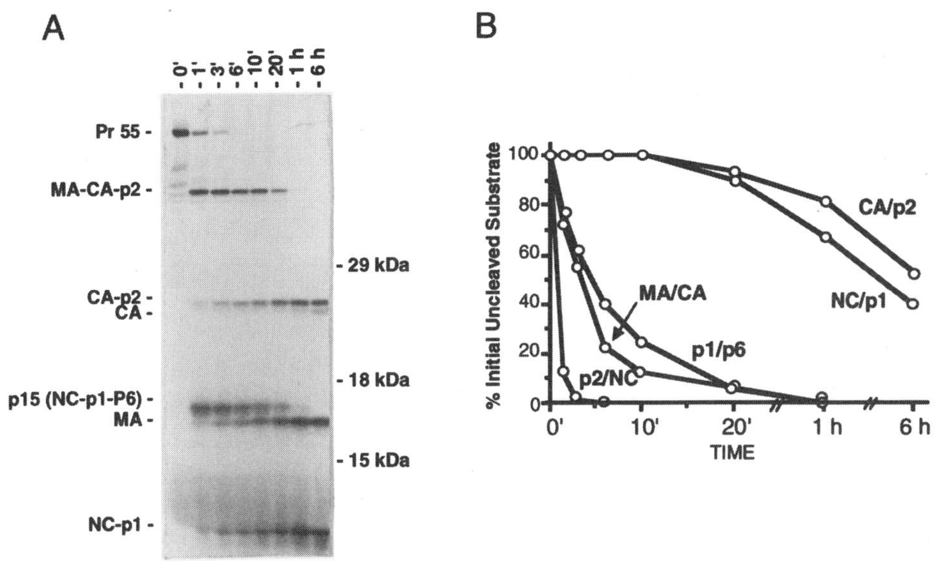

Figure 4. Sequential processing of HIV-1 Gag in vitro at pH 7. (A) Labeled Gag was produced by in vitro transcription followed by translation in rabbit reticulocyte lysate. Purified HIV-1 protease was added, aliquots taken at the times indicated, and the products of the reaction separated by SDS-polyacrylamide gel electrophoresis (SDS-PAGE). The composition of the transient processing intermediates and final products are shown to the left. (B) The time course of cleavage of each of the Gag processing sites. The percent of each of the uncleaved Gag processing sites over time was determined as previously described (Pettit et al., 1994).

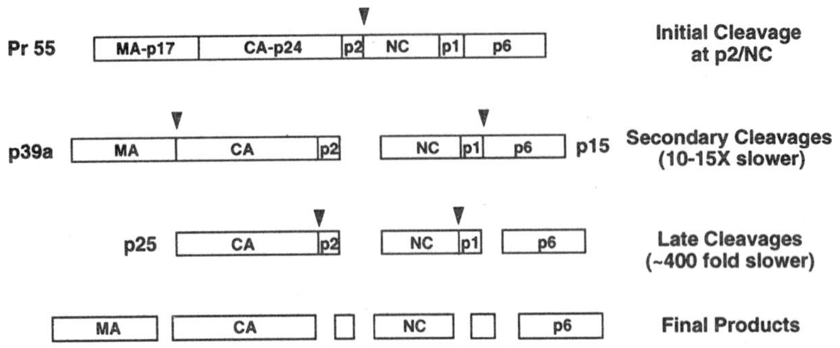

Figure 5. Schematic of the sequential processing cascade of HIV-1 Gag in vitro at pH 7. The composition of processing intermediates are shown. Based on their relative rates of cleavage, the Gag sites were separated into three general classes. Initial cleavage occurred at the p2/NC site followed by two secondary cleavages at the MA/CA and p1/p6 sites (~10-fold slower). In the final cleavage events (~400 fold slower than the p2/NC site), the small peptide domains of p1 and p2 were removed from NC and CA, respectively. The order of cleavages and the composition of processing intermediates were similar to that observed in the virion.

cleavage site can easily be estimated from the composition and quantity of the intermediates during the timecourse of the assay. The order of cleavages in the Gag precursor in vitro at pH 7 is p2/NC > p1/p6 ~ MA/CA > NC/p1 > CA/p2 (Figure 4B), with a maximal difference in cleavage rate of 400-fold at neutral pH (Tritch et al., 1991; Pettit et al., 1994). At pH 5.0, near the kinetic optimum for PR, the difference in cleavage rates is only 15-fold, which suggests that processing site sequences are optimized to maximize differences in cleavage rates at physiological pH. The two Gag sites that differed the most flanked the 14 amino acid p2 domain, suggesting a possible role for p2 in regulating sequential processing of Gag.

Having characterized the relative rate of cleavages of the Gag sites, we probed for additional levels of regulation beyond processing site sequence alone by examining the independence of the Gag cleavage events. Using mutagenesis, cleavage at each site was blocked by substitution of a P1 Ile in individual mutant clones. We next examined the effect of inhibiting cleavage at the mutated site on the rate of cleavage of the other four (unmutated) Gag sites. We found that cleavage at the Gag processing sites generally occurred independently; however, cleavage at the slow CA/p2 site was accelerated 20-fold when initial cleavage of the downstream p2/NC site was blocked (Pettit et al., 1994). This experimental evidence strongly suggests that the p2 domain of HIV-1 Gag functions as part of a regulatory mechanism to slow the rate of cleavage at the CA/p2 site during sequential processing.

We further investigated the role of p2 in regulating sequential processing by deleting the p2 domain from the Gag precursor. The deletions were constructed to allow an evaluation of the rate of cleavage for the sites flanking the p2 domain without the mitigating influence of p2. Three deletions were made that retained either the seven amino acid sequence of the slow CA/p2 site (Δp2-slow), the seven amino acid sequence of the initial cleavage site at p2/NC (Δp2-fast), or created a composite CA/NC processing site sequence (Δp2-composite). When the two processing sites flanking p2 were removed from the regulatory effect of p2, the difference in the rate of cleavage was reduced from 400-fold to roughly 15-fold (Figure 6), primarily through a 20-fold acceleration of cleavage of the CA/p2 site. Moreover, through the combined effects of a p2 deletion and lower pH (5.0) we observed a 100-fold acceleration of cleavage of the CA/p2 site. From these results it is

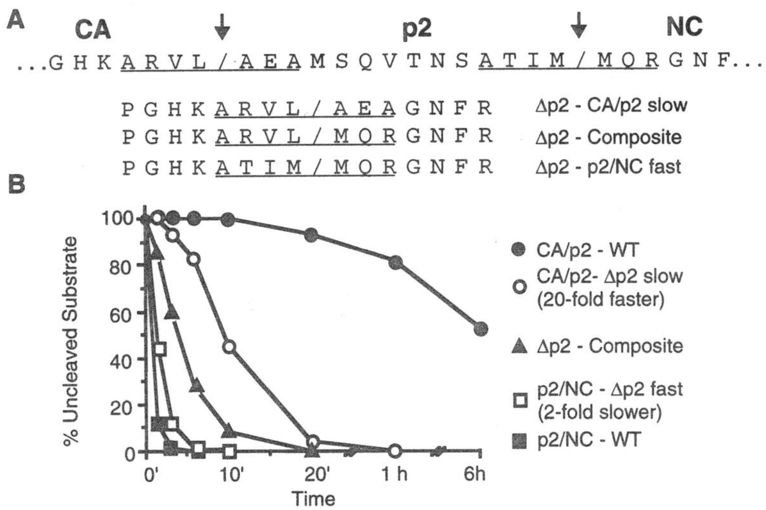

Figure 6. The influence of the p2 domain of Gag cleavage rate of the flanking sites in vitro at pH 7. (A) The p2 domain was deleted to retain the 7 amino acid processing site sequence (underlined) of the slower CA/p2 site (Δp2-slow), the faster p2/NC site (Δp2-fast), or a composite CA/NC site (Δp2-composite). (B) Time course of the cleavages of the CA/p2 site and p2/NC site without the presence of p2 (open symbols). The time course of cleavage in the two sites in wild type Gag is represented with corresponding closed symbols. Deletion of p2 resulted in 20-fold faster cleavage of the CA/p2 site and 2-fold slower cleavage of the p2/NC site (right). The time course of cleavage of the composite CA/NC (Δp2-composite) is also shown.

evident that cleavage at the CA/p2 site is downregulated significantly during the sequential processing cascade. Thus, the p2 domain of Gag contributes significantly in the regulation of sequential processing in vitro by increasing the differences in the rate of CA/p2 and p2/NC site to a level greater than that obtainable by sequence variation alone.

We examined whether the p2 domain of Gag provided a necessary function in regulation of Gag processing during virion morphogenesis. The p2 domain was deleted from the Gag precursor and the effects on processing, virion morphogenesis, and infectivity determined (Pettit et al., 1994). We chose for this analysis the Δp2-composite construction, as it removes the influence of the p2 domain on Gag processing without affecting the primary amino acid sequence of either the mature CA or NC. Virus deleted in p2 was not deficient for release from HeLa cells or in the generation of the final processed Gag products as judged by western blot. However, virions produced were not infectious, and when examined by thin section electron microscopy showed a diffuse, malformed core structure. Thus, the p2 domain of Gag has a necessary role in the regulation of sequential processing and in the formation of the normal cone-shaped core structure of the virus. This evidence strongly supports a role for regulated, sequential processing in providing the final proteins at the proper time during virion morphogenesis.

The mechanism of how the p2 domain regulates the rate of cleavage at the CA/p2 site is not known. Between HIV-1 and HIV-2, p2 is conserved in location, but is poorly conserved in sequence. Amongst HIV-1 isolates, the sequence of p2 is highly variable in composition and length (Louwagie et al., 1993). The only residues that are conserved in all isolates are the sequence of the CA/p2 processing site and the core sequence (P2 to P2′) of the p2/NC processing site. This implies that initial cleavage at the p2/NC site is an important aspect of p2 function.

Regulation of Gag Cleavage by Interaction with Viral RNA

We have also examined whether processing can be influenced by factors other than those encoded in Gag; i.e., by external factors. The NC domain, when embedded in the Gag precursor, functions in the capture of genomic RNA (Gorelick et al., 1990, Dorfman et al., 1993, Darlix et al., 1995). Once freed from Gag as a mature product, NC has several functions which include compacting the two genomic RNAs into the viral core and promoting strand-transfer (Darlix et al., 1995). NC also has activities to support reverse transcription of the genomic RNA early in the viral life-cycle, functioning as an processivity factor similar to E. coli single-stranded binding protein (You and McHenry, 1993; Ji et al., 1996).

During the sequential processing cascade of Gag, there are at least four forms of NC generated: NC in full length Gag, in the p15 early intermediate consisting of NC–p1–p6, in NC–p1, and as mature NC. We investigated whether the interaction of Gag with RNA influenced the rate at which the Gag sites were cleaved in vitro. Synthetic mRNA present during translation of Gag in the rabbit reticulocyte lysate was removed prior to the addition of PR by treatment with pancreatic RNase A. Removal of RNA resulted in roughly a 5-fold decrease in the rate of cleavage of the p15 intermediate (Figure 7) (Sheng and Erickson-Viitanen, 1994). The upregulation effect of RNA was localized to sites within the p15 intermediate (NC/p1 and p1/p6). The rates of cleavage of the early p2/NC cleavage and the upstream MA/CA cleavage were not affected. This result suggests interaction of RNA with the NC domain of Gag and p15 provides an additional level of regulation, in this case positive, during sequential processing.

We examined which residues of NC were necessary for the up-regulation of NC–p15 cleavage upon interaction with RNA by mutagenesis of NC–p15. The mature NC protein is composed of 55 amino acids and is highly basic in amino acid composition. There are two zinc finger motifs carrying the characteristic C–C–H–C motif that are flanked in between and at each end by basic residues. The mutants were designed to test which areas of NC are necessary for the up-regulation of p15 cleavage. CS1 and CS2 mutations disrupt the first and second zinc fingers, respectively, by substituting Ser for the two first Cys resides it the finger. The P3BE mutations substitutes three glutamic residues for a basic RKK motif located in the seven amino acid linker between the fingers (residues 32–34). Of the three NC mutations, only the P3BE mutation abolished upregulation of p15 cleavage in vitro by RNA (Sheng and Erickson-Viitanen, 1994). This result was somewhat surprising given the prior determination by others that the zinc fingers are absolutely

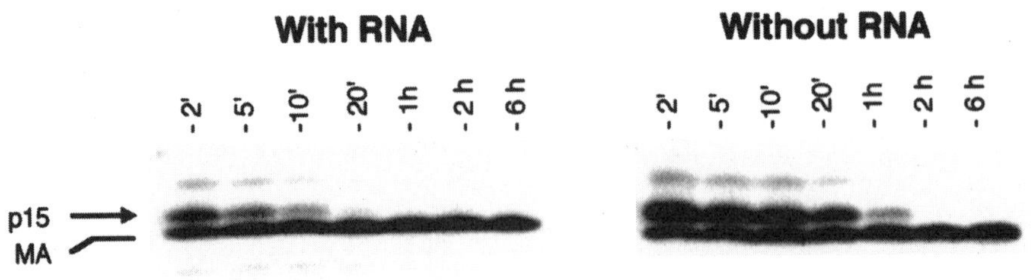

Figure 7. The stimulating effect of RNA on cleavage of the NC-p15 intermediate in vitro. In vitro HIV-1 Gag processing assays were performed as described (Pettit et al., 1994). Aliquots of the processing reaction were taken and separated by SDS-PAGE (Laemmli, 1970). The two reactions shown are identical except that RNA in the reaction on the right was removed prior to the assay by preincubation for 30′ with RNase A (100μg/ml). The position of the p15 intermediate is shown by the arrow. The lower band is the matrix protein (MA) which is generated by a separate cleavage event.

essential in the selection and capture of viral RNA in the budding virion (Gorelick et al., 1990; Dorfman et al., 1993; Gorelick et al., 1993; Dannull et al., 1994).

We examined whether the basic motif between the zinc fingers of NC functioned in the regulation of p15 cleavage during Gag processing in the context of the virion. The P3BE and CS2 mutations were transferred to an infectious clone of viral DNA and characterized for their effects on viral release, NC–p15 generation and cleavage, infectivity, and virion morphology. The CS2 mutation serves as a control, since that mutation creates a lesion in NC closer to the affected downstream processing sites than the P3BE mutations. The phenotype of the two NC mutations were similar in that there was no significant effect on release of virions from transfected HeLa cells. The two mutations were also similarly defective in the selective packaging of viral RNA and had similar significant defects in infectivity (at least 1000-fold) (data not shown). However, when the stability of the p15 intermediate was examined by pulse-chase, only the P3BE mutation was observed to significantly delay further cleavage of p15 compared to wild type (data not shown). Conversely, the CS2 mutation actually reduced NC–p15 stability. Thus, the basic RKK motif located between the zinc fingers is necessary for the up-regulation of p15 cleavage upon interaction with RNA in vitro and in the virion.

We further examined the effect of the P3BE mutation, and the loss of p15 cleavage upregulation, on virion morphology. The CS2 mutant resulted in little disruption of virion morphology compared to wild type. This is similar to the finding of Gorelick et al., who noted that virions with mutations in the zinc coordinating residues of the first or second fingers had wild type morphology (Gorelick et al., 1990). The P3BE mutation, however, resulted in virions that were severely defective in core formation. P3BE virions had a open, donut-shaped morphology similar to immature or PR deficient virions; however, these virions were different in that they contained fully processed precursors (data not shown). These observations demonstrate the importance of the basic RKK motif of NC in regulating the cleavage of p15 and in formation of the virion core.

The specificity of the interaction between RNA and wild type NC–p15 that resulted in the upregulation of cleavage of the downstream processing sites was further examined. We found the interaction to be reversible, as returning synthetic Gag mRNA to RNase treated p15 again upregulated p15 cleavage to similar levels similar to p15 not treated with RNase (Sheng and Erickson-Viitanen, 1994). The interaction responsible for upregulation appears to be specific for RNA and not other charged species. Synthetic Gag mRNA, synthetic p15 RNA, rRNA and brome mosaic virus RNA were equally able to enhance cleavage of p15 by PR. In contrast, most DNA species examined, and poly-amino acids of glutamic acid or lysine failed to upregulate cleavage of p15 in vitro, even when added in significantly higher amounts. To further characterize the nucleotide specificity of the p15–RNA interaction, the 400 nucleotide mRNA initially identified was studied as a series of smaller RNA and DNA oligonucleotide species, ranging in size from 24 to 200 nucleotides. We found that only a small fraction of synthetic oligonucleotide species tested were able to enhance p15 cleavage by the viral protease to levels comparable to that observed with synthetic mRNA encoding p15. We found further that the ability to promote cleavage was correlated to the rank order of binding as measured by a sensitive filter binding assay.

These results demonstrate that the interaction between RNA and p15 is highly specific and dependent on particular sequences or secondary structures within the RNA. In addition, the specificity of the interaction is an important aspect of core formation in the virion. Thus, it appears that the specific interaction of NC with RNA serves as a regulatory trigger to promote further cleavage of NC–p15 during the sequential processing cascade in addition to providing a necessary role in core formation during virion morphogenesis.

CONCLUSION

In this report we have presented data exploring the basis of the diversity of retroviral processing sites sequences. We started with a characterization of that diversity, and followed with an examination of the role of the sequence diversity in sequential processing of the HIV-1 Gag precursor. We found that the processing sites differ significantly in their rates of cleavage. In addition to the influence of sequence diversity in controlling the rate of cleavages, there exists a minimum of two additional mechanisms that regulate the proteolytic processing of Gag. Evidence presented here indicates that these control mechanisms are important for proper virion morphogenesis and in establishing the infectivity of virions. Taken together, these results imply that the interaction of HIV-1 PR with the Gag precursor does not occur fortuitously, but rather, is a designed, regulated, and critical aspect of the viral life cycle.

ACKNOWLEDGMENTS

We thank Lora Cavallo and Dr. Jack Giffith of the Lineberger Comprehensive Cancer Center Electron Microscopy Core Facility for assistance with electron microscopy. We also thank Marlene Rayner for helpful discussions and valuable assistance. This work was supported in part by NIH grant R01-AI 25321 to R. S.

REFERENCES

Black, P. L., Downs, M. B., Lewis, M. G., Ussery, M. A., Dreyer, G. B., Petteway, S. J. and Lambert, D. M. (1993) *Antimicrob. Agents Chemother.*, ger, R. M., Veber, D. F. and Sigal, I. S. (1988) *Biochem. Biophys. R*, **37,** 71–77.

Bryant, M. and Ratner, L. (1990) *Proc. Natl. Acad. Sci. U S A*, **87,** 523–527

Crawford, S. and Goff, S. P. (1985) *J. Virol*, **53,** 899–907.

Dannull, J., Surovoy, A., Jung, G. and Moelling, K. (1994) *Embo J.*, **13,** 1525–1533.

Darke, P. L., Nutt, R. F., Brady, S. F., Garsky, V. M., Ciccarone, T. M., Leu, C. T., Lumma, P. K., Freidin*es. Commun.*, **156,** 297–303.

Darlix, J. L., Lapadat, T. M., de, R. H. and Roques, B. P. (1995) *J. Mol. Biol.*, **254,** 523–537.

Dorfman, T., Luban, J., Goff, S. P., Haseltine, W. A. and Gottlinger, H. G. (1993) *J. Virol.*, **67,** 6159–6169.

Erickson-Viitanen, S., Manfredi, J., Viitanen, P., Tribe, D. E., Tritch, R., Hutchison III, C. A., Loeb, D. D. and Swanstrom, R. (1989) *AIDS Res. Hum. Retroviruses*, **5,** 577–591.

Gorelick, R. J., Chabot, D. J., Rein, A., Henderson, L. E. and Arthur, L. O. (1993) *J. Virol.*, **67,** 4027–4036.

Gorelick, R. J., Nigida, S. J., Bess, J. J., Arthur, L. O., Henderson, L. E. and Rein, A. (1990) *J. Virol.*, **64,** 3207–3211.

Gottlinger, H. G., Sodroski, J. G. and Haseltine, W. A. (1989) *Proc. Natl. Acad. Sci. USA*, **86,** 5781–5785.

Gowda, S., Stein, B. and Engleman, E. (1989) *J. Biol. Chem*, **264,** 8459–8462.

Griffiths, J. T., Phylip, L. H., Konvalinka, J., Strop, P., Gustchina, A., Wlodawer, A., Davenport, R. J., Briggs, R., Dunn, B. M. and Kay, J. (1992) *Biochemistry*, **31,** 5193–5200.

Hansen, M., Jelinek, L., Whiting, S. and Barklis, E. (1990) *J. Virol.*, **64,** 5306–5316.

Henderson, L. E., Benveniste, R. E., Sowder, R., Copeland, T. D., Schultz, A. M. and Oroszlan, S. (1988) *J. Virol.*, **62,** 2587–2595.

Henderson, L. E., Bowers, M. A., Sowder, R. 2., Serabyn, S. A., Johnson, D. G., Bess, J. J., Arthur, L. O., Bryant, D. K. and Fenselau, C. (1992) *J. Virol.*, **66,** 1856–1865.

Henderson, L. E., Copeland, T. D., Sowder, R. C., Schultz, A. M. and Oroszlan, S. (1988) In *Human retroviruses, cancer, and AIDS: approaches to prevention and therapy.* (Ed, Bolognesi, D.) Alan R. Liss, Inc., New York, pp. 135–147.

Jacks, T., Power, M. D., Maslarz, F. R., Luciw, P. A., Barr, P. J. and Varmus, H. E. (1988) *Nature*, **331,** 280–283.

Ji, X., Klarmann, G. J. and Preston, B. D. (1996) *Biochemistry*, **35,** 132–143.

Kaplan, A. H., Zack, J. A., Knigge, M., Paul, D. A., Kempf, D. J., Norbeck, D. W. and Swanstrom, R. (1993) *J. Virol.*, **67,** 4050–4055.

Katoh, I., Yoshinaka, Y., Rein, A., Shibuya, M., Odaka, T. and Oroszlan, S. (1985) *Virology,* **145,** 280–292.

Kohl, N. E., Emini, E. A., Schleif, W. A., Davis, L. J., Heimbach, J. C., Dixon, R. A., Scolnick, E. M. and Sigal, I. S. (1988) *Proc. Natl. Acad. Sci. USA,* **85,** 4686–4690.

Kraussiich, H. G., Ingraham, R. H., Skoog, M. T., Wimmer, E., Pallai, P. V. and Carter, C. A. (1989) *Proc. Natl. Acad. Sci. USA,* **86,** 807–811.

Laemmli, U. K. (1970) *Nature,* **277,** 680–685.

Louwagie, J., McCutchan, F. E., Peeters, M., Brennan, T. P., Sanders-Buell, E., Eddy, G. A., van der Groen, G., Fransen, K., Gershy-Damet, G.-M., Deleys, R. and Burke, D. S. (1993) *AIDS,* **7,** 769–780.

Mervis, R. J., Ahmad, N., Lillehoj, E. P., Raum, M. G., Salazar, F. H., Chan, H. W. and Venkatesan, S. (1988) *J. Virol.,* **62,** 3993–4002.

Miller, M., Schneider, J., Sathyanarayana, B. K., Toth, M. V., Marshall, G. R., Clawson, L., Selk, L., Kent, S. B. H. and Wlodawer, A. (1989) *Science,* **246,** 1149–1152.

Moore, M. L., Bryan, W. M., Fakhoury, S. A., Magaard, V. W., Huffman, W. F., Dayton, B. D., Meek, T. D., Hyland, L., Dreyer, G. B., Metcalf, B. W., Strickler, J. E., Gorniak, J. G. and Debouck, C. (1989) *Biochem. Biophys. Res. Comm.,* **159,** 420–425.

Partin, K., Krausslich, H. G., Ehrlich, L., Wimmer, E. and Carter, C. (1990) *J. Virol.,* **64,** 3938–3947.

Pearl, L. H. and Taylor, W. R. (1987) *Nature,* **328,** 482.

Pettit, S. C., Moody, M. D., Wehbie, R. S., Kaplan, A. H., Nantermet, P. V., Klein, C. A. and Swanstrom, R. (1994) *J. Virol,* **68,** 8017–8027.

Pettit, S. C., Simsic, J., Loeb, D. D., Everitt, L., Hutchison III, C. A. and Swanstrom, R. (1991) *J. Biol. Chem.,* **266,** 14539–14547.

Poorman, R. A., Tomasselli, A. G., Heinrikson, R. L. and Kezdy, F. J. (1991) *J. Biol. Chem.,* **266,** 14554–14561.

Ratner, L., Fisher, A., Jagodzinski, L. L., Mitsuya, H., Liou, R.-S., Gallo, R. C. and Wong-Staal, F. (1987) *AIDS Res. Hum. Retroviruses,* **3,** 57–69.

Rhee, S. S. and Hunter, E. (1987) *J. Virol.,* **61,** 1045–1053.

Ross, E. K., Fuerst, T. R., Orenstein, J. M., ONeill, T., Martin, M. A. and Venkatesan, S. (1991) *AIDS Res. Hum. Retroviruses,* **7,** 475–483.

Schatzl, H., Gelderblom, H. R., Nitschko, H. and K., v. d. H. (1991) *Arch. Virol,* **120,** 71–81.

Sheng, N. and Erickson-Viitanen, S. (1994) *J. Virol.,* **68,** 6207–6214.

Sommerfelt, M. A., Petteway, S. R. J., Dreyer, G. B. and Hunter, E. (1992) *J. Virol.,* **66,** 4220–4227.

Swain, A. L., Gustchina, A. and Wlodawer, A. (1991) In *Structure and function of the aspartic proteinases,* Vol. 306 (Ed, Dunn, B. M.) Plenum Press, New York, pp. 433–441.

Tozser, J., Gustchina, A., Weber, I. T., Blaha, I., Wondrak, E. M. and Oroszlan, S. (1991) *FEBS Lett.,* **279,** 356–360.

Tritch, R. J., Cheng, Y. E., Yin, F. H. and Erickson-Viitanen, S. (1991) *J. Virol.,* **65,** 922–930.

Veronese, F. D., Copeland, T. D., Oroszlan, S., Gallo, R. C. and Sarngadharan, M. G. (1988) *J. Virol,* **62,** 795–801.

Veronese, F. D., Rahman, R., Copeland, T. D., Oroszlan, S., Gallo, R. C. and Sarngadharan, M. G. (1987) *AIDS Res. Hum. Retroviruses,* **3,** 253–264.

Wagner, R., Fliessbach, H., Wanner, G., Motz, M., Niedrig, M., Deby, G., von, B. A. and Wolf, H. (1992) *Arch. Virol.,* **127,** 117–137.

You, J. C. and McHenry, C. S. (1993) *J. Biol. Chem.,* **268,** 16519–16527.

3

A CELLULAR ANTI-APOPTOSIS PROTEIN IS CLEAVED BY THE HIV-1 PROTEASE

Bruce D. Korant, Peter Strack, Michelle W. Frey, and Christopher J. Rizzo

DuPont Merck Pharmaceutical Co.
Experimental Station
Wilmington, Delaware 19880-0336

1. ABSTRACT

Cleavage of non-viral proteins is rarely observed with the HIV-1 protease (HIV pr). One such cleavage event occurs with bcl-2, an important cytoprotective protein. The loss of bcl-2 has biological consequences, leading to enhanced HIV replication and programmed death of the host cell. A strategy is proposed to suppress HIV with non-cleavable mutants of bcl-2.

2. INTRODUCTION

The essential role of HIV pr in processing viral precursor proteins has been under intense study, culminating recently in three important new antiviral drugs which are potent, selective, competitive inhibitors of viral protein cleavages (reviewed in 1). Also reported are cleavages of various cellular polypeptides by the viral protease (reviewed in 2), but those proteolytic events are poorly understood with respect to their relevance in virus replication.

Interest in cellular proteins as HIV pr substrates grew from our earlier efforts to construct mammalian cell-based assays dependent on the action of the viral protease. We reported previously the splitting of firefly luciferase by HIV pr (3). The luciferases are potentially ideal reporters for cell-based assays, because the light-emitting reactions they catalyze are performed extremely efficiently (quantom yield approaching 1.0 for the firefly enzyme), and because measurement is quantitative, rapid and specific, with negligible background activity in animal cells. The only drawback is the initial cost of the luminometer instrument.

During construction of cell-based assays based on splitting of firefly luciferase by the HIV pr, we found that cytocidal action of HIV pr (4) was based on induction of apoptosis or programmed death in protease expressing cell (5). We searched for potential substrates and found that the cytoprotective protein bcl-2 was selectively cleaved by HIV pr, and here propose a model for HIV replication based upon proteolysis-initiated suicide of HIV infected lymphocytes.

3. HIV pr CLEAVES THE CYTOPROTECTIVE PROTEIN bcl-2

Cell death initiated by HIV closely parallels the course of disease in infected people. The mechanism is unclear, but likely involves a direct effect of one or more viral products on the cells rather than immuno-modulation or growth factor deprivation (see 5 for a recent review). We observed the direct splitting of the cytoprotective protein bcl-2 in cells or cell-free extracts exposed to small quantities of HIV pr, primarily at a site between F112 and A113 located in a sequence in bcl-2 which is similar to an optimal HIV pr processing site. Although this could be incidental, it subsequently leads to a marked increase in oxidative stress in the cells and a lowering of intracellular pH; both conditions which favor HIV replication. Moreover, these conditions are also antecedents of death of cells by apoptosis or programmed suicide.

Why should a virus adopt such a lethal strategy? We propose that it accomplishes two goals of the virus. First, it activates transcription of viral RNA, and second it provides an additional way for the virus to ensure its spread in the human host, since apoptotic cell fragments, after being scavenged, lead to infection of macrophages, and transport to distant organs of the host, including possibly the nervous system (Figure 1).

4. CONCLUSIONS

The recent introduction of three new HIV pr inhibitors as antiviral treatments for AIDS is an important milestone in treating this catastrophic infectious disease. Because the viral protease may be the central actor in the virus strategy to kill host cells as it replicates, the new drugs are predicted to be highly effective at protecting the immune and nervous systems after HIV infection, as well as suppressing "viral load," i.e., the quantity of infectious virus in circulation. A separate option we are pursuing is to modify the gene

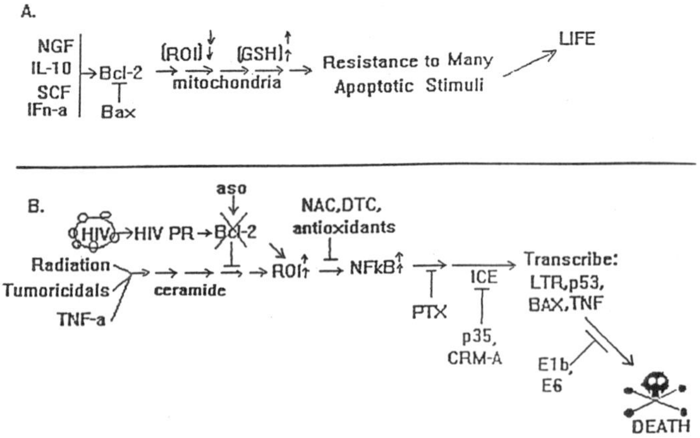

Figure 1. Scheme for regulation of HIV replication and apoptosis by bcl-2.

for bcl-2 so that the protein is less susceptible to cleavage by the HIV pr. This is not completely straightforward; altered bcl-2 must also retain its cytoprotective properties, and there are many amino acids in bcl-2 conserved throughout evolution from nematode to man, in particular in a portion of the molecule where the primary HIV cleavage site is located. However, if the proper mutein can be identified and its gene delivered in a suitable vector, it may make available an antiviral therapy to many HIV-positive people who cannot use the current anti-HIV drugs.

ACKNOWLEDGMENTS

Key contributions to this study were made by P. Strack and M. Frey, using support provided by the DuPont Merck postdoctoral fellowship program. J. Corman carried out protein sequence analysis.

REFERENCES

1. J. Bartlett. Protease inhibitors for HIV infection. Ann. Intern. Med. **124**, (1996), 1086–1087.
2. A. Tomasselli and R. Heinrikson. Specificity of retroviral protease: An analysis of viral and non-viral protein substrates, Meth. Enzymol. **241**, (1994), 279–301.
3. B. Korant, Z. Lu, P. Strack and C. Rizzo. HIV protease mutations leading to reduced inhibitor sensitivity, Adv. Exp. Med. Biol., **389** (1996), 241–246.
4. H.-G. Kraüsslich. Human immunodeficiency virus proteinase dimer as a component of the viral polyprotein prevents particle assembly and viral infectivity, Proc. Natl. Acad. Sci. USA **88**, (1991), 3213–3217.
5. P. Strack et al. Apoptosis mediated by the HIV protease is preceded by cleavage of bcl-2, Proc. Natl. Acad. Sci. USA **93**, (1996), 9571–9576.

4

ENGINEERING CATALYTICALLY DEFECTIVE FORMS OF HIV PROTEASE TO MODULATE ITS ACTIVITY

Charles S. Craik, Lilia M. Babé, Deborah Dauber, Fiona McPhee, Jason Rosé, and Ayce Ünal

Department of Pharmaceutical Chemistry
University of California at San Francisco
San Francisco, California 94143

INTRODUCTION

The protease encoded in the HIV-1 genome plays a crucial role in the life cycle of the virus, cleaving the $p55^{gag}$ and $p160^{gag\text{–}pol}$ precursors into their mature and functional forms Since either genetic inactivation or chemical inhibition of the protease results in the production of non-infectious viral particles, the development of inhibitors of the HIV-1 protease has been pursued as a means of preventing viral replication. Although many inhibitors now exist that can block viral maturation and slow viral replication *in vivo*, resistance to the inhibitors over a prolonged period of time is often observed [2–5]. We are interested in defining the role of the protease in the viral life cycle by engineering variant forms of the enzyme with known properties. This approach has led us to consider an alternate method of regulating the activity of HIV protease for potential therapeutic intervention.

The retroviral proteases are members of the aspartyl protease family [1,6], but unlike other mammalian proteases of this family, they consist of two identical subunits. Dimerization is a prerequisite for activity, since each monomer contributes one catalytic aspartyl residue to the active site of the enzyme [7,8]. The protease dimerization process is driven by the formation of a four-stranded β-sheet, composed of the five N- and five C-terminal residues from each monomer. This interface is a very stable structure, with a dissociation constant of 5–50 nM [9–11]. Particle assembly, protease activation, precursor processing and virion budding must occur in a coordinated fashion to yield infectious particles [12].

Assembly of viral capsids from the $p55^{gag}$ and $p160^{gag\text{–}pol}$ polyprotein precursors occurs at the plasma membrane, and cleavage to their mature forms occurs during particle release from the host cell [13]. Autoproteolytic activation of the protease follows the dimerization of two $p160^{gag\text{–}pol}$ polyproteins during the assembly process and is a prerequisite for

Aspartic Proteinases, edited by James.
Plenum Press, New York, 1998.

all subsequent processing events [14]. Viral particles in which the protease has been mutated to remove the active site aspartic acid residues will assemble in the absence of proteolytic activity, but the resultant virions are non-infectious and morphologically aberrant [1,15,16]. Thus, assembly of a viral protease dimer where one monomer is catalytically defective will result in an inactive protease dimer and immature non-infectious virions.

DEFINING THE LEVEL OF HIV-1 PROTEASE ACTIVITY

Although it is known that elimination of PR activity prevents viral maturation, detailed information regarding the minimal proteolytic activity required for formation of an infectious virion was not available. It has been reported that in the presence of limiting concentrations of protease inhibitors it is possible to partially inhibit the protease, resulting in virions with small defects in processing and large decreases in infectivity [17]. We wanted to analyze the effect of decreased levels of protease activity on viral maturation and infectivity, and determine if a correlation exists between the level of protease activity and infectivity.

We examined the requirement for protease activity in the maturation process by studying protease mutants with specifically diminished catalytic activity. Three mutations were engineered in the active site region of the protease that resulted in enzymes with reduced proteolytic activity, Thr-26 to Ser (T26S), Ala-28 to Ser (A28S), Asp-25 to Asn (D25N). These variant enzymes were expressed in *E. coli* using a bicistronic expression system, and analyzed kinetically for effects on substrate binding and catalytic efficiency [18]. The catalytic efficiency (k_{cat}) of HIV-1 protease is decreased 4-fold in the T26S variant and approximately 50-fold in the A28S variant. To analyze the effect of the protease mutants on virion maturation and infectivity, genes containing the single protease mutations were cloned into an HIV proviral vector, pHIV-gpt (Figure 1), which produced all of the viral gene products except envelope [19]. Transfection of this vector into mammalian cells permitted the formation of viral capsids which were analyzed with regard to RNA content, polypeptide composition, and degree of maturation [19]. By co-transfection with an envelope-encoding plasmid (Figure 1) it was possible to generate infectious particles to analyze the effects of altered polyprotein processing on viral infectivity. Virions containing the T26S protease variant, in which there is 25% protease activity are very similar to wild-type virions, although slight reductions in infectivity are observed. In contrast, virions containing the A28S protease variant are not infectious, even though a limited amount of polyprotein processing does occur. There appears to be a linear correlation between the level of protease activity and particle infectivity as shown in Figure 2 (infectivity vs k_{cat}/K_m). This suggests that a threshold of protease activity exists between 4-fold and

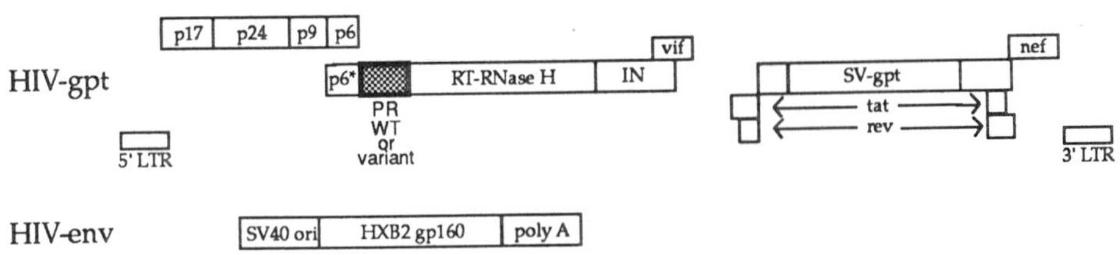

Figure 1. Schematic of plasmid constructs encoding the HIV-1 genome. See text for description. LTR, long terminal repeat.

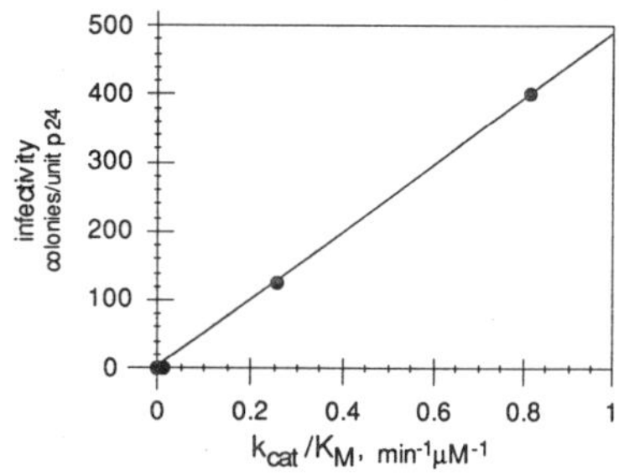

Figure 1. Correlation between HIV-1 infectivity and catalytic efficiency (k_{cat}/K_m) of HIV-1 protease. Values plotted for HIV PR Q7K, D25N, T26S, A28S are from reference 18.

50-fold reduction, below which processing is insufficient to yield infectious particles. This data also suggests that a reduction of protease activity by 50-fold or greater is sufficient to prevent the formation of infectious particles. Our results indicate that although lowering the protease activity does not appear to affect the viral capsid assembly process, reduction of activity by 50-fold or greater is sufficient to prevent the formation of infectious virions.

CATALYTICALLY DEFECTIVE HIV-1 PROTEASE MONOMERS PREVENT PROTEASE ACTIVATION AND VIRION MATURATION BY TRANS-DOMINANT INHIBITION

It has been shown that HIV particle assembly and protease activation occur in a coordinated fashion and the cascade of proteolytic events leading to viral maturation could be blocked by protease inhibition. Thus, we proposed that protease inhibition could also be achieved by forming catalytically defective protease dimers. The term "dominant negative" was applied to describe the ability of defective (wt) gene products to interfere with the function of wild-type gene products *in vivo* [20]. The use of dominant negative inhibition has been proposed for a number of HIV proteins [21] and demonstrated for variants of gag [22] and rev [23]. The inhibition due to variant rev was shown to occur through the formation of inactive protein complexes [24]. More recently, T-lymphocytes stably transfected with mutant *rev* [25] or *gag* genes [26] were shown to inhibit the growth of HIV-1 following infection. A similar strategy may be more effective at a stage of the viral life cycle dependent upon a catalytic function such as proteolytic processing.

Initial *in vitro* experiments with purified proteins suggested that mixing of wt and mutant protease subunits could lead to inactivation in a concentration-dependent fashion [27]. It was assumed that this loss of activity was due to the formation of inactive heterodimers through dynamic subunit exchange. We reasoned that assembly of inactive heterodimers *in vivo* would also result in the loss of protease function and a block in viral maturation. This event could be monitored by the subsequent accumulation of unprocessed polyproteins and the secretion of non-infectious virions as diagramed in Figure 3. The addition of defective protease monomers could also serve to titrate proteolytic activity over a broad range and thus provide information about the threshold of activity needed to produce infectious virus.

We have assayed the formation of inactive protease heterodimers by co-expressing proteins from the HIV-gpt proviral DNA carrying either the wt genome [19] or one with a mutation in the protease gene that results in an inactive enzyme. The mutation that converts the viral protease active site Asp25 to Asn (D25N) has been well characterized *in vitro* and *in vivo* [1] and was used in these studies [28]. The trans-dominant inhibitory activity of a

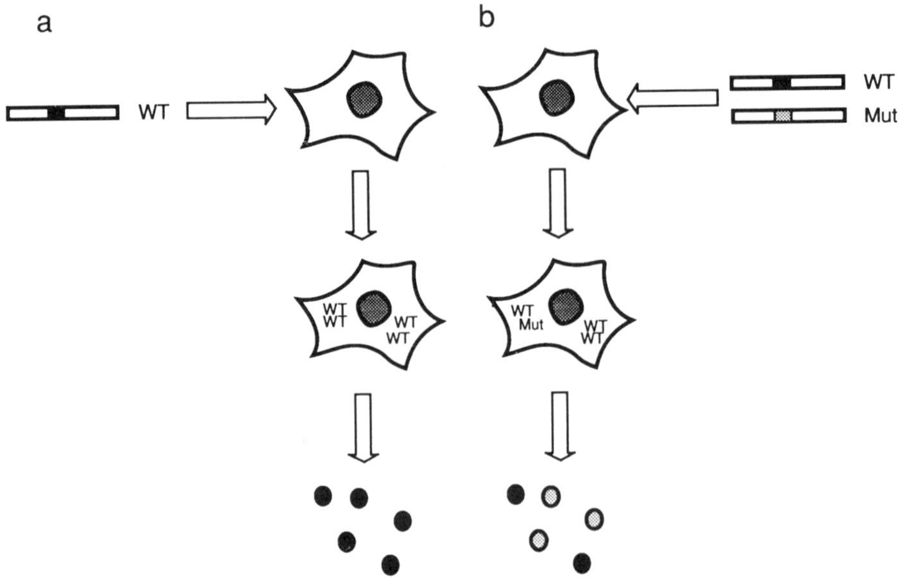

Figure 3. Schematic for transfection experiments with wild-type and trans-dominant HIV-1 protease inhibitors. (a) A plasmid encoding the HIV-1 genome (see Figure 1) is transfected into tissue culture cells (NIH 293T, HeLaT4 or CEM-SS) and infectious particles (dark circles) result. (b) Plasmids encoding the HIV-genome and a wild-type protease or an engineered trans-dominant inhibitor variant are co-transfected into tissue culture cells. Infectious particles (dark circles) or immature particles (gray circles) result.

defective protease monomer with the D25N mutation was measured by transient transfection (Figure 3). A proviral plasmid that included the drug-selectable *E. coli* gpt gene was used to deliver the wild-type or mutant proteases to cultured cells. Co-expression of the wt proviral DNA (HIV-gpt) with increasing amounts of the mutant proviral DNA (HIV-gpt D25N) resulted in a concomitant decrease in proteolytic activity as monitored by *in vivo* viral polyprotein processing (Figure 4). The viral particles resulting from inactivation of the protease were mostly immature, consisting predominantly of unprocessed $p55^{gag}$ and $p160^{gag\text{-}pol}$ polyproteins. In the presence of HIV-1 gp160 env, the number of secreted non infectious particles correlated with the presence of increasing amounts of the defective protease. Greater than 97% reduction in infectivity was observed at a 1:6 ratio of wt to defective protease DNA. This provides an estimate of the level of inhibition required for effectively preventing virion processing that is greater than 95% of HIV protease activity. Stable expression of the defective protease in monkey kidney cells reduced the yield of infectious particles from these cells by 90% upon transfection with the wt proviral DNA [28]. These results show that defective subunits of the viral protease exert a trans-dominant inhibitory effect resulting from the formation of catalytically compromised heterodimers *in vivo*, ultimately yielding non infectious viral particles.

ENGINEERING HIV-1 PROTEASE HETERODIMERS AS MACROMOLECULAR INHIBITORS OF VIRAL MATURATION

Crystallographic structures of HIV-1 protease complexed with peptide mimetics and non-peptide inhibitors have played a major role in obtaining highly potent and specific

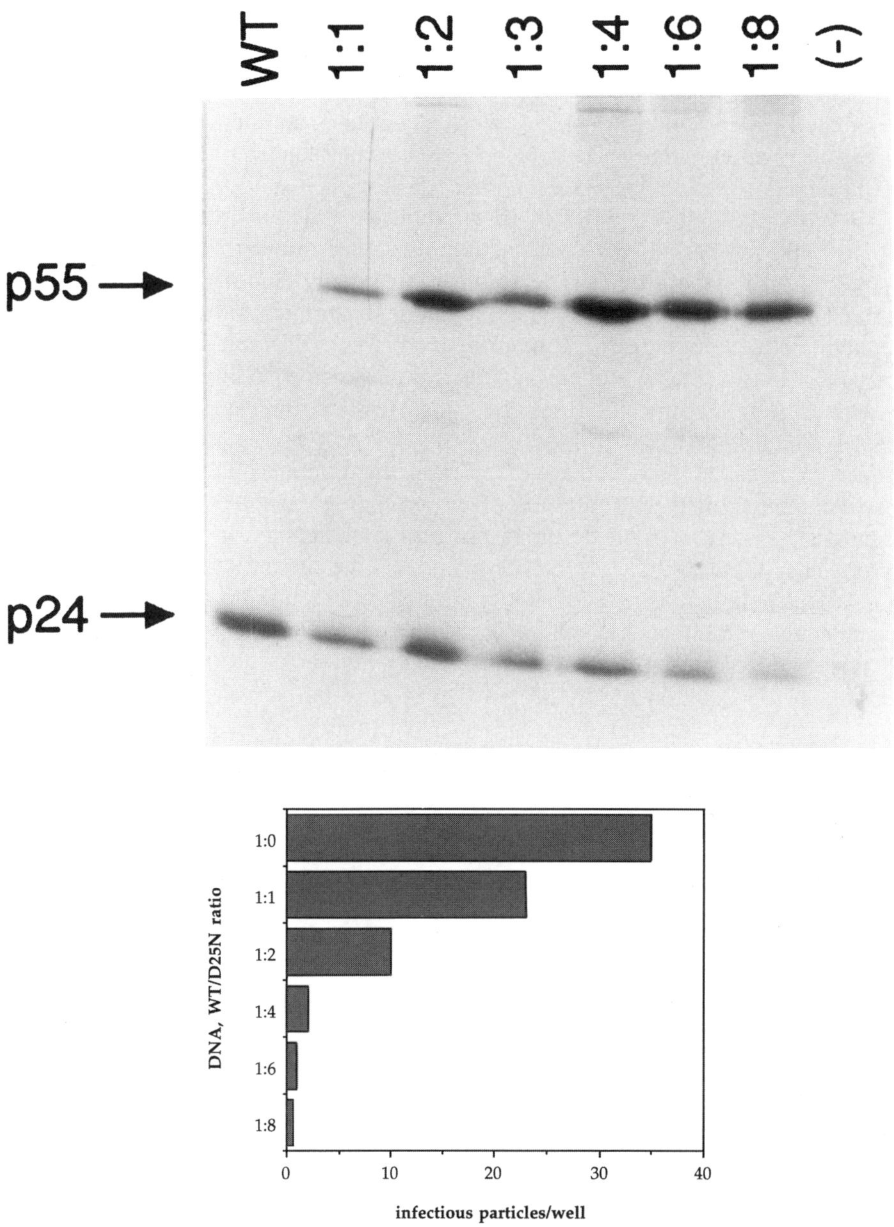

Figure 4. Inhibition of polyprotein processing by co-transfection of wt HIV-gpt and increasing amounts of mutant HIVD25N-gpt proviral DNAs. COS-7 cells were infected with proviral DNAs in the ratios shown and supplemented with the HIV-env plasmid encoding HIV-1 gp160. Culture supernatant was collected 48 post transfection. At that time, fresh culture media containing ^{35}[S]-methionine was added to label newly synthesized proteins over an 8h period. The supernatant containing radioactive virions was subjected to sucrose sedimentation gradients (as described by Page et al., 1990 (ref. 19)). (a) The separation of radiolabeled viral proteins on 12% SDS-PAGE followed by autoradiography is shown. (b) The number of infectious particles in the respective culture supernatants. These values were obtained by infecting HeLaT4 cells and applying mycophenolic acid to select infected cells. Cell colonies develop which can be stained and counted as described in reference 19.

small molecule inhibitors for this protease. Although numerous inhibitors have demonstrated efficacy in cell culture and clinical studies, resistant variants of HIV-1 PR towards the various structurally diverse protease inhibitors is now well documented [29, and references therein]. Nonetheless, the initial antiviral efficacy of these inhibitors still validates the protease as a therapeutic target. While strategies involving multiple protease inhibitors may eventually block the emergence of resistant variants, an alternative approach that is less sensitive to point mutations may be preferable. One such alternative is a gene therapy treatment targeting the viral protease dimerization requirement.

We have modeled a number of HIV-1 protease structures to aid in the design of mutations in the protease dimer interface to promote heterodimer formation and decrease the stability of the mutant protease homodimer interaction. This structure-based approach identified amino acid substitutions [25,49,50] at the dimer interface of HIV-1 protease that could facilitate preferential association of heterodimers and inhibit self-association of the defective monomers [30,31] (Figure 5). Proviral plasmids encoding the designed, defective proteases were constructed. Transient co-transfection of human kidney 293 cells with equimolar amounts of the wild type plasmid (HIV-gpt) and the plasmids carrying the designed proteases showed decreased particle maturation and infectivity. The defective proteases exhibited different inhibitory effects depending upon the engineered amino acid substitution (Figure 6). These designed protease subunits appear to be *trans*-dominant inhibitors since the activity of wild type dimers can be titrated out in a dose-dependent fashion [31]. These results show that it is possible to design HIV protease monomers as macromolecular inhibitors that can be used in conjunction with or as an alternative to small molecule inhibitors for controlling the enzymatic activity of HIV-1 protease. The

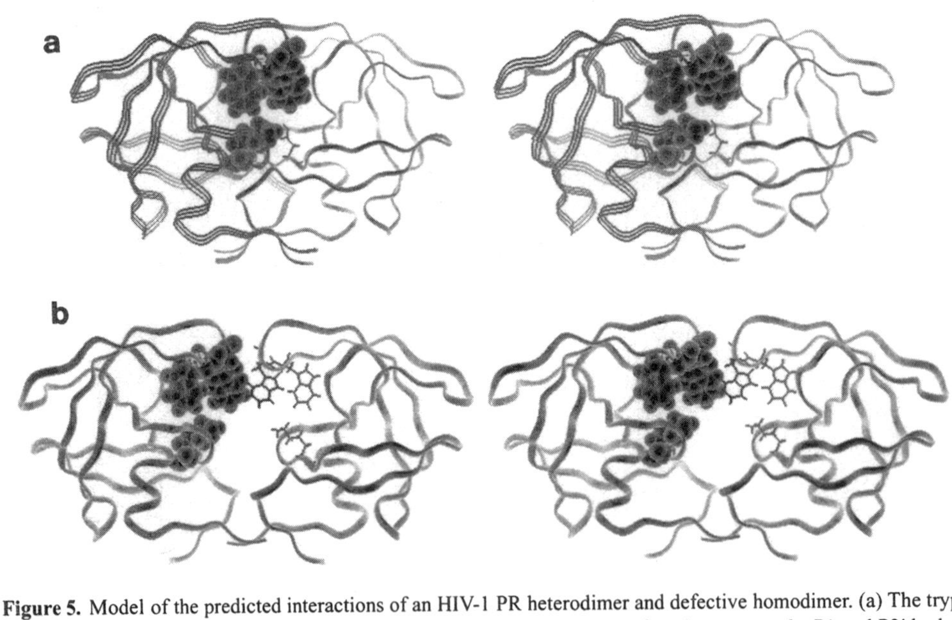

Figure 5. Model of the predicted interactions of an HIV-1 PR heterodimer and defective homodimer. (a) The tryptophan side chains at positions 49 and 50 of the defective PR monomer are found to occupy the P1 and P2′ hydrophobic sites, respectively, of the opposing wild-type monomer. The lysine at position 25 is shown interacting with the aspartic acid of the opposing wild-type monomer. These additional interactions between the variant and the wild-type monomer suggest that the Asp25Lys, Gly49Trp, Ile50Trp (KWW) substitutions could enhance binding affinity on dimerization and enhance the dominant negative effect. (b) Model of a defective protease homodimer showing the repulsive interactions that would decrease self association.

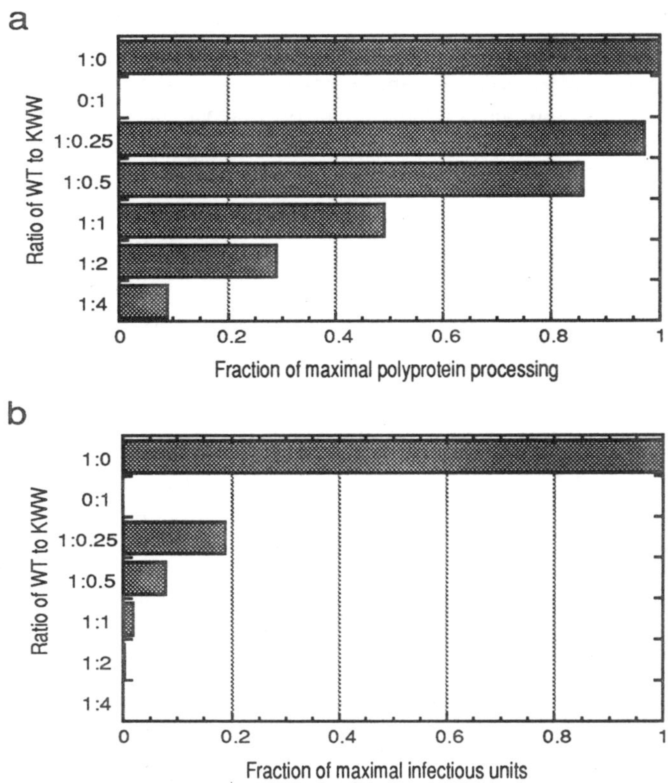

Figure 6. Dose-dependent inhibition of polyprotein processing and infectivity by co-transfection of wild-type and engineered proviral DNA. Analysis of the inhibition of HIV-1 PR polyprotein processing and infectivity by co-transfection of plasmids encoding wild-type and engineered PR monomers at varying ratios. (a) Quantitation of the effects of HIV-1 PR KWW on HIV-1 polyprotein processing. Purified viral capsids were quantitated by p24 ELISA before and after digestion with exogenous viral PR. The p24 values obtained for untreated samples indicated the amount of processed capsid protein (and partially processed p25). The p24 values obtained after *in vitro* processing indicated the amount of total capsid protein (unprocessed and processed). (b) Supernatants of the experiments discussed in (a) were used to infect 293-T cells to determine the titer of infectious particles present. See reference 31 for experimental procedures.

potency of defective HIV protease monomers as macromolecular antivirals may reflect the essential nature of the multiple catalytic steps this enzyme must perform for proper viral maturation and assembly. Furthermore, unlike a small molecule inhibitor, a macromolecular inhibitor can have a greater interacting surface area with its target. This greater surface area can be an advantage in overcoming the effects of amino acid substitutions that result from the error prone replication cycle of HIV.

As mentioned above, dominant negative inhibition of HIV replication has already been shown for defective forms of gag [22] and rev [23]. We show here that a dominant negative inhibitor can be made that targets an enzyme already shown to be a sensitive point in the HIV life cycle, the protease. We then use molecular modeling to design more potent versions of the dominant negative inhibitor and demonstrate that this strategy is a valid approach for designing macromolecular inhibitors. A similar strategy could be applied to other multimeric targets including reverse transcriptase and integrase. Therefore, obstacles could be set at different stages of the viral life cycle. Although it is promising that several different defective gene products have shown antiviral effects when expressed in HIV-1 infected T cells [25,26,32,33], significant advances have to be made regarding the delivery of such macromolecular inhibitors to $CD4^+$ cells in a therapeutic setting. However, the use of exceptionally potent defective protease monomers as trans-dominant inhibitors offers a unique possibility for a single or multifaceted gene therapy approach in the clinical treatment of HIV-1 infection.

ACKNOWLEDGMENTS

C.S.C. is a recipient of NIH grant GM39552 and D.S.D. is supported by an NSF Graduate Fellowship.

REFERENCES

1. Kohl NE, Emini EA, Schleif WA, Davis LJ, Heimbach JC, Dixon RA, Scolnick EM and Sigal IS, Active human immunodeficiency virus protease is required for viral infectivity. *Proc Natl Acad Sci U S A* **85**(13): 4686–90, 1988.
2. Martin LN, Soike KF, Murphey-Corb M, Bohm RP, Roberts ED, Kakuk TJ, Thaisrivongs S, Vidmar TJ, Ruwart MJ, Davio SR and et al., Effects of U-75875, a peptidomimetic inhibitor of retroviral proteases, on simian immunodeficiency virus infection in rhesus monkeys. *Antimicrob Agents Chemother* **38**(6): 1277–83, 1994.
3. McQuade TJ, Tomasselli AG, Liu L, Karacostas V, Moss B, Sawyer TK, Heinrikson RL and Tarpley WG, A synthetic HIV-1 protease inhibitor with antiviral activity arrests HIV-like particle maturation. *Science* **247**(4941): 454–6, 1990.
4. Tomasselli AG, Howe WJ, Sawyer TK, Wlodawer A and Heinrikson RL, The complexities of AIDS: an assessment of the HIV protease as a therapeutic target. *Chim. Oggi* : 6–274, 1991.
5. Wlodawer A and Erickson JW, Structure-based inhibitors of HIV-1 protease. *Annu Rev Biochem* **62**: 543–85, 1993.
6. Pearl LH and Taylor WR, A structural model for the retroviral proteases. *Nature* **329**(6137): 351–4, 1987.
7. Wlodawer A, Miller M, Jaskolski M, Sathyanarayana BK, Baldwin E, Weber IT, Selk LM, Clawson L, Schneider J and Kent SB, Conserved folding in retroviral proteases: crystal structure of a synthetic HIV-1 protease. *Science* **245**(4918): 616–21, 1989.
8. Navia MA, Fitzgerald PM, McKeever BM, Leu CT, Heimbach JC, Herber WK, Sigal IS, Darke PL and Springer JP, Three-dimensional structure of aspartyl protease from human immunodeficiency virus HIV-1. *Nature* **337**(6208): 615–20, 1989.
9. Babé LM, Rosé J and Craik CS, Synthetic "interface" peptides alter dimeric assembly of the HIV 1 and 2 proteases. *Protein Science* **1**(10): 1244–53, 1992.
10. Cheng YS, Yin FH, Foundling S, Blomstrom D and Kettner CA, Stability and activity of human immunodeficiency virus protease: comparison of the natural dimer with a homologous, single-chain tethered dimer. *Proc Natl Acad Sci U S A* **87**(24): 9660–4, 1990.
11. Zhang ZY, Poorman RA, Maggiora LL, Heinrikson RL and Kezdy FJ, Dissociative inhibition of dimeric enzymes. Kinetic characterization of the inhibition of HIV-1 protease by its COOH-terminal tetrapeptide. *J Biol Chem* **266**(24): 15591–4, 1991.
12. Krausslich HG, Human immunodeficiency virus proteinase dimer as component of the viral polyprotein prevents particle assembly and viral infectivity. *Proc Natl Acad Sci U S A* **88**(8): 3213–7, 1991.
13. Stephens EB and Compans RW, Assembly of animal viruses at cellular membranes. *Annu Rev Microbiol* **42**: 489–516, 1988.
14. Debouck C, Gorniak JG, Strickler JE, Meek TD, Metcalf BW and Rosenberg M, Human immunodeficiency virus protease expressed in Escherichia coli exhibits autoprocessing and specific maturation of the gag precursor. *Proc Natl Acad Sci U S A* **84**(24): 8903–6, 1987.
15. Gottlinger HG, Sodroski JG and Haseltine WA, Role of capsid precursor processing and myristoylation in morphogenesis and infectivity of human immunodeficiency virus type 1. *Proc Natl Acad Sci U S A* **86**(15): 5781–5, 1989.
16. Peng C, Ho BK, Chang TW and Chang NT, Role of human immunodeficiency virus type 1-specific protease in core protein maturation and viral infectivity. *J Virol* **63**(6): 2550–6, 1989.

17. Kaplan AH, Zack JA, Knigge M, Paul DA, Kempf DJ, Norbeck DW and Swanstrom R, Partial inhibition of the human immunodeficiency virus type 1 protease results in aberrant virus assembly and the formation of noninfectious particles. *J Virol* **67**(7): 4050–5, 1993.
18. Rosé JR, Babe LM and Craik CS, Defining the level of human immunodeficiency virus type 1 (HIV-1) protease activity required for HIV-1 particle maturation and infectivity. *J Virol* **69**(5): 2751–8, 1995.
19. Page KA, Landau NR and Littman DR, Construction and use of a human immunodeficiency virus vector for analysis of virus infectivity. *J Virol* **64**(11): 5270–6, 1990.
20. Herskowitz I, Functional inactivation of genes by dominant negative mutations. *Nature* **329**(6136): 219–22, 1987.
21. Feinberg MB and Trono D, Intracellular immunization: trans-dominant mutants of HIV gene products as tools for the study and interruption of viral replication. *AIDS Res Hum Retroviruses* **8**(6): 1013–22, 1992.
22. Trono D, Feinberg MB and Baltimore D, HIV-1 Gag mutants can dominantly interfere with the replication of the wild-type virus. *Cell* **59**(1): 113–20, 1989.
23. Malim MH, Bohnlein S, Hauber J and Cullen BR, Functional dissection of the HIV-1 Rev trans-activator--derivation of a trans-dominant repressor of Rev function. *Cell* **58**(1): 205–14, 1989.
24. Hope TJ, Klein NP, Elder ME and Parslow TG, trans-dominant inhibition of human immunodeficiency virus type 1 Rev occurs through formation of inactive protein complexes. *J Virol* **66**(4): 1849–55, 1992.
25. Bahner I, Zhou C, Yu XJ, Hao QL, Guatelli JC and Kohn DB, Comparison of trans-dominant inhibitory mutant human immunodeficiency virus type 1 genes expressed by retroviral vectors in human T lymphocytes. *J Virol* **67**(6): 3199–207, 1993.
26. Smythe JA, Sun D, Thomson M, Markham PD, Reitz MS, Jr., Gallo RC and Lisziewicz J, A Rev-inducible mutant gag gene stably transferred into T lymphocytes: an approach to gene therapy against human immunodeficiency virus type 1 infection. *Proc Natl Acad Sci U S A* **91**(9): 3657–61, 1994.
27. Babé LM, Pichuantes S and Craik CS, Inhibition of HIV protease activity by heterodimer formation. *Biochemistry* **30**(1): 106–11, 1991.
28. Babé LM, Rosé J and Craik CS, Trans-dominant inhibitory human immunodeficiency virus type 1 protease monomers prevent protease activation and virion maturation. *Proc Natl Acad Sci U S A* **92**(22): 10069–73, 1995.
29. Condra JH, Schleif WA, Blahy OM, Gabryelski LJ, Graham DJ, Quintero JC, Rhodes A, Robbins HL, Roth E, Shivaprakash M and et al., In vivo emergence of HIV-1 variants resistant to multiple protease inhibitors. *Nature* **374**(6522): 569–71, 1995.
30. Dauber D, McPhee F, Ünal A and Craik CS, Optimization of a macromolecular inhibitor of HIV-1 protease. In: *Structure and Function of Aspartic Proteinases: Retroviral and Cellular Enzymes* (Ed. James MNG). Plenum Publishing Corporation, New York, 1997.
31. McPhee F, Good AC, Kuntz ID and Craik CS, Engineering HIV-1 protease heterodimers as macromolecular inhibitors of viral maturation. *Proc. Natl. Acad. Sci.* **93:** 11477–81, 1996.
32. Escaich S, Kalfoglou C, Plavec I, Kaushal S, Mosca JD and Bohnlein E, RevM10-mediated inhibition of HIV-1 replication in chronically infected T cells. *Hum Gene Ther* **6**(5): 625–34, 1995.
33. Junker U, Escaich S, Plavec I, Baker J, McPhee F, Rosé JR, Craik CS and Böhnlein E, Intracellular expression of HIV-1 protease variants inhibits replication of wild-type and protease inhibitor resistant HIV-1 strains in human T cell lines. *J. Virology* **70:** 7765–72, 1996.

5

THE ASPARTIC PROTEINASE FROM EQUINE INFECTIOUS ANAEMIA VIRUS

D. J. Powell,[1] D. Bur,[2] A. Wlodawer,[3] A. Gustchina,[3] B. M. Dunn,[4] and J. Kay[1]

[1]School of Molecular and Medical Biosciences
University of Wales
Cardiff CF1 3US, Wales, United Kingdom
[2]F. Hoffmann La Roche AG
CH-4070 Basel, Switzerland
[3]Macromolecular Structure Laboratory
NCI-Frederick Cancer Research and Development Center
Frederick, Maryland 21702
[4]Department of Biochemistry and Molecular Biology
University of Florida Medical School
Gainesville, Florida 32610

Swamp fever or equine infectious anaemia is an infectious disease caused by a non-oncogenic retrovirus. The primary mode of transmission between horses is via a blood-sucking insect. Equine infectious anaemia virus (EIAV) is closely related to other lentiviruses of ruminants and primates, such as HIV, SIV and FIV (feline immunodeficiency virus). Just as with these other viruses, EIAV encodes an aspartic proteinase which is essential for processing *gag* and *gag-pol* polyproteins in order to generate new infectious viral particles.[1] The sequences of EIAV and HIV-1 proteinases have ~30% identity overall. With the large number of HIV-1 PR inhibitors that have been generated,[2] including those now licensed by the FDA (Invirase™, Norvir™ and Crixivan™), it was important to establish whether any of these might be effective against EIAV PR. Since no effective treatment currently exists for affected horses, this might be a useful approach for management of this wasting disease.

A systematic evaluation of more than 20 potent HIV-proteinase inhibitors, including Invirase™, Norvir™ and Crixivan™, revealed that almost all of these compounds were completely unable to block the activity of purified recombinant EIAV proteinase.[3] In this respect, the proteinase from EIAV behaves like a drug-resistant mutant of HIV proteinase that has arisen under selective drug pressure.[4] Only one inhibitor (HBY-793) of HIV proteinase showed comparable potency towards the EIAV enzyme.[3] Compounds closely related to HBY-793 such as A-76889 and A-77003, which differ from each other only in their stereo-

Aspartic Proteinases, edited by James.
Plenum Press, New York, 1998.

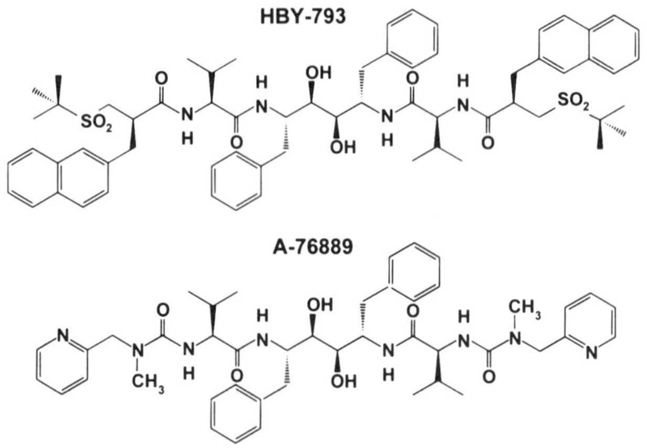

Figure 1. Structures of the symmetrical inhibitors HBY-793 and A-76889.

chemistry, and which have substituents from P_2 through P_2' that are identical in structure to those of HBY-793 (Fig. 1), did not inhibit EIAV proteinase.[3] The differences in specificity of EIAV and HIV-1 proteinases would thus appear to be at least partially explained by interactions between enzyme and inhibitor (or substrate) at the periphery of the active site cleft.

Insight into these aspects was gained from an X-ray structure that was solved at 1.8Å resolution and refined to a crystallographic R-factor of 0.136 of a complex between HBY-793 and a mutant form of EIAV proteinase.[5] In this mutant, Ile54 was replaced by glycin.e[3] The reciprocal change (G48V) which occurs at the equivalent position in HIV-1 proteinase has been detected in the resistant virus that emerges in some patients in response to drug treatment.[6] All of the amino acids which contact a bound substrate or inhibitor from S_3/S_3' to S_1/S_1' are identical in EIAV and HIV proteinases with only three exceptions. These are located at residue 30 (Thr in EIAV; Asp in HIV), residue 56 in EIAV proteinase = residue 50 in HIV-1 proteinase and the previously mentioned Ile54 in EIAV which corresponds to Gly48 in the HIV-1 enzyme.[5]

The potency of the symmetrical inhibitor HBY-793 towards the I54G mutant EIAV PR was not changed from that against the wild-type enzyme (Table 1).

In contrast, the ability of the I54G mutant EIAV proteinase to cleave the pseudo-symmetrical peptide substrate

Ac–Tyr–Arg–Ala–Arg–Val–Phe∗Nph–Val–Arg–Ala–Ala–Lys

Table 1. Kinetic constants for the interaction of ligands with wild-type and mutant EIAV proteinases

	Substrate 1			HBY-793
Proteinase	K_m (μM)	k_{cat} (sec^{-1})	k_{cat}/K_m ($sec^{-1}mM^{-1}$)	K_i (nM)
Wild-type	15	7	470	0.1
I54G	250	7	28	0.1
I54G, Y48H	240	3	10	6

Values were measured at pH 4.7 using substrate 1 =

Ac–Tyr–Arg–Ala–Arg–Val–Phe∗Nph–Val–Arg–Ala–Ala–Lys

was more than an order of magnitude lower than that of the wild-type EIAV proteinase, primarily as a consequence of the ability of the enzyme to bind the substrate, as reflected in the K_m value (Table 1). Even larger differences between the wild-type and I54G mutant enzymes were measured when other substrates were used for the comparison.[3] Thus, the presence/absence of the two 4-carbon side-chains (in the homodimer) of isoleucine at residue 54 has a considerable influence on the positioning of substrate and its turnover but does not appear to disrupt the binding of the one inhibitor (HYB-793; Table 1) that is effective against EIAV proteinase. The precise nature and fit between ligand and enzyme is thus of crucial significance and has been described in detail previously.[3,5]

In contrast to these findings measured with the I54G single mutant enzyme, a different effect was observed when a second mutation (Tyr→His) was introduced at position 48. The resultant double mutant (I54G, Y48H) EIAV proteinase had a specificity constant that was, at most, only 3-fold lower than that measured for the single mutant (I54G) enzyme towards the long, pseudo-symmetrical substrate

Ac–Tyr–Arg–Ala–Arg–Val–Phe*Nph–Val–Arg–Ala–Ala–Lys

whereas the potency of the (shorter) HBY-793 inhibitor was reduced by 60-fold compared to the K_i values measured against the wild-type or I54G mutant proteinases (Table 1). Residue 48 can be seen in the crystal structure solved for EIAV PR to be situated at the periphery of the β-hairpin loops or flaps of the enzyme[5] in a position where it would not be able to contact the inhibitor and thus influence its binding directly (Fig. 2).

In the EIAV structure, a four stranded β-sheet consisting of two inverted hairpin loops are evident in the region of position 48, one strand stretching from Phe63 to Val82,

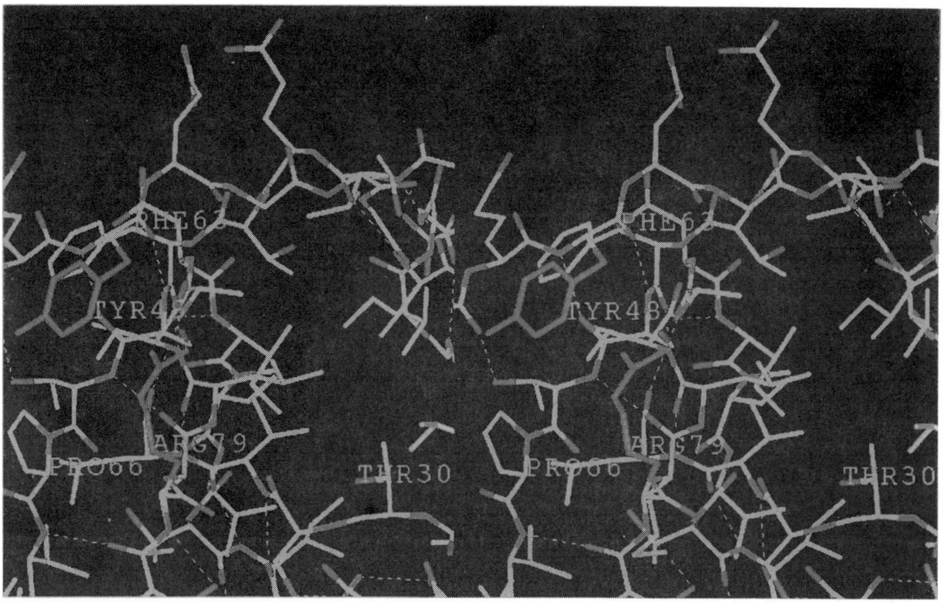

Figure 2. Stereo diagram depicting the location of Tyr48 in EIAV proteinase. The guanidino group of Arg79 is positioned immediately to the right, and the ring of Pro66 is evident below and behind the ring of Tyr 48. The residues contributing to the two strands of the β-hairpin loop (flap) are evident in the upper right hand quadrant. The side-chain of the P_2Val residue of the inhibitor sits in the S_2 pocket. Hydrogen bonds are marked with dashed lines.

the other, with Asn16 at its apex, extends from Pro9 to Leu23. This Asn residue is perfectly pinned down by H-bonding, from the side-chain amide bonds to the main chain amide of Thr68 in one direction, and to the main chain carbonyl of Tyr43. This very stable 4-stranded sheet thus positions Arg79 at the top of the right-hand strand (Fig. 2), such that the 3 methylene groups of its side-chain can make hydrophobic interactions with the ring of Pro66 (and potentially an H-bond between the side-chain amido group and the side-chain oxygen of Ser64). The guanidino moiety is stacked with the ring of the Tyr48 side-chain such that the P_i cloud of electrons from the Tyr ring makes an electrostatic interaction with the positive charge on Arg79 (Fig. 2). Thus, this Tyr residue is in a position where it may be predicted to be important for the mechanism that opens and closes the flaps to enable substrate and inhibitors to gain access to the active site of retroviral proteinases.[7] In the I54G/Y48H double mutant, a His residue occupies this position. At acidic pH values, such as 4.7 at which the assays (Table 1) were conducted, the His side-chain would be expected to be positively charged, causing electrostatic repulsion between itself and the similarly charged guanidino moiety of Arg79. Since the main-chain and most of the side-chain of Arg79 is strictly defined (see above), energy would have to be expended for it to move. This is unlikely, and so it is more probable that it is the His residue that has to seek an alternative location. This relocation could relay itself along the length of the strand, resulting in an alternative flap conformation in the homodimer, and the anomalous binding of substrates or inhibitors.

Since this Tyr48His replacement occurs on both flaps of the homodimeric enzyme, it is not unreasonable to conclude that the interaction of a potent inhibitor, HBY-793, with the I54G, Y48H double mutant proteinase would be particularly susceptible to any minute changes in the conformation of the two flaps. In contrast, the longer and more flexible substrate (Table 1) may itself be able to make suitable, reciprocal adjustments in the positioning of its peripheral residues such that the scissile peptide bond is still located in an appropriate juxtaposition to the catalytic apparatus to enable catalysis to proceed comparably by the I54G, Y48H double mutant and I54G single mutant enzymes.

The crucial role of retroviral proteinases in viral replication has led to the development of an enormous variety of inhibitors targetted against these enzymes. Treatment with proteinase inhibitors does, however, result in viral mutation and many drug-resistant mutations of HIV-1 proteinase have been described.[4,8] The replacement amino acids that are introduced into the mutant HIV-1 virus sometimes occur naturally at the equivalent positions in the wild-type proteinases from other retroviruses. EIAV proteinase might thus be thought of as a particularly drug-resistant mutant of HIV-1 proteinase since it readily cleaved HIV-1 substrates but was resistant to all but one of the potent HIV-1 inhibitors tested.[3] Contacts between residues of substrate/inhibitor with components of the enzyme distant from the catalytic apparatus appear to have a crucial role, yet most of the inhibitors developed as anti-AIDS drugs are usually designed to be as short as possible for pharmacokinetic and bioavailability purposes. It is evident then that consideration of these facets may allow the realisation of more effective therapies for the future treatment of these debilitating viral diseases in both humans and horses.

ACKNOWLEDGMENTS

This work was supported by awards from the UK AFRC (for DJP), by the National Cancer Institute's Dept. of Health and Human Services, U.S.A. (to A.W.) under contract with ABL and from the NIH (Grant AI 18571 to BMD). The valuable contributions made to this work by Dr. J. Knolle (Hoechst AG, Frankfurt, Germany), Drs S. L. Payne and S. F. J. Le Grice (Case Western Reserve University School of Medicine, Cleveland, OH, U.S.A.) in providing HBY-793, viral DNA clones and inspiration and encouragement, respectively, are gratefully acknowledged and much appreciated.

REFERENCES

1. Rushlow, K., Peng, X. X., Montelaro, R. C. & Shih, D. S. (1992). *Virology* **188**, 396–401.
2. Wlodawer, A. & Erickson, J. E. (1993). *Annu. Rev. Biochem.* **62**, 543–85.
3. Powell, D. J., Bur, D., Wlodawer, A., Gustchina, A., Payne, S. L., Dunn, B.M. & Kay, J. (1996). *Eur. J. Biochem.* **241**, 664–674.
4. Erickson, J. W. (1995). *Nature Struct. Biol.* **2,** 523–529.
5. Gustchina, A., Kervinen, J., Powell, D. J., Zdanov, A., Kay, J. & Wlodawer, A. (1996). *Protein Sci.* **5**, 1453–1465.
6. Jacobsen, H., Hanggi, M., Ott, M., Duncan, I.B., Owen, S., Andreoni, M., Vella, S. & Mous, J. (1996). *J. Infect. Dis.* **173**, 1379–1287.
7. Collins, J. R., Burt, S. K. & Erickson, J.W. (1995). *Nature, Struct. Biol.* **2**, 334–338.
8. Condra, J.H., Schleif, W.A., Blahy, O.M., Gabryelski, L.J., Graham, D. J., Qunitero, J. C., Rhodes, A., Robbins, H.L., Roth, E., Shivaprakash, M., Titus, D., Yang, T., Teppler, H., Squires, K. E., Deutsch, P. J. & Emini, E. A. (1995). *Nature* **374**, 569–571.

6

THE EFFECT OF SUBSTRATES ON THE KINETICS AND THE *IN VIVO* THRESHOLD ACTIVITY OF MUTANT HIV-1 PROTEASES

Jacques Ermolieff, Xinli Lin, and Jordan Tang

Protein Studies Program
Oklahoma Medical Research Foundation
University of Oklahoma Health Sciences Center
Oklahoma City, Oklahoma 73104

INTRODUCTION

The role of the human immunodeficiency virus (HIV) protease in the processing of *gag* and *gag-pol* polyproteins is essential for the maturation and the life cycle of the virus. The great potential of HIV protease as a therapeutic target for treating the acquired immunodeficiency syndrome, AIDS, has now been confirmed by the recent success of HIV inhibitor drugs to suppress HIV propagation in clinical trials (Ho *et al.*, 1995; Wei *et al.*, 1995; Coffin, 1995). In spite of the success, a serious problem associated with the clinical use of HIV protease inhibitor drugs is the rapid development of drug resistance by HIV. The basis of this drug resistance is the rapid propagation of HIV in the human body and the selection of resistance strains of the virus containing mutant proteases in the presence of inhibitor drugs (Coffin, 1995). The molecular basis of the resistance, therefore, resides on the properties of mutant proteases of the resistant HIV strains, which must have sufficient catalytic activity to support the viral propagation, yet a reduced sensitivity toward the inhibitors. This separate modulation of catalytic and inhibition activities in mutant HIV-1 proteases have now been demonstrated in many laboratories.

In order to predict the viability and resistant potential of different mutant HIV-1 proteases, it is important to directly compare the catalytic and inhibition activities of wild-type and mutant HIV-1 proteases. We have proposed that the value of k_{cat}/K_m can be used for direct comparison because under the intravirion processing conditions, the effective substrate concentrations should be below the K_m values of the substrates (Tang and Hartsuck, 1995). Knowing the inhibition constants, K_i values, the intravirion proteolytic processing activities (called 'Mutation Modulated Activity' or MMA) of any HIV-1 mutant can be calculated in the presence or absence of an inhibitor (Tang and Hartsuck, 1995). By comparing k_{cat}/K_m value of the wild-type HIV-1 protease to those of the HIV-1 mutants

Aspartic Proteinases, edited by James.
Plenum Press, New York, 1998.

from *in vitro* or *in vivo* experiments, the 'threshold activity' of different mutant HIV-1 proteases can be calculated as percentage of that of the wild-type enzyme (Tang and Hartsuck, 1995). In order to assess the viability and resistance potential of different mutant proteases, it is essential to accurately determine their relative k_{cat}/K_m values.

However, for assay convenience, nearly all the kinetic parameters of mutant HIV-1 proteases have been obtained by using synthetic chromogenic substrates. Our previous calculations of MMA and 'threshold' values were derived from kinetic data using a nine-residue substrate which is cleaved at a Leu-paranitrophenylalanyl bond (Lin *et al.*, 1995). How do different substrates, especially the substrates derived from the native cleavage sites, affect the relative values of k_{cat}/K_m for different mutant HIV-1 proteases have not been clearly demonstrated.

We report in this paper the kinetic data of selected mutant HIV-1 protease measured using peptide substrates derived from natural processing sites in *gag-pol*. Our results indicate that the low solubility of many of these substrates with natural processing sequence post a serious handicap to the kinetic measurement. We have also found considerable variations for the relative k_{cat}/K_m values among mutant enzymes.

EXPERIMENTAL PROCEDURES

Peptide substrates are synthesized in the Molecular Biology Resource Center on this campus. The preparation of the wild-type and some mutant enzymes are as previously described (Ido *et al.*, 1991; Lin *et al.*, 1995). New mutant enzymes, L90M, G48V, I84V and G48V/L90M were prepared in a similar way as previously described (Lin *et al.*, 1995).

The sequences of three peptide substrates are as follows: Substrate A, derived from the cleavage site between MA and CA is VSQNY*PIVQ (* represents the cleavage site). Substrate D, derived from the processing site between NC and p6, is KGRPGNF*LQSRP. Substrate E, derived from the cleavage site at the N-terminus of the protease, is RQGTVSFNF*PQITL. Some charged residues are added to positions outside of P4 and P4′ in order to improve the solubilities.

The assays for the hydrolysis of substrates A, D and E were carried out in 0.1 M Na acetate containing 0.1 M NaCl and 1% dimethylsulfoxide, pH 5.0 at 37°C. The reaction was stopped by the addition of guanidine-HCl to a final concentration of 7 M. The hydrolysates were separated on reverse-phase HPLC in a Beckman Gold instrument using a 4.6 × 250 mm C-18 column. The solvent system was H_2O/acetonitrile. The product peaks were quantitated using the integration software provided by the HPLC manufacturer. The kinetic parameters were calculated by fitting the data to Michaelis-Menten equation using Enzfitter program (Leatherbarrow, 1987).

RESULTS AND DISCUSSION

k_{cat} and K_m Values of Wild-Type and Mutant HIV-1 Proteases Measured Using Peptides Derived from the *gag-pol* Processing Sites

The kinetic constants of the wild-type and 8 mutant HIV-1 proteases are shown in Table 1. Substrate A is sufficiently soluble for the collection of k_{cat} and K_m data from all 9 enzymes. The kinetic parameters for 4 of the enzymes, the wild-type, G48H, V82A and V82E, had previously been determined using a chromogenic substrates (Lin *et al.*, 1995). Substrates D and E have very low solubilities under the assay conditions. The kinetic parameters can only be determined for the wild-type enzyme but can not be determined accurately for the mutant enzymes (Table 1). Our k_{cat}/K_m data of substrates A and D for the wild-type enzyme are similar to the value obtained by Tozser *et al.* (1991). However, the values for peptide E hydrolyzed by the wild-type enzyme are very different from two works.

Relative k_{cat}/K_m Values of Wild-Type and Mutant HIV-1 Proteases from Different Substrates

The relative k_{cat}/K_m values (R values) of wild-type and 3 mutant HIV-1 proteases for substrate A are shown in Table 2. For comparison, the R values for the hydrolysis of 7 chromogenic substrates by the same enzymes are also shown in Table 2.

There is apparently a great deal of variations among the R values determined from different substrates. Considering the trend, the following general conclusions can be made. (1) Even though the specific numbers do not agree well for the R values from a given enzyme, the numerical values are more consistently in the order of W.T. > G48H > V82A > V82E. (2) While small changes in substrate structure can bring larger changes in R values, substrates with very different structures can have close R values. For example, the R values for G48H hydrolysis of substrate Chr. and 5 are 1.68 and 3.59, the values for substrates 2, 3, 4 and 6, which have structures similar to that of substrates Chr. and 5, are near 0.2–0.9. On the other hand, the R value of substrate A (0.18) is closer to substrates 1 and 4 than the values between these two groups of chromogenic substrates. The discrepancy among the chromogenic substrates is mainly the 2–3 fold activity increase of G48H over the wild-type enzyme when substrates Chr. and 5 are used. The molecular basis of this activity increase is discussed in a separate chapter (Hong *et al.*) in this volume. The large discrepancy for G48H may be an exception because the R values for V82A and V82E are more internally consistent, including the comparison between values from substrate A and those from the chromogenic substrates.

From the R value comparison above, it appears that to assess the relative catalytic activity of mutant HIV-1 proteases, a single substrate may not be sufficient to provide an accurate estimation. The present results suggest the use of several substrates in order to obtain a trend of relative activities. Further search of other peptide substrates derived from the processing sites would be necessary to simplify the assay. It would also be helpful to use the peptide derived from the site of the rate-limiting step in the processing of *gag-pol*, which has not been identified currently.

In Vivo Threshold Activity for HIV-1 Mutant Viability

The '*in vivo* threshold activity', which is the level of enzymic activity required by HIV-1 protease mutants to support viable HIV-1 propagation inside human body, has not be determined. Based on the activity (defined by k_{cat}/K_m) of mutant V82A, the *in vitro* threshold activity is 26% of the activity wild-type enzyme (Tang and Hartsuck, 1995). However, previous data was based on the assay using a chromogenic substrate. The current data using the k_{cat}/K_m of V82A assayed by peptide A, the *in vitro* threshold is 14%.

The *in vivo* threshold activity can be estimated from the k_{cat} and K_m values for mutants G48V, L90M and G48V/L90M (Table 1), since these mutants have been observed in AIDS patients (Jacobsen *et al.*, 1996). The k_{cat}/K_m values of these three mutants in the same order as above are 9%, 8% and 1.6% of that of the wild-type HIV-1 protease. The lowest value from the double mutant suggest that 1.6% to be sufficient for the viability of the virus *in vivo*.

Table 1. Kinetic constants, k_{cat} and K_m, of wild-type and mutant HIV-1 proteases

	W.T.	G48V	G48H	V82A	V82E	I84V	A28S	L90M	G48V/L907
Substrate = peptide A									
K_m (mM)	1.5 ± 0.4	9.8 ± 2.7	3.3 ± 0.9	4.2 ± 0.9	10.3 ± 2.0	4.8 ± 2.2	≥70.0	1.15 ± 0.35	3.8 ± 0.7
k_{cat} (s^{-1})	51.8 ± 4.9	31.3 ± 5.2	20.1 ± 2.8	20.7 ± 1.9	30.9 ± 4.2	9.7 ± 2.8	≤19.2	3.14 ± 0.46	2.2 ± 0.2
Substrate = peptide D									
K_m (mM)	6.4 ± 2.6	≥ 40							
k_{cat} (s^{-1})	12.7 ± 4.2	≤ 0.18							
Substrate = peptide E									
K_m (mM)	0.54 ± 0.15	≥ 1.6		≥ 1.7	≥ 1.8				
k_{cat} (s^{-1})	62.7 ± 14.0	≤ 372.8		≤ 47.0	≤ 191.2				

Assay condition: 0.1 NaOAc, 0.1 M NaCl, 1% DMSO, pH 5 and 37°C.

Table 2. Relative k_{cat}/K_m values of wild-type and four mutant HIV-1 proteases[a]

		W.T.		G48H		V82A		V82E	
Substrates		k_{cat}/K_m	R values	k_{cat}/K_m	R values	k_{cat}/K_m	R values	k_{cat}/K_m	R values
A	VSQNY*PIVQ	34.5	1.0	6.09	0.18	4.93	0.14	3.00	0.087
Chr.	KARVL*XEAM	275	1.0	462	1.68	72	0.26	113	0.41
1	KARVL*XEAZG	1400	1.0	377	0.24	165	0.12	104	0.08
2	KARVZ*XEAZG	370	1.0	324	0.88	155	0.42	94	0.26
3	KARVF*XEAZG	775	1.0	639	0.83	175	0.27	176	0.23
4	KARAF*XEAZ	605	1.0	167	0.28	65	0.11	27	0.05
5	KARLZ*XEAZ	69	1.0	248	3.59	26	0.39	pc	—
6	KARIZ*XEAZ	1211	1.0	783	0.65	691	0.57	174	0.14

[a]The relative kcat/Km values, RV, were calculated for each substrate by setting the kcat/Km values of the wild-type HIV-1 protease as 1.0 and express the mutant values as ratio to the wild-type. The sequences of the peptide substrates are shown in single amino acid codes, where the non-conventional codes for norleucine is Z and for paranitrophenylalanine is X. Asterisk denotes the site of cleavage. The data for substrate A is taken from Table 1. The kinetic parameters determined from 7 chromogenic substrates are taken from Lin et al. (1995).

ACKNOWLEDGMENTS

This work is supported by NIH grant AI38189. The authors wish to thank Marcus Dehdarani for excellent assistance in this work.

REFERENCES

Coffin, J.M. (1995) *Science* **267:**483–489.

Ho, D.D., Neumann, A.U., Perelson, A.S., Chen, W., Leonard, J.M. and Markowitz, M. (1995) *Nature* **373:**123–126.

Hong, L., Zhang, C., Hartsuck, J.A., Foundling, S. and Tang, J. (1997) *Aspartic Protease* Meeting in Banff, Canada - October, 1996.

Ido, E., Han, H.-P., Kezdy, F.J. and Tang, J. (1991) *J. Biol. Chem.* **266:**24359–24366.

Jacobsen, H., Hanggi, M., Ott, M., Duncan, I.B., Owen, S., Anderoni, M., Vella, S. and Mous, J. (1996) *J. Infect. Diseases* **173:**1379–1387.

Leatherbarrow, R.J. (1987) *Enzfitter (A Nonlinear Regression Data Analysis Program for the IBM PC* Elsevier Science Publishers, Amsterdam.

Lin, Y., Lin, X., Hong, L., Foundling, S., Heinrikson, R.L., Thaisrivongs, S., Leelamanit, W., Ratherman, D., Shah, M., Dunn, B.M. and Tang, J. (1995) *Biochemistry* **34:**1143–1152.

Tang, J. and Hartsuck, J.A. (1995) *FEBS Letters* **367:**112–116.

Tozser, J., Blaha, I., Copeland, T.D., Wondrak, E.M. and Oroszlan, S. (1991) *FEBS Lett.* **281:**77–80.

Wei, X., Ghosh, S.K., Taylor, M.E., Johnson, V.A., Emini, E.A., Deutsch, P., Lipson, J.D., Bonhoeffer, S., Nowak, M.A., Hahn, B.H., Saag, M.S. and Shaw, G. M. (1995) *Nature* **373:**117–122.

7

A COMPARISON OF *gag–pol* PRECURSOR CLEAVAGE IN NATURALLY ARISING HIV VARIANTS

Gregory Bloom,[1] Elena Perez,[4] Shefal Parikh,[1] John Kay,[2] John Mills,[3] Maureen Goodenow,[4] and Ben M. Dunn[1]

[1]Department of Biochemistry and Molecular Biology
[2]Department of Biochemistry
University of Wales
Cardiff CF1 1ST
[3]Department of Molecular Virology
Roche Products Limited
Hertfordshire, AL7 3AY United Kingdom
[4]Department of Pathology
University of Florida
Gainesville, Florida 32610-0245

1. INTRODUCTION

Currently, three inhibitors of HIV-I protease are showing promise in clinical use against HIV infection. However, the development of resistance continues to be an important consideration in protease inhibitor therapy. Mutations in protease which lead to resistance have been observed in both inhibitor treated and untreated individuals (Borman et al., 1996; Condra et al., 1995). Variations have also been observed to a higher degree in areas of the *gag* region including p7, and the cleavage site sequences (Barrie et al., 1996).

The variations, occurring in the *gag* region, may be concomitant with variations occurring in the protease region. The combination of these changes may allow an otherwise inactive *gag–pol* region to be processed. Elucidation of how changes in the *gag* region affect processing of the *gag–pol* precursor is therefore important in understanding the extent of mutations possible within protease while still allowing precursor polyprotein

Aspartic Proteinases, edited by James.
Plenum Press, New York, 1998.

processing to occur. This knowledge should prove useful in future inhibitor design and in evaluation of current therapies.

To this end, we have constructed an *E. coli* based expression system which allows the production and subsequent processing of a *gag–pol* precursor by its own protease or by a protease isolated from a different variant to be analyzed in a time dependent manner.

2. MATERIALS AND METHODS

2.1. Expression

M15 *E.coli* cells containing a pDS56 plasmid with the variant gag-pol regions isolated from PBMCs of HIV-1 positive inhibitor naive patients and containing the pDMI.1 plasmid were grown in LB (100 mg/ml Amp, 25 μg/ml Kana) until an $O.D_{600}$ of 0.8 was reached. The culture was induced with 400 μg/ml IPTG. Time points were taken at 0, 5, 10, 15, 20, 30 and 60 minutes. At each time point, 300 μl of the culture was removed and an $O.D_{600}$ was taken. Seven hundred μl was also remove and centrifuged at 14,000 RPM on a table top microcentrifuge for 5 minutes to pellet the cells. The supernatant was removed and the pellet resuspended in 200 μl of LSB.

2.2. Electrophoresis and Western Blot

Equivalent amounts of cells were and electrophoresed using a SDS-PAGE gel (10% separator, 3% stacker Tris- tricine). The proteins were blotted onto a PVDF membrane for subsequent use in a Western blot assay. A monoclonal HIV-I p25/24 Gag antibody was used for visualization of antigen.

3. RESULTS AND DISCUSSION

The purpose of this study was to address factors which affect processing of the HIV-1 *gag–pol* polyprotein precursor. To this end, an expression vector was designed to allow replacement of a region of *gag–pol* from HXB10 with *gag–pol* segment isolated from patients infected with HIV-1. This region was expressed using an *E. coli* based system production and processing of the polyprotein precursor was followed in a time dependent manner.

Figure 1 illustrates the results of the first experiments. In this case, three different *gag-pol* segments were expressed and the processing observed. The rate and completeness of processing of the polyprotein precursor varied considerably among the three constructs. Additionally, six other constructs were analyzed. These constructs also showed variations in processing (data not shown).

To determine what role if any gag plays in its own processing, a series of chimeric constructs was created. Figure 2 shows the segments used from each of the variants to form the chimeras. Figure 3 illustrates the results of the chimera expression. It is clear from these data that the D1.10 variant processes *gag* regions from the three other variants differently from its own *gag* region. When D1.21 is expressed with its own *pol* region, processing of the precursor does not occur (Figure 1 lane 1). However, when its *pol* region is replaced with the D1.10 *pol* region, as seen in Figure 3, processing does occur.

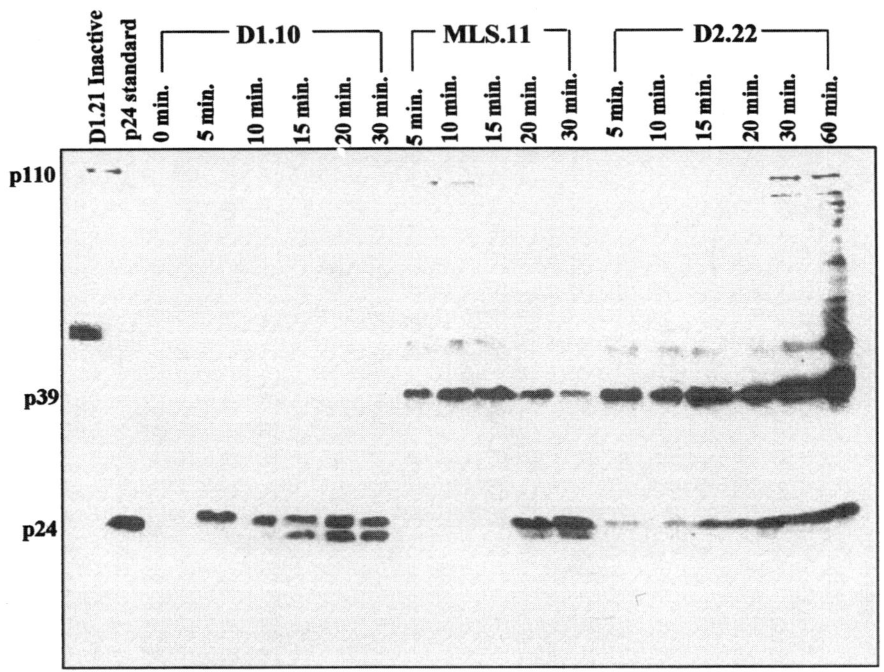


```
LAI      PQITLWQRPLVTIKIGGQLKEALLDTGADDTVLEEMSLPGRWKPKMIGGIGGFIKVRQYDQILIEICGHKAIGTVLVGPTPVNIIGRNLLTQIGCTLN
 D1.10   ------------------------------------T-T----------------------VP------------------------------------
 D1.21   --------------------------R---------T-T------------------R---VP------------------------------------
 D1.22   ------------------------------------T-T-K---------------------P-----R------------------------------
MLS.11   ----------------------------------D-N-------------------------P------------------------------------
```

Figure 1. Western blot using p24 antibody showing expression and variation in processing among three *gag–pol* polyprotein precursors. Note levels of products p110, p39 and p25/24 at different time points during the experiment. D1.10 is termed fast/complete, MLS.11 intermediate/complete and D2.22 intermediate/incomplete. These designations are given for the rate and completeness of processing of the *gag–pol* polyprotein precursor by protease. Lane 1 represents an inactive precursor. The protease amino acid sequence is given for each of the variants.

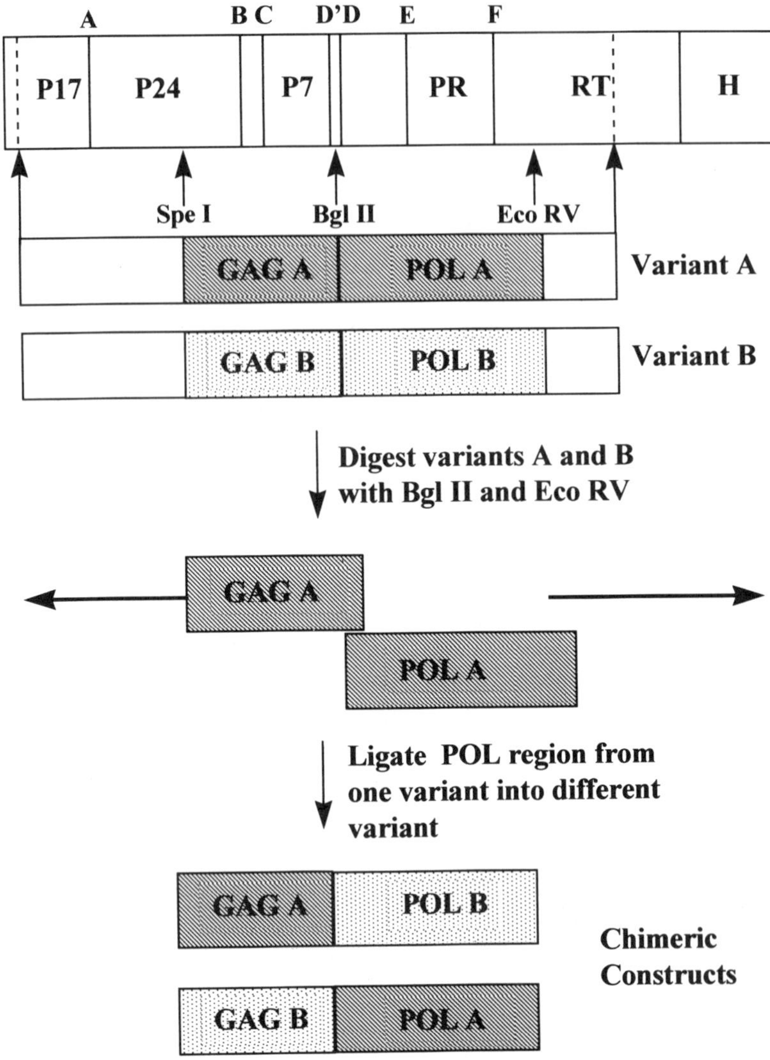

Figure 2. Diagram of the generation of chimeric protein constructs illustrating the region of *gag–pol* exchange between variants. Letters represent cleavage sites within the polyprotein.

This processing is not, however, the same as that observed for the D1.10 variant (Figure 1). This type of pattern is also found in the two other chimeras, wherein the rate and completeness of processing change when their own *pol* regions are replaced by D1.10 *pol* but not to the same rate and completeness observed in the original D1.10 variant. These data indicate that while protease does have a role in determining rate and completeness of *gag–pol* precursor processing, regions within the *gag* domain also influence processing. Future studies will attempt to map the site of these determining factors and ascertain how these factors interact with the protease region to regulate the processing of the *gag–pol* polyprotein precursor.

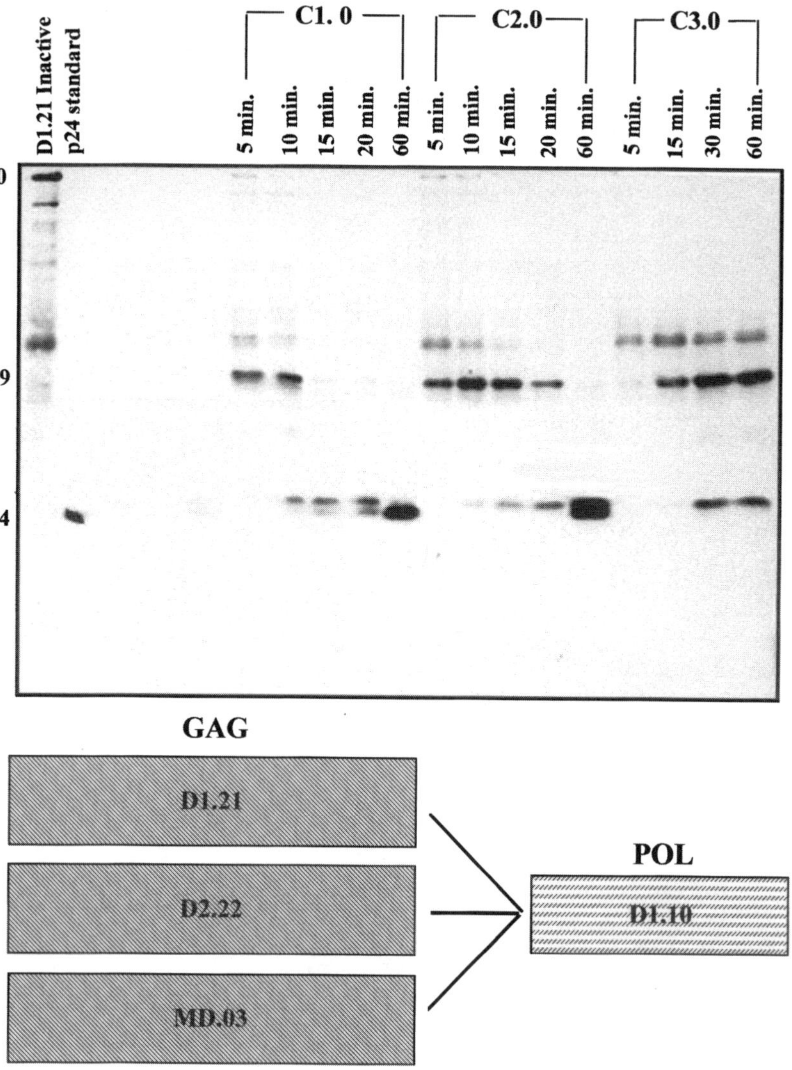

Figure 3. Western blot using p24 antibody showing expression and variation in processing among three chimeric *gag–pol* polyprotein precursors. Note levels of products p110, p39, and p25/24 at the different time points during the experiment. Lane 1 represents an inactive precursor. Diagram showing source of *gag* and *pol* regions used in each chimera.

ACKNOWLEDGMENTS

This work was supported by grant AI 28571 from NIH to Ben M Dunn.

REFERENCES

Barrie, K. A., Perez, E., Lamers, S., Farmerie, W., Dunn, B., Sleasman, J., Goodenow, M. 1996 Virology 219:407–416.

Borman, A., Paulous, S., Clavel, F. 1996 J. Gen. Virol. 77:419–426.

Condra, J., Schleif, W., Blahy, O., Gabryelski, L., Graham, D., Quintero, J., Rhodes, A., Robbins, H., Roth, E., Shivaprakash, M., Titus, D., Yang, T., Toppler, H., Squires, K., Deutsch, P., Eminl, E. 1995 Nature 374:569–571

8

X-RAY CRYSTALLOGRAPHIC STUDIES OF THE STRUCTURE-FUNCTION RELATIONSHIPS OF HIV-1 PROTEASE

Lin Hong, Cai Zhang, Jean A. Hartsuck, Steve Foundling, and Jordan Tang

Protein Studies and Crystallography Programs
Oklahoma Medical Research Foundation
University of Oklahoma Health Sciences Center
Oklahoma City, Oklahoma 73104

INTRODUCTION

Human immunodeficiency virus (HIV) protease is responsible for the processing of gag and gag-pol polyprotein precursors to produce structural proteins and enzymes for the mature virus. Hence, it is essential for the maturation and infectivity of the virus and is thus a major target for the development of inhibitor drugs. But these efforts have been handicapped by the development of drug resistant viral strains.

Therefore, the structural changes in HIV protease mutants are of interest for understanding the drug resistance. Here we report the X-ray crystallographic structures of two mutant HIV-1 proteases, G48H and A28S, in complex with a peptidic inhibitor U-89360E (structure shown below) and compare these structures to that of the wild-type protease complexed with the same inhibitor (1).

Position 48 in the primary structure of the protease has been identified as one of the critical residue for conferring resistant to inhibitors (2,3). G48H mutant has been shown to have greater catalytic efficiency for certain substrates than that of the wild-type enzyme (4).

A28S mutation in HIV-1 protease potentially places a hydrogen bond between Ser-28 and active-site Asp-25, as exists in pepsin. Comparison of kinetic data for A28S mutant and the wild-type enzymes shows that the mutation results in a decrease of catalytic efficiency with little or no change of active-site pKa values (5) (Table 1). Determination and study of the crystal structure of A28S HIV-1 protease should lead to a better understanding of the catalytic mechanism and the structure-function relationship of HIV-1 protease.

Aspartic Proteinases, edited by James.
Plenum Press, New York, 1998.

Table 1. Kinetics parameters of wild type and mutant HIV-1 protease with U89360E

	kcat (sec^{-1})	Km (mM)	kcat/Km	Ki (nM)
Wild type*	11	40	0.275	20
G48H*	30	65	0.462	119
Wild type‡	9.8	12.8	0.766	20
A28S‡	0.037	17.4	0.002	1520

Two different substrates were used for G48H and A28S in the kinetic measurements.
*References (4) and (6). Chromogenic substrate A was used.
‡Reference (5). Radiometric substrate [^{3}H]Ac-SQNYPVVR was used.

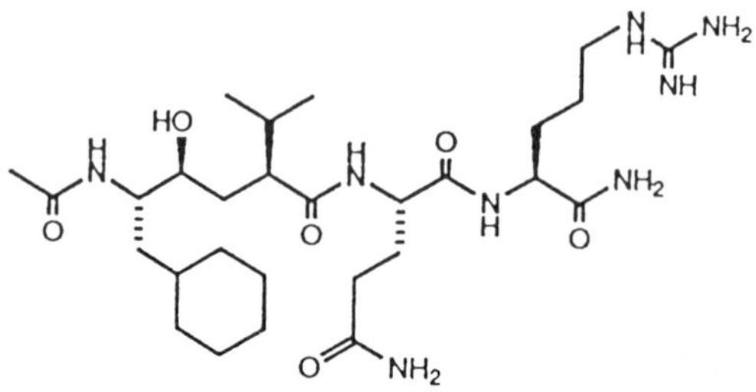

U89360E

RESULTS AND DISCUSSION

The Structure of G48H HIV-1 Protease

The G48H/U89360E complex was crystallized in space group I222. The unit cell parameters were: a=58.5 Å, b=88.2 Å, c=94.2 Å. The crystallographic refinement was carried out using the refinement program package TNT (7). Since this is a new space group for HIV-1 protease, the initial phases were solved by molecular replacement (8). The wild type HIV-1 protease structure previously solved (1) was used as the search model. The calculation was carried out with the molecular replacement program package MRCHK (9). Molecular graphics program O was used for map display and model fitting (10). The final structure has a R-factor of 0.193. Diffraction data in the resolution range of 20–2.3 Å was used for the refinement. The overall B factor is 39.

Clear and continues electron density for the inhibitor was observed in the active site of the G48H HIV-1 protease. In the wild type structure, two inhibitor orientations were observed with equal occupancy related by the 2-fold symmetry axis of the protease dimer (1). Whereas only one inhibitor orientation was observed in the active site of G48H. Unlike the wild type HIV-1 protease which had a space group of $P6_1$, the two monomers of G48H are not related by a pseudo crystallographic axis.

The inhibitor U-89360E does not have C2-symmetry. It makes contacts with the protease residues in P1, P1′, P2′, and P3′ subsites. Part of the inhibitor arginine residue in the P3′ pocket extends outside the binding cleft. Different crystal packing interactions of this part of the inhibitor in different space groups are expected, which may contribute to the fact that two inhibitor orientations in the crystal structure of the wild type HIV-1 protease/U89360E complex were observed (space group $P6_1$), yet only one orientation is present in the G48H mutant complex (space group I222).

The flaps of HIV-1 protease (minimally from residue Lys-45 to Lys-55) cover the substrate binding cleft and are assumed to play a very important role in capturing and correctly positioning of the substrate in the active site (11). The dynamic mobility of these residues compared to the rest of the protein is supported by the fact that the flaps of the inhibitor bound protease adopt a closed conformation versus an open one for that of the HIV-1 protease with no bond inhibitor (12). The tip of the flaps are located between residue 50 and 51. The amino acid sequence of the wild type HIV-1 protease between residue 48 and 51 is GIGG which is considered to be the consensus amino acid sequence for the flap regions of aspartic proteases.

G48H crystal structure shows that introduction of a histidine at position 48 allows several interactions which are not present in the wild type protease.

1. In the G48H mutant protease, His-48 side chain is located at the outside face of the flap region. It has close contacts with Phe-53 in both monomers of the protease (Figure 1). This interaction and the larger side chain of the histidine would result in less conformational freedom of the flap. Flap rigidity may be responsible for the increased catalytic activity of the mutant enzyme for some substrates (4).
2. The arginine side chain of the inhibitor at the S3′ position extends outside the binding cleft. Van der Waals contacts between this part of the inhibitor and the side chain of His-B48 were observed. A least square, rigid body superposition between the wild type and G48H structure using main chain protein atoms was

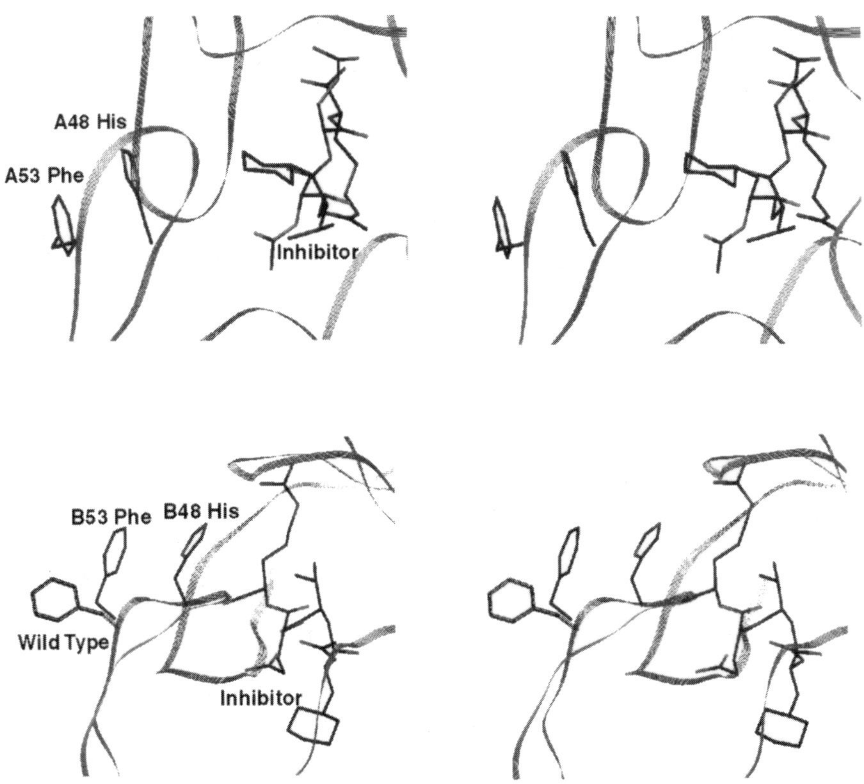

Figure 1. In the G48H mutant protease, His-48 side chain extends from the flap and contacts the solvent. It interacts with Phe-53. The upper picture is for the A monomer of the protease, and the lower one is for the B monomer.

performed, the inhibitor conformation bound to the mutant enzyme was compared to that of the wild type enzyme. Of the two inhibitor orientations present in the wild type, the one which is the same as that of the mutant structure was used for the comparison. The result showed a inhibitor conformation change presumably due to the interaction mentioned above (data not shown). The extent of this interaction is substrate-dependent. The corresponding interaction is not present for His-A48, because P2 and P3 subsites are not occupied by the inhibitor.

Upon the rigid body superposition, it was also observed that the mutation causes significant protein back-bone deviation from that of the wild type in the flap regions around residue Pro-A44, which may contribute to the change of catalytic efficiency of the mutant protease.

The Structure of A28S HIV-1 Protease

For A28S mutant, The space group was determined to be P61, which is isomorphous with the wild type structure (1). The latter was used to calculate the initial phase of the refinement. The unit cell parameters are: a=b=63.6 Å, c=83.7 Å. The program packages described above were used to solved the A28S structure. The final R-factor was 0.194 for the data range of 20–2.0 Å. The overall B factor is 37.6. Not surprisingly, we observed two equally occupied inhibitor orientations related by the 2-fold symmetry axis of the protease dimer as we did for the wild type enzyme (1).

The hydrogen bond pattern introduced by the mutation was unexpected. In A28S, the Ser-28 side chain assumes two orientations in both monomers of the protease (Figure 2), resulting in hydrogen-bond either to Asp-25 catalytic carboxyl group or Asp-30 main chain carbonyl.

In the mutant structure where there is a hydrogen bond between Ser-28 and Asp-25, there is an apparent rotation of the catalytic carboxyl group of the latter in the A monomer compared to that of the wild type protease binding to the same inhibitor (Figure 3). A more modest re-orientation of Asp-B25 side chain was also observed. Conceivably, these

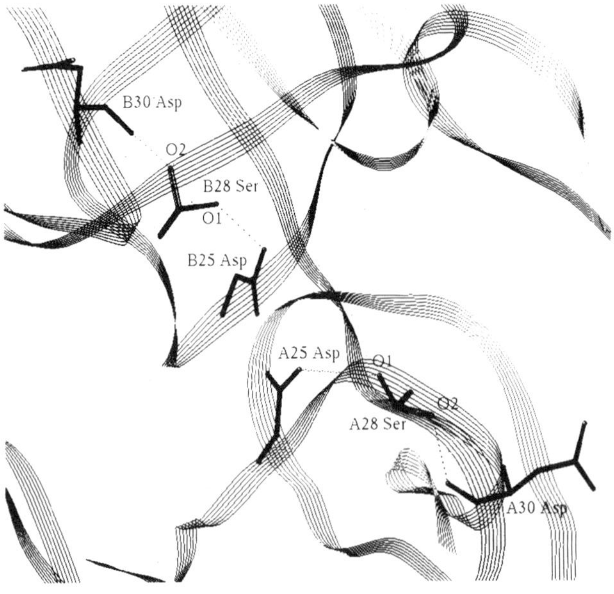

Figure 2. In A28S mutant, the Ser-28 side chain assumes two orientations, resulting in a hydrogen-bond either to Asp-25 or Asp-30.

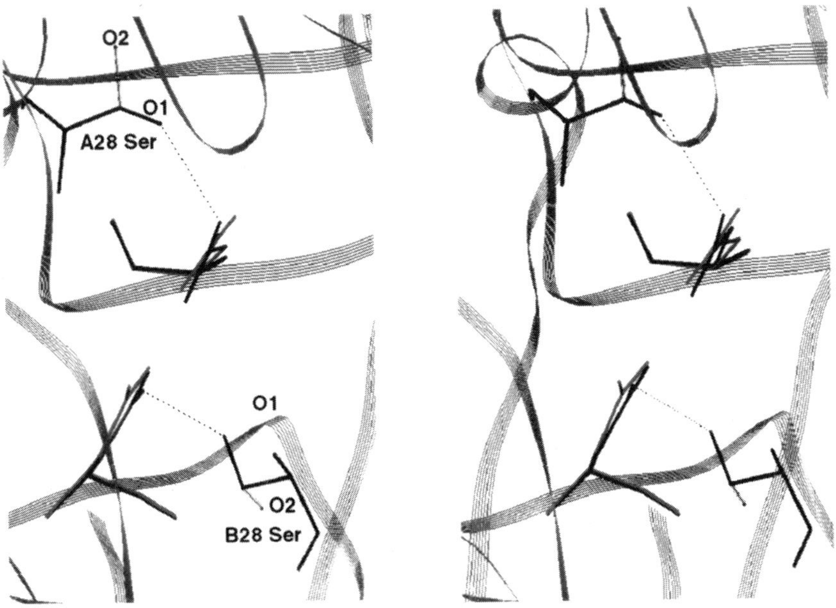

Figure 3 Hydrogen bond formation between Ser-28 and Asp-25 causes a rotation of the catalytic carboxyl group. The black model represents the A28S mutant structure, and the gray model represents the wild type.

new hydrogen bonds would introduce certain conformational restraints to the active site. The ability for the catalytic carboxyl groups to rotate was also predicted by A. M. Silva et al. in a quantum mechanics simulation of the catalytic process (13).

If there is a dynamic equilibrium between the two observed side chain orientations of Ser-28 in solution, it would cause the hydrogen bonds to the Asp-25 side chain to be constantly formed and broken, and induce a continual reorientation of the side chain. This may well have an adverse effect on the HIV-1 protease active site configuration with a concomitant decrease in catalytic capability. The fact that these structural perturbations decrease the protease activity (kcat) indicates that coordinated conformational changes in the active site are necessary for the catalytic process of HIV-1 protease.

REFERENCES

1. L. Hong, A. Treharne, J. A. Hartsuck, S. Foundling, and J. Tang, Biochemistry, 35:10627–10633 (1996)
2. T. Ridky and J. Leis, J. Biol. Chem., 270:29621–29623 (1995)
3. H. Jacobsen, M. Hanggi, M. Ott, I. B. Duncan, S. Owen, M. Andreoni, S. Vella, and J. Mous, J. Infect. Dis., 173:1379–1387 (1996)
4. Y. Lin, X. Lin, L. Hong, S. Foundling, R. L. Heinrikson, S. Thaisrivongs, W. Leelamanit, D. Raterman, M. Shah, B. M. Dunn, and J. Tang, Biochemistry, 34:1143–1152 (1995)
5. E. Ido, H.-P. Han, F. J. Kezdy, and J. Tang, J. Biol. Chem., 266:24359–24366 (1991)
6. A. D. Richards, L. H. Phylip, W. G. Farmerie, P. E. Scarborough, A. Alvarez, B. M. Dunn, P. -H. Hirel, J. Konvalinka, P. Strop, L. Pavlickova, V. Kostka, and J. Kay, J. Biol. Chem., 265:7733–7736 (1990)
7. D. E. Tronrud, L. F. Ten Eyck, B. W., Matthews, Acta Cryst., A43:489–503 (1987)
8. M. G. Rossmann, Editor. The Molecular Replacement Method. New York: Gordon and Breach (1972)
9. X. J. Zhang, and B. W. Matthews, Acta Cryst., D50:675–686 (1994)
10. T. A. Jones, J.-Y. Zou, S. W. Cowna, and M. Kjeldgaard, Acta Cryst., A47:110–119 (1991)
11. J. Tang, and G. Koelsch, Prot. & Pept. Lett., 2:257–266 (1995)
12. A. Wlodawer, and J. W. Erickson, Ann. Rev. Biochem., 62:543–585 (1993)
13. A. M. Silva, R. E. Cachau, H. L. Sham, and J. W. Erickson, J. Mol. Biol., 255:321–346 (1996)

9

OPTIMIZATION OF A MACROMOLECULAR INHIBITOR OF HIV-1 PROTEASE

Deborah S. Dauber, Fiona McPhee, Ayçe Ünal, and Charles S. Craik

Department of Pharmaceutical Chemistry
University of California
San Francisco, California 94143

1. INTRODUCTION

HIV protease (HIV PR) has been identified as a key target in the treatment of HIV infection, and antiviral drugs that block its proteolytic activity have been developed. However, there are numerous ways in which a drug can lose effectiveness during long-term therapy. In particular several protease-mediated mechanisms result in reduced antiviral sensitivity of the various anti HIV protease inhibitors. As with inhibitors of reverse transcriptase, protease inhibitors have shown a sensitivity to variations in the protease that weaken drug affinities, both in cell culture and *in vivo* [1,2]. These include: 1) mutations of active site residues that are direct substrate specificity determinants; 2) mutations of non-active site residues that indirectly interfere with inhibitor binding through long range structural changes; 3) mutations that increase enzyme catalysis; 4) mutations that increase enzyme stability and half-life; and 5) mutations that provide new substrate specificities with corresponding mutations in the gag–pol polyprotein precursor substrate. Since the binding modes of both the natural substrates for HIV protease and the substrate analog class of inhibitors are similar, amino acid substitutions that occur in the substrate binding pocket are among the more common resistance mutations that arise in response to HIV protease inhibitors. However, the structural basis for the loss of sensitivity of a viral protease to a drug is not always obvious or predictable. This inability to anticipate the variety of mechanisms of viral resistance to new protease inhibitors complicates the task of designing more effective drugs. In this work, we consider the possibility of creating a new class of inhibitors that can pose a greater challenge to the development of protease-mediated viral resistance.

In the presence of substrate analog inhibitors, mutant proteases are generally selected in which the shape of the active site has changed enough to weaken the strength of inhibitor binding without irreparably damaging its affinity for the physiological ligand. There is a limit, however, to the mutability of the protease: certain regions of the protein are faithfully conserved in all clinical isolates [3]. An inhibitor that interacts specifically with these regions of the protease, which are presumably essential to maintaining structure

Aspartic Proteinases, edited by James.
Plenum Press, New York, 1998.

and/or catalytic activity, may present a greater barrier to the development of viral resistance. With this in mind, several groups have developed compounds that disrupt the dimer interface and subsequently block the formation of the protease dimer [4–6]. However, these inhibitors currently suffer from a lack of potency and/or specificity. Ideally, an agent that disrupts the dimer interface would be able to inhibit dimer formation while increasing both affinity and specificity by binding to other regions of the protease. This increased surface area between inhibitor and wild type monomer can be attained by increasing the size of the inhibitor to that of a protein. Thus, a catalytically inactive variant of HIV PR is expected to inhibit specifically wild type PR by precluding activity through its own inability to catalyze proteolysis while binding through interactions normally essential to the formation of wild type homodimers.

The concept of creating a defective protein to block the activity of a multicomponent assembly has precedence. Yeast geneticists have used this "dominant negative" approach to determine the functions of various cloned gene products through loss-of-function mutations [7]. Our initial studies used the catalytically inactive variant Asp25Asn (D25N) as a dominant negative inhibitor of wild type (WT) HIV protease [8]. The D25N variant inhibited HIV PR WT in transiently transfected cells, as demonstrated both by evaluation of gag polyprotein processing as well as by measuring the yield of infectious virions produced by infected cells. In addition, a stable cell line constitutively expressing HIV PR D25N blocked viral maturation and infectivity when challenged with a WT viral genome.

A classical approach for the iterative development of a potent small molecular inhibitor generally starts with an initial candidate, which is varied in a directed fashion to search for more potent inhibitors. Analogously, the initial success of HIV PR D25N as a macromolecular inhibitor was encouraging as an initial compound. In engineering a second generation dominant negative inhibitor, our goal was to increase the affinity of the inhibitor for the wild type monomer while decreasing the ability of the inhibitor to form homodimers. The combination of these effects would lead to a higher concentration of heterodimer and a smaller population of wild type homodimers. In an excess of inhibitor, the concentration of WT HIV PR homodimers would drop below the threshold necessary for virion maturation and infectivity.

2. METHODS

2.1. Establishment of Transiently Transfected and Stable Variant-Expressing Cell Lines

Vectors were constructed that expressed the HIV genome with variant forms of the protease. The HIV-gpt form of the virus was chosen for our experiments because it permitted quantitation of the virus [9]. HIV-gpt DNA constructs containing WT and variant protease genes were co-transfected as described [8,10] and virus particles were assayed for proteolytic maturation and infectivity.

A cell line that stably expressed the entire HIV genome containing HIV PR KWW was developed as previously described [8]. Cells producing the HIV PR KWW variant were then "challenged" by transfecting with the WT HIV-gpt DNA construct.

2.2. Maturation and Infectivity Assays

Following either co-transfection or challenge with the WT genome, sucrose gradient purified viral particles were analyzed by either ELISA or immunoassay using anti-p24 antibodies. Virion maturation was measured as the ratio of processed p24 to total p55 gag production.

To produce infectious virions, cells were co-transfected with the HXB2-env DNA construct [9] along with HIV-gpt plasmids. The virions produced by these cells were used to infect HeLa-T4 cells. The number of viral plaques formed on monolayers of HeLa-T4 cells under mycophenolic acid selection correspond to the number of infectious particles. The number of particles produced by cells expressing WT, variant, or both forms of the protease were then compared.

2.3. Molecular Modeling

Design of the dominant negative variants was conducted using the SYBYL modeling package [11]. Using the SYBYL match algorithm, seven high resolution structures of HIV-1 PR bound to various ligands were chosen from the Brookhaven Protein Data Base. These structures were aligned as described [10] and the bound ligands overlaid to map out a substrate binding site. Substitutions were then chosen for one monomer that filled the binding site without creating any steric repulsion within the heterodimer.

3. RESULTS AND DISCUSSION

Seven structures of protease bound to different ligands were overlaid and the HIV PR binding pocket was mapped (Figure 1a) as described above. Substitutions in the variant monomer were suggested based on amino acids that were able to fill the substrate binding pocket on the opposing monomer. The following modifications to the HIV PR monomer were suggested by the modeling: Asp25Arg, Asp25Lys, Gly49Trp, Ile50Trp. The advantage of Arg and Lys at position 25 is two-fold: aliphatic regions help to fill the binding pocket while a positively charged group can form a salt bridge with the catalytic aspartic acid on the wild type monomer (Figure 1b,c). Tryptophans at positions 49 and 50 are predicted to fill the P1 and P2′ pockets, respectively. An energy-minimized model using the triple mutant Lys25/Trp49/Trp50 (KWW) shows that these substitutions fill the binding pocket of an inhibitor-WT heterodimer [10,12]. Conversely, the strength of the inhibitor homodimer is expected to be severely weakened relative to the heterodimer due to steric repulsion between the tryptophans as well as ionic repulsion between the positively charged residues at position 25 (Figure 1d). Experimental assessment of the above substitutions was done using singly-, doubly- and triply-mutated protease genes.

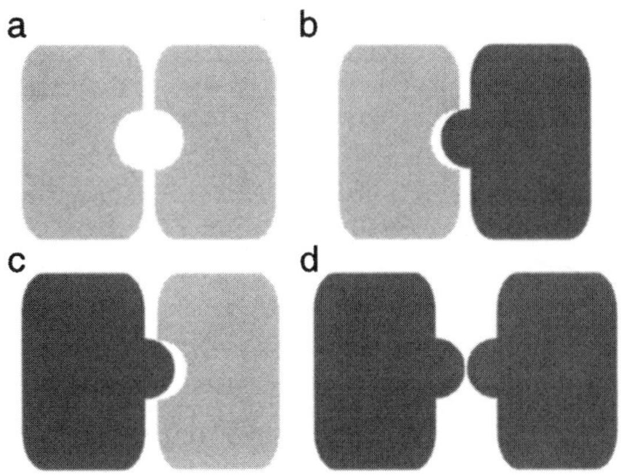

Figure 1. The four different species that are formed using HIV PR KWW and HIV PR WT monomers. Intrusions signify the active site; extrusions signify the charged lysine and bulky tryptophan mutations. Relative concentrations of a) HIV PR WT homodimer, b,c) HIV PR KWW/HIV PR WT heterodimer, and d) HIV PR KWW homodimer will depend on the K_d values for each species.

Studies on the HIV PR heterodimers were conducted by co-transfecting dominant negative inhibitor genes and HIV PR WT into COS-7 cells at a 1:1 ratio. Inhibition of HIV PR was analyzed as a function of gag polyprotein processing as well as of the yield of infectious virions (Figure 2a,b, [10]). A comparison of the various inhibitors suggests that the HIV PR KWW variant inhibits proteolytic processing the most effectively (Figure 2a), although the margin of difference is not large. The strength of the inhibition, however, appears much more marked when measured by the yield of infectious virions (Figure 2b). At ratios of 1:1 for inhibitor to wild type DNA, nearly all of the engineered variants show at least 90% inhibition of production of infectious virions. The best inhibitor, the HIV PR KWW variant, showed 95% inhibition in these conditions.

Inhibition of polypeptide maturation by the HIV PR KWW variant was shown to be dose-dependent by transiently transfecting increasing ratios of HIV PR KWW to HIV PR WT DNA into COS-7 cells [10,12]. Again, inhibition as monitored by gag processing is marked but modest. However, at ratios as low as 1:2 of HIV PR WT DNA to KWW HIV PR DNA the level of infectious virions produced has dropped to undetectable levels. Compared to the 75% inhibition demonstrated by HIV PR D25N in the comparable experiment [8], the HIV PR KWW variant is clearly demonstrating a higher level of inhibition.

We wished to determine whether stable expression of HIV PR KWW could protect cells from challenge with WT HIV genome. The B7 cell line, which constitutively expresses the HIV-gpt gag and gag/pol polypeptides containing the HIV PR KWW variant, was established and transfected with WT HIV-gpt DNA. Immunoblots using antibodies against p24 show that gag polyprotein processing has been almost completely eliminated (Figure 3a). In addition, production of infectious virions in the B7 cell line requires addition of five times the amount of wild type HIV-gpt DNA needed to produce a comparable level of virions in the HIV PR D25N-expressing 12A2 cell line (Figure 3b, [8]). Further cell culture studies have shown that expression of the HIV PR KWW variant both delays susceptibility of cells to clinical isolates of HIV and lowers the ability of the virus to

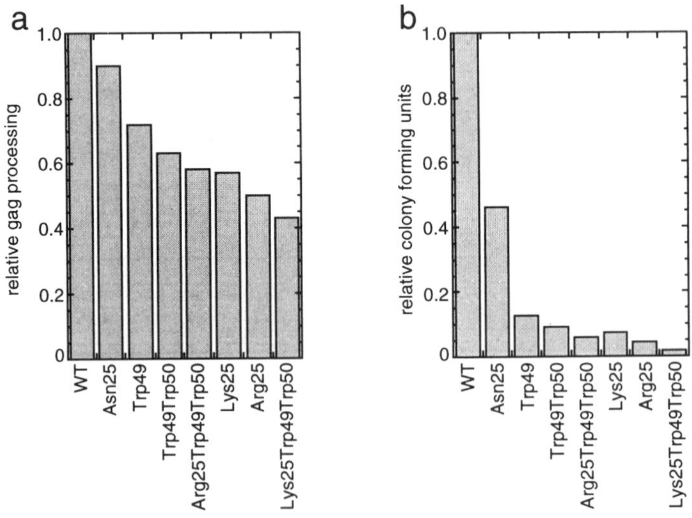

Figure 2. A comparison of a) gag polyprotein processing, as measured by anti-p24 ELISA, and b) colony forming units produced in cells transiently co-transfected with variant and HIV PR WT DNA. Values are normalized for culture volumes and are represented as fractions relative to values for uninhibited HIV PR WT.

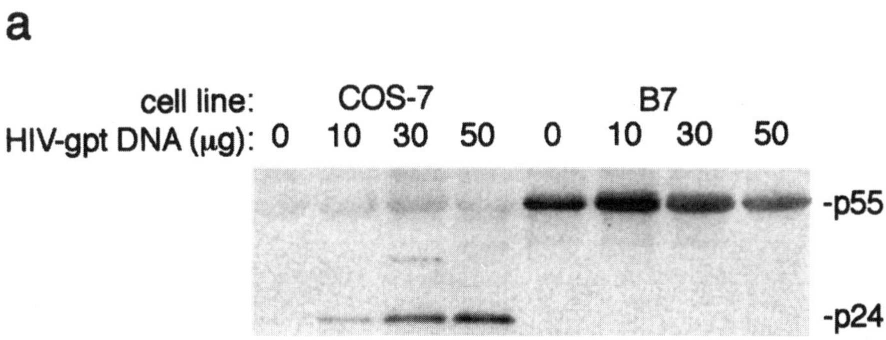

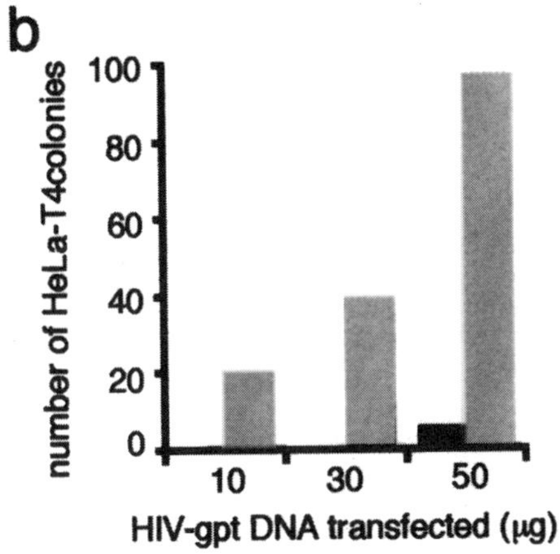

Figure 3. A comparison of gag polyprotein processing and viral infectivity for B7 and naive COS-7 cell lines. a) An immunoblot using anti-p24 antibodies shows that even with increased WT genome being transfected the virus is unable to properly cleave the gag polyprotein. b) Infectivity assays on Hela-T4 cells show that B7 cell lines produce a low level of infectious virions (black bars) only at the highest level (50 μg) of transfected WT HIV-gpt DNA. Naive COS-7 viral production (gray bars) was used as a control.

reproduce as measured by p24 production [13]. In addition, a combination of KWW-expression and protease inhibitor administration further inhibits viral production.

4. CONCLUSIONS

In this work, we have explored the development of a new class of macromolecular inhibitors of multimeric proteins. The essential role of HIV protease in the life cycle of HIV makes it an ideal test case for development of an engineered dominant negative inhibitor. We have demonstrated the potential of HIV PR KWW to be a prototype of this class of inhibitors; future work will focus on further characterization of the HIV PR KWW/HIV PR WT heterodimer, design of increasingly more potent inhibitors, and determination of the sensitivity of dominant negative inhibitors to viral resistance.

ACKNOWLEDGMENTS

C.S.C. is a recipient of NIH grant GM39552, D.S.D. is supported by an NSF Graduate Fellowship, and A.U. by a grant from CFAR.

REFERENCES

1. Condra JH, Schleif WA, Blahy OM, Gabryelski LJ, Graham DJ, Quintero JC, Rhodes A, Robbins HL, Roth E, Shivaprakash M and et al., In vivo emergence of HIV-1 variants resistant to multiple protease inhibitors. *Nature* **374**(6522): 569–71, 1995.

2. Mellors J, Schinazi R and Larder B, Mutations in HIV-1 RT and protease. In: *Human Retroviruses and AIDS* (Eds. Myers G, Wain-Hobson S, Henderson LE, Korber B, Jeang K-T and Pavlakis GN). Theoretical Biology & Biophysics, Los Alamos, New Mexico, updated 1995.
3. Winslow DL, Stack S, King R, Scarnati H, Bincsik A and Otto MJ, Limited sequence diversity of the HIV type 1 protease gene from clinical isolates and in vitro susceptibility to HIV protease inhibitors. *AIDS Res Hum Retroviruses* **11**(1)**:** 107–13, 1995.
4. Babé LM, Rosé J and Craik CS, Synthetic "interface" peptides alter dimeric assembly of the HIV 1 and 2 proteases. *Protein Science* **1**(10)**:** 1244–53, 1992.
5. Schramm HJ, Billich A, Jaeger E, Rucknagel KP, Arnold G and Schramm W, The inhibition of HIV-1 protease by interface peptides. *Biochem Biophys Res Commun* **194**(2)**:** 595–600, 1993.
6. Zhang ZY, Poorman RA, Maggiora LL, Heinrikson RL and Kezdy FJ, Dissociative inhibition of dimeric enzymes. Kinetic characterization of the inhibition of HIV-1 protease by its COOH-terminal tetrapeptide. *J Biol Chem* **266**(24)**:** 15591–4, 1991.
7. Herskowitz I, Functional inactivation of genes by dominant negative mutations. *Nature* **329**(6136)**:** 219–22, 1987.
8. Babé LM, Rosé J and Craik CS, Trans-dominant inhibitory human immunodeficiency virus type 1 protease monomers prevent protease activation and virion maturation. *Proc Natl Acad Sci U S A* **92**(22)**:** 10069–73, 1995.
9. Page KA, Landau NR and Littman DR, Construction and use of a human immunodeficiency virus vector for analysis of virus infectivity. *J Virol* **64**(11)**:** 5270–6, 1990.
10. McPhee F, Good AC, Kuntz ID and Craik CS, Engineering HIV-1 protease heterodimers as macromolecular inhibitors of viral maturation. *Proc. Natl. Acad. Sci.* **93:** 11477–81, 1996.
11. Tripos Associates Inc., SYBYL Molecular Modeling Package. Tripos Associates, St. Louis, Version 6.0, 1994.
12. Craik CS, Babé LM, Dauber D, McPhee F, Rosé J and Ünal A, Engineering catalytically defective forms of HIV protease to modulate its activity. In: *Structure and Function of Aspartic Proteinases: Retroviral and Cellular Enzymes* (Ed. James MNG). Plenum Publishing Corporation, New York, 1997.
13. Junker U, Escaich S, Plavec I, Baker J, McPhee F, Rosé JR, Craik CS and Böhnlein E, Intracellular expression of HIV-1 protease variants inhibits replication of wild-type and protease inhibitor resistant HIV-1 strains in human T cell lines. *J. Virology* **70:** 7765–72, 1996.

ABSTRACTS

SAQUINAVIR: DESIGN AND DEVELOPMENT

Joseph A. Martin

Roche Research Centre, 40 Broadwater Road, Welwyn Garden City, Herts AL5 2BN, England

The elucidation of molecular events involved in the replication of human immunodeficiency virus (HIV) provided a range of targets for medicinal chemists to design novel therapeutic agents. One target, the HIV proteinase, was characterised as a member of the aspartic family of proteinases from its amino acid sequence, active-site mutation studies and its inhibition by pepstatin. This enzyme is required for post-translational modification of the viral polyproteins, Pol and Gag–Pol, to afford both the structural proteins and enzymes necessary for the production of new infectious virus particles. The search for inhibitors of this essential enzyme has been one of the most widely studied approaches in the search for an effective treatment of HIV infection.

We have designed potent and selective inhibitors of HIV proteinase. Our strategy was based on the concept of transition-state mimetics and a detailed knowledge of peptide sequences recognised and specifically cleaved by the viral enzyme. This presentation will describe studies that led to saquinavir, the first HIV proteinase inhibitor to be marketed. Structure-activity relationships and biological properties will be discussed.

CRYSTALLOGRAPHY AND THE DESIGN OF POTENT NON-PEPTIDIC HIV PROTEASE INHIBITORS: WHAT IT HAS TAUGHT US

K. D. Watenpaugh and M. N. Janakiraman

Structural Analytical & Medicinal Chemistry, Pharmacia & Upjohn, Kalamazoo, Michigan 49001

In the development of potent HIV protease inhibitors with good pharmacokinetic properties about 80 crystal structures of protein/inhibitor complexes were determined. From an initial weak inhibitor lead (phenprocoumon) with good pharmacokinetic characteristics, a classic structure-based drug design paradigm was carried-out with synthetic chemistry, competitive binding assays, crystallography, modeling and biological testing. Crystallography proved invaluable since modeling did not always predict the binding details correctly at key points during the process. Starting with the 4-hydroxy coumarins, the central template bound into the catalytic pocket of the protease evolved to pyrones and then dihydropyrones. Through substitutions including various aromatic sulfonamides off the central template into additional binding pockets, the binding of the inhibitors was greatly improved. Later design efforts were directed towards maintaining the tight binding while improving the pharmacokinetic properties and drug efficacy. As a result, orally bioavailable, non-peptidic HIV protease inhibitors were discovered with *in vitro* activity comparable to the peptide-derived inhibitors. Various methods have been used to produce a drug candidate that would be less effected by mutations of the protease. For example, many crystal structures were determined using HIV-2 protease which has only a 50% sequence identity to HIV-1 protease. The crystal structures lend insight as to why these inhibitors are less effected by the usual mutations in the substrate binding region and are highly specific for HIV proteases. As a consequence, HIV-1 isolates highly resistant to the peptide-derived inhibitors remain sensitive to these 'third generation' inhibitors.

Rather than describing the structure-based drug discovery process, we will focus on the crystallographically determined structural information. From the unusual binding of some of the inhibitors and from the accidental binding of auto-catalytic peptide fragments further information on the aspartyl protease mechanism of action can be derived. Octapeptides were synthesized to mimic the accidentally bound peptides and were co-crystallized. Peptide-containing crystals provide further insight into the initial binding of proteins in the catalytic pocket of aspartyl proteases in both active and inactive conformations.

A MODULAR APPROACH TO HIV-1 PROTEINASE INHIBITOR DESIGN

J. Konvalinka,[1] T. Uhlíková,[1] M. Hradilek,[1] I. Pichová,[1] M. Souèek,[1] H.-G. Kraeusslich,[2] and J. Vondrásek[1]

[1]Institute of Organic Chemistry and Biochemistry, Academy of Science of Czech Republic, Flemingovo, n. 2, 166 10 Praha 6, Czech Republic
[2]Abteilung für Zellbiologie und Virologie, Heinrich-Pette Institut, Hamburg, Germany, D-20251

HIV-1 proteinase (PR) represents an important target for antiviral chemotherapy. Most inhibitors of HIV proteinase reported to date are transition state mimics and bind to the active site of the enzyme. The only alternative to active-site inhibition explores the fact that PR is an obligatory homodimer, each monomer subunit contributing one aspartate residue to the active site. It has been shown that the N- and C-termini of the PR sequence are responsible for approximately 50% of the intersubunit ionic and hydrogen bond interactions. Indeed, several in vitro studies with inhibitors targeted to the dimerization domain have been reported using peptides derived from the amino acid sequence of the N- and C-termini of the PR.

We have designed, synthesized and tested bifunctional, "modular" inhibitors of HIV PR combining an active-site inhibitor tethered to a structure targeted to the dimerisation domain of the enzyme via a short hinge. We have observed that combination of the dimerisation and active-site inhibitors shows synergistic effect. Moreover, the modular inhibitor has significantly lower IC_{50} value than the equimolar mixture of the two parent inhibitors. The pH dependence of the modular inhibitor activity is flatter than that of the parent active-site targeted inhibitor, probably due to the increase of dimerisation constant with increasing pH.

We hypothesize that these modular inhibitors might have several advantages over the simple mixture of inhibitors. They may be able to bind to the PR domain in uncleaved Gag/Pol polyproteins prior to viral particle assembly. Modular inhibitors may be thus transported to the viral particle together with viral proteins via their dimerisation inhibitor segment. The relatively high local concentration of the inhibitor within the particle may then improve the inhibitory potential of the active-site-directed segment. Other possible advantage of modular inhibitors as well as mixtures of active-site and dimerisation inhibitors represents the fact that development of resistance to such compounds might be restricted because nonlethal parallel mutations in two essential domains of the enzyme might be less probable. The analysis several of these hypotheses, the mechanism of action of these inhibitors and their activity in mammalian cells infected by a laboratory strain of HIV will be discussed.

DATABASE OF THE STRUCTURES OF HIV PROTEASE

J. Vondrasek and A. Wlodawer

Macromolecular Structures Laboratory, NCI-FCRDC, ABL - Basic Research Program, Frederick, Maryland 21702

The inhibitors of HIV-1 protease, one of the most important enzymes in the life cycle of the HIV, are currently the most powerful agents for blocking HIV infection. The initial and very crucial step of the process of drug design was the determination of the structure of HIV-1 protease in complex with inhibitors. Estimated number of crystallographic structures of HIV PR is over 400, but most of them have not been published due to proprietary concerns of pharmaceutical companies. We have now embarked on a project sponsored by NIGMS to assemble all the existing structures of HIV-1, HIV-2, and SIV proteases into a single, unique database. We are motivated by the fact that with the completion of the design stage of HIV-1 protease-inhibiting drugs, some of the structural information might soon be lost. We are also convinced that, with several drugs on the market, the need for confidentiality of these structures is gone, but that this information may be very useful for other projects in the future, as well as for the design of the next generation of PR inhibitors.

The basic idea of the database is to be as informative as possible. The database is available to the general scientific community through the WWW service. The main part is organized as a spreadsheet and involves information about structure, X-ray measurement conditions, resolution, R factor, type and schematic formula of the inhibitor, dimension of the unit cell, some biological data and details of the K_i measurements, the journal where the paper was published, and many others. Using some adapted spreadsheet functions everyone will be able to use sorting, listing and obtaining the chosen subset of the data.

The second part of the database provides more scientific insight into the phenomena of the binding of particular inhibitors. All of the collected structures are transformed into a common frame of reference in the same way, using the alignment of all of the C_α atoms of the enzyme. These data are in PDB-like format, which means that all of the information from the PDB file will be retained. Moreover, the annotation of all structures is uniform. A set of transformation matrices will be provided for the various alignments of the chosen structures. The transformation itself is controlled by pressing a button in the Web page. A subset of inhibitor structures is put together as an independent part of the database to provide insight into the structural rules of the inhibitor binding. We also intend to compute and release some additional data, such as the volume of the active site, its surface, related data from simulations, etc. The database will be flexible enough to accommodate new information as it becomes available.

HIV	Human Immunodeficiency Virus
SIV	Simian Immunodeficiency Virus
PR	Protease
WWW	World Wide Web
PDB	Protein Database Brookhaven

Research sponsored by the National Cancer Institute, DHHS, under contract with ABL.

10

IN VITRO SELECTION AND CHARACTERIZATION OF VX-478 RESISTANT HIV-1 VARIANTS

S. Pazhanisamy, Judith A. Partaledis, B. G. Rao, and David J. Livingston

Vertex Pharmaceuticals Incorporated
130 Waverly Street
Cambridge, Massachusetts 02139

1. ABSTRACT

VX-478 (141W94), a potent inhibitor of HIV protease, is in late stage clinical trials for the treatment of HIV infection and AIDS. Resistant viruses were raised *in vitro* by passage of HIV-1_{IIIB} in the presence of increasing concentrations of VX-478 and the related hydroxyethylamino sulfonamide inhibitor VB-11,328. By direct PCR analysis of selected viruses, a number of mutations were identified (L10F, M46I, I47V, I50V and I84V) in the protease gene. These mutations were introduced into recombinant HIV-1 protease and the mutant enzymes assayed against a panel of inhibitors of diverse chemical structure. For VX-478, significant increases in IC_{90} and K_i were observed for virus or protease, respectively, containing I50V single mutation or an M46I/I47V/I50V triple mutation. The mutant proteases were also characterized for their kinetic competence to process substrates representing cleavage sites of gag-pol viral polypeptide. The kinetic data were interpreted with the aid of molecular modeling to understand the effect of mutations on inhibitor binding and processing of the gag-pol polypeptide to generate infective virions.

2. INTRODUCTION

During the life cycle of HIV, the viral protease plays a vital role in the maturation of many structural proteins and enzymes from the gag-pol polypeptide. Therefore, it is an attractive drug discovery target for the control of HIV infection and AIDS. It has been shown by mutagenesis that Asp 25 at the active site of HIV protease is essential for its catalytic activity and hence it belongs to the aspartyl protease class of enzymes.[1] Unlike the mammalian and fungal aspartyl proteases, which are monomeric enzymes, the aspartyl

Aspartic Proteinases, edited by James.
Plenum Press, New York, 1998.

proteases of retroviral origin are homodimers—each subunit contributing one aspartyl residue for catalysis. The HIV protease three-dimensional structure has been determined in numerous laboratories.[2] A wealth of knowledge available on the drug discovery effort in Renin, a related mammalian aspartic protease, paved the way for designing very potent inhibitors for HIV protease in relatively short time.[3–5] In late 1995 and early 1996, three HIV protease inhibitors, saquinavir, indinavir, and ritonavir, were approved for the treatment of HIV infection and AIDS, and more are being evaluated in the clinic. Several protease inhibitors have been claimed to be more clinically effective than any single reverse transcriptase (RT) inhibitor in reducing viral load.[6] VX-478 (Fig. 1) is a protease inhibitor being jointly evaluated in phase III clinical trials by Vertex Pharmaceuticals, Glaxo Wellcome and Kissei Pharmaceuticals. VX-478 is a potent inhibitor of HIV-1 and HIV-2 proteases with K_i values of 0.6 nM and 19 nM, respectively. This inhibitor is highly specific to HIV protease (K_i > μM for mammalian aspartyl proteases), is water soluble (0.19 mg/ml in PBS) and has an excellent oral bioavailability (30–80%)[7] in animal models. In phase I clinical trials,[8–9] VX-478 was shown to be well absorbed and well tolerated. Phase II dose escalating trials[10] demonstrated that the viral load could be reduced by 99% over three weeks when VX-478 was administered at 1200 mg twice daily.

The rapid mutation rate of HIV can limit the efficacy of antiretroviral agents. Under drug pressure resistant mutations emerge even more rapidly. Since the HIV protease is a homodimer of subunits, each mutation in the subunit results in modifying interactions with the ligand at two locations in the active site (one at the prime side and the other at the non-prime side[11]). Therefore, an ideal protease inhibitor would not only be potent and bioavailable but would also induce few resistant mutations of consequence in the protease. We were interested in identifying the possible mutations in HIV protease gene that might cause clinical resistance to VX-478. Early identification of *in vitro* resistance profile may be valuable for proper design of therapeutic regimens using VX-478. Here we summarize the results of *in vitro* selection experiments that led to the identification of VX-478 resistant viruses and the characterization of such resistant viruses against a panel of inhibitors[12] (Fig. 1). The mutations identified in the HIV protease gene were expressed either individually or in combination in recombinant HIV protease and were further characterized by their effect on inhibitor binding and catalytic efficiency of substrate processing by the protease.[13]

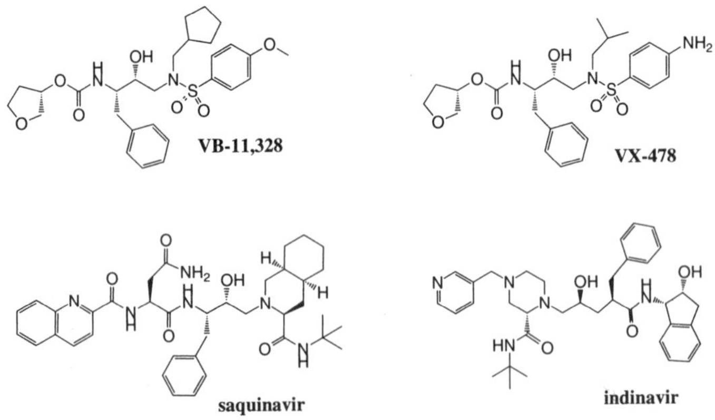

Figure 1. Structure of protease inhibitors used in this study to determine the antiviral activity and inhibition constant against mutants.

3. PHENOTYPIC AND GENOTYPIC CHANGES IN HIV-1_{IIIB} VIRUS SELECTED IN THE PRESENCE OF INCREASING CONCENTRATIONS OF VX-478 OR VB-11328

HIV-1_{IIIB} was serially passaged in CEM-SS cells in the presence of increasing concentrations of VX-478 and VB-11,328, two inhibitors of the hydroxyethylamino sulfonamide class (Fig. 1). Once the rate of viral production attained (or exceeded) that in the presence of two fold lower concentration of the drug, the concentration of the drug was further doubled in the next passage. Otherwise, the same concentration was maintained in the subsequent passages. At every passage, a small sample of the virus was saved for characterization. Direct PCR sequencing was performed to determine the mutations in the protease gene. The order of appearance of mutations in the protease gene is similar (Table 1) for VX-478 and VB-11,328. In Fig. 2 location of the mutations are indicated with respect to the VX-478 binding at the active site of the HIV protease dimer. The first mutation to appear was L10F, a residue approximately 10 Å away from the active site. As the selecting concentration of the drug was increased, the I84V mutation appeared in the background of L10F at 160 nM and ² 800 nM of VB-11328 and VX-478, respectively. After several passages (at selecting concentrations of ² 640 nM and 1600 nM of VB-11328 and VX-478, respectively), 84V reverted to 84I and a new mutation appeared at residue 50 giving rise to I50V in the background of L10F. Further passage of the virus at higher concentrations of drug resulted in adding M46I and I47V to the already present mutations L10F and I50V. At 3.2 μM of VX-478, the selected virus contained D60V which was not observed in the presence of VB-11,328. However, in the latter case the highest concentration of drug used was 2.5 μM.

The viral stocks saved at various passages were assayed against a panel of structurally diverse inhibitors (Fig. 1) — VX-478, VB-11328, saquinavir and indinavir. In the increase in IC_{90} for these compounds with respect to that for wild type virus is shown as a function of the mutations which appeared in the protease gene (Table 1). The double mutation L10F/I84V had no effect on the IC_{90} of saquinavir and had only a 2-fold effect on indinavir IC_{90}. The I50V containing viruses from either passage experiment had only a 2- and 6-fold effect on the IC_{90} of saquinavir and indinavir, respectively. However for VX-478 and VB-11328, the I84V mutation affected the IC_{90} moderately (38- and 8-fold increase,

Table 1. Antiviral activity of protease inhibitors against passaged variants of HIV-1

Selecting compound	Passage no.	Selecting conc. (nM)	Mutations in protease	Fold increase in IC_{90} (nM)			
				VB,11328	VX-478	Saquinavir	Indinavir
VB,11328	0	0	None	1 (23)	1 (26)	1 (22)	1 (27)
	10	80	L10F	1	ND[a]	ND	ND
	12	160	L10F, I84V	4	ND	ND	ND
	15	640	L10F, I50V	19	53	1	ND
	17	1280	L10F, M46I, I50V	65	100	1	4
	20	2560	L10F, M46I, I47V, I50V	100	240	2	4
VX-478	0	0	None	1	1	1	1
	7	800	L10F, I84V	8	38	1	2
	8	1600	L10F, I50V	ND	87	ND	ND
	9	1,600	L10F, M46I, I47V, I50V	108	150	1	3
	10	3200	L10F, M46I, I47V, I50V, D60V	46	170	1	6

[a]Not determined.

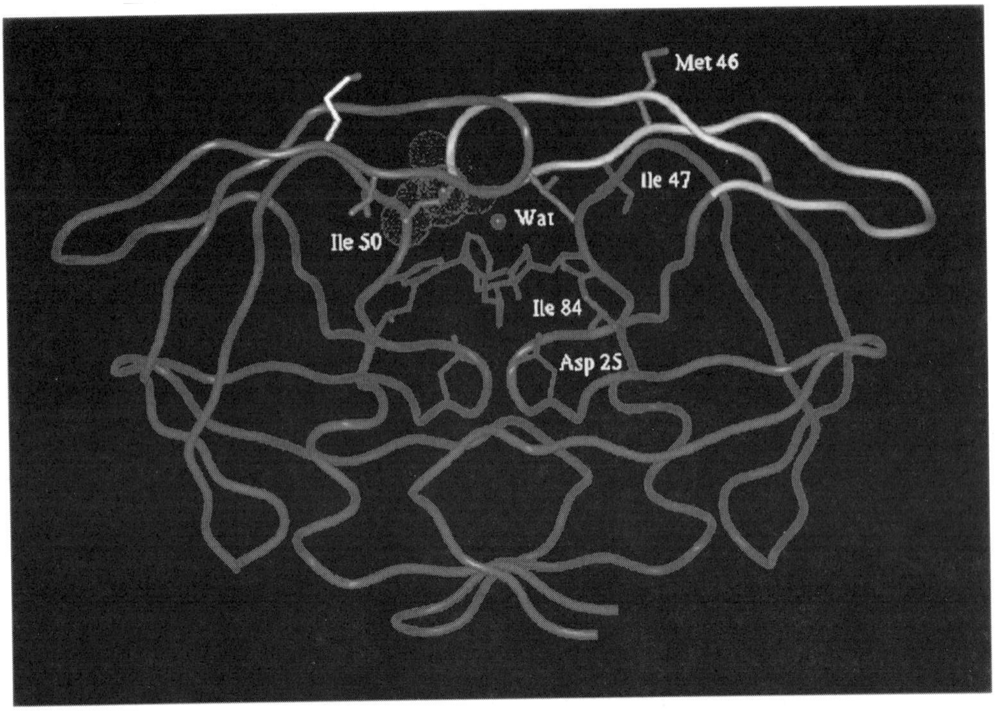

Figure 2. Ribbon diagram of HIV-1 protease dimer with VX-478 at the active site. The residues that mutate during the *in vitro* selection of virus in presence of VX-478 and the active site aspartic acid are identified.

respectively) whereas the I50V containing viruses show a much larger increase in IC_{90} (100- to 240-fold).

4. MOLECULAR INTERACTIONS OF MUTANTS WITH INHIBITORS

The mutations identified in the *in vitro* selection experiment were expressed, individually and in combination, in recombinant HIV-1 protease. The purified mutant enzymes were assayed against the same panel of inhibitors (Fig. 1) used to characterize the mutant viruses. The increase in K_i values relative to the wild type are reported in Table 2. Leu 10 is located about 10 Å from the active site, hence the Phe 10 mutation has very little effect on the binding of any of the compounds. Residue 46 is on a long loop called the 'flap' that closes down on the active site upon binding a ligand. There are two flaps, comprising residues 35 to 57 in each chain, per active dimer (Fig. 2). Molecular dynamics simulation studies[14] suggest that the M46I mutation results in locking the flap portion of the enzyme in the closed conformation. Residue 47, also on the flap, is part of the S_2/S_2' binding pocket. Among the inhibitors used in this study, only indinavir makes any interaction with this residue. Therefore, only a moderate effect on inhibitor binding (1–4 fold change in K_i) against the single mutations M46I and I47V was observed. Residues Ile 50 and Ile 84 are at the center of the active site and their side chains are at the interface of subsites 1 and 2, and therefore, are capable of interacting with residues at P_2 through P_2'. Hence, it is not

Table 2. Inhibition constant, K_i, for protease inhibitors against mutants of HIV-1 protease

HIV PR mutant	Fold increase in K_i (nM) for			
	VB-11328	VX-478	Saquinavir	Indinavir
Wild type	1 (<0.1)[a]	1 (0.6)	1 (0.8)	1 (1.0)
L10F	>1	1	1	4
M46I	>1	1	1	4
I47V	>2	1	1	3
I50V	**>30**[b]	**83**	**21**	10
I84V	>10	**23**	12	**20**
L10F, I84V	**>22**	**20**	**21**	**24**
M46I, I47V, I50V	**>140**	**270**	**41**	**29**

[a]The number in parentheses is the inhibition constant, K_i (nM), measured against wild type HIV-1 protease.
[b]Ratio of K_i for mutant protease over that of the wild type value rounded to the nearest whole number. Ratios ≥ 20 are highlighted.

surprising to find a greater than 10-fold effect on the binding of these compounds vs. the I50V or I84V mutants. In the case of the I84V mutation, the effect is 12- to 23-fold on K_i on all inhibitors examined. On the contrary, I50V affects the binding of VX-478 by 83-fold which is a much higher effect than that observed for saquinavir and indinavir with only 21- and 10-fold increases in K_i, respectively. This confirms that the I50V mutation appeared due to selection pressure applied by VX-478 in the medium. Modeling studies, including solvent accessible surface area calculations, molecular dynamics and free energy perturbation, suggest that the increase in K_i is due to the loss of hydrophobic interactions, but not due to loss of any hydrogen bond interactions of the 'flap water' (water molecule hydrogen bonded with amide groups of Ile 50 and Ile 50′ of the flaps and the inhibitor). The Cγ methyl group of Ile 50 makes hydrophobic interactions with all of the carbon atoms of the P_2' phenyl ring of VX-478. On the other hand, in saquinavir and indinavir, only two of the three methyl groups of t-butyl at P_2' are in direct contact with the Cγ methyl group of Ile 50. On the non-prime side, the P_2 indanol group of indinavir is making a direct hydrophobic interaction with the entire side chain of Ile 50′, not just with the terminal Cγ group. Therefore, absence of the Cγ methyl group of Ile50 / Ile50′ residues by the I50V mutation causes a greater loss in binding of VX-478 than for indinavir or saquinavir. Addition of M46I and I47V mutations to I50V reduces the binding of all of the inhibitors by a factor of 2- to 4-fold.

5. MOLECULAR INTERACTIONS OF MUTANTS WITH SUBSTRATES

The substrate specificity of HIV protease is diverse, recognizing at least eight different cleavage sites on the gag–pol polypeptide.[15] Therefore, a single substrate cannot be used to completely describe the effect of the mutation on catalysis. We used four peptide substrates representing three different cleavage sites to determine the catalytic competence of the mutant proteases. The identity of the peptide substrates, the cleavage sites they represent and catalytic parameters obtained with wild type HIV-1 protease are presented in Table 3. The catalytic efficiency represented by the peptides spans three orders of magnitude. Substrate 4 is the least efficiently cleaved peptide. If the gag–pol polypeptide has to be processed completely for the virion to be infectious, then the most slowly processed

Table 3. Catalytic parameters for the wild type HIV-1 protease-catalyzed hydrolysis of peptide substrates

Cleavage site	Substrate[a]	K_m (μM)	k_{cat} (s^{-1})	k_{cat}/K_m (mM^{-1} s^{-1})
p24–X	1 (Leu–pNP)	17	16.3	940
p24–X	2 (Leu–Ala)	69	1.2	17.5
p17–p24	3 (Tyr–Pro)	740	8.6	11.6
PR–RT	4 (Phe–Pro)	1465	0.7	0.5

[a]Substrate 1 is His–Lys–Ala–Arg–Val–Leu–(NO_2)Phe–Glu–Ala–Nle–Ser–NH_2, substrate 2 is His–Lys–Ala–Arg–Val–Leu–Ala–Glu–Ala–Met–Ser–NH_2, substrate 3 is Ser–Gln–Asn–Tyr–Pro–Ile–Val–OH and substrate 4 is Ac–Thr–Leu–Asn–Phe–Pro–Ile–Ser–Pro–OH. Residues flanking the cleavage site were shown in parentheses. pNP is p-nitrophenylalanine.

cleavage site would be the most likely one to determine the rate of production of infectious virion and should be the most important one to be measured.

The catalytic efficiencies for the mutants measured against the four peptides are tabulated (Table 4) in comparison to the wild type protease. The catalytic efficiency for processing the substrate decreases in most cases but the decrease is ≤ 3-fold for the L10F, M46I, I47V and I84V mutants. The I50V mutation results in about 5-fold decrease in catalytic efficiency for substrates 1 and 2 but nearly a 25-fold decrease for substrates 3 and 4. The substrate models show that Ile 50 and Ile 50′ side chains are in contact with P_2, P_1, P_1' and P_2' side-chains of the substrates, as in the case of inhibitors noted earlier. The degree of contact of the Ile 50 and Ile 50′ residues with these different substrates varies, however. Using a 5.0Å distance cut-off, we found that the C_γ methyl groups of Ile 50 and Ile 50′ side chains make the largest number of contacts (four) with the Ile side chains at P_2' of substrates 3 and 4, but only one contact with the P_2' Glu side chain of substrates 1 and 2 (Fig. 3). On the non-prime side, there are two contacts with the Val side-chain of substrates 1 and 2, and one with the Asn side-chain of substrates 3 and 4. The larger number of interactions of the P_2' Ile is due to the fact that Ile side chain is buried deeper into the hydrophobic region of the S_2' pocket whereas the Glu side chain is swung away from the hydrophobic region, but towards the solvent-exposed portion of the enzyme making

Table 4. Effect of mutations in the HIV-1 protease on the catalytic efficiency of substrate processing

	Fold increase in k_{cat}/K_m ($mM^{-1}s^{-1}$)			
Enzyme	Substrate 1	Substrate 2	Substrate 3	Substrate 4
Wild type	1.00 (940)[a]	1.00 (17.5)	1.00 (11.6)	1.00 (0.5)
L10F	0.66	ND[b]	ND	ND
M46I	0.83	ND	ND	ND
I47V	0.50	1.18	0.60	0.68
I50V	0.32	0.17	**0.04**[c]	**0.04**
I84V	0.34	0.21	0.44	0.53
L10F/I50V	0.36	0.16	**0.03**	**0.03**
M46I/I47V/I50V	0.20	0.22	**0.06**	**0.08**

[a]Number in parentheses represents the catalytic efficiency, k_{cat}/K_m ($mM^{-1}s^{-1}$), for the wild type HIV-1 protease.

[b]Not determined.

[c]Changes of >10-fold in k_{cat}/K_m are in bold.

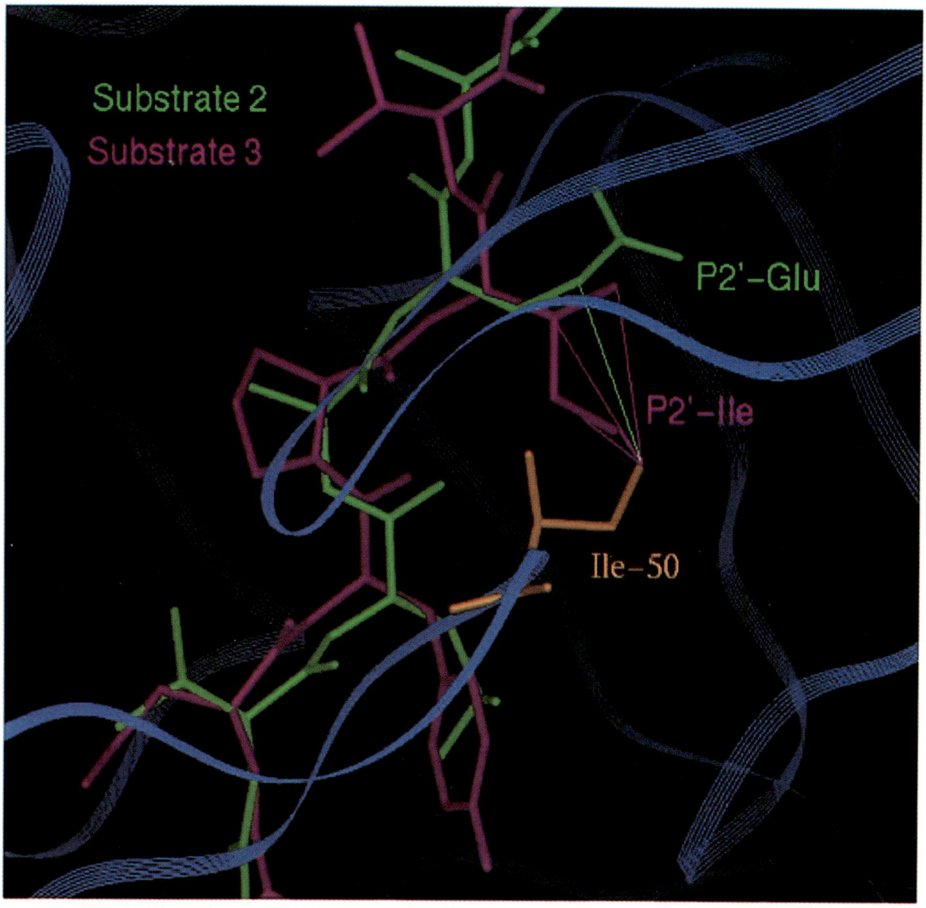

Figure 3. Models of substrates 2 (green) and 3 (purple) in the active site of HIV-1 protease (blue ribbon). For clarity, only the P_2' residues of these substrates (Glu and Ile for substrates 2 and 3, respectively) and Ile 50 residue of the enzyme are labeled. The dashed lines indicate close contacts (5 Å cut-off) between C_δ of Ile 50 and the P_2' residues of the substrates.

hydrogen bonds with the main chain and side chain of Asp 30. The effect of the I50V mutation, therefore, is more severe for processing substrates 3 and 4, than substrates 1 and 2.

6. EFFECT OF MUTATIONS ON APPEARANCE OF RESISTANCE

During selection of resistant virus in presence of VX-478 the following mutations appeared in the protease gene: L10F, M46I, I47V, I50V and I84V. Among these mutations L10F, M46I and I47V have marginal effects on the catalysis of the gag–pol polypeptide cleavage sites and inhibitor binding. I84V and I50V have a significant effect on inhibitor binding. Such mutations are also capable of effecting the catalytic efficiency of the polpypeptide processing which is vital for the generation of infectious virion. The virus has to carefully balance the gain (increase in K_i value of inhibitor) over the loss (decrease in catalytic efficiency) in selecting any mutations to confer resistance. Hence it is useful to determine the vitality[16] values for the mutations which confer a significant effect on the inhibitor binding. The vitality value, defined in Eq. 1, slightly rearranged from the original definition, is a ratio reflecting the decrease in the inhibition constant over the decrease in the catalytic efficiency.

$$\text{Vitality} = \frac{(K_i)_{mut} / (K_i)_{wt}}{(k_c / K_m)_{wt} / (k_c / K_m)_{mut}} \tag{1}$$

This ratio is much greater than one for a mutation to be considered a resistant mutation. In Fig. 4 vitality values for I84V, I50V and the triple mutant M46I/I47V/I50V are plotted as a function of substrate identity for VX-478, saquinavir and indinavir. For I84V the vitality values for VX-478, saquinavir and indinavir are within a factor of two (Fig. 4, panel A). If the mutation were truly due to the presence of VX-478, the vitality value for VX-478 would have been selectively higher than that against saquinavir or indinavir. The I50V mutation on the other hand, decreases the binding of VX-478 more selectively than that for saquinavir or indinavir and hence the vitality values are much higher for VX-478 for any of the substrates (Fig. 4, panel B). This is consistent with the observation that during the selection of resistant virus in the presence of VX-478 the I84V mutation which initially appears is replaced with the I50V mutation. However, among the substrates that were assayed against I50V mutant, substrate 4 (the most slowly processed of the substrates by wild type protease) has a vitality

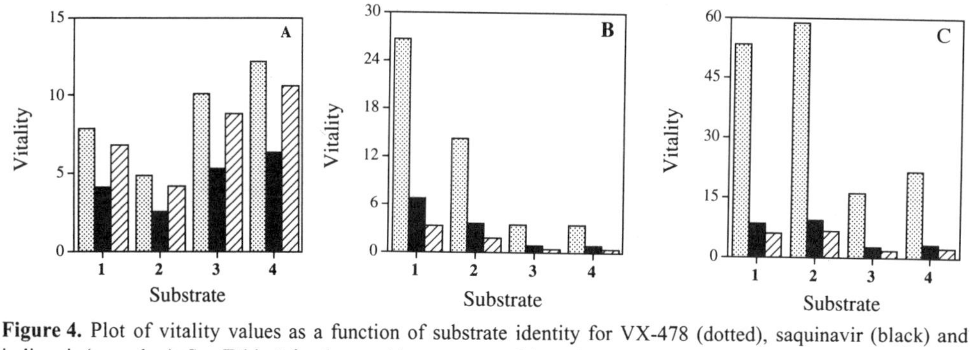

Figure 4. Plot of vitality values as a function of substrate identity for VX-478 (dotted), saquinavir (black) and indinavir (cross bar). See Table 1 for the peptide sequence of the substrates 1–4. Panels A, B and C represent data for the mutants I84V, I50V and the triple mutant M46I/I47V/I50V, respectively.

value of only 3 for VX-478. For the same substrate and inhibitor, this value increases to 21 for the triple mutant M46I/I47V/I50V (Fig. 4, panel C). The triple mutant, therefore, is preferred for survival of the virus in presence of VX-478. The added mutations, M46I and I47V, decrease the binding affinity of VX-478 in addition to increasing the catalytic efficiency. Thus they are compensatory mutations for viral survival in the presence of VX-478. The vitality is low (≤3) for saquinavir and indinavir suggesting that the triple mutant will not protect the virus in the presence of these inhibitors.

7. CONCLUSIONS

A number of mutations (L10F, M46I, I47V, I50V and I84V) appeared in the protease gene of the resistant HIV selected in presence of hydroxyethylamino sulfonamide inhibitors, VX-478 and VB-11328. *In vitro* kinetic characterization of the recombinant HIV protease mutants with a panel of inhibitors and substrates led to the following conclusions: 1) L10F has very little effect on inhibitor binding and I84V is less selective in effecting the binding of the panel of inhibitors of diverse chemical structure, 2) the single and the triple mutant I50V and M46I/I47V/I50V, respectively, affect the binding of VX-478 and VB-11328 more selectively, 3) the catalytic competence of the triple mutant is higher than that of the single mutant by a factor of 1.5 to 2.0 and, therefore, M46I and I47V are compensatory mutations, and 4) for the triple mutation, the affinity for VX-478 is decreased by a factor of 3 while the catalytic competence of the protease for substrate 4 is increased by a factor of 2 compared to the I50V mutation. Adding M46I and I47V, therefore, increases the survival of the virus in presence of VX-478. The I50V mutation substantially reduces the binding affinity for VX-478 as well as the catalytic competence of the protease to process the gag–pol polypeptide as revealed by the model substrates. This severely compromised catalytic competence of the protease to process the gag–pol polypeptide at specific cleavage sites may reduce the likelihood of the appearance of I50V mutation in causing clinical resistance to VX-478.

ACKNOWLEDGMENTS

We gratefully acknowledge the contributions of our colleagues Dr. Steve Chambers, John Fulghum, Dr. Olga Futer, Nara Margolin, and Cameron M. Stuver whose publications are reviewed here. Our sincere thanks to Dr. Scott Raybuck for his critical reading of the manuscript. We would also like to thank the HIV drug discovery team at Vertex and Dr. Vicki Sato for her continued support and inspiration.

REFERENCES

1. Meek, T. D. (1992) *J. Enz. Inhib.* **6**, 65–98.
2. Wlodawer, A., and Erickson, J. W. (1993) *Annu. Rev. Biochem.* **62**, 543–585.
3. Darke, P. L., and Huff, J. R. (1994) *Adv. Pharmacol.* **25**, 399–454.
4. Pillay, D., Getman, M. B. D., and Richman, D. D. (1995) *Rev. Med. Virology* **5**, 23–33.
5. De Clercq, E. (1995) *J. Med. Chem.* **38**, 2491–2517.
6. XI International Conference on AIDS, Vancouver, Canada, 1996. Abstract # Mo. B. 1137 and # Mo. B. 170.
7. Studenberg, S. D., Dhal, R. A, and Wooley, J. L. on " Pharmacokinetics, Excretion, and Mass Balance Studies in Dogs with 141W94 (VX-478), an HIV-1 Protease Inhibitor." Presented at the 7th International Society for the Study of Xenobiotics, San Diego, CA, 1996. Abstract # 327.

8. Painter, G. R., Ching, S., Reynolds, D., St.Clair, M., Sadler, B. M., Elkins, M., Blum, R., Dornsife, R., Livingston, D., Partaledis, J. A., Pazhanisamy, S., Tung, R. and Tisdale, M. (1996) *Drugs of the Future* **21**, 347–350.
9. St. Clair, M.H., Millard, J., Tisdale, M., Parry, N., Sadler, B.M., Blum, M. R., and Painter, G. on "*Invitro* antiviral activity of 141W94 (VX-478) in combination with other antiviral agents." Presented at the Consensus Symposium on Combined Antiviral Therapy, Lisbon, Portugal, July 26, 1995.
10. Schooley, R. on "Preliminary data on the safty and antiviral efficacy of the novel protease inhibitor 141W94 in HIV-infected patients with 150 to 400 CD4+ cells/mm^3" Presented at the 36th Interscience Conference on Antimicrobial Agents and Chemotherapy, New Orleans, La. Abstract # LB7a.
11. Schechter, I., and Berger, A. (1967) *Biochem. Biophys. Res. Commun.* **27**, 157–162.
12. Partaledis, J. A., Yamaguchi, K., Tisdale, M., Blair, E. E., Falcione, C., Maschera, B., Myers, R. E., Pazhanisamy, S., Futer, O., Cullinan, A. B., Stuver, C. M., Byrn, R. A., and Livingston, D. J. (1995) *J. Virology*. **69**, 5228–5235.
13. Pazhanisamy, S., Stuver, C. M., Cullinan, A. B., Margolin, N., Rao, B. G. and Livingston, D. J. (1996) *J. Biological Chem.* **271**, 17979–17985.
14. Collins, J. R., Burt, S. K., and Erickson, J. W. (1995) *Nature Struct. Biol.* **2**, 334–338.
15. Debouk, C. (1991) *Adv. Exp. Med. Biol.* **306,** 407–415.
16. Gulnik, S. V., Suvorov, L. I., Liu, B., Yu, B., Anderson, B., Mitsuya, H., and Erickson, J. W. (1995) *Biochem.* **34**, 9282–9287.

11

SENSITIVITY TO INHIBITION AND CATALYTIC EFFICIENCY OF HIV PROTEINASE MUTANTS

S. I. Wilson,[1] L. H. Phylip,[1] S. V. Gulnik,[2] J. S. Mills,[3] D. Bur,[4] B. M. Dunn,[5] and J. Kay[1]

[1]School of Molecular and Medical Biosciences
University of Wales
Cardiff CF1 3US, Wales, United Kingdom
[2]NCI-FCRDC
Frederick, Maryland 21702
[3]Roche Research Centre
Welwyn Garden City, AL7 3AY, United Kingdom
[4]F. Hoffmann La Roche AG
CH-4070 Basel, Switzerland
[5]Department of Biochemistry and Molecular Biology
University of Florida
Gainesville, Florida 32610

The first potent inhibitor of HIV proteinase to be licensed by the FDA for anti-AIDS therapy was Ro31–8959 (Invirase™). Approval followed shortly thereafter for ABT-538 (Norvir™) and L-735,524 (Crixivan™). Virus variants have been found to emerge however in patients being treated with these compounds and concerns have been voiced that treatment with one drug might generate resistance not only against that inhibitor but also against the other compounds.[1] In patients treated with Ro31–8959, virus variants with decreased sensitivity are selected with the mutations G48V and L90M in the proteinase component of the viral genome,[2] whereas with ABT-538 and L-735,524 different resistant mutations have been detected,[1,3] e.g. at residues 82 (such as V82F) and 84 (I84V).

The ability of these three licensed compounds to inhibit homogeneous preparations of each of these recombinant mutant proteinases was determined *in vitro* and compared to the values obtained with wild-type proteinase (Table 1). Ro31–8959 was almost unchanged in its effect on V82F mutant proteinase and was only 3-fold less effective against the L90M mutant proteinase compared to wild-type. In contrast, the I84V and G48V mutant proteinases were ~20-fold and 30-fold less sensitive to inhibition by this compound (Table 1). The glycine to valine substitution at position 48 and the Ile to Val replacement at position 84 both occur at residues that can make direct contact with inhibitors bound in

Aspartic Proteinases, edited by James.
Plenum Press, New York, 1998.

Table 1. Interaction of several inhibitors with wild-type and mutant HIV-1 proteinases

Proteinase	Ro31-8959 K_i (nM)	L-735,524 K_i (nM)	ABT-538 K_i (nM)	HBY-793 K_i (nM)
Wild-type	0.3	2	0.1	0.1
G48V	8	12	1.7	0.1
V82F	0.5	15	9	1.7
I84V	5	3	1.4	0.1
L90M	0.8	3	0.5	0.1

the active site cleft of HIV proteinase. In contrast, leucine at position 90 in each subunit is located almost immediately underneath the polypeptide loop that contains the active site Asp25 residue that each subunit contributes to the catalytic mechanism of HIV proteinase.[4] As such, the L90M mutation occurs at a location where the newly-introduced Met side chain cannot make contact directly with a bound inhibitor and thus any changes observed for inhibitor binding in response to this mutation require more subtle explanations. This will necessitate elucidation of the crystal structure of the L90M mutant proteinase complexed with the inhibitor.

Introduction of a valine side chain into position 48 of HIV proteinase by molecular modelling shows that the Val48 side chain can make too close contacts with the rigid quinoline ring at P_3 of the Ro31–8959 inhibitor (Fig. 1). Examination of the inhibitory potency of a systematic series of compounds with structures closely related to Ro31–8959 in which the P3 and P1′ substituents were systematically varied, confirmed that the ~30-fold increase in resistance measured for the G48V mutant proteinase against Ro31–8959 required *both* the quinoline ring at P_3 and the decahydro isoquinoline to be present at P_1' in Ro31–8959.[5] Inhibitor structures are shown in Fig. 2.

All of the mutant proteinases showed resistance against the compound ABT-538 (Table 1). A five-fold loss in potency was measured against the L90M mutant proteinase and even larger losses in efficacy, up to 90-fold for V82F, were detected against the proteinases with mutations at residues 48, 82 and 84. In contrast, the potency of L-735,524 towards the I84V and L90M mutants was not significantly changed in relation to that measured for wild-

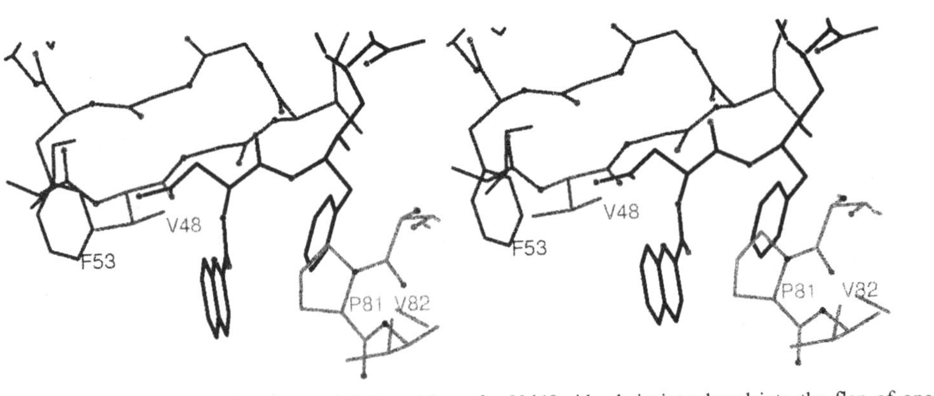

Figure 1. Stereodiagram depicting the modelled position of a Val48 side-chain introduced into the flap of one subunit in the structure solved for HIV-1 proteinase complexed with the inhibitor Ro31–8959. The P_3 quinoline ring and P1 phenylalanine residue of the inhibitor are depicted adjacent to Pro81and Val82 from the other subunit.

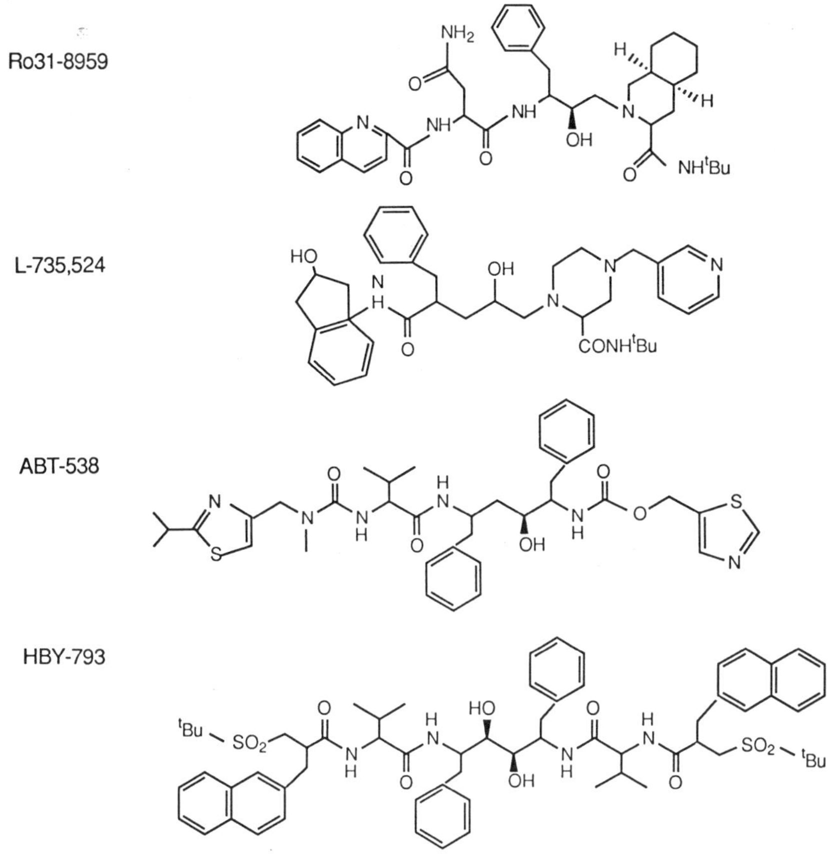

Figure 2. Structures of several inhibitors of HIV-1 proteinase.

type proteinase, although losses in efficacy of six and seven-fold were detected against the G48V and V82F mutant proteinases (Table 1). Since HIV proteinase is a symmetrical enzyme, the potency of a symmetrical inhibitor (HBY-793; Fig. 2) towards the mutant enzymes was compared to that measured for the wild-type enzyme (Table 1). This hydrophobic compound was comparably effective against all of the enzymes with the exception of the V82F mutant which showed a 17-fold increase in resistance (Table 1). The most noticeable feature of this symmetrical compound is the presence of two naphthylalanine substituents in the P_3 and P_3' positions (Fig. 2). Structures solved for this inhibitor complexed with retroviral proteinases[7] show that these large, hydrophobic side-chains nestle very closely together with the aromatic rings of the Phe side-chains in the P_1/P_1' positions respectively of the inhibitor and each fits very snugly into a pocket (in the symmetrical dimer) formed by the side-chains of Arg8′, Asp29, Pro81 and Val82. Thus, if the side-chain of Val82 (and Val 82′) contributing to each of these pockets is increased in size to that of a Phe residue, the two bulkier Phe side-chains will occupy some of the space required by the P_3/P_3' naphthylalanine substituents. A reduction in potency of HBY-793 towards the V82F mutant proteinase would arise inevitably as a consequence of this conflict.

Reduction in inhibitor sensitivity of a mutant proteinase is not the only aspect of viral resistance that needs to be considered however. In order for a virus encoding a mutant prote-

Table 2. Hydrolysis of peptide substrates by wild-type and mutant HIV-1 proteinases.

	Substrate A			Substrate B			Substrate C		
	K_m (μM)	k_{cat} (sec^{-1})	k_{cat}/K_m ($mM^{-1}sec^{-1}$)	K_m (μM)	k_{cat} (sec^{-1})	k_{cat}/K_m ($mM^{-1}sec^{-1}$)	K_m (μM)	k_{cat} (sec^{-1})	k_{cat}/K_m ($mM^{-1}sec^{-1}$)
Wild-type	13	14	1,080	190	10	55	15	23	1,530
G48V	12	5	420	190	0.5	3	90	2.5	30
V82F	35	6	170	225	4.5	20	260	13	50
I84V	35	10	285	250	3.6	15	85	10	120
L90M	10	10	1,000	440	4	9	25	5	200

Values were measured at pH 4.7.
A = Lys–Ala–Arg–Val–Leu*Nph–Glu–Ala–Nle–NH_2
B = Tyr–Val–Ser–Gln–Asn–Phe*Pro–Ile–Val–Gln–Asn–Arg
C = Ac–Tyr–Arg–Ala–Arg–Val–Phe*Nph–Val–Arg–Ala–Ala–Lys

inase to be able to replicate and remain viable, the mutant proteinase has to operate at a minimal level of enzymic competence. This aspect was examined by comparing the relative abilities of wild-type and these selected mutant proteinases to hydrolyse synthetic peptide substrates representative of the –Hydrophobic*Hydrophobic–, –Aromatic*Pro– and pseudo-symmetrical types of cleavage junction that are known to be present within the HIV polyproteins.[6]

For the –Hydrophobic*Hydrophobic– type substrate (A in Table 2), only minor changes in specificity constant (k_{cat}/K_m) were detected for most of the mutant enzymes compared to the value for wild-type proteinase. Only the V82F mutant showed a more significant drop (~6-fold) in catalytic efficiency. It is readily evident that wild-type proteinase is much less efficient in its ability to hydrolyse the peptide substrate (B in Table 2) representative of the –Aromatic*Pro– type of cleavage junction compared to the –Hydrophobic*Hydrophobic– type of cleavage junction peptide mimic (A in Table 2). A ~20-fold difference in specificity constants was measured. Against substrate B, the mutant proteinases showed a progressive decrease in their specificity constants, mostly because of lowered k_{cat} values, in the order wild-type > V82F, I84V > L90M > G48V. Perhaps not surprisingly, the pseudo-symmetrical substrate (C in Table 2) has the highest specificity constant in absolute terms for the symmetrical, wild-type HIV proteinase. All of the specificity constants measured for the mutant proteinases using substrate C were considerably lower than that of the wild-type enzyme, ranging from 7- to 12-fold for the L90M and I84V mutants to 30- to 50-fold for the G48V and V82F mutant proteinases. Since each of these mutations occurs twice (one in each subunit) in the symmetrical dimeric HIV proteinase, it would seem that any subtle repositioning of a substrate in the active site cleft in order to attempt to circumvent these two altered enzyme side-chains is more difficult to achieve with a pseudo-symmetrical substrate such as C. Re-adjustments elsewhere in the orientation of side-chains/backbone of the substrate may not position the scissile peptide bond in optimal juxtaposition to the catalytic apparatus for cleavage by the mutant proteinases. In contrast to the large losses in enzymic efficiency measured against the pseudo-symmetrical substrate C (Table 2), it was discussed earlier that the symmetrical inhibitor HBY-793 did not have a reduced potency (Table 1) against the mutant enzymes (except V82F). This apparent discrepancy may be explained by the difference in length between substrate C and HBY-793, suggesting that interactions on the periphery of the enzyme may have a considerable influence on the binding and positioning of the longer substrate.

When the ability of the wild-type and mutant proteinases to cleave a viral polyprotein (recombinant *gag*) was compared, similar patterns of processing were observed in all cases.[5] Thus the order in which the various cleavage junctions in *gag* are processed by wild-type proteinase[6] was maintained for the mutant enzymes, confirming that the mutant proteinases have little difference in selectivity for cleavage at any of the individual junctions. Rather, they have a reduced overall efficiency compared to the wild-type proteinase.

From these data, it appears that the G48V mutation introduces the most disabling change into HIV-1 proteinase. However, this mutation has never been observed in virus from HIV-positive individuals who have had no treatment with proteinase inhibitors and it arises only in response to treatment of patients with Ro31–8959 (Invirase™[8]) and not with the other inhibitors. Furthermore, in Invirase-treated patients, it is a very rare occurrence. In contrast, the L90M mutation arises in virus in ~50% of HIV-positive individuals undergoing lengthy treatment with Ro31–8959 alone, or in combination with other drugs. The L90M mutant proteinase was the least affected enzyme in terms of its catalytic efficiency and while it was only slightly more resistant to Ro31–8959, it was still just as susceptible as wild-type proteinase to inhibition by a variety of other inhibitors.[5] Indeed, viruses carrying the L90M mutation in their proteinase do exhibit comparable sensitivities to wild-type virus for a number of the inhibitors that have been tested.[3,9]

ACKNOWLEDGMENTS

Supported by an MRC-Industrial Collaborative Studentship (for SIW) jointly with Roche Research Centre, and by grants from the MRC AIDS Directed Programme (to JK) and from the NIH (Grant AI-28571 to BMD).

REFERENCES

1. Condra, J.H., Schleif, W.A., Blahy, O.M., Gabryelski, L.J., Graham, D.J., Quintero, J.C., Rhodes, A., Robbins, H.L., Roth, E., Shivaprakash, M., Titus, D., Yang, T., Teppler, H., Squires, K.E., Deutsch, P.J. & Emini, E.A. (1995). *Nature*, **374**, 569–571.
2. Jacobsen, H., Yasargil, K., Winslow, D.L., Craig, J.C., Kröhn, A., Duncan, I.A. & Mous, J. (1995). *Virology*, **206**, 527–534.
3. Molla, A., Korneyeva, M., Gao, Q., Vasavanonda, S., Schipper, P.J., Mo, H.-M., Markowitz, M., Chernyavskiy, T., Niu, P., Lyons, N., Hsu, A., Granneman, G.R., Ho, D.D., Boucher, C.A.B., Leonard, J.M., Norbeck, D.W. & Kempf, D.J. (1996). *Nature Med.*, **2**, 760–765.
4. Erickson, J.W. (1995). *Nature Struct. Biol.*, **2**, 523–529.
5. Wilson, S.I., Phylip, L.H., Mills, J.S., Gulnik, S.V., Erickson, J.W., Dunn, B.M. & Kay, J. (1997). *Biochim. Biophys. Acta*, in press.
6. Pettit, S.C., Moody, M.D., Wehbie, R.S., Kaplan, A.H., Nantermet, P.V., Klein, C.A. & Swanstrom, R. (1994). *J. Virol.*, **68**, 8017–8027.
7. Gustchina, A., Kervinen, J., Powell, D.J., Zdanov, A., Kay, J. & Wlodawer, A. (1996). *Protein Sci.*, **5**, 1453–1465.
8. Jacobsen, H., Hanggi, M., Ott, M., Duncan, I.B., Owen, S., Andreoni, M., Vella, S. & Mous, J. (1996). *J. Infect. Dis.*, **173**, 1379–1387.
9. Tisdale, M., Myers, R.E., Maschera, B., Parry, N.R., Oliver, N.M. & Blair, E.D. (1995). *Antimicrobial Agents Chemother.*, **39**, 1704–1710.

12

COMPUTER ASSISTED EVALUATION OF INHIBITOR SENSITIVITY TO HIV-1 PR MUTANTS

Fredy Sussman,[1,2*] Luis A. Martínez,[1,2] and M. Carmen Villaverde[1,2†]

[1]Protein Studies Program
Oklahoma Medical Research Foundation
Oklahoma City, Oklahoma 73104
[2]Research and Education Computing
Oklahoma University Health Sciences Ctr.
Oklahoma City, Oklahoma 73104

One of the most promising targets in the development of an anti-viral therapy against the AIDS virus has been the protease (HIV-1 PR) that the HIV virus uses for the cleavage of the *gag* and *gag–pol* complex during viral replication.[1] Many research laboratories have developed HIV-1 PR inhibitors that have become lead compounds for anti-viral drugs. Some of these compounds have been found to stop very effectively the replication of the HIV-1 virus *in vitro*.[2–4] However they seem to have limited clinical benefit, due in part to the development of viral resistance. The HIV virus is able to generate drug resistance because of the high mutational frequency in the replication of its genome. Under the 'evolutionary pressure' from an anti-viral agent a drug resistant mutant is selected. The primary sequence modifications that are generated in this way are called 'resistance' mutations and have been observed *in vitro*[5–9] and *in vivo*.[10,11] They maintain the level of catalytic activity, allowing the virus to replicate, but they lower the affinity of the HIV-1 PR for the anti-viral drug, reducing the drug's potential therapeutic power. At present, the molecular mechanisms that underlie drug resistance are not fully understood. Their elucidation will represent a step towards the formulation of novel drugs that could overcome this hurdle.

Here we review the methods developed and used in our laboratory for the determination of the sensitivity of an inhibitor towards a mutant strain of the HIV-1 PR. The first method is based on a solvent accessibility protocol for the prediction of the free energies of binding recently developed in our laboratory.[12] Our solvent accessibility algorithm should be ideally suited to predict binding sensitivity towards mutations that involve the

* Corresponding author: Protein Studies Program, Oklahoma Medical Research Foundation, Oklahoma City, Oklahoma 73104.
† Permanent address: Departamento de Química Orgánica, Facultad de Química, 15706 Santiago de Compostela (Spain).

Aspartic Proteinases, edited by James.
Plenum Press, New York, 1998.

replacement of one hydrophobic residue for another. A vast number of the observed resistant mutants seem to affect hydrophobic and van der Waals interactions and could be considered to produce defects in the binding pockets of the enzyme analogous to those generated by hydrophobic mutations in the core of a protein.[5–11] In this paper we apply our prediction algorithm to study the effect of the resistant mutant V82A on the affinity towards several inhibitors. This mutant strain has been identified as a resistant mutant in vitro,[5] as well as in clinical isolates.[11] Our results indicate that the of use of molecular mechanics based algorithms are sufficient to predict three dimensional structures with many of the features of the mutated complex. The calculated differences between the affinities of the native and mutant (A82) strains towards these inhibitors are in good agreement with experimental data. Partition of the free energy of binding into the contribution of single binding pockets reveals location of the binding pocket defects, and point towards design strategies for inhibitors that are immune to resistance mutations.[19]

The second method used in our lab for the study of the effect resistance mutations on inhibitor sensitivity is a modified Poisson-Boltzmann Finite difference method (mPBFD). We have shown it to be ideally suited for the study of resistance mutations involving change of polarity. With this method we have studied the effect of the replacing Val-82 for Asp, Glu or Gln residues. Below we give a brief description of both methods and a summary of some of the results obtained.[14]

1. SOLVENT ACCESSIBILITY FREE ENERGY PREDICTOR (SAFE_p)[12]

The method's basic assumption is that the free energy of an inhibitor transfer from water to the binding pockets of HIV-1 PR is analogous to the process of transfer from a medium of higher polarity to a one of lower polarity, since most residues that line the active site and specificity pockets of this enzyme are hydrophobic. To investigate this hypothesis we have generated a simple algorithm which fitted closely the observed free energies of binding for a series of known inhibitors.[12] The additive nature of this approach enabled us to partition the free energy of binding into the contributions of single fragments. The resulting analysis allowed us to rank the importance of the enzyme's subsites for binding: although all the enzyme's pockets contribute to binding, the ones that bind the P2–P2′ span of the inhibitor are in general the most critical for high inhibitor potency. Moreover, perusal of the energy contributions of single side chains has shown a broad specificity for some of the inhibitor fragments located in the central portion of the HIV-1 PR inhibitors. These observations are in agreement with experimental data, providing a validation for the physical relevancy of our method.[12]

Our calculations were performed on the complexes between HIV-1 PR and inhibitors A-77003, A-78791 and Ro-31–8959.[13,14] The starting point for most of our calculations were the wild type structures of the HIV-1 PR-Inhibitor complexes available from the Brookhaven data bank.[13,14] Prior to any calculation the hydrogens were added and their position was regularized using the Quanta suite of programs.[15] For the most part the structures of the mutant strains were obtained from the native complexes by replacing the V82 side chain by Ala. The complex was surrounded by ~1600 water molecules and a molecular dynamics (MD) trajectory of more than 110 picoseconds was performed for each of the structures. The software package used for the molecular dynamics calculations was CHARMm.[16] The time step used was 0.5 femtoseconds. In the first 30 picoseconds of the trajectory only the water molecules were allowed to move, in order to achieve a better solvent distribution around the solute. The free energies of binding were obtained as an average calculated over the MD frames collected during the last 20 picoseconds of the trajectory. For the sake of consistency the same protocol was applied to the bound native structure as well. The only known structure of the complex between an inhibitor and a mutant is that of A-77003 bound to the V82A strain.[17] In this case the SAFE_p algorithm was applied directly to the X-ray crystallographic structures (native and mutant), allowing for the comparison of the sensitivity of binding between the experimentally determined structures and the modeled ones.

One of the most useful analytical tools in SAFE_p is the division of the free energy of binding into the contributions of single inhibitors fragments. Fig. 1 shows such a free energy partition for all HIV-1 PR inhibitor complexes studied. Perusal of this figure shows the location of the 'hydrophobic defects' generated by the mutation without the need of surveying directly the structures. In the case of the A-77003 inhibitor, (see Fig. 1A–B), the V82A mutant further breaks down the almost symmetric distribution of free energies found for the S1 and S1′ sites, by reducing the affinity of the enzyme only against the P1 fragment. This feature was already noticed by direct inspection of the X-ray structures.[17] There is also a reduction in the affinity of the P3′ fragment. The crystallographic free energy profile patterns are similar to those of the modeled ones. Nevertheless, the effect of the mutation upon the inhibitor affinity is higher for the modeled structures. The calculated free energy difference between the native and the mutant enzymes for the modeled structure is 1.0 ± 0.3 Kcal/mol, while the one obtained from the X-ray crystallographic structures gives a difference of only 0.4 Kcal/mol. The experimental values are either 0.7[17] or 1.6[18] Kcal/mol (depending on the experimental conditions), in closer agreement with the result of the modeled structures. The free energy profiles for this inhibitor clearly indicate a strategy for the design of inhibitors

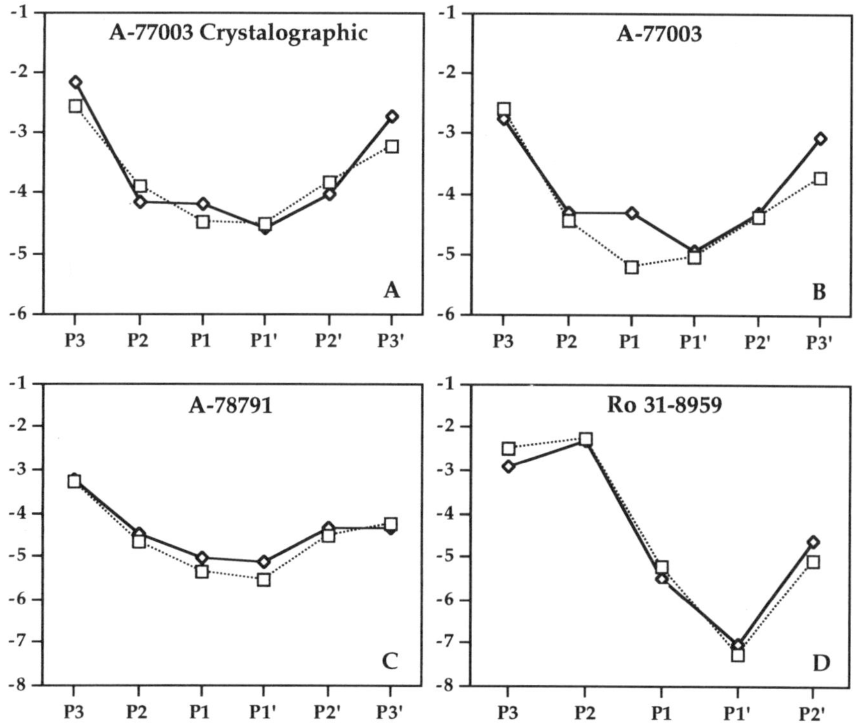

Figure 1. Partition of the free energy of binding to the wild type and V82A strains of HIV-1 PR into the contributions of the inhibitor's P fragments (as defined in ref. 12). Energy is given in dimensionless units ($\Delta G/RT$)
WT Muta

insensitive to this mutation. For this inhibitor, the groups at either P1 and/or P3′ should be redesigned to fit their respective pockets. One possible alternative would be the introduction of bulkier groups at any of the inhibitor fragments in contact with binding sites sensitive to the mutation.[5] Our results for the inhibitor Ro-31–8959 indicate that it is immune to the V82A mutation in agreement with experimental data[5] (the calculated difference in the free energy of binding is 0.0 ± 0.3 Kcal/mol). Perusal of the free energy profile (Fig. 1D) shows only very small differences in the free energy contributions of the binding pockets for the native and mutant complexes with this inhibitor. We surmise that the lack of sensitivity is due to the placement of a bulky decahydroisoquinoline moeity in P1′. Our calculations suggest other alternatives for inhibitors with lower sensitivity to resistance mutations. We predict that A-78791, (an inhibitor with lower symmetry than A-77003, due to the abstraction of an hydroxyl group in its isostere), has a free energy difference of binding between the native and the wild type of only 0.6 ± 0.4 Kcal/mol.

2. POLAR RESISTANCE MUTATIONS

Some of the resistance mutants in vivo and in vitro involve the replacement of an apolar residue for a polar one. We have shown that the mechanism of resistance is different to the one mutations in which apolar residues are reduced in size. While the mechanism for the latter resistance mutations involves a reduction in the van der Waals contact between inhibitor and enzyme upon contact, the mutations to polar residues lower their inhibitor affinity through differential solvent effects.[20] To evaluate the inhibitor sensitivity of these resistance mutations we have used a modified Poisson-Boltzman finite differences method (mPBFD). We have studied in this case the effect of some polar mutants of the protease at position 82 (V82D, V82E and V82N) on the affinity towards inhibitor U-85548E. The chemical structure of this inhibitor is shown in Fig. 2.

In the framework of the mPBFD method, the differential free energy of binding of two HIV-1 PR variants (i.e. V82 and V82X) can be expressed as:

$$\Delta\Delta G_{\text{bind}} = \Delta G_g^{(u\to h)}(V82 \to X82) + \Delta G_{\text{solv}}^{(h)}(V82 \to X82) - \Delta G_{\text{solv}}^{(u)}(V82 \to X82) \quad (1)$$

where $\Delta G_g^{(u\to h)}$ is the energy difference for the mutation V82→X in the 'pseudo gas-phase' between the bound and unbound HIV-1 PR. The ΔG_{solv} terms are the difference in solvation between the mutant and native molecules in the bound and ligand free states.

In this approach the energy of mutating the unbound and bound forms of the HIV-1 PR in the pseudo-gas phase (i.e., $\Delta G_g^{(u)}$, $\Delta G_g^{(h)}$) corresponds to the difference in coulom-

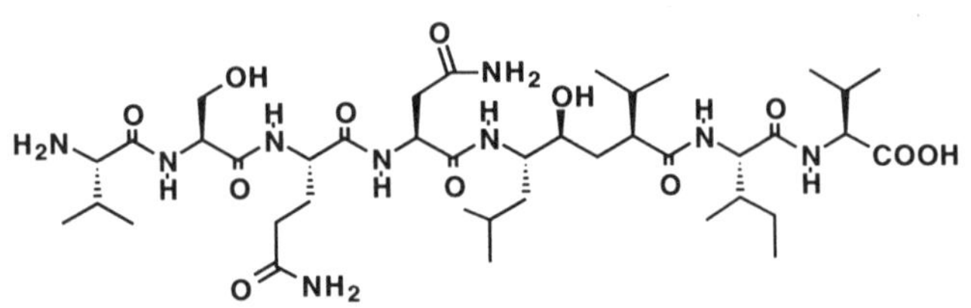

U-85548E

Figure 2. Chemical structures of the HIV-1 PR inhibitor used in our mPBFD calculations.

bic energies between the mutant and native variants of the molecule (i.e., $\Delta G_g^{(u)} = \Delta G_{\mathrm{coul}}^{(u)}$, $\Delta G_g^{(b)} = \Delta G_{\mathrm{coul}}^{(b)}$ and $\Delta G_g^{(u\rightarrow b)} = \Delta G_{\mathrm{coul}}^{(u\rightarrow b)}$).

In the FDPB approach the solvation terms in eq. 1 correspond to the electrostatic component as determined by the reaction field component. We have modified them so as to include a correction term depending on the solvent accessibility:

$$\Delta G_{\mathrm{solv}} = \Delta G_{\mathrm{solv,elec}} + \Delta G_{\mathrm{solv,acc}} \tag{2}$$

where $\Delta G_{\mathrm{solv,elec}}$ is the electrostatic component of the solvation energy calculated directly from the reaction field energy. $\Delta G_{\mathrm{solv,acc}}$ contains the energy of van der Waals interactions between solvent and solute, the entropy penalty for reorganizing the solvent molecules around a solute and the energy devoted to create a cavity in the solvent for the solute molecule. All the components of this free energy term depend directly on the size and shape of the first solvation shell and hence indirectly on the solvent accessibility surface of the solute.

Table 1 lists the calculated differential free energies of binding as well as their components derived from the analysis presented above. As seen from this table the coulombic contributions and those from the solvation and its entropic related correction term have opposite trends. While $\Delta G_{\mathrm{coul}}^{(u\rightarrow b)}$ favors the binding of the inhibitor to the mutated strains of HIV-1 PR, both $\Delta G_{\mathrm{solv,elec}}$ and $\Delta G_{\mathrm{solv,acc}}$ oppose it. Hence, the calculated decrease in the binding affinity could be considered as driven by solvation and its entropic related corrections.

In the ligand-free state, the difference in the contribution of the electrostatic solvation energy, between the mutant and native molecules, $\Delta G_{\mathrm{solv,elec}}^{(u)}(V82 \rightarrow X82)$, is negative (see Table 1). That is, the V82D, V82N and V82E strains are favored by the solvation energy over the wild-type molecule. The increased solvation energy of the mutant strains will result in a lowering of their inhibitor binding affinity, because these molecules have to

Table 1. Free energy contributions (in Kcal/mol) to the differential binding affinity of the native and mutant HIV-1 PR strains for inhibitor U-85548E

	Mutations (V82 → X)		
Free energy[a]	V → D	V → E	V → N
$\Delta G_{\mathrm{coul}}^{(u)}$	−23.1	−44.2	−66.0
$\Delta G_{\mathrm{coul}}^{(b)}$	−78.8	−65.4	−79.6
$\Delta G_{\mathrm{coul}}^{(u\rightarrow b)}$	−55.7	−21.2	−13.6
$\Delta G_{\mathrm{solv,elec}}^{(u)}$	−38.9	−15.9	−34.9
$\Delta G_{\mathrm{solv,elec}}^{(b)}$	24.1	0.1	7.6
$\Delta G_{\mathrm{solv,elec}}^{(u\rightarrow b)}$	63.0	16.0	42.5
$\Delta\Delta G_{\mathrm{bind,elec}}$	7.3	−5.2	28.9
$\Delta G_{\mathrm{solv,acc}}$	26.8 (8.9)[b]	5.4 (1.8)	26.8 (8.9)
$\Delta\Delta G_{\mathrm{bind}}$	34.1 (16.2)[b]	0.2 (−3.4)	55.7 (37.8)

[a]Most of the terms have the same meaning as in Equations 1–6. $\Delta G_{\mathrm{solv,elec}}^{(u\rightarrow b)} = \Delta G_{\mathrm{solv,elec}}^{(b)} - \Delta G_{\mathrm{solv,elec}}^{(u)}$. $\Delta G_{\mathrm{bind,elec}}$ is the electrostatic component of the binding energy, without the solvent accessibility correction.

[b]The entropy corrections were calculated for k = 75 (cal mol^{-1}/Å^2) and k = 25 (cal mol^{-1}/Å^2) (in parentheses).

pay, upon binding, a penalty equal to the surplus solvation energy of the unbound molecules. Moreover, when an inhibitor binds to the HIV-1 PR, the residues at position 82 and 82′ are directly involved in the binding of the peptidic fragments P1′/P3′ and P1/P3 respectively.[4,12] Hence, binding involves a reduction in the solvent accessibility of this side chain, a process that is energetically adverse for polar mutants, and which translates ultimately into unfavorable (positive) contributions to the differential free energy of solvation of the mutant HIV-1 PR strains, $\Delta G^{(b)}_{\text{solv,elec}}(V82 \rightarrow X82)$ (see Table 1).

As seen from Table 1, if the proportionality factor for the surface accessibility term used in our calculations is 75 cal/Å², the binding order predicted by our scoring function for the affinity of the various HIV-1 PR strains to inhibitor U-85548E:

$$V82 > E82 > D82 > N82$$

which corresponds to the observed binding order for several inhibitors studied in this work, including U-85548E.[21] Calculations with the lower value of k, do not agree with the experimental order. This result is in line with the growing body of evidence that indicates that k has a value of 50 cal/Å² or higher. Although the decreased inhibitor affinity of these mutants is driven by the solvent and entropic related contributions, the binding order shown above not always is determined by them. For polar charged mutants (D82, E82), the affinity ranking is dominated by solvent contributions (see Table 1). Differential coulombic interactions serve only to modulate the free energy scoring of these mutants. The affinity ranking of the N82 HIV-1 PR strain has large solvent and entropic contributions, but other terms (i.e., coulombic) play a part. This strain shows the largest sensitivity to binding for U-85548E, an effect that is brought about in part not only by a sizable solvent related component but by a small contribution of the difference in coulombic energies, the one term that favors binding to the polar mutants (see Table 1).

ACKNOWLEDGMENTS

We thank Dr. A. Wlodawer for the coordinates of the HIV-1 PR/inhibitor Ro-31-8959 complex used in this study. This work was supported by NIH Grant GM-49553 and O.C.A.S.T. Grant HS3-011 4588. We would like to thank NCI for access to their supercomputer facilities.

REFERENCES

1. Kohl, N. E.; Emini, E. A.; Schleif, W. A.; Davis, L. J.; Heimbach, J. C.; Dixon, R. A. F.; Scolnick, E. M.; Sigal, I. S. *Proc. Natl. Acad. Sci. U.S.A.* **1988**, *85*, 4686–4690.
2. Tomaselli, A. G.; Howe, W. J.; Sawyer, T. K.; Wlodawer, A.; Heinrikson, R. L. *Chimicaoggi-Chemistry Today* **1991**, *9*, 6–27.
3. Debouck, C.; Metcalf, B. W. **1990**, *Drug. Devel. Res. 21*, 1–17.
4. Wlodawer A.; Erickson, J. W. *Annu. Rev. Biochem.* **1993**, *62*, 543–585.
5. Otto, M. J.; Garber, S.; Winslow, D. L.; Reid, C. D.; Aldrich, P.; Jadhav, P. K.; Patterson, C. E.; Hodge, C. N.; Cheng, Y.-S. E. *Proc. Natl. Acad. Sci. U.S.A.* **1993**, *90*, 7543–7547.
6. El-Farrash, M. A.; Kuroda, M. J.; Kitazaki, T.; Masuda, T.; Kato, K.; Hatanaka, M., Harada, S. *J. Virol.* **1994**, *68*, 233–239.
7. Ho, D. D.; Toyoshima, T.; Mo, H.; Kempf, D. J.; Norbeck, D.; Chen, C.-M.; Wideburg, N. E.; Burt, S. K.; Erickson, J. W.; Singh, M. K. *J. Virol.* **1994**, *68*, 2016–2020.
8. Kaplan, A. H.; Michael, S. F.; Wehbie, R. S.; Knigge, M. F.; Paul, D. A.; Everitt, L.; Kempf, D. J.; Norbeck, D. W.; Erickson, J. W.; Swanstrom, R. *Proc. Natl. Acad. Sci. U.S.A.* **1994**, *91*, 5597–5601.

9. Markowitz, M.; Mo, H.; Kempf, D. J.; Norbeck, D. W.; Bhat, T. N.; Erickson, J. W.; Ho, D. D. *J. Virol.* **1995**, *69*, 701–706.
10. Winslow, D. L.; Stack, S.; King, R.; Scarnati, H.; Bincsik, A.; Otto, M. J. *AIDS Res. Hum. Retroviruses* **1995**, *11*, 107–113.
11. Condra, J. H.; Schleif, W. A.; Blahy, O. M.; Gabryelski, L. J.; Graham, D. J.; Quintero, J. C.; Rhodes, A.; Robbins, H. L.; Roth, E.; Shivaprakash, M.; Titus, D.; Yang, T.; Teppler, H.; Squires, K. E.; Deutsch, P. J.; Emini, E. A. *Nature* **1995**, *374*, 569–571.
12. Nauchitel, V.; Villaverde, M. C.; Sussman, F. *Protein Sci.*, **1995**, *4*, 1356–1364.
13. Hosur, M. V.; Bhat, T. N.; Kempf, D. J.; Baldwin, E. T.; Liu, B.; Gulnik, S.; Wideburg, N. E.; Norbeck, D. W.; Appelt, K.; Erickson, J. W. *J. Am. Chem. Soc.* **1994**, *116*, 847–855.
14. Krohn, A.; Redshaw, S.; Ritchie, J. C.; Graves, B. J.; Hatada, M. H. *J. Med. Chem.* **1991**, *34*, 3340–3342.
15. Molecular Simulations, Inc., Waltham, MA, Quanta version 4.0.
16. Brooks, B. R.; Bruccoleri, R. E.; Olafson, B. D.; States, D. J.; Swaminathan, S.; Karplus, M. *J. Comput. Chem.* **1983**, *4*, 187–217.
17. Baldwin, E. T.; Bhat, T. N.; Liu, B.; Pattabiraman, N.; Erickson, J. W. *Nature Struct. Biol.* **1995**, *2*, 244–249.
18. Gulnik, S. V.; Suvorov, L. I.; Liu, B.; Yu, B.; Anderson, B.; Mitsuya, H.; Erickson, J. W. *Biochemistry* **1995**, *34*, 9282–9287.
19. Sussman, F. Martinez, L, and Villaverde M.C. (submitted to *Biochemsitry*).
20. Sussman, F., Villaverde, M.C. and Davis, A. (Protein Science, in press).
21. Lin Y, Lin X, Hong L, Foundling S, Heinrikson RL, Thaisrivongs S, Leelamanit W, Raterman D, Shah M, Dunn BM, Tang J. *Biochemistry* **1995**, *34* , 1143–1152.

13

INVESTIGATION OF AN ALLOSTERIC SITE OF HIV-1 PROTEINASE INVOLVED IN INHIBITION BY Cu^{2+}

Helena Danielson, Maria T. Lindgren, Per-Olof Markgren, and Ulrika Nillroth

Department of Biochemistry
University of Uppsala
BMC, Box 576, S-751 23 Uppsala, Sweden

1. INTRODUCTION

Development of drugs against HIV is dependent on a detailed understanding of the enzymatic properties of the target viral enzymes and how different ligands can interfere with their catalytic properties. The viral proteinase is responsible for processing of viral polyproteins and is essential for viral particle maturation. Inhibition of the proteinase is therefore an efficient means of preventing viral replication, a principle that is used by the most efficient drugs currently available for treatment of AIDS. These drugs are transition-state analogues that bind to the active site of the enzyme with high affinity and specificity. However, there is a major problem with rapid resistance development. In the case of these proteinase inhibitors, they become less active as a result of point mutations in the proteinase, reducing the affinity for the inhibitor. It is therefore of importance to identify new types of inhibitors, with different mechanisms and/or binding sites.

It has previously been reported that metal ions and metal ion complexes inhibit HIV-1 proteinase.[1–4] The residues responsible for binding of metals and the inhibition mechanisms involved have not been fully elucidated, although several features have been revealed. Binding directly to the active site has been proposed for zinc and Cu^{1+} complexes.[2–4] In contrast, the inhibition by Cu^{2+} occurs by a cysteine dependent mechanism, unless an exogenous thiol is present.[1] There are two cysteine residues in each subunit (67 and 95), Cys67 has been suggested to be of particular importance for Cu^{2+}-inhibition, since it is very reactive and readily modified.[5,6] Also, it is adjacent to a sequence that occurs in a tripeptide hormone that binds copper in human plasma (–Cys67–Gly68–His69–Lys70–).[7] Figure 1 shows the structure of HIV-1 proteinase and the residues discussed are displayed.

The aim of the present work was to identify the residues responsible for the binding of Cu^{2+} and to determine if inhibition is due to reversible interaction with an allosteric site

Aspartic Proteinases, edited by James.
Plenum Press, New York, 1998.

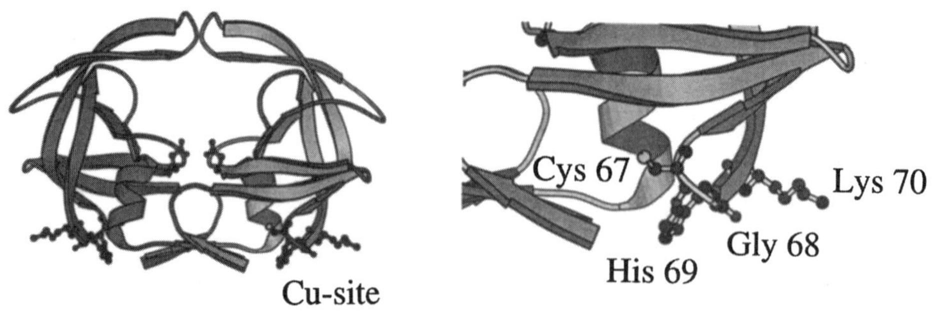

Figure 1. HIV-1 proteinase. Representation of native dimeric HIV-1 proteinase generated with InsightII/Biosym from coordinates 3phv.ent[11] and displayed using Molscript.[12]

or if it is a result of irreversible inactivation, e.g., cysteine oxidation. By using immobilized metal affinity chromatography (IMAC)[8] as an analytical tool we have shown that His69 is important for reversible high affinity binding to immobilized Cu^{2+}. Inhibition studies reveal that the same residue is also essential for inhibition by copper, suggesting that it is present in an allosteric site that indirectly affects catalysis. It was also found that mutation of Cys67 had no effect on either binding of copper or inhibition.

2. MATERIALS AND METHODS

2.1. Cloning, Expression, and Purification of HIV-1 Proteinase

The gene for wild type HIV-1 proteinase was amplified by PCR from plasmid pn10e15[9] containing the *pol* gene for HIV-1 strain HXB. The fragment was inserted into plasmid pET11-a for expression in *E. coli* strain BL21(DE3) (Novagen, Madison, WI, USA). Mutations at C67S and H69A were introduced by site directed mutagenesis of the gene in M13mp18, using Amersham RPN 1526 kit (Amersham, Bucks., U.K.). The protein was isolated from inclusion bodies and purified by a procedure based on that described by Taylor et al.[10]

2.2. Binding Studies Using IMAC

Cu^{2+} was immobilized on iminodiacetic acid (Novarose IDA_{low} 1000/49, Innovata, Bromma, Sweden). The column was loaded with 50 mM $CuSO_4$ until completely blue, washed with 300 mM imidazole and equilibrated with 20 mM Na_2HPO_4, pH 7.0, 1 M NaCl. After application of protein, loosely bound proteins were removed by rinsing with 5 and 10 mM imidazole. A step-wise gradient, with 20, 60 and 300 mM imidazole, in the same buffer as used for equilibration, was used for elution. Finally, the column was washed with 200 mM EDTA to desorb Cu^{2+} and remaining protein.

2.3. Inhibition Studies

The kinetic constants were determined in an assay where enzyme (approx. 10–20 nM) was added to 0.01–0.10 mM HIV-substrate III (Bachem, Bubendorf, Switzerland) in reaction buffer (50 mM glycine, 50 mM Na-acetate, 50 mM MES, 50 mM TRIS, 1 mM EDTA,

1 mM DTT and 1 M NaCl, pH 5,5) to a final volume of 120 μl. The reaction was monitored spectrophotometrically at 300 nm and 37°C during 5 min. The substrate is a chromophoric peptide: H–His–Lys–Ala–Arg–Val–Leu–Phe(pNO_2)–Glu–Ala–Nle–Ser–NH_2 which is hydrolyzed between Leu and Phe(pNO_2). The extinction coefficient for the reaction at 300 nm is 1800 $M^{-1} cm^{-1}$.

Inhibition of wild type and mutant HIV-1 proteinase was measured using an internally quenched fluorescent peptide substrate, DABCYL–g–Abu–Ser–Gln–Asn–Tyr–Pro–Ile–Val–Gln–EDANS (Bachem, Bubendorf, Switzerland). Measurements were performed in 96-well microtiter plates using a Fluoroskan Ascent plate reader (Labsystems, Finland). Excitation and emission wavelengths were 355 and 500 nm, respectively. The assay buffer contained 0.1 M NaAc, pH 5.0, 1 M NaCl, 2.3% DMSO and 5 μM substrate. Enzyme concentrations were 1.7 nM for wild type, 3.7 nM for C67S and 6.3 nM for H69A, copper was added as $CuCl_2$.

3. RESULTS

3.1. Binding Studies

Wild type HIV-1 proteinase binds tightly to Cu^{2+}-IDA sepharose, requiring 300 mM imidazole for elution. Complete elution requires addition of EDTA, whereby the metal is also desorbed. The C67S mutant binds with similar affinity, while the H69A mutant showed a weaker affinity. Although it bound to the column, it was completely eluted with 60 mM imidazole. Figure 2 shows the elution profiles for wild type and the H69A enzyme. Importantly, the enzyme eluted with imidazole was active.

3.2. Inhibition Studies

The kinetic properties of the mutants were similar to that for the wild type enzyme (Table 1).

However, they were not equally sensitive towards Cu^{2+}, Figure 3. Most notably, the H69A mutant was essentially not inhibited by copper. Also, the C67S mutant was at least as sensitive towards copper as the wild type enzyme.

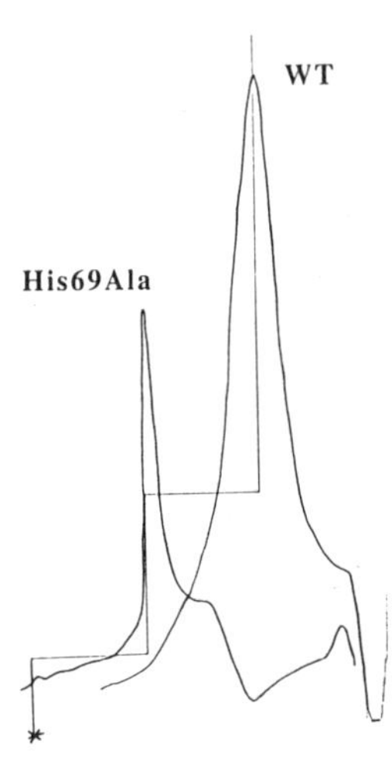

Figure 2. Elution of wild type and H69A HIV-1 proteinase from Cu^{2+}-IDA sepharose. The elution profiles, A_{280}, of IMAC experiments with wild type and H69A mutant HIV-1 proteinase have been overlaid. The corresponding profile for C67S overlaps with that for wild type enzyme. The marked line represents the imidazole gradient, 20, 60 and 300 mM.

Table 1. Kinetic properties of HIV-1 proteinase[a]

Enzyme	k_{cat} (s^{-1})	K_M (μM)	k_{cat}/K_M ($mM^{-1}s^{-1}$)
wt	6.1 ± 0.3	11.2 ± 1.7	545 ± 86
C67S	14.7 ± 1.3	18.7 ± 5.3	786 ± 233
H69A	11.3 ± 2.0	13.5 ± 5.2	837 ± 354

[a]Values and standard errors from non-linear regression analysis of experimental data are shown.

4. DISCUSSION

The present study was performed using wild type and mutant HIV-1 proteinase, in order to determine the role of key residues in binding and inhibition by copper. Mutation of Cys67 and His69 was based on the potential importance of these residues for the binding of Cu^{2+}. Both residues are located in a site distinct from the active site, but suggested to be involved in the inhibition by copper since the inhibition has been shown to be cysteine dependent and a peptide hormone containing the same sequence binds copper with high affinity, using histidine as a ligand.

The results indicate that the mutants did not differ significantly from the wild type enzyme in the absence of copper. Therefore, the mutations do not directly affect the catalytic mechanism or the binding of substrate. Since they show different sensitivities towards copper, they differ in their ability to bind copper and/or, as a result of binding copper, indirectly interfere with the catalytic cycle. It was unexpected that the cysteine mutant appeared to have similar affinity towards immobilized copper as the wild type enzyme and had a similar or even increased sensitivity towards copper inhibition. It had previously been reported that cysteine mutants were not inhibited by copper, unless exogenous thiols were present. Since there were no thiols in the current experiments there must be an additional explanation. A major difference between our experiments and those using cysteine mutants, is that the mutants were double mutants, i.e., cys67 and cys95 were both replaced by

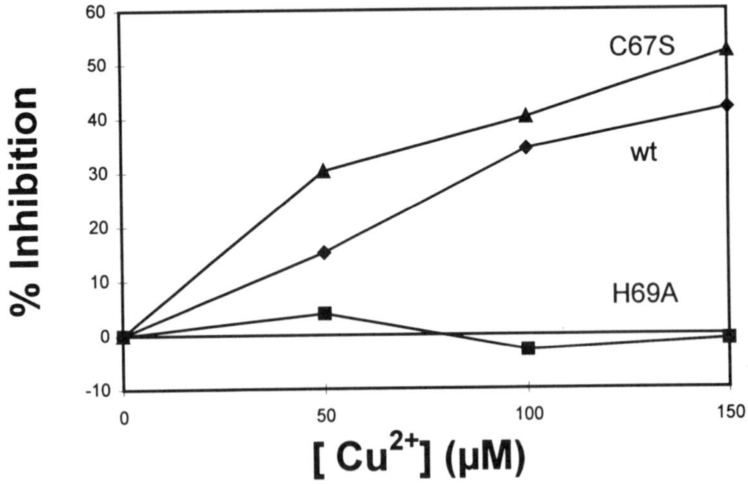

Figure 3. Inhibition of wild type and mutant HIV-1 proteinase by Cu^{2+}. Enzymatic activity was measured in acetate buffer pH 5.0, 1 M NaCl with an internally quenched fluorescent substrate, copper was added as $CuCl_2$.

α-aminobutyric acid. The importance of histidine as a ligand for copper was verified by both IMAC and inhibition studies. This provides evidence for an allosteric site, where His69 is included. There are no other histidines present in the enzyme.

In addition to identification of an allosteric site the present result show that interaction with copper is a reversible process. Since enzyme eluted from immobilized copper was active, the interaction did not involve an irreversible inactivation, e.g., as a result of oxidation of cysteine. This is also supported by the fact that cysteine mutants are inhibited by copper. It is therefore not evident why binding of copper to a site distinct from the active site, results in reduced catalysis. There are no obvious intramolecular contacts where copper cross links different secondary structures or domains, or interferes with other intramolecular interactions, that would explain the inhibition. Further studies adressing this question are therefore necessary before the inhibition mechanism can be fully described.

REFERENCES

1. Karlström, A.R. and Levine, R.L. (1991) Proc. Natl. Acad. Sci. USA **88**, 5552–5556
2. Zhang, Z.-Y., Reardon, I.M., Hui, J.O., O'Connell, K.L., Poorman, R.A., Tomaselli, A.G. and Heinrikson, R.L. (1991) Biochemistry **30**, 8717–8721
3. Woon, T.C., Brinkworth, R.I. and Farlie, D.P. (1992) Int. J. Biochem. **24**, 911–914
4. Davis, D.A., Branca, A.A., Pallenberg, A.J., Marschner, T.M., Patt, L.M., Chatlynne, L.G., Humphrey, R.W., Yarchoan, R. and Levine, R.L. (1995) Arch. Biochem. Biophys. **322**, 127–134
5. Karlström, A.R., Shames, B.D. and Levine, R.L. (1993) Arch. Biochem. Biophys. **304**, 163–169
6. Davis, D.A., Dorsey, K., Wingfield, P.T., Stahl, S.J., Kaufman, J., Fales, H.M. and Levine, R.L. (1996) Biochemistry **35**, 2482–2488
7. Pickart, L. and Lovejoy, S. (1987) Meth. Enzymol **147**, 314–328
8. Porath, J., Carlsson, J., Olsson, I. and Belfrage, G. (1975) Nature **258**, 598–599
9. Unge, T., Ahola, H., Bhikhabhai, R., Bäckbro, K., Lövgren, S., Fenyö, E.M., Honigman, A., Panet, A., Gronowitz, J.S. and Strandberg, B. (1990) AIDS Research and Human Retroviruses **6**, 1297–1303
10. Taylor, A., Brown, D.P., Kadam, S., Maus, M., Kohlbrenner, W.E., Weigl, D., Turon, M.C. and Katz, L. (1992) Appl. Microbiol. Biotechnol. **37**, 205–210
11. Lapatto, R., Blundell, T., Hemmings, A., Overington, J., Wilderspin, A., Wood, S., Merson, J.R., Whittle, P.J., Danley, D.E., Geoghegan, K.F., Hawrylik, S.J., Lee., S.E., Scheld, K.G. and Hobart, P.M. (1989) Nature **342**, 299–302.
12. Kraulis, P. (1991) J. Appl. Cryst. **24**, 946–950.

14

ANALYSIS OF AUTOPROCESSING OF MASON-PFIZER MONKEY VIRUS PROTEINASE *IN VITRO*

Three Active Forms of Proteinase

Iva Pichová,[1] Aleš Zábranský,[1] Iveta Košťálová,[2]
Olga Hrušková-Heidingsfeldová,[1] Martin Andreansky,[3]
Eric Hunter,[3] and Tomáš Ruml[2]

[1]Department of Biochemistry
Institute of Organic Chemistry and Biochemistry
Academy of Sciences of the Czech Republic
Prague, Czech Republic
[2]Department of Biochemistry and Microbiology
Institute of Chemical Technology
Prague, Czech Republic
[3]University of Alabama at Birmingham
Birmingham, Alabama

1. INTRODUCTION

Mason-Pfizer Monkey Virus (M-PMV, also called SRV-3) is a primate retrovirus that was first isolated from a spontaneous breast carcinoma of a female rhesus monkey *Macaca mulatta*.[1,2] Infection of macaque species with M-PMV causes an AIDS-like disease.[3–6] M-PMV is characterized by the self-assembly of the Gag, Gag–Pro and Gag–Pro–Pol precursors into intracytoplasmic particles (procapsids) within the infected cell cytoplasm. These morphogenetic properties of the virus are typical for D-type retrovirus.[7,8]

The M-PMV genomic organization, 5′–LTR–*gag*–*pro*–*pol*–*env*–LTR–3′, is reminiscent of Human T-cell Leukemia (H-TLV-1) and Bovine leukemia viruses[9–11], in that the viral proteinase is encoded by the second ORF, denoted *pro*, which overlaps *gag* by 55 codons at its 5′ end and *pol* by 5 codons at its 3′ end. The amino acid sequence bears marked homology to the retroviral proteinase sequences of Simian AIDS retrovirus (SRV-1),[12] H-TLV-I[9] and Visna.[13] The proteinase is expressed by a frameshifting mechanism. The frameshifts occur within the *gag/pro* and *pro/pol* overlaps to generate fused polyproteins Gag–Pro

Aspartic Proteinases, edited by James.
Plenum Press, New York, 1998.

(Pr95) and Gag–Pro–Pol (Pr180). The *gag* gene encodes a precursor, Pr78 (Gag), containing six capsid proteins. The order of Gag precursor is NH_2–p10(MA)–pp24/18–p12–p27(CA)–p14(NC)–p4–COOH.[15,16] Proteinase precursors preassembled into procapsids are not activated while the capsids remain in the cytoplasm.[17] During or shortly after particle release, the viral proteinase cleaves the viral polyproteins to the structural proteins and viral enzymes. This results in a morphological change of the virus particle into the mature virion.[18] Cleavage of the type D virus capsid precursors is a prerequisite for virus infectivity.

Recently, we have shown that the M-PMV proteinase (M-PMV PR) precursor expressed in *E. coli* undergoes *in vitro* self-processing to yield two proteolytically active proteins of molecular masses 17 and 12 kDa.[19] We now describe the presence of a third form of M-PMV-PR with a molecular mass of 13 kDa. To study the possible roles of these different forms of the M-PMV PR, we have performed a comparative analysis of the 17, 13 and 12 kDa M-PMV proteinase forms. We have developed a process for isolation of individual autocatalytically truncated products, and have determined factors affecting stability for all these enzymes. Substrate specificities of these three forms were compared using synthetic peptides corresponding to processing sequences in M-PMV Gag, Gag–Pro and Env polyproteins.

2. RESULTS AND DISCUSSION

2.1. Purification and Processing

Expression of the 26 kDa proteinase precursor was described by Hrušková-Heidingsfeldová et al.[19] The M-PMV PR precursor accumulated during the expression period in insoluble cytoplasmic inclusion bodies. After disruption of cells by X-press and sonication, inclusion bodies (IB) were separated from soluble cytoplasmic fractions by centrifugation. IB were solubilized in TRIS-HCl buffer pH 7.0 containing 9M urea. Dialysis was then carried out either against pH 7.0 buffer or against 50 mM sodium acetate buffer pH 5.2. Processing of the 26 kDa proteinase precursor occurs during renaturation of the protein by dialysis. SDS PAGE analysis of samples after dialysis showed that both dialysis procedures yielded a mixture of 17 and 13 kDa forms of the M-PMV PR. Under acidic conditions, an autocatalytic cleavage of 17 and 13 kDa forms occurred to generate the mature 12 kDa form. Further experiments showed that higher ionic strengths accelerated autoprocessing of M-PMV proteinase into the truncated form. At pH 5, in the presence of 2 M NaCl and 4°C, the complete processing of p17 and p13 into the mature 12 kDa form was complete in six days. However, at pH 7 without sodium chloride, 17 and 13 kDa proteinases have been stable for about one year. These results demonstrate that pH and the ionic strength significantly influence the processing of protein precursors into truncated molecules.

The separation of the 17 and 13 kDa forms was performed on SE-Sephadex. The 13 kDa form flowed through the support at pH from 5 to 7, the 17 kDa form bound on SE-Sephadex at pH ranging from 5 to 5.5. It was eluted with 1M sodium chloride.

2.2. Proteolytic Activity

All forms were tested for their proteolytic activities with the peptide substrate ATPQVYNpHVRKA followed by evaluation of the cleavage products by RP-HPLC on a Vydac C_{18} RP column in methanol/H_2O. The activity assay was performed as described by

Hrušková-Heidingsfeldová et al.[9] The specific activities of the individual forms, expressed as nanomols of converted substrate by 1 μg of proteinase at 37°C in 2h, differ significantly. The 17 kDa form displayed the highest specific activity, while the 13 and 12 kDa forms exhibit only 58 and 12% of that of the longest form of M-PMV PR, respectively.

2.3. Isoelectric Points

The isoelectric point, determined by isoelectric focusing, of p17 is 9.1, those of p13 and p12 are 6.4 and 8.1, respectively.

2.4. Amino Acid Composition, N- and C-Terminal Sequences

Amino acid analyses, N-terminal sequencing and determination of the molecular mass by ESI mass spectrometry confirmed the sequence of 17 and 12 kDa forms published by Hrušková-Heidingsfeldová et al.[17] The new form of M-PMV PR (p13) is 7 amino acids longer than the mature M-PMV PR (p12). The N-termini of all three forms are identical. The C-terminal sequence of p17 is –GFGNF, that of 13 kDa form is –DIVTA and 12 kDa form has at the C-terminus the sequence –IMMCS.

2.5. pH Optimum

All forms of the M-PMV proteinase exhibit maximal proteolytic activity, measured with the chromogenic peptide substrate ATPQVYNpHVRKA, in the pH range from 3 to 7, at pH 4.5.

2.6. Substrate Specificities

To study the possible roles of the three forms of M-PMV proteinase, we have compared their substrate specificities using peptides spanning the processing sites within the M-PMV Gag, Gag–Pro and Env polyproteins. The analysis of results, summarized in Table 1, revealed that all forms of M-PMV proteinase have similar substrate specificities. However, specific activities and preferences of cleavage sequences in polyprotein precursors were different. While all forms cleave preferentially the sequence corresponding to N-terminus of proteinase, for the cleavage site within the cytoplasmic domain of the transmembrane glycoprotein gp 22 (TM) they exhibit the lowest affinity. This sequence was

Table 1. Kinetic parameters of the peptide substrates derived from the processing sites in M-PMV polyproteins

	17 kDa			13 kDa			12 kDa		
Origin of cleavage site	K_M [mM]	k_{cat} [s^{-1}]	k_{cat}/K_M [s^{-1} mM^{-1}]	K_M [mM]	k_{cat} [s^{-1}]	k_{cat}/K_M [s^{-1} mM^{-1}]	K_M [mM]	k_{cat} [s^{-1}]	k_{cat}/K_M [s^{-1} mM^{-1}]
Gag MA/PP	0.2	0.8	4.0	0.23	1.0	4.35	0.19	0.6	3.16
GagPP/p12	0.84	0.3	0.36	0.43	0.2	0.46	0.28	0.15	0.54
Pro N terminus	0.10	3.5	35.00	0.08	1.2	15.00	0.02	0.7	35.00
Env gp22–gp20	0.36	0.04	0.11	0.40	0.04	0.10	0.54	0.05	0.09

The assays were performed in 0.05 M sodium acetate buffer pH 5.3 containing 0.3 M NaCl, 0.002M EDTA, 0.05% 2-mercaptoethanol at 37°C. The concentration of proteins in the assay was 0.2 ∝M, and the concentration of peptides varied between 5–700 ∝M. The course of the cleavage reaction was followed by reverse-phase HPLC. The kinetic parameters were determined at less than 20% substrate turnover by a nonlinear regression program Enzfitter.

predicted from *in vivo* studies,[20] and we have verified the location of the cleavage for the first time through the *in vitro* studies described here. The results presented here are in agreement with the data from preliminary *in vivo* experiments showing the presence of several forms of M-PMV proteinase in virions (M. Andreanský, personal communication).

3. CONCLUSIONS

When expressed in bacteria as a 26 kDa precursor, the proteinase is first processed at the N-terminus, yielding a proteolytically active 17 kDa form of the proteinase. Apparently, this initial cleavage is followed by a self-processing that leads to the emergence of an active 13 kDa form of M-PMV PR. Depending on conditions, the autoprocessing can continue to yield the mature 12 kDa form of M-PMV proteinase. The truncation of proteinase molecule occurs only at the C-terminal part of the molecule and results in the loss of proteolytic activity. The substrate specificity studies show that the available cleavage sites in viral precursors are cleaved by M-PMV PR forms with different rates. The results described here allow us to hypothesize that the individual forms of M-PMV proteinase may be responsible for cleavage at different parts of the budding virion, or at different stages of maturation.

This work was supported by grant 203/97/0416 from the Grant Agency of the Czech Republic.

REFERENCES

1. Chopra, H. C. & Mason, M. M. (1970) Cancer Res. 30, 2081–2086.
2. Jensen, E.M., Zelljadt, I., Chopra, H.C. & Mason, M.M. (1970) Cancer Res. 30, 2388–2393.
3. Bryant, M.L., Gardner, M.B., Marx, P.A., Maul, D.H., Lerche, N.W., Osborn, K.G., Lowenstein, L.J., Bodgen, A., Arthur, L.O. & Hunter, E. (1986) J. Natl. Cancer Inst. 77, 957–965.
4. Fine, D. L., Landon, J. C., Pienta, R. J., Kubicek, M. T., Valerio, M. J., Loeb, W. F. & Chopra, H. C. (1975) J. Natl. Cancer Inst. 54, 651–658.
5. Daniel, M.D., King, N.W., Letwin, N.L., Hunt, R.D., Sehgal, P.K. & Desroisiers, R.C. (1984) Science 223, 602–605.
6. Daniel, M. D., Letvin, N. L., King, N. W., Kannagi, M., Sehgal, P. K., Hunt, R. D., Kanki, P. J., Essex, M., & Desrosiers, R. C. (1985) Science 228, 1201–1204.
7. Kramarsky, B., Sarkar, N.A. & Moore, D.H. (1971) Proc. Natl. Acad. Sci. USA 68, 1603–1697.
8. Fine, D., & Schochetman, G. (1978) Cancer Research 38, 3123–3139.
9. Seiki, M., Hattori, S., Hirayama, Y. & Yoshida, M. (1983) Proc. Natl. Acad. Sci. USA 80, 3618–3622.
10. Shimotohno, K., Takahashi, Y., Shimizu, N., Gojobori, T., Golde, D. W., Chen, I. S. Y., Miwa, M. & Sugimura, T. (1985) Proc. Natl. Acad. Sci. USA 82, 3101–3105.
11. Sagata, N., Yasunaga, T., Tsuzuku-Kawamura, J., Ohishi, K., Ogawa, Y. & Ikawa, Y. (1985) Proc. Natl. Acad. Sci. USA 82, 677–681.
12. Power, M.D., Marx, P.A., Bryant, M.L., Gardner, M.B., Barr, P.J., and Luciw, P. (1986) Science 231, 1567–1572.
13. Sonigo, P., Alizon, M., Staskus, K., Klatzmann, D., Cole, S., Danos, O., Retzel, E., Tiollais, P., Haase, A. & Wain-Hobson, S. (1985) Cell 42, 369–382
14. Sonigo, P., Barker, Ch., Hunter, E. & Wain-Hobson, S. (1986) Cell 45, 375–385.
15. Bradac, J. & Hunter, E. (1986) Virology 150, 491–502
16. Henderson, L. E., Sowder, R., Smythers, G., Benveniste, R. E. & Oroszlan, S. (1985) J. Virol. 55, 778–787.
17. Rhee, S.S. & Hunter E. (1987) J. Virol. 61, 1045–1053.
18. Sommerfelt, M. A., Petteway, S. R., JR., Dreyer, G., B. & Hunter, E.(1992) J. Virol. 7, 4220–4227.
19. Hrušková-Heidingsfeldová, O., Andreanský, M., Fábry, M., Bláha, I., Štrop, P. & Hunter, E. (1995) J. Biol. Chem. 25, 15053–15058
20. Brody, B.A., Rhee, S.S., Sommerfelt, M.A. & Hunter, E. (1992) Proc. Natl. Acad. Sci.USA 89, 3443–3447.

15

SUBSEQUENT SELFPROCESSING OF BOVINE LEUKEMIA VIRUS PROTEINASE *IN VITRO*

Iva Pichová,[1] Anna Teplitsky,[2] Romana Cubínková,[1] Aleš Zábranský,[1]
Gil Shoham,[2] and Olga Hrušková-Heidingsfeldová[1]

[1]Institute of Organic Chemistry and Biochemistry
Czech Academy of Science
Flemingovo náměstí 2, 166 10, Prague 6
Czech Republic
[2]Department of Inorganic Chemistry
Hebrew University of Jerusalem
Israel

1. INTRODUCTION

Bovine leukemia virus (BLV) is a B-cell lymphotropic retrovirus closely related to human T-cell leukemia viruses (HTLV-1 and -2)[1] BLV can naturally infect cattle in which it induces a disease complex termed enzootic bovine leukosis. Experimental infection of rabbits, however, causes severe depression in. immune function.[2] Although BLV and HTLV are grouped with C-type oncoviruses, during assembly they can be barely distinguished from lentiviruses, forming thus a specific group.[3]

The genome of these viruses is organized into *gag*, *pro*, *pol*, *env* and *pX* genes. Proteinase is encoded by the *pro* reading frame located between *gag* and *pol* genes which code for the viral structural proteins and enzymes, respectively. In order to express the *pro* and *pol* gene products, a ribosomal frameshift occurs within the overlap of *gag*/*pro* and *pro*/*pol* reading frames.[4] Thus, the *pro* gene is translated as a part of the Gag–Pro and Gag–Pro–Pol polyprotein precursors.

Proteinases of retroviruses process the viral precursor polyproteins to produce the mature structural proteins and the enzymes essential for the viral replication. Therefore, they are considered as one of the targets for chemotherapy of retroviral diseases. Substrate specificity of retroviral proteinases has been characterized to a detail. Nevertheless, the process of proteinase activation is not fully understood as yet. PR is believed to be cleaved autocatalytically from the polyprotein precursor. However, each precursor molecule contains only one subunit of proteinase, and to yield the active enzyme, the dimerization of subunits is required. The mechanism of the proteinase dimerization and processing has

Aspartic Proteinases, edited by James.
Plenum Press, New York, 1998.

been studied using artificially truncated proteinase precursors, in order to explore the role of the regions adjacent to PR in the regulation of the PR activation. Our previous studies have revealed, that the retroviral proteinases which are encoded by an independent *pro* reading frame may serve as a good model for the study of selfprocessing events.[5]

BLV PR has been isolated from virions and its N- and C-terminus have been determined.[6] The monomers of BLV PR were found to consist of 126 amino acid residues (Fig. 1), and thus, BLV PR has been considered to be the largest retroviral proteinase studied. Furthermore, the BLV PR consisting of 126 amino acids was chemically synthesized by the solid phase method. After renaturation, the synthetic enzyme was fully active and able to cleave appropriate substrates.[7] Despite of these results, existence of a shorter form of BLV PR has been predicted. The hypothesis has been confirmed by structural modeling and chemical synthesis of the BLV PR consisting of only 116 residues.[8] The enzyme was active and cleaved oligopeptides derived from processing sites on the BLV Gag polyprotein.

In our previous work, we have cloned and expressed in *E. coli* a part of the BLV *pro* reading frame, coding for a 22 kDa proteinase precursor.[9] Proteinase cleaved itself out of the precursor yielding a proteolytically active species whose subunits consisted of 126 amino acid residues. Furthermore, the N- and C-termini of the recombinant proteinase were identical to those of the BLV PR isolated from virions. During *in vitro* studies, however, we observed that the purified BLV PR undergoes a selfprocessing to a shorter form, that retains a proteolytic activity. The secondary selfprocessing occurred particularly during the experiments in which a high protein concentration is achieved (e.g., crystallization trials). In the present study we examined the selfprocessing of purified BLV PR as a function of different conditions.

2. RESULTS

In our previous work we observed that active fractions of BLV PR may be stored in various buffers at 4°C for several years, without any changes in the specific activity or electrophoretic homogenity.[10] This refers, however, only to purified or partially purified samples, whose concentration is lower than approximately 0.7μmol l^{-1}. In more concentrated samples, the BLV PR begins to be cleaved to a species of a lower molecular weight. Since this event occurred reproducibly in various preparations of BLV PR, most of which underwent purification using affinity chromatography, we assume, that the cleavage of the 126 amino acid form ("long sequence") is an autocatalytical process that results in the 116 amino acid form ("short sequence") that has been predicted by others.[8] The peptide ASISI*PEEV, which corresponds to the boundary between the long and the short sequence was found to be cleaved by BLV PR. The rate of the long sequence processing depends on conditions of storing and treatment.

```
L S I P L A R S R P S V A V Y L S G P W L Q P S Q N Q A
L M L V D T G A E N T V L P Q N W L V R D Y P R I P A A
V L G A G G V S R N R Y N W L Q G P L T L A L K P E G P
F I T I P K I L V D T S D K W Q I L G R D V L S R L Q A
S I S I * P E E V R P P M V G
```

Figure 1. Sequence of BLV PR. The boundary between the long and short sequence is indicated by asterisk.

2.1. Effect of pH on the Selfprocessing

Identical aliquots of the purified long form of BLV PR were transferred into buffers of pH ranging from 5 to 8. The changes of pH value did not cause any change of the electrophoretic pattern of the samples whose concentration was up to 0.7 μmol l^{-1}. Then the samples were concentrated to approximately 2.5 μmol l^{-1} and screened by 15% SDS-PAGE. The processing of the long form occurred most efficiently at pH 6–6.5. This is in agreement with previous studies, in which BLV PR was reported to be active from pH 3.5 to 8, with maximum activity in the range of pH 5 to 6.5.[11] Concentrated BLV PR precipitates more rapidly at pH 5. Thus, the concentration of soluble enzyme at pH 5 may become lower and the selfprocessing is slowed down.[10] Although the activity of BLV PR decreases at pH values above 7, the long sequence cleavage occurs with time even in concentrated samples (approx. 3 μmol l^{-1}) stored at pH 8.

2.2. Effect of Salt Concentration

The activity of BLV PR was found to be stimulated by high salt concentrations. Raising the ionic strength from 100mM to 750mM NaCl affected positively the cleavage of HIV-1 Pr^{55}Gag by BLV PR.[12] Oligopeptide substrates are hydrolyzed most rapidly in presence of 1.5 M NaCl.[11] High salt concentration, however, does not affect the BLV PR subsequent selfprocessing in diluted samples (enzyme concentration bellow approx. 0.7 μmol l^{-1}). Processing of the BLV PR long sequence was enhanced particularly in concentrated samples that were incubated with cation exchanger and then eluted by 1–2 M NaCl. Similar effect has been observed with Mason Pfizer monkey virus proteinase (Pichová et al., "Analysis of Autoprocessing of Mason-Pfizer Monkey Virus Proteinase *in vitro*," this book).

2.3. Concentrating of the BLV PR in Presence of Inhibitor

An active-site targeted inhibitor BocNle[CH_2NH]Nle–Ile–Ala–NH_2 that was initially designed for the studies of HIV-1 PR inhibits BLV PR in micromolar concentration (unpublished results). We used this compound in order to examine, whether the processing of the long sequence may be slowed down in presence of a BLV PR inhibitor. The inhibitor was added to a final concentration of 25 μmol l^{-1} to several aliquots of purified BLV PR samples. Then the samples were concentrated using microcon concentrator to various extent and screened using 15% SDS PAGE. In presence of inhibitor, no short sequence has been detected up to approximately 8 μmol l^{-1} concentration of the PR. By contrast, the control sample without inhibitor yielded a mixture of the both proteinase forms. Further concentrating, however, lead to the long sequence cleavage even in presence of the inhibitor used.

2.4. Temperature-Dependent Loss of Activity

The activity of BLV PR was monitored at various time points and temperatures. Storing of BLV PR at a room temperature or at 40°C revealed striking stability of this enzyme. As opposed to HIV-1 PR, incubation of BLV PR samples at a room temperature or 37°C causes almost no decrease of the enzyme activity within the first 48 hours. Diluted proteinase samples showing a loss of activity after long incubation periods were compared to the original BLV proteinase stock solutions using 15% or 18% SDS-PAGE. Surprisingly, no change in the molecular weight was observed.

On the other hand, the secondary selfprocessing can occur when the samples are stored at 4°C and it can be enhanced by incubating the BLV PR at a temperature optimum (40°C), provided the enzyme concentration exceeds approximately 4 μmol l^{-1}. In such samples, prolonged incubation at a temperature above 37°C causes a loss of activity characterized by occurrence of cleavage products of a molecular weight lower than that of the short BLV PR sequence. It suggests that the 116 residue species that has been designated the short sequence may be the minimal form of active BLV PR. Thus, it appears that of the various factors studied, the enzyme concentration is the most critical for the secondary selfprocessing.

3. DISCUSSION

The fact that BLV PR can occur in (at least) two active forms *in vitro*, raises several questions: 1) may the short form of the PR be present also *in vivo*? If so, what is the biological significance of such a stepwise selfprocessing? 2) May the C-terminal decapeptide play any role in the regulation of proteolysis within the viral particles? 3) In most of our experiments, the long sequence processing was incomplete and resulted in a mixture of the long and short sequences. Do the monomers consisting of 126 and 116 residues associate only with subunits of a similar molecular weight, or is it reasonable to expect existence of a "heterodimer" composed of one short and one long subunit?

Up to now, the retroviral PR stepwise processing has been observed only in Mason-Pfizer monkey virus (M-PMV).[5] While the BLV PR subsequent cleavage is driven by the increased enzyme concentration, the M-PMV PR selfprocessing depends rather on ionic strength (Pichová et al., Analysis of autoprocessing of Mason-Pfizer Monkey Virus proteinase *in vitro*, this book). Interestingly, proteinases of both BLV and M-PMV are encoded by the *pro* reading frame, separated from *gag* and *pol* genes. In M-PMV, which is a D-type, immunosuppressive retrovirus, three forms of virus encoded proteinase have been observed both *in vivo* and *in vitro*. Results of virological studies were in a good agreement with the data obtained with the purified M-PMV PR. This allows us to believe that also the short sequence of BLV PR is a biologically relevant species. It is attractive to speculate that the stepwise selfprocessing plays a role in the regulation of the PR activation or in the regulation of the maturation of viral particles. Although we have not observed any decrease in the substrate hydrolysis rate caused by occurrence of the short sequence, possible differences in proteolytic activities between the two forms of BLV PR should be studied more extensively in future.

This work was supported by grant A 4055503 from the academy of Sciences of the Czech Republic.

REFERENCES

1. Hrušková-Heidingsfeldová, O. (1995) Folia Biologica (Prague) 41, 201–212.
2. Altanerová, V., Ban, J., and Altaner, C. (1989) AIDS 3, 755–758.
3. Gelderblom, H.R. (1991) AIDS 5, 617–638.
4. Nam, S.H., Copeland, T.D., Hatanaka, M., and Oroszlan, S. (1993) J. Virol. 67, 196–203.
5. Hrušková-Heidingsfeldová, O., Andreánsky, M., Fábry, M., Bláha, I., Štrop, P., and Hunter, E. (1995) J. Biol. Chem. 270, 15053–15058.

6. Yoshinaka, Y., Katoh, I., Copeland, T.D., Smythers, G.W., and Oroszlan, S. (1986). J. Virol. 57, 826–832.
7. Bláha, I., Tozsér, J., Kim, Y., Copeland, T.D., and Oroszlan, S. (1992) FEBS Lett. 309, 389–393.
8. Précigoux, G., Geoffre, S., Léonard, R., Llido, S., Dautant, A., Langlois d'Estaintot, B., Picard, P., Ménard, A., Guillemain, B., and Hospital, M. (1993). FEBS Lett. 326, 237–240.
9. Andreánsky, M., Hrušková-Heidingsfeldová, O., Sedláèek, J., Konvalinka, J., Bláha, I., Jeèmen, P., Hořejší, M., Strop, P., and Fábry, M. (1991) FEBS Lett. 287, 129–132.
10. Hrušková-Heidingsfeldová, O., Piskáèek, M., Cubínková, R., Zábranský, A., and Pichová, I. (1996). Collect. Czech. Chem. Commun 61, 1254–1260.
11. Ménard, A., Mamoun, R.Z., Geoffre, S., Castroviejo, M., Raymond, S., Précigoux, G., Hospital, M., and Guillemain, B. (1993) Virology 193, 680–689.
12. Konvalinka, J., Heuser, A.-M., Hrušková-Heidingsfeldová, O., Vogt, V.M., Sedláèek, J., Štrop, P., and Krausslich, H.-G. (1995) Eur. J. Biochem. 228, 191–198.

16

MECHANISM OF ACTION OF ASPARTIC PROTEASES

I. V. Kashparov, M. E. Popov, and E. M. Popov

Shemyakin-Ovchinnikov Institute of Bioorganic Chemistry
Russian Academy of Sciences
Moscow, Russia

Analysis of crystallographic structures of aspartic proteases allows to conclude that nonbonded interactions of various enzymes of this family with respective specific substrates produce in every case practically identical unstable electronic systems which include catalytic groups of the active center, a cleaved peptide bond and a water molecule. The idea of a similar mutual orientation of these components in all nonbonded complexes of aspartic proteases [1] can reliably be verified by an a priori semi-empirical computation of the enzyme-substrate interactions for two representatives of this group which differ essentially in their amino acid sequences and three-dimensional structures. As the objects of theoretical investigations of catalytic mechanism of aspartic proteases, we have chosen human immunodeficiency virus protease (HIV1-PR) and fungal protease rhizopuspepsin. Because of the restricted volume of the publication, here we only present the results for rhizopuspepsin and its interaction with the substrate Ac–Pro^1–Phe^2–His^3–Lys^4–Phe^5–Val^6–OMe and the substrate-like inhibitor Ac–Pro^1–Phe^2–His^3–$Phe^4(\Psi)$–CH_2–Phe^5–Val^6–OMe which forms a stable complex, investigated by X-ray diffraction [2]. The results for the HIV1-PR are to be published in a separate communication. A quantitative study of the rhizopuspepsin catalytic act is based on the theory and computation method described recently [3]. According to this scheme, the conformational properties of free molecules of the substrate, inhibitor and of the enzyme active center were analyzed first, and then a formation of the enzyme-inhibitor and enzyme-substrate complex was modeled.

1. CONFORMATIONAL PROPERTIES OF THE ASP^{35} AND ASP^{218} SIDE CHAIN

Conformational properties of the two catalytically active aspartic residues were studied using technique described in [4]. Figure 1 shows alignment of the X_1–X_2 conformational maps of Asp^{35} and Asp^{218} in the potential field of rhizopuspepsin active site (623

Aspartic Proteinases, edited by James.
Plenum Press, New York, 1998.

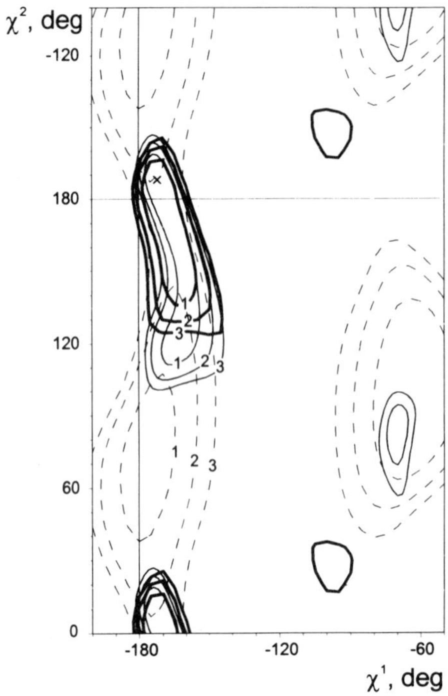

Figure 1. A cross-section of the X_1–X_2 potential energy surfaces of Asp^{35} (thin line), Asp^{218} (bold line) and of methyl amide of N–acetyl–L–Asp.

atoms of 48 amino acid residues according to the selected model) and X_1–X_2 plot of the N–acetyl–L–aspartic acid methyl amide.

The maps are plotted for the experimental φ–ψ values of the native enzyme [5]. It is noteworthy that both Asp^{35} and Asp^{218} side chains have almost identical conformational freedom in terms of their X_1 and X_2 dihedral angles. The energetically allowed region for both residues occupies on the plot a relatively small area with calculated minimum corresponding to the experimental X_1, X_2 crystallographic values; this minimum also lies within allowed region for a free aspartic amino acid residue. This fact indicates that aspartate side chains in the rhizopuspepsin active site structural environment do not undergo any steric tension, yet have substantially restricted conformational freedom. A special analysis of Asp^{35} and Asp^{218} nearest neighborhood demonstrates that their mutual orientation and conformational possibilities are directly determined by the next three residues in the sequence: Thr, Gly, Ser (Thr) which are conservative in all enzymes of this family. All types of nonbonded interactions: Van der Waals, electrostatic, torsional and hydrogen bonding contribute to the active $Asp^{35}(^{218})$ conformational stability with dominating contribution made by the dispersion component.

2. THE INHIBITOR AND SUBSTRATE BINDING MODES

In calculations of the enzyme–inhibitor complex we took into account 623 atomic coordinates of 48 residues of the active center and a nitrogen atom of the inhibitor Pro^1 pirrolydine ring. This nitrogen atom was rigidly fixed in its position, but Pro^1 residue was allowed to adopt any orientation about it. All side chain orientations were also fixed in the model except those which undergo a conformational change upon inhibitor binding. Evaluation of conformational properties of the Pro^1 residue revealed its single energetically favorable orientation in the enzyme potential field. The next step of the binding pro-

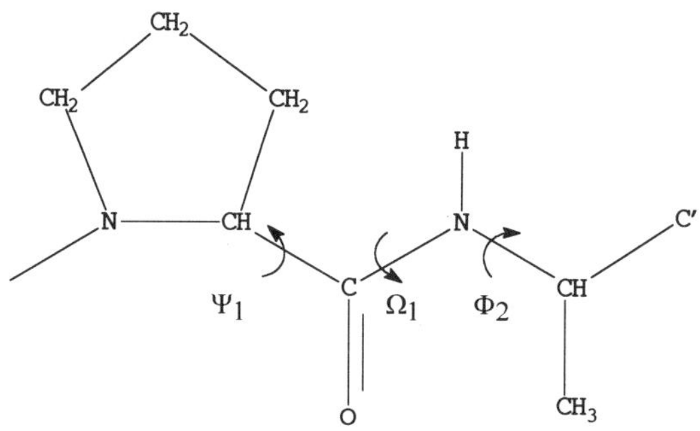

Figure 2. Fragment of the inhibitor with the variable conformational parameters.

cedure involved analysis of conformational freedom of the peptide group between Pro[1], which was assumed to be fixed during this step, and Phe[2] residue with $–CH_2–C_6H_5$ fragment replaced by methyl group (Figure 2).

Preferred orientations of this fragment in the enzyme potential field were determined from allowed regions of the $\psi1–\varphi2$ plot, which in its turn served as the starting values for the energy minimization procedure varying $\psi1$, $\omega1$ and $\varphi2$ dihedral angles. Further study of the conformational state of Phe[2] residue followed two directions supplementing each other. The first involved calculations of the inhibitor Phe[2] residue with CH_3 group replaced by $–CH_2–C_6H_5$ and was aimed at the preferred side chain orientation. Within the second, conformational possibilities of the Phe[2] main chain and of the second peptide link were determined by plotting $\varphi2–\psi2$ map followed by the energy refinement for the fragment. At this point second residue was replaced by Ala. As the next step, Phe[2] side chain is returned to its place, the $X_1–X_2$ energy map is plotted and the $\varphi2–\psi2$ crossection is analyzed again. The computation has revealed two possible Phe[2] side chain orientation with almost orthogonal and approximately isoenergetic positions of the side chain phenolic ring. The similar steps involving alternating evaluation of conformational surfaces and energy minimization procedure was used for the subsequent residue-by-residue building up of the inhibitor molecule and fitting it in the enzyme potential field. The results of theoretical prediction for the enzyme–inhibitor complex are compared with the experimental crystallographic data in Table 1.

A close agreement between the theoretical and experimental structures is also demonstrated in Figure 3.

The root mean square difference between predicted and experimental structures is 0.23 Å. As it is obvious from Table 1, conformational states of all residues of the sorbed hexapeptide inhibitor correspond to low energy states of the free residues, and the complete inhibitor molecule to the optimal β-structure of the free inhibitor, which provides the best possible opportunities for stabilizing nonbonded interactions with the enzymic active site. Quantitative estimations of the contributions of each inhibitor residue to the total energy of the enzyme–inhibitor interactions is given in the last right column of Table 1.

The substrate molecule differs from the analyzed inhibitor only by a single replacement of the Phe[4][Ψ]–CH_2 fragment by the standard Lys[4] residue which is connected to rhizopuspepsin primary specificity [6]. The computational procedure of the substrate binding in the enzyme active site followed the scheme described above and was performed independently from the analogous study of the enzyme–inhibitor complex. The rhizopuspepsin "flap" region including Ile[75]–Ser[81] residues was not fixed, but allowed to move during all optimization stages. However, its conformational parameters have not changed during substrate binding. Table 2 shows dihedral angles in the predicted structure of the

Table 1. Experimental (upper string) and theoretically predicted (bottom string) values of dihedral angles (deg.) and stabilization energies of rhizopuspepsin inhibitor Ac–Pro^1–Phe^2–His^3–Phe^4[ψ] in its complex with enzyme

	Main chain			Side chain		U
Residue	Φ	Ψ	Ω	X_1	X_2	kcal/mol
Pro^1	−60	162	179	—	—	−17.1
	−60	167	177	—	—	
Phe^2	−114	157	180	−66	−41	−32.6
	−112	163	180	−65	−55	
His^3	−140	65	−179	−49	−117	−47.8
	−143	57	180	−52	−112	
Phe^4[Ψ]	−99	59	129	−48	−33	−35.8
	−95	68	121	−44	−40	
Phe^5	−83	164	176	67	97	−54.5
	−87	167	177	73	98	
Val^6	−113	152	178	64	—	−17.1
	−111	131	180	67	—	

Michaelis complex, which according to our theory of biocatalysis [1] is stable with respect to nonbonded interactions and unstable with respect to chemical transformations.

In the rhizopuspepsin active site, the substrate backbone, as in case with substrate-like inhibitor, adopts an extended conformation (Figure 4).

Since conformational states of all substrate residues correspond to low-energy regions, the stabilizing enzyme–substrate interactions (Table 2) do not entail any steric tensions. In the predicted conformation of the substrate, a carbonyl oxygen of the scissile bond points toward Asp^{35} and Asp^{218} side chains, forcing out the conservative water molecule Wat-2 (Wat507 in rhizopuspepsin numbering) (Figure 5). This situation promotes the hydration of the substrate facilitating the nucleophilic attack on carbonyl carbon atom. The second water molecule (W858 in rhizopuspepsin numbering) located in the native enzyme active site also cannot retain its position: it is supplanted by $C\beta H_2$ of Lys^4 side chain. The position of water molecule which could have served as a nucleophile in catalytic reaction was found by potential energy calculations for the multitude of points between Asp^{35}, Asp^{218} and a scissile bond. In this way we have determined first the oxygen atomic coordinates, corresponding to the minimum energy, and then orientations of two hydroxyl groups. This position of water molecule is shown in Figure 5 (see also Figure 2 in [3]).

In the nonbonded complex water oxygen is at 2.8 Å against the carbonyl carbon; the proton of one hydroxyl group points toward nitrogen atom of the scissile bond, while the proton of other hydroxyl forms a hydrogen bond (at 1.7 Å distance) to the carboxyl oxygen of Asp^{35}. In this situation the plane of the substrate peptide group is nearly orthogonal

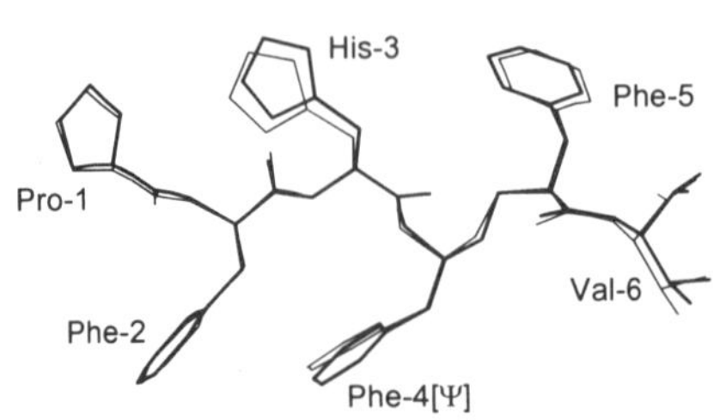

Figure 3. Superposition of the experimental (thin line) and *a priori* calculated (bold line) structures of the inhibitor in the rhizopuspepsin active site.

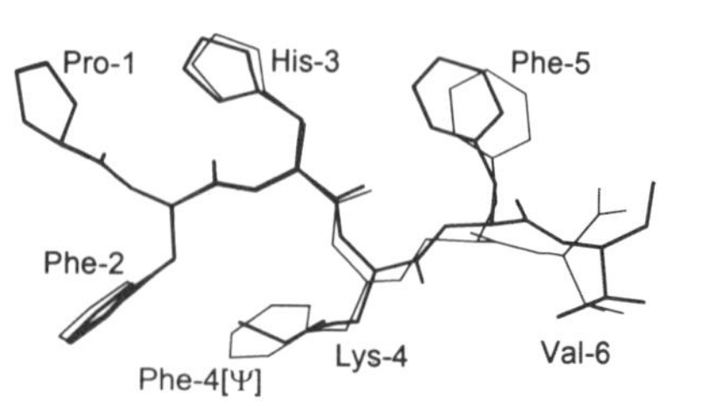

Figure 4. Superposition of *a priori* calculated structures of the inhibitor (thin line) and the substrate (bold line) in the rhizopuspepsin active center.

Table 2. Theoretically predicted values of dihedral angles and stabilization energies of the substrate molecule in the enzyme–substrate nonbonded complex in the absence of water molecule WAT507

	Main chain			Side chain				U
Residue	Φ	Ψ	Ω	X_1	X_2	X_3	X_4	kcal/mol
Pro^1	−60	167	177	—	—	—	—	−13.5
Phe^2	−110	163	180	−70	−55	—	—	−32.4
His^3	−148	72	185	−47	−110	—	—	−49.4
Lys^4	−95	81	161	−53	−60	−162	−56	−56.4
Phe^5	−128	143	182	48	86	—	—	−26.3
Val^6	−125	94	176	183	61	60	—	−19.8

to the plane formed by N–C′ bond and water oxygen. Mutual orientation of the substrate, water molecule and catalytic groups of aspartate residues was automatically produced during the procedure of selection of low-energy states of the nonbonded complex. In the course of computation no assumption was made about chemical aspects of the enzyme–substrate interactions. However a possible productive nature of the predicted position of P_1–P_1' substrate bond, i.e. a chemical instability of the complex is quite evident. In fact, the H_2O hydroxyl group is in position necessary, on the one hand, for a nucleophilic attack on the carbonyl carbon of the substrate and, on the other hand, for an interaction of the H_2O hydrogen with the amide nitrogen of the cleaved bond. It is noteworthy that the calculated substrate and nucleophile mutual orientation, form a polyfunctional system working on an autocatalysis principles, when each reaction facilitates a following one, (see [4] for more details). Moreover, the surroundings act in the same direction: interactions of O^δ of Asp^{35} with H_2O hydrogen (at distance 1.7 Å) enhance nucleophilicity of water oxygen while interaction of carbonyl carbon atom of the hydrolyzed bond with hydrogen atom of Asp^{218} hydroxyl group increases the oxygen negative charge thus facilitating a formation of the intermediate tetrahedral adduct. Here, as in case of α-chymotrypsin and β-trypsin [4] the very essence of biocatalytic process consists in creation of the polyfunctional system, i.e. in reducing a polymolecular reaction, unthinkable in usual organic chemistry, to a monofunctional, essentially intramolecular reaction.

In conclusion, we examined an alternative scheme proposed in a number of publications on the mechanism of catalytic reaction of aspartic proteases [2,7–9], in which the

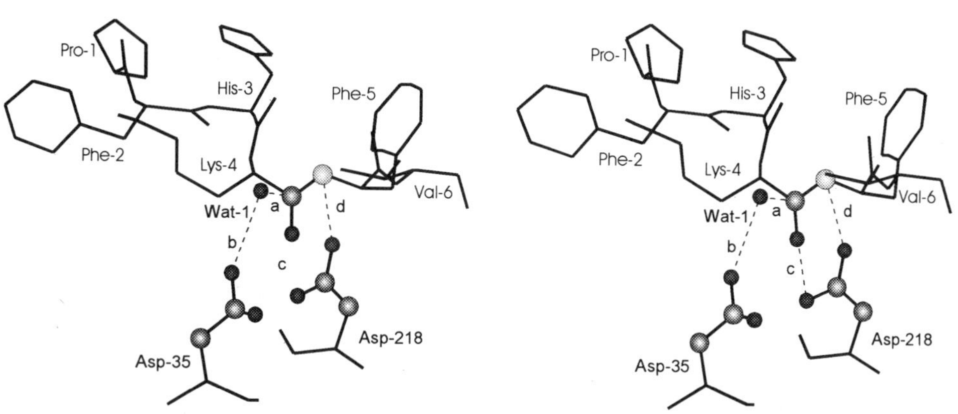

Figure 5. Stereoscopic view of mutual arrangement of the scissile bond, Asp^{35}, Asp^{218} and Wat-1 molecule in rhizopuspepsin active site. The distances are (in Å): a, 2.7; b, 2.8; c, 2.9; d, 3.7.

Table 3. Theoretically predicted values of dihedral angles and stabilization energies of the substrate molecule in the enzyme–substrate nonbonded complex in presence of water molecule WAT507

	Main chain			Side chain				U
Residue	Φ	Ψ	Ω	X_1	X_2	X_3	X_4	kcal/mol
Pro^1	−60	166	177	—	—	—	—	−13.7
Phe^2	−110	155	179	−68	−54	—	—	−32.8
His^3	−150	76	−166	−43	−107	—	—	−49.5
Lys^4	−95	14	173	−47	181	−159	81	−54.7
Phe^5	−53	142	181	72	84	—	—	−23.7
Val^6	−111	88	180	182	62	60	—	−16.9

role of a nucleophile is ascribed to the crystal water molecule that resides between the two aspartic active carboxyl groups. Taking the position of this water molecule fixed, a multi-stage energy optimization has been performed to determine the geometry and a binding energy of the new conformationally equilibrium substrate binding mode in the enzyme potential field. The substrate fragment that undergo the most significant conformational change is dipeptide Lys^4–Phe^5 which contains the scissile bond (Table 3).

During the optimization procedure certain changes in the dihedral angles of both the substrate backbone and side chain are to take place for preserving a complementarity of potential surface of two interacting molecules. This particularly is evident for the Lys^4 side chain (which adopts a new conformational state in S_1 enzymic locus) and seven flap residues (Ile^{75}–Ser^{81} in rhizopuspepsin numbering) which retire slightly (approx. at 0.5 Å by Ca atoms) from the active aspartates. The finale optimization results in the 7.0 kcal/mol difference in the stabilization energy in favor of the conformation with water molecule W507 absent from the active center. This energy difference involves only Lys^4–Val^6 substrate fragment. Geometry of the enzyme–substrate sensitive region including active Asp^{35} and Asp^{218}, a scissile bond and the water molecule Wat-2 (W507 in rhizopuspepsin numbering) is shown in Figure 6.

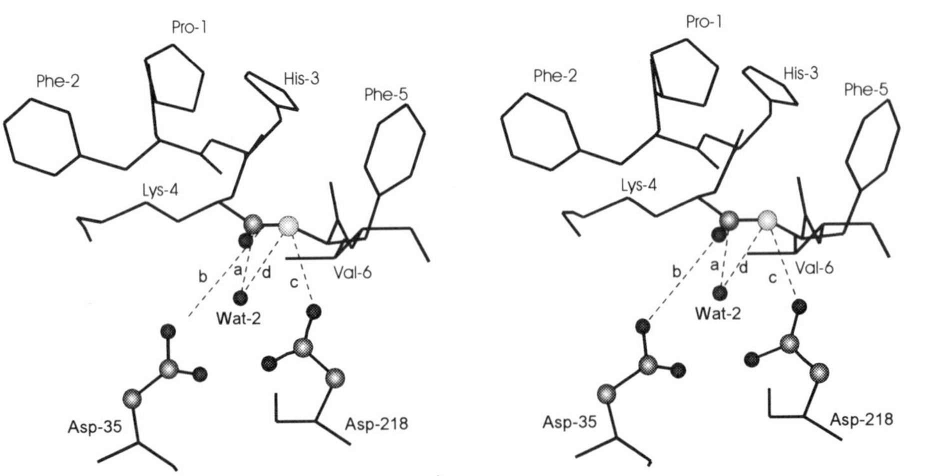

Figure 6. Stereoscopic view of mutual arrangement of the scissile bond, Asp^{35}, Asp^{218} and crystallographic Wat-2 (W507) molecule in rhizopuspepsin active site. The distances are (in Å): a, 2.4; b, 3.7; c, 3.5; d, 2.6.

The most significant feature of this structure is the distance 3.7 Å between carbonyl oxygen and a nearest atom O^{δ} of Asp^{35}, which is obviously too large for an effective electrophilic activation involving Asp^{35}. In this aspect this structure seems to be less preferred compared to that represented at Figure 5 which involves Asp^{218} residue as general acid. The chemical aspects of aspartic protease catalysis will be studied quantitatively by quantum chemistry methods.

REFERENCES

1. Popov, E.M., Kashparov, I.V., & Popov M.E. (1994). Uspekhi Biologicheskoy Khim. (rus). 34: 40–82
2. Suguna, K et al. (1987) Proc. Natl. Acad. Sci. U.S.A. 84: 7009–7013
3. Popov, M.E., Kashparov, I.V., Popov E.M. This issue.
4. Popov, E.M. (1992). Structural-functional organization of proteins. Moskva, Nauka, 359 p. (rus.)
5. Suguna, K. et al. (1987). J. Mol.Biol. 196: 877–900
6. Hofmann, T., Hodges, R.S., James, M.N.G. (1984) Biochemistry, 23: 635–643
7. Pearl, L. (1987) FEBS Lett. 214: 8–12
8. Jaskolski, M. et al. (1991) Biochemistry 30: 1600–1609
9. James, M.N.G. et al. (1992) Biochemistry 31: 3872–3886

THEORY AND METHOD OF A PRIORI COMPUTATION OF CATALYTIC ACTS OF ASPARTIC AND SERINE PROTEINASES

M. E. Popov, I. V. Kashparov, and E. M. Popov

Shemyakin-Ovchinnikov Institute of Bioorganic Chemistry
Russian Academy of Sciences

Theory of enzymatic catalysis based on the new concept of structure-functional organisation of proteins is developed. Previously it has been shown [1], that dynamic conformational properties of a polypeptide chain, manifesting in the folding process cannot be described using current ideas of classical equilibrium thermodynamics. Novel principles of non-linear nonequilibrium thermodynamics (Prigogine physics) have been introduced, giving for the first time explanation to all characteristic features of protein folding process (stochastic mechanism, short time and precision of self-assembling) and of protein interaction with other molecules [1–3]. These principles formed the base for a quantitative approach to the investigations of mechanism of enzymatic catalysis. In a spontaneous enzymatic reaction, decisive contribution is made by irreversible fluctuations (bifurcations), characteristic for non-equilibrium dissipative ensembles. These fluctuations drastically change the evolution of the system, producing new energetically favourable states due to the formation of specific stabilising contacts with other parts of a protein molecule in the folding process or with a substrate molecule in enzymatic reaction. It was shown [4], that the process of spontaneous transformations of thermal energy of irreversible fluctuations into conformational and electronic energy of enzymatic catalysis cannot have counterparts in the model chemical (congruent) reactions governed by completely different mechanisms of equilibrium thermodynamics and statistical physics. During a spontaneous process of the formation of protein globule, the highly organised physiologically active structure emerges due to bifurcations, and during a spontaneous catalytic cycle there emerges a still more highly organised nonbonded complex.

The important assumption of the proposed theory is that conformational and electronic rearrangements follow each other and occur independently. This makes possible to analyse electronic states of chemical bonds participating in a catalytic reaction separately from conformational rearrangement of the substrate and enzyme active site, and vice versa. That is, the validity of the Frank–Condon principle is presumed for all chemical stages of a catalytic act, and the validity of the Born–Oppenheimer principle for all inter-

Aspartic Proteinases, edited by James.
Plenum Press, New York, 1998.

mediate conformational stages. Since hundreds of atoms participate in the nonbonded substrate-enzyme interactions while only tens take part in a purely electronic enzyme-substrate chemical reaction, the latter can be treated using classical approach with semi-empirical atom-atomic potentials and the former by quantum-chemistry methods.

Analysis of all conformational rearrangements of enzyme active site and of the substrate along the reaction path so as to reveal the low-energy structures and their potentiality to transformations is possible only in case of co-ordination and consistency of all participating bifurcations, i.e. fluctuations which bring about the energetically most favourable structures for the investigated fragment. The idea of the highest degree of correlation between nonbonded intra- and inter-residue interactions in protein molecule, first suggested in 1971 [4], is now confirmed by many experimental facts.

Figure 1 shows experimental conformational (φ,ψ) points of all residues except Gly and Pro of pepsin [5], rhizopuspepsin [6], endothiapepsin [7] and HIV-1 proteinase [8] plotted on the conformational map of N-acetyl-L-alanine, representing the potential surface of a free element of protein backbone. It is evident, that in all four aspartic proteinases conformational points of amino acid residues are located in low-energy regions. The majority of points are within 1 kcal/mole contour, and all residues without any exception are within 2 kcal/mole contour. Thus, conformations of all residues in the native states of aspartic proteinases as well as other proteins, which structures are now known with a sufficiently high resolution (less than 2 Å), are completely defined by short-range interactions. Indicating a high-degree harmony between all types of interactions in enzyme molecule, this fact is also an indisputable evidence against the concept of strain and induced deformations as driving forces of enzymatic catalysis.

Thus, general principles of the proposed theory may be formulated:

1. The biocatalytic reaction is a non-linear nonequilibrium process, that does not have counterparts in model chemical reactions.
2. The enzyme and substrate molecules which enter into interactions, are completely (electronically and conformationally) non-stained structures.

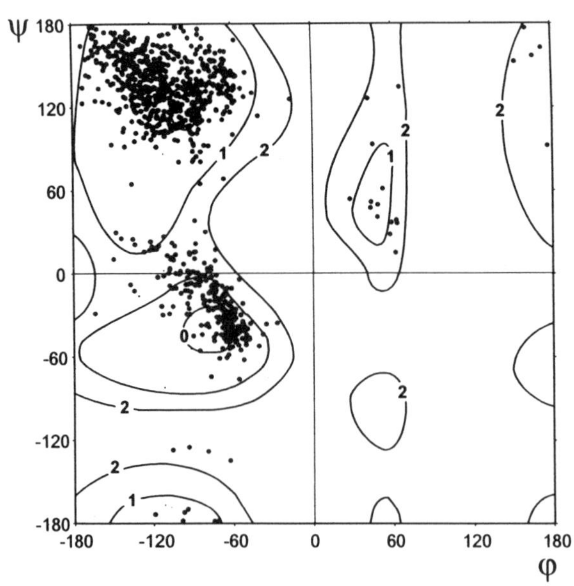

Figure 1. Conformational map φ-ψ on N-acetyl-L-alanine methylamide and conformational points of HIV-1 PR, pepsin, rhizopuspepsin and endotiapepsin amino acid residues (with the exception of Gly and Pro). Contours of the equipotential sections are given for 0.0, 1.0 and 2.0 kcal/mole.

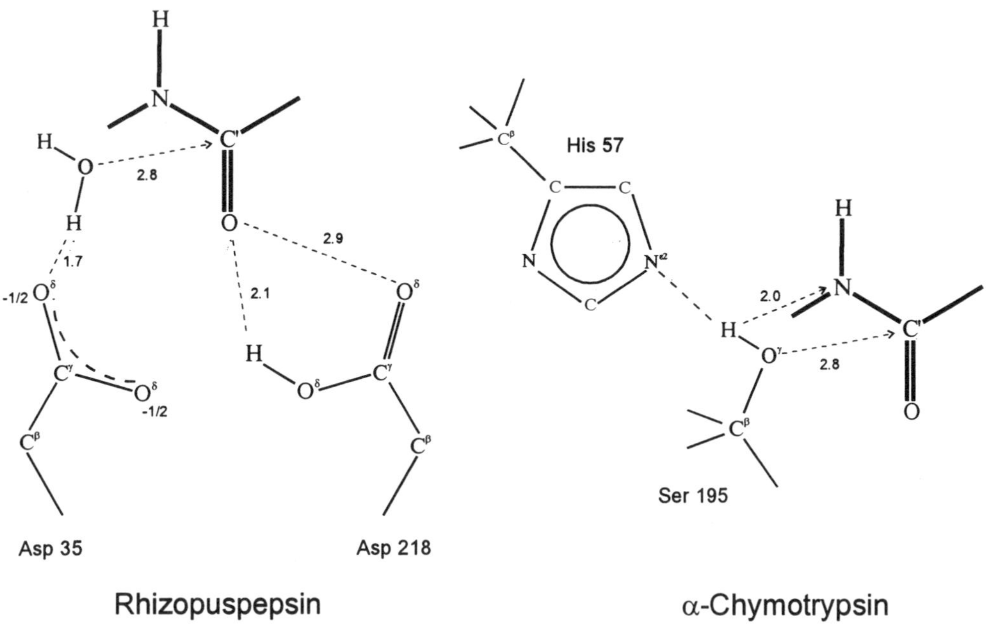

Figure 2. The computed model of the scissile bond productive position in the nonbonded complexes of Aspartic and Serine proteinases. The distances are indicated in broken lines.

3. For a native enzyme, a high degree of correlation between all types of intra- and inter-residue interactions is characteristic.
4. During an enzymatic cycle, conformational rearrangements of enzyme and substrate can be treated independently of electronic changes according to the Born–Oppenheimer principle. The rearrangement of chemical bonds occurs without changing conformational states of interacting molecules according to the Frank–Condon principle.
5. Conformational changes within an enzyme-substrate complex result in formation of electronically unstable state when a minimum nonbonded interaction energy is reached; consequent rearrangements of chemical bond lead to the formation of conformationally unstable structure with two lowest energy of valence interactions.

The proposed theory was tested in a priori calculations of catalytical acts of α-chymotrypsin and β-trypsin [2]. In this issue we analyse catalytic cycles of aspartic proteinases, human immunodeficiency virus type 1 proteinase (HIV-1 PR) and rhizopuspepsin. The completely independent a-priory calculations of serine and aspartic proteinases produced essentially the same orientation of the substrate cleavable bond about catalitically important residues of the enzyme active site (Fig. 2).

The numerical results of these studies are presented in a separate communication.

REFERENCES

1. Popov, E.M. (1989). Structural organization of Proteins (in Russian), Moskwa, Nauka.
2. Popov, E.M. (1992). Structure-Functional Organization of Proteins (in Russian), Moskwa, Nauka.

3. Popov, E.M. (1993). Priroda (in Russian), 6: 59–68
4. Popov, E.M., Lipkind, G.M. (1971) 5: 624–636; 667–679
5. Sielecki, A.R. et al. (1990) J. Mol. Biol. 214: 148–170
6. Suguna, K. et al. (1987) J.Mol.Biol 196: 877–900
7. Blundell, T. et. al. (1990) J. Mol.Biol. 211: 919–941
8. Wlodawer, A. et al. (1989) Science. 245: 616–621

ABSTRACTS

STRUCTURAL MECHANISMS OF DRUG RESISTANCE FOR HIV-1 PROTEASE MUTANTS

J. W. Erickson, T. N. Bhat, B. Yu, S. Gulnik, and L. Suvorov

Structural Biochemistry Program, SAIC, NCI-FCRDC, Frederick, Maryland 21702

A major limitation to the efficacy of antiviral therapy for AIDS has been the emergence of resistance to drugs directed against both HIV-1 reverse transcriptase and HIV-1 protease (HIV PR). X-ray crystallography and enzymology have been employed to study the effects of mutations in HIV PR on enzyme function and inhibitor binding in order to understand the structural mechanisms of drug resistance. Crystal structures have been determined for several clinically-tested inhibitors in complexes with a variety of single and double mutants of HIV PR that include both active site and non-active site mutants. These structures reveal unexpected conformational changes for both protein and inhibitor, and illustrate key roles played by water molecules in modulating the structural effects of mutations. This analysis underscores the significance of enzyme *and* inhibitor flexibility as well as solvent structure in HIV PR/inhibitor interactions. This structural information can be used to aid the design of second generation inhibitors targeted against drug resistant variants of HIV PR, and suggests structure-based clinical strategies to combat resistance.

IN VIVO EMERGENCE OF HIV-1 ESCAPE MUTANTS RESISTANT TO MULTIPLE PROTEINASE INHIBITORS

Jon H. Condra, Daniel J. Holder, William A. Schleif, Olga M. Blahy, Robert M. Danovich, Lori J. Gabryelski, Donald J. Graham, Julio C. Quintero, Audrey Rhodes, Helen L. Robbins, Elizabeth Roth, Malathi Shivaprakash, Donna Titus, Tao Yang, Jeffrey A. Chodakewitz, Paul J. Deutsch, Randi Y. Leavitt, Ferdinand E. Massari, and Emilio A. Emini

Merck Research Laboratories, West Point, Pennsylvania 19486

Indinavir (IDV, CRIXIVAN®, MK-639, L-735,524) is a potent and selective inhibitor of the human immunodeficiency virus type 1 (HIV-1) protease. During early clinical trials, in which patients initiated therapy with suboptimal doses of IDV, we followed the emergence of viral resistance to the inhibitor by genotypic and phenotypic characterization of primary HIV-1 isolates. Development of resistance coincided with variable patterns of multiple substitutions among at least eleven protease amino acid residues. No single substitution was present in all resistant isolates, indicating that resistance evolved through multiple genetic pathways. Measurable resistance to IDV (≥4-fold) was found to require the co-expression of at least three or more specific amino acid substitutions in the protease, and thus IDV-resistant mutants may not pre-exist in natural viral populations. Despite this complexity, alterations of residues V82 (to A, F, or T) or M46 (to I or L) were shown to be approximate indicators of resistance selection. We also extended our previous finding that IDV-resistant viral variants exhibit varying patterns of cross-resistance to a diverse panel of HIV-1 protease inhibitors. Every viral isolate examined that exhibited resistance to IDV (29 isolates from 19 patients) was also cross-resistant to ritonavir and XM412. Between 60–80% of these IDV-resistant isolates were cross-resistant to saquinavir, VX-478, and/or SC-52151, and the spectrum of cross-resistance varied from isolate to isolate.

Finally, we noted a direct association between the number of protease amino acid substitutions and the observed level of IDV resistance. The evolution of this resistance was cumulative, indicating the need for ongoing viral replication. This finding is supported by the results of later clinical studies, in which optimal doses of IDV were used alone or in combination with reverse transcriptase inhibitors. When replication was efficiently inhibited by these optimized therapies, the emergence of resistance to IDV and to the RT inhibitors was suppressed. In contrast, when therapy was begun at suboptimal levels of viral inhibition, the development of resistance limited the benefit of later optimal therapy. These observations strongly suggest that therapy should be *initiated* with the most efficacious regimen available, both to block viral spread and to inhibit the replication that drives the evolution of resistance.

SECOND LOCUS IN HIV-1 RESISTANCE TO PROTEASE INHIBITORS: CONSEQUENCES AND INPUT ON DRUG DESIGN

D. Lamarre, L. Doyon, G. Croteau, D. Thibeault, G. McKercher, and L. Pilote

Bio-Méga/Boehringer Ingelheim Research Inc., 2100 Cunard Street, Laval (Québec) H7S 2G5, Canada

Protease inhibitor therapies against the human immunodeficiency virus type 1 (HIV-1) could maintain benefits despite the development of resistance if emerging viruses show decreased viral fitness. To explore this possibility, potent substrate analog protease inhibitors have been used to select HIV-1 variants with high-level resistance in culture. These resistant variants not only contain multiple genotypic changes in the protease locus, resulting in up to four mutations within the active site of the enzyme, but also contain modifications in one or two protease cleavage sites. These latter mutations are the first evidence of the involvement of a second locus in HIV-1 resistance to protease inhibitors. The accumulation of mutations located in the active site of the protease can lead to a dramatic increase (10^5-fold) in Ki towards protease inhibitors and is accompanied by a substantial loss (10^3-fold) of enzyme activity towards peptide substrates. When introduced in viral clones, protease mutations impaired polyprotein processing, consistent with a decrease in enzyme activity in virions. Although viruses with high-level resistance to protease inhibitors are selected and thus able to replicate *in vitro*, an impaired growth phenotype is observed as a consequence of this functional constraint. Multiple active-site mutations in the protease indeed confer a significant growth reduction of viral clones in peripheral blood mononuclear cells and T-cell lines. The presence of second-locus mutations which promote and even rescue viral growth without affecting resistance to inhibitors is further evidence of the impaired fitness of these highly resistant viruses. By giving rise to better peptide substrates for the protease and to better polyprotein processing by mutant proteases in virions, these compensatory cleavage site mutations indicate a rate-limiting step in polyprotein processing in highly resistant viruses. These second-locus mutations also suggest a limit in the mutability of the protease and should therefore have important consequences on the design of more efficient antiviral strategies against HIV-1.

SITE-DIRECTED MUTAGENESIS OF FIV PROTEASE: EXPRESSION, ACTIVATION, PURIFICATION, AND CHARACTERIZATION

Li Zhang and Ben M. Dunn

Department of Biochemistry and Molecular Biology, University of Florida, College of Medicine, Gainesville, Florida 32610-0245

Feline immunodeficiency virus encodes a protease (FIV-PR) of a length (116 amino acids) which serves the same polyprotein processing function as the related enzyme from human immunodeficiency virus (HIV-PR, 99 amino acids).

The biological properties of FIV are quite similar to primate lentiviruses, and the viral infection of domestic cats leads to clinical symptoms which parallel some of those observed in human AIDS, including a decline in $CD4^+$ lymphocytes, an inversion of the CD4/CD8 lymphocyte ratio, and a hypergammaglobulinemia. For these reasons, the feline system may provide an excellent small-animal model for human immunodeficiency virus infection. Despite the strong structural homology, binding of inhibitors to the two enzymes shows extreme diversity. Potent HIV-PR inhibitors, developed as potential anti-AIDS therapeutics, bind very weakly to FIV-PR. Many sub-nanomolar inhibitors of HIV-PR exhibit greater than 1000-fold reduction in binding strength to FIB-PR. The possible structural origins of the differences in affinity for active site binding may be related to sequence differences at critical positions in the two proteins.

Using site-directed mutagenesis, the isoleucine in FIB-PR (position 48, HIV-PR numbering) was replaced with glycine to mimic the specificity of HIV-PR. The resulting mutant protease was overexpressed in *E. coli* in the form of inclusion bodies. The protein was solubilized, refolded, concentrated and purified using gel filtration chromatography. The enzymatic specificity of the mutant protease was determined with different chromogenic substrates, derived from Lys–Ala–Arg–Val–Leu*Nph–Glu–Ala–Nle–Gly–NH_2, with systematic substitution of the amino acid residues in P_5, P_2, P_1, and P_2' at different pH values. Whereas most of the peptides used in this study were hydrolyzed extremely slowly by the mutant enzyme at pH 4.7, at higher pHs the mutant enzyme was more efficient with some substrates. The spectrophotometric assay has been used to characterize the protease further with respect to pH dependence, ionic strength dependence, and the effect of competitive inhibitors.

This work was supported by NIH grant AI28571 to BMD.

ANTI-HIV PROTEINASE ANTIBODIES THAT INHIBIT ENZYME ACTIVITY

J. Sedlácek,[1] J. Lescar,[2] R. Stouracová,[1,2] M.-M. Riottot,[2] V. Chitarra,[2] J. Brynda,[1,2] M. Fábry,[1] M. Horejsi,[1] and G. A. Bentley[2]

[1]Department of Gene Manipulation, Institute of Molecular Genetics, Academy of Sciences of the Czech Republic, 166 42 Prague, Czech Republic
[2]Unite d'Immunologie Structurale, Institut Pasteur, Paris, France

The current knowledge suggests that, in addition to the active site, two other parts of the protein molecule of the proteinase of human immunodeficiency virus (HIV PR) are directly involved in the enzyme function: (i) the flap region involved in ligand binding during each catalytic cycle, and (ii) the C- and N-terminal parts involved in obligatory dimerization of protomers. We reasoned that the enzyme would be inactivated by interfering with the said regions, and, further, that design of such alternative (non-active-site) inhibitors could be preferably aided by structure of monoclonal antibodies (mAbs): the flaps and the terminal strands have no known natural *sensu stricto* ligands that could exemplify a prototype inhibitor structure and serve productively as "lead compounds" (as non-cleavable substrate analogs can for the active-site inhibitors).

Potent inhibition with a nanomolar range of Ki values was achieved by mouse mAbs (IgGs or their derived Fab or Fab′ fragments) reactive to the dimerization and flap regions of HIV PR, termed mAb 10C8 (binding epitope in N-terminal segment, positions 1–7) and mAb F11 (binding epitope in positions 36–46). The Ki for mAb 10C8 thus surpasses by four orders of magnitude the Ki values found for interface peptides reactive to the same enzyme region (e.g. Schramm *et al.*, BBRC 194:595, 1993). A true dissociative mechanism for inhibitory mAbs has been indicated by much poorer inhibition of the single-chain form of HIV PR. Crystals of the Fab F11 have been obtained both in the free state and as complexes formed with the HIV PR flap-epitope peptides, 36–46 and 36–57 (Lescar *et al.*, Prot. Sci. 5:966, 1996). Solved crystal structures of the complexes show that the interaction between HIV PR and Fab F11 is mainly contributed by one of the six complementarity defining regions of the mAb. Protein crystallography studies with Fab′ 10C8 are also progressing.

Construction of potent inhibitors of a lower molecular weight based on structures of inhibitory antibodies will be critically dependent on chances to preserve a substantial part of macroscopic tight-binding properties while eliminating parts of the "framework" of mAb. Practical advantages may be assumed for thus constructed alternative inhibitors acting at the enzyme regions distinct from its active site, e.g. that they would react with the polyprotein precursor of proteinase prior to the enzyme activation (Darke *et al.*, Biochemistry 33:98, 1994), and that additional constraints would prevent development of drug resistance by mutational changes in the N- and C-terminal strands.

STRUCTURE OF EQUINE INFECTIOUS ANEMIA VIRUS PROTEINASE COMPLEXED WITH AN INHIBITOR

Jukka Kervinen,[1] Alla Gustchina,[1] David Powell,[2] Alexander Zdanov,[1] John Kay,[2] and Alexander Wlodawer[1]

[1]Macromolecular Structure Laboratory, NCI-Frederick Cancer Research and Development Center, ABL-Basic Research Program, Frederick, Maryland 21702
[2]University of Wales, Cardiff, Wales, United Kingdom

Equine infectious anemia virus (EIAV) belongs to the lentiviral family of retroviruses and it is a causative agent of an infectious anemia (swamp fever) in horses. EIAV proteinase (PR) processes viral polyproteins into functional molecules. X-ray structure of a complex of recombinant EIAV PR with the inhibitor HBY-793 has been solved at 1.8 Å resolution and refined to a crystallographic R-factor of 0.136. The overall fold of EIAV PR is very similar to the fold of other retroviral proteinases. However, the appearance of the second ([alpha])-helix in the monomer is a feature not previously reported for retroviral PRs. Despite their strong structural homology, different retroviral PRs show extreme diversity in the binding of substrates and inhibitors. The diversity in affinity may be explained by the structural differences caused by sequence diversity at critical positions in the active site cleft and nearby regions. Here, comparison of the high resolution structures of EIAV PR, feline immunodeficiency virus PR, HIV PR, and Rous sarcoma virus PR are used to explain some of those differences.

18

COMPARISON OF THE SPECIFICITY OF THE ASPARTIC PROTEINASES TOWARDS INTERNALLY CONSISTENT SETS OF OLIGOPEPTIDE SUBSTRATES

Ben M. Dunn,[1] Kohei Oda,[2] John Kay,[3] Chetana Rao-Naik,[1] W. Todd Lowther,[1] Brian M. Beyer,[1] Paula E. Scarborough,[1] and Marina Bukhtiyarova[1]

[1]Department of Biochemistry and Molecular Biology
University of Florida College of Medicine
Gainesville, Florida 32610-0245
[2]Department of Applied Biology
Kyoto Institute of Technology
Kyoto 606, Japan
[3]School of Molecular and Medical Biosciences
University of Wales College of Cardiff
Cardiff, CF1 3US, Wales, United Kingdom

At the 1993 conference in Gifu, the concept of direct comparison of catalytic specificity between enzymes of the aspartic proteinase class was introduced.[1] To achieve this, sets of oligopeptide substrates were used for precise kinetic studies. The parent peptide was: Lys–Pro–Ala–Lys–Phe*Nph–Arg–Leu, with cleavage occurring between the Phe and Nph (p-nitrophenylalanine) residues in all cases. In that report, we presented the set of peptides created by varying the P5, P4, P3, P2, P2′ and P3′ residues of the parent sequence. We substituted Ala, Ser, Asp, Leu, and Arg in each position, thereby exploring a range of potential interactions. In our previous discussion, the critical role of the residues in the P3 and P2 positions of the substrate was emphasized. Three characteristic behaviors were noted for the enzymes; pig pepsin, human cathepsin D, and Rhizopuspepsin. Pepsin showed a broad specificity, but could not tolerate an Arg residue in P3 at low pH; cathepsin D demonstrated a much tighter specificity, being unable to cleave any peptide with the Lys residue in P2, but showing strong activity versus peptides with Leu or Ala in that position. Rhizopuspepsin possessed the broadest specificity of all, with rapid cleavage of all ten substrates in the set compared. In all cases, the highest activity measured for each enzyme, against the best substrate for that enzyme, was comparable: in the range of $0.8–2.4 \times 10^6$ M^{-1} sec^{-1} for k_{cat}/K_m.

Aspartic Proteinases, edited by James.
Plenum Press, New York, 1998.

In the years since the previous meeting, we have completed our analysis of the 28 peptides in the full series described above (P5–P2, and P2′–P3′ variation), and have also utilized these peptides to analyze the effects of mutagenesis in selected positions of the enzymes.[2,3] In this report, we summarize our findings for the parent peptide series on a selection of enzymes. In addition, we introduce a second series of peptides in which the offending Lys residue in the P2 position, which makes most of the first series less useful for many members of the aspartic proteinase family, is replaced by a Glu residue.[4–6] In this new series, we have varied P2, P2′, and P3′, but have also included a wider selection of amino acids to more fully explore the binding preferences in these three subsites.

We will present the results in a graphical format in this chapter; this provides a visual method to distinguish between residues of differing effectiveness in the substrate/enzyme interaction. The value of k_{cat}/K_m is presented on the vertical axis, and the substitution is plotted on the horizontal axis. In each plot, the value for the parent peptide is actually plotted several times, to reflect the substitution present in each of the six positions. A horizontal dotted line is provided at the level observed with the parent peptide. Thus, the effect of a substitution can be readily assessed at a glance, by deciding whether the new peptide provides a value that falls above or below the dotted line (Figure 1).

For the data presented with pepsin, the parent peptide sequence provided an excellent substrate, with the highest values of the specificity constant (k_{cat}/K_m), except for the substitution of Leu in P2′. This was expected, as the parent peptide was chosen based on earlier work by Pohl and Dunn[7] that showed strong activity of pepsin vs. this peptide. However, many other substitutions provide substrates that were cleaved at readily-measurable rates. For example, Leu in P5 was nearly as good as Lys, and Ala in P2 as good as Lys, while Leu was close behind. In contrast, Pro in P4 was by far the best substituent, and Leu in P3′ was clearly better than any of the others tested.

In contrast, the plot for human cathepsin E (Figure 2) showed a much higher selectivity. Only the peptides in which substitution of the Lys residue in P2 has occurred showed easily measurable rates of cleavage. The maximum value observed, however, 1.6 $\times\ 10^6\ M^{-1}\ sec^{-1}$ for the peptide with Leu in P2, was in the same range as the best substrate for pepsin. This result was similar to that seen with human cathepsin D earlier.[1]

In Figure 3 we present the corresponding data for Rhizopuspepsin. This fungal enzyme was the most non-specific enzyme we have encountered in this family. All the substitutions in the parent peptide lead to substrates with good values of k_{cat}/K_m, with only two exceptions: Arg in P4 and Asp in P2′. These charged substituents were not able to be

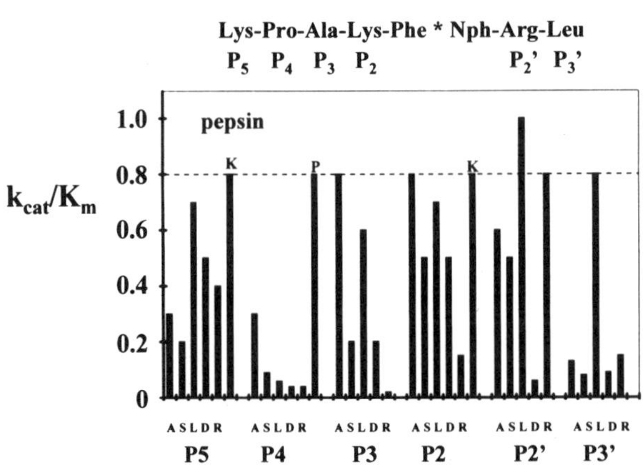

Figure 1. Values of k_{cat}/K_m measured at pH 3.5 and 37° for the pepsin-catalyzed cleavage of peptides based on the parent sequence shown above the graph. The five substitutions (Ala, Ser, Leu, Asp, Arg) are given in that order for each position. Where an extra value is determined, due to other substitution present in the parent sequence, e.g., Lys in P5, this is given following the five standard substitutions, and identified by a one-letter code above the bar for that peptide. The horizontal dashed line represents the value obtained with the parent peptide.

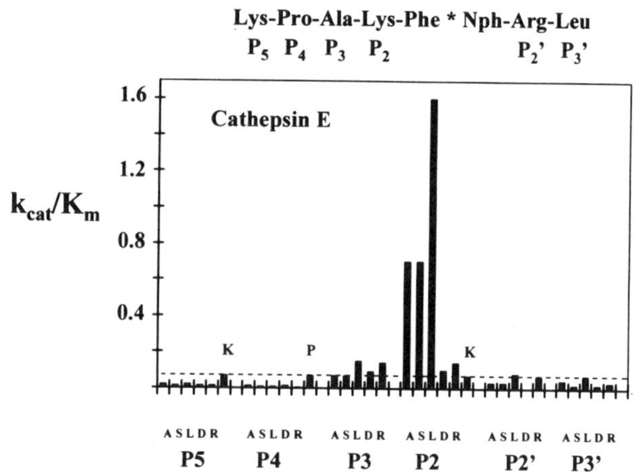

Figure 2. Values of k_{cat}/K_m measured at pH 4.5 and 37° for the Cathepsin E-catalyzed cleavage of peptides based on the parent sequence shown above the graph. All other features are as described in Figure 1.

accomodated in the largely hydrophobic S4 and S2′ binding subsites in the active site cleft. In several cases, substitutions in the parent sequence result in oligopeptides that were cleaved more efficiently, seen in Figure 3 as bars that reach above the dashed line representing the value of k_{cat}/K_m obtained with the parent peptide. As the parent sequence was chosen based on results obtained with pig pepsin, it was not surprising that some peptides would be better for a different enzyme, as seen here as well as in Figure 2 for cathepsin E.

We have now analyzed 12 different enzymes of the aspartic class with this series of substrates, and find that the results fall into two different groups: some enzymes were able to efficiently process the parent sequence and many of the substituted peptides, such as seen with pig pepsin and Rhizopuspepsin; many others, however, were only able to process the oligopeptides where the Lys residue in P2 was replaced by a more hydrophobic amino acid, such as seen with cathepsin D[1,2] and E.[6] Because many interesting enzymes fall into the latter group, we have created a new series of oligopeptides based on a parent peptide derived from our earlier work,[4] Lys–Pro–Ile–Glu–Phe*Nph–Arg–Leu.

Data derived from the cathepsin E-catalyzed cleavage of oligopeptides with changes in P3, P2′ and P3′ are presented graphically in Figure 4. In the case of this series, we have utilized more substituents in the various positions, in order to explore additional features

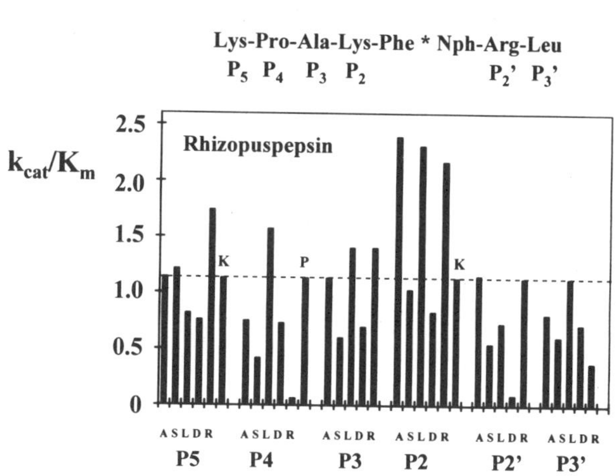

Figure 3. Values of k_{cat}/K_m measured at pH 3.5 and 37° for the Rhizopuspepsin-catalyzed cleavage of peptides based on the parent sequence shown above the graph. All other features are as described in Figure 1.

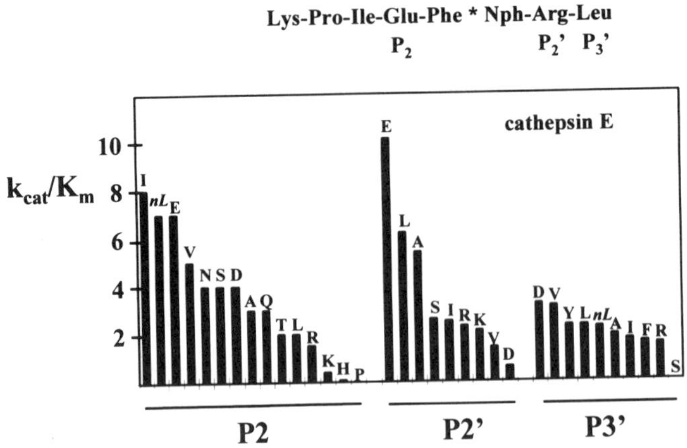

Figure 4. Values of k_{cat}/K_m measured at pH 4.5 and 37° for the cathepsin E-catalyzed cleavage of peptides based on the parent sequence shown above the graph. Here the substitutions are arranged in order of decreasing efficiency of cleavage of the resulting peptide, and are identified by a one-letter code above the bar for the peptide with that substitution. The substitution of norleucine is represented by *nL*.

of binding and catalysis. In particular, we have incorporated the non-natural amino acid Norleucine, because many enzyme prefer hydrophobic substituents.

Considering first the P2 substitutions, cathepsin E had a clear preference for hydrophobic residues. The positively-charged substituents Arg, Lys, and His made very poor substrates, although they can be cleaved at reduced rates. The substitution of Pro in P2 resulted in a peptide that was not cleaved by cathepsin E. In this series of peptides, we have discovered several sequences that are cleaved at high rates, leading to the highest values of k_{cat}/K_m we have observed in this family. However, the highest value, $8 \times 10^6\ M^{-1}sec^{-1}$, is still only about four times that seen in Figure 3 for the best substrates with Rhizopuspepsin. Thus, the catalytic efficiency of the aspartic proteinases are all in the same range with these substrates, confirming the similarity of the catalytic mechanism throughout this family.

Turning to the P2′ substitutions, the peptide with Glu in P2′ gave the best substrate, while the substitution of Asp lead to a depression of the efficiency of turnover, confirming the observation made with Rhizopuspepsin in Figure 3. We believe that this difference is due to the shorter length of the Asp side chain (B. M. Beyer and B. M. Dunn, unpublished observations). In the P3′ substitutions, Asp provided the highest value of k_{cat}/K_m; however, the range of values observed due to substitution in this position was very small. One notable exception was seen for the substrate with Ser in P3′. This peptide was not cleaved at measurable rates under conditions where the other nine peptides are good substrates. The origin of this effect is not known at this time. It should be pointed out that the preferences for each subsite are clearly different. As an example, the Asp in P2′ is the worst substitution, while Asp in P3′ gives the best value of k_{cat}/K_m.

The kinetic parameter k_{cat}/K_m, often called the specificity constant for an enzyme-substrate pair, reports on all events occurring from binding of the two components up to the rate limiting step. The k_{cat} parameter, or turnover number, gives a measure of the entire catalytic process. To help identify the origin of the changes seen for substrate variation in the cathepsin E results, we plot in Figure 5 the values of k_{cat} and in Figure 6 the values of K_m in the same order as presented in Figure 4.

In comparing Figures 4–6, it can be seen that, in general, the substrates that showed the highest values of k_{cat}/K_m also had higher values of k_{cat} and lower values of K_m. How-

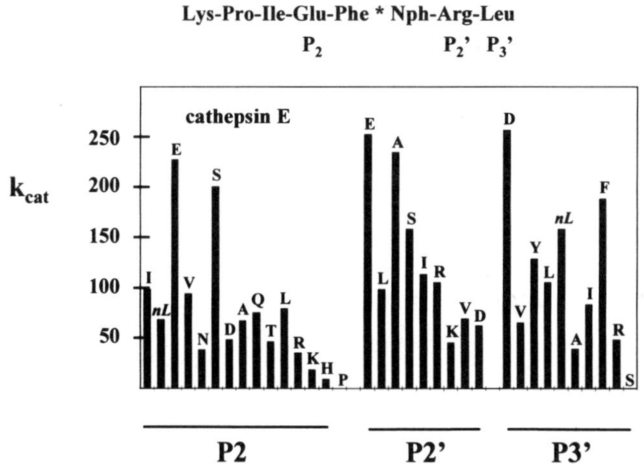

Figure 5. Values of k_{cat} measured at pH 4.5 and 37° for the cathepsin E-catalyzed cleavage of peptides based on the parent sequence shown above the graph. All other features are as described in Figure 4. Pro in P2 and Ser in P3′ are represented as zero in this plot, as no catalysis was observed in either case.

ever, it was also clear that there are many exceptions to this generalization; for example, the substitution of Asn in P2 yielded a low value of k_{cat} which was accompanied by a low value of K_m, and the substitition of Phe in P3′ lead to a high value of k_{cat}, but also a high value of K_m. This illustrates a compensation between the k_{cat} parameter and the K_m value. In the case of an enzyme such as a protease with an extended active site cleft, some forces that lead to tighter binding might also lead to poorer catalysis by creating non-productive conformations. Conversely, apparently weaker binding might be accompanied by more efficient turnover as product release steps might be faster.

It should also be noted that the optimal substrate for a particular application can vary. For example, for the most sensitive assay during purification steps, the k_{cat} parameter

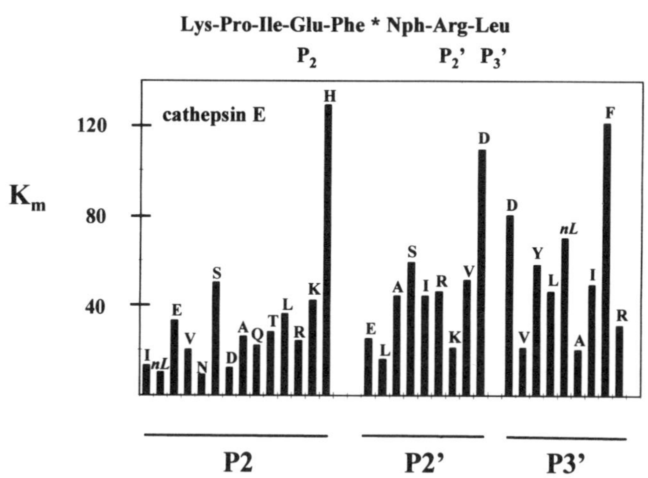

Figure 6. Values of K_m measured at pH 4.5 and 37° for the cathepsin E-catalyzed cleavage of peptides based on the parent sequence shown above the graph. All other features are as described in Figure 4. The substitution of Pro in P2 and Ser in P3′ gave peptides that showed no cleavage; therefore, no values are plotted in this figure for those peptides.

is most critical; however, for assay of competitive inhibitors, a substrate with an acceptable value of k_{cat} (10–100 sec^{-1}) combined with a relatively high value of K_m (greater than 50 μM) is better.

In conclusion, we have presented a graphical method to compare the catalytic efficiency of enzymatic activity of various aspartic proteinases against two series of oligopeptide substrates. In creating these peptides we have sought to provide the capability to analyze different members of the family with a constant series of peptides in order to permit cross-comparisons. We have also incorporated systematic variation in several positions flanking the cleavage point to enable an analysis of the forces involved in binding at each sub-site within the active site cleft. This has provided a wealth of information that reports on the specificity and selectivity of each enzyme. In several cases, collaborators have visited the Dunn laboratory at the University of Florida, bringing their favorite enzyme and conducting these specificity studies. This laboratory will be pleased to continue these efforts.

ACKNOWLEDGMENTS

The authors would like to thank our many collaborators who have enriched our knowledge of the aspartic proteinase world through their interactions. This work was supported in the Florida laboratory through NIH grant DK18865 (to B.M.D.).

REFERENCES

1. Ben M. Dunn, Paula E. Scarborough, W. Todd Lowther, and Chetana Rao-Naik (1995) Comparison of the Active Site Specificity of the Aspartic Proteinases Based on a Systematic Series of Peptide Substrates. "Aspartic Proteinases: Structure, Function, Biology, and Biomedical Implications" (K. Takahashi, ed.) Plenum Press, New York, pp 1–9
2. Paula E. Scarborough and Ben M. Dunn (1994) Redesign of the Substrate Specificity of Human Cathepsin D: The Dominant Role of Position 287 in the S2 Subsite. Protein Engineering, 7, 495–502.
3. W.Todd Lowther, Pavel Majer and Ben M. Dunn (1995) Engineering the Substrate Specificity of Rhizopuspepsin: The Role of Asp 77 of Fungal Aspartic Proteinases in Facilitating the Cleavage of Oligopeptide Substrates with Lysine in P1, Protein Science 4, 689–702.
4. Ben M. Dunn, Melba Jimenez, Benne F. Parten, Martin J. Valler, Carole E. Rolph and John Kay, (1986) Systematic Series of Synthetic Chromophoric Substrates for Aspartic Proteinases, Biochem. J., 237, 899–906.
5. Paula E. Scarborough, Kunchur Guruprasad, Thomas L. Blundell, Gary R. Richo and Gregory E. Conner (1993) Exploration of Subsite Binding Specificity of Human Cathepsin D through Kinetics and Knowledge-based Molecular Modeling, Protein Science, 2, 264–276.
6. Chetana Rao-Naik, Kunchur Guruprasad, Brian Batley, Stephen Rapundalo, Jeffrey Hill, Tom Blundell, John Kay and Ben M. Dunn (1995) Exploring the Binding Preferences/Specificity in the Active Site of Human Cathepsin E, Proteins: Structure, Function, and Genetics 22, 168–181.
7. Jan Pohl and Ben M. Dunn (1988) Secondary Enzyme-Substrate Interactions: Use of Kinetic Evidence for Ionic Interaction Between Substrate Side Chains and The Pepsin Active Site. Biochem., 27 4827–4834.

19

EXPRESSION OF CHIMERIC HUMAN ASPARTIC PROTEINASES

John M. Chirgwin,[1,2] Suzanne Schultz,[1,2] and Deepali Sachdev[1,2]

[1]Audie L. Murphy Veterans Administration Research Service
[2]Departments of Medicine and Biochemistry
The University of Texas Health Science Center at San Antonio
San Antonio, Texas 78284-7877

ABSTRACT

The aspartic proteinases form a gene family whose sequences and three-dimensional structures are strongly conserved. These monomeric proteases are initially translated as prepro- precursors into the lumen of the endoplasmic reticulum (ER), whence they are post-translationally modified and sorted. We have studied three members of the gene family which are expressed in man. Procathepsin D carries N-linked oligosaccharides that are modified to contain mannose 6-phosphate residues. Mannose 6-phosphate receptors in the *trans* Golgi and on the plasma membrane bind and deliver the proenzyme to an acidic prelysosomal compartment and then recycle. Procathepsin D is processed to the mature enzyme and active at low pH (~4.5). Pepsinogen is the abundantly secreted precursor to gastric pepsin and is generally not glycosylated. Propeptide cleavage and activity occur extracellularly at acid pH (>3). Unlike cathepsin D and pepsin, renin shows exquisite specificity towards its physiological substrate, angiotensinogen, and is active at neutral pH in the circulation. Its glycosylation is variable, and cleavage of the propeptide takes place near neutrality at a paired basic residue site. These three aspartic proteinases share certain common features. Procathepsin D and pepsinogen are processed and active at acid pH. Procathepsin D and prorenin usually carry N-linked glycosylation sites. Procathepsin D is targeted to the lysosome, while pepsinogen and prorenin are sorted into secretory vesicles. Sorting into regulated secretory granules occurs in cells which express this specialized pathway.

We were interested in the practicality of expressing chimeras between the human proenzyme forms of cathepsin D, pepsin and renin. The studies were aimed at testing whether such chimeras would be successfully synthesized by transfected mammalian cells and whether they would be sorted through the ER-Golgi pathway and either secreted or targeted to lysosomes. Kornfeld and colleagues[1] have pioneered the use of pepsinogen as a framework for the substitution of segments from procathepsin D.

Aspartic Proteinases, edited by James.
Plenum Press, New York, 1998.

INTRODUCTION

The aspartic proteinase gene family is characterized by a high degree of sequence[2] and structural[3] conservation. The active enzymes, from both fungal and vertebrate sources, are characterized by similarly folded N- and C-terminal lobes. Each domain contributes one of the two essential catalytic aspartic acid residues to the active site and may have evolved from an ancestral tandem gene duplication.[2] The two domains are connected by an antiparallel β-sheet whose six strands are contributed by segments from the N and C domains and the amino terminus of the protein.[3] All of the proteases have an amino terminal propeptide of roughly 45 amino acids, in addition to the cotranslationally removed ER signal peptide. The propeptide sequences are less conserved than the mature proteins. Those proteases which are active at acid pH have a conserved propeptide lysine which may be bound to the active site aspartyl carboxyl groups. Detailed comparison of the crystal structures of pepsin and pepsinogen,[4,5] suggests that the conformation of the mature protein (N and C lobes) is little changed within the pro-form. A major difference is the arrangement of the amino terminus. The first strand (N1) of the six-stranded β-sheet connecting the two lobes is contributed by the amino terminus in the mature protein. In the pro-form this same strand is contributed by the amino terminus of the propeptide, while what is to become the mature amino terminus is about 30 Å away, connecting the propeptide to the body of the protein.

The large size of the propeptides of the aspartic proteinases may reflect the substantial dimensions of the active site clefts which they fill. Alternatively, they could serve as armatures around which the body of the protein folds. Such a propeptide function is known for the subtilisin protease gene family.[6] Several groups have shown that prorenin can be expressed, with substantially reduced efficiency, after deletion of its propeptide.[7] We found that the cathepsin D is only modestly reduced in expression relative to the undeleted pro-form.[8] The data indicate that the propeptides of these aspartic proteinases are not essential for their expression, although they may facilitate folding in the ER or subsequent progress through intracellular compartments. We have recently tested the necessity of the propeptide for the expression of pepsin in mammalian cells. In transiently transfected human 293 cells, the expression of pepsin relative to pepsinogen was reduced ~10 fold within the cells, as indicated by pulse-chase experiments (Schultz and Chirgwin, unpublished). The secretion of the expressed pepsin was further reduced ~4–20 fold compared to pepsinogen. Similar results were obtained with a variant of pepsin/pepsinogen carrying an N-linked glycosylation site on each lobe at the positions equivalent to N70 and N199 of procathepsin D. The results indicate that the propeptide is increasingly important for the cellular expression of aspartic proteinases in the order cathepsin D < renin < pepsin. In no case, however, is the propeptide absolutely essential as an aid to intramolecular folding.

The pro-forms of the aspartic proteinases are the substrates for the cellular machinery in the ER and Golgi which sorts the nascent proteins to secretory granules or lysosomes. Recent experiments with prorenin suggest that the proteolytic cleavage site at the pro:mature junction carries the information for targeting to the regulated granules of mouse pituitary AtT20 cells.[9] Specific targeting could be transferred to a test protein by fusion with the cleavable pro sequence. In *Saccharomyces cerevisiae* vacuolar targeting of proproteinase A, the yeast homologue of cathepsin D, is mediated by the pro-sequence in a man6P-independent mechanism. The proteinase A propeptide confers vacuolar targeting to fusion proteins[10] and binds to the Vps10p sorting receptor, which also targets carboxypeptidase Y to the vacuole.[11] Our data[8] indicate that the propeptide is necessary for the delivery of procathepsin D to the lysosome, suggesting that there may be a yeast-like lysosomal targeting system conserved in mammals which utilizes a propeptide sorting receptor. This idea is supported by the identification of a procathepsin L propeptide receptor in mammalian endosomes[12] and of a plasma membrane procathepsin D receptor.[13] Conservation of the yeast mechanism in mammalian cells is also suggested by the observation of Nishimura et al.[14] that rat procathepsin D, but not a pro-deleted mutant, was delivered to the vacuole of transformed yeast cells.

Inconsistent with a central role for the propeptide of cathepsin D in lysosomal targeting are the results from the Kornfeld laboratory.[1] Major determinants for the addition of man6P to glycopepsinogen chimeras expressed in *Xenopus* oocytes were found to be contributed by sequences rich in lysine residues on the surface of the C-terminal lobe of cathepsin D. These data have been substantially extended.[15] However, it was found that many of the chimeras tested in oocytes at 19°C were not readily expressed in mammalian cells because of problems with folding.[16] We have therefore constructed a series of pairwise chimeras between procathepsin D, pepsinogen, and prorenin. These have been tested for expression, secretion, targeting to the lysosome, and potential proteolytic activity.

METHODS AND MATERIALS

Chimeric cDNAs were constructed by PCR as 5′ (pro+N domain) Hind III to Sal I ~700bp and 3′ (C domain) Sal I to Nco I ~550bp fragments. The Sal I (GTCGAC) site within the coding sequence specifies Val-Asp. The digested and purified fragments were ligated to a Hind III + Nco I digested vector, which drives expression with the CMVie promoter and adds a C-terminal 13aa myc epitope extension,[17] which was not assayed in these experiments. The human pepsinogen full length cDNA was a generous gift from Dr. Tom Taggart. Glycosylation sites corresponding to the N70 and N199 sites of cathepsin D were added by mutagenesis. The resultant glycopepsinogen is nearly identical to that described by Baranski et al.[1] DNAs were transiently transfected into the cell lines 293 (human embryonic kidney) and CHO-L76 (rodent) as previously described.[8] Two days after transfection the cells were labeled overnight with [^{35}S] Cys + Met. Conditioned media and cell lysates were prepared and immunoprecipitated with rabbit polyclonal antisera directed against the sequences contributed by the 5′ DNA fragment, followed by protein A sepharose. Antibodies against human pepsin and recombinant prorenin were generously provided by Drs. Mike Samloff and Peter Hobart. Immunoprecipitates were resolved on reducing, denaturing 12.5% polyacrylamide gels and visualized by fluorography with Enhance (New England Nuclear) following the manufacturer's instructions.

Proteolytic activity was determined on nonradioactive cell lysates and serum-free conditioned media of tissue culture wells prepared in duplicate with the labeled wells. The cathepsin D assay was a micromodification of the standard hemoglobin hydrolysis assay.[18] Hemoglobin was radiolabeled by reductive methylation with [^{14}C]-formaldehyde and $NaBH_3CN$ and purified by gel filtration. Reactions were carried out in V-bottom 96-well microtiter dishes for 4 hr at 37°C in a volume of 200 μl of 0.1M NaOAc pH 4.5 with ~25,000 cpm of substrate. After the addition of cold carrier hemoglobin and TCA to a final concentration of 3%, the plates were centrifuged and the radioactivity of the supernatants determined by liquid scintillation.

DESIGN OF ASPARTIC PROTEINASE CHIMERAS

The six-stranded β-sheet connecting the N and C domains was selected as the least disruptive site to fuse the domains from different proteases. The third (L1) and fourth (L2) strands of this sheet form the bonded interface between the sequences of N and C domains. Within the sheet, there is a long connecting strand between strands 2 (M1) and 6 (N2), whose arrangement can be seen clearly in figure 5b of Andreeva et al.[19] By splicing the chimeras within this strand, only the L1:L2 bonding will be chimeric. Since the propeptide as well as the mature N terminus are contributed by the 5′ cDNA fragment, the N1:M1 β-strand interactions (which are different in pro and mature forms) will be unaltered in the chimeras relative to the wild-type encoded proteins. The site of the junction in cathepsin D is Leu–Gly–Gly–Thr[Val]–Asp–Ser–Lys. The bracketed Val indicates the new residue created by the mutagenic introduction of the Sal I site at procathepsin D position 186. The corresponding sequence in pepsinogen is Phe–Gly–Gly–Ile[Val]–Asp–Ser–Ser and Leu–Gly–Gly–Ser[Val]–Asp–Pro–Gln in prorenin.

If one assumes that the folded conformations of the propeptides are unaltered in the chimeras, then the major novel domain interactions expected will be those between that side of the active site cleft contributed by the C domain and the propeptide (whose sequence is that of the N domain protease). Šali et al.[20] analyzed a series of inhibitor-complexed structures of endothiapepsin. The data supported a model in which the two domains move as separate, independent rigid bodies, and this model was extrapolatable to other aspartic proteinases. Thus N:C domain interactions, outside of the β-sheet, were expected to be minimal.

RESULTS

We constructed 5′ (prepro + N lobe) and 3′ (C lobe) cassettes for all three human proteinases to share a common Sal I [Val186-Asp] junction site. In addition to the six possible chimeric combinations, we also assembled the three control nonchimeric proteins, each referred to as 186V, for comparison with the unmutated plasmids. All twelve plasmids (6 chimeras, 3 186Vs, and 3 parentals) were constructed with identical untranslated regions and all carried a C-terminal myc epitope extension of 13 amino acids.[21] All twelve DNAs were expressed by transient transfection in human 293 and rodent CHO-L76 cells. Proteins were analyzed by continuous labeling for 16 hrs followed by specific immunoprecipitation. The results are shown in Figure 1. The experiment has been repeated several times. We noted some variability in the degree of secretion of some of the chimeras into the medium, particularly with CHO cells, which we attribute to the known selective handling of human procathepsin D in rodent cells.[22]

The three 186V point mutant proteins behaved similarly to their wild-type partners. The constructs with the C lobe of cathepsin D replaced by pepsin [CD-PG] or renin [CD-RN] sequences were little altered in their levels of expression or secretion relative to the procathepsin D [CD, CD186V] controls. Replacing the C lobe of pepsin with cathepsin D [PG-CD] or renin [PG-RN] sequences dramatically reduced the expression levels, and in addition PG-RN was poorly secreted. Replacement of the C lobe sequence of prorenin with that of cathepsin D [RN-CD] resulted in substantially decreased expression but efficient secretion. A converse effect was seen with substitution of the C lobe of prorenin with pepsin sequence [RN-PG]. This chimera was expressed at a level comparable to that of pepsinogen but was only slightly secreted.

The cell lysates immunoprecipitated with anti-cathepsin D showed a band just above the 30kDa molecular weight marker. This corresponds to the endogenous heavy chain of the two-chain form of human lysosomal cathepsin D. The band was not substantially altered in any of the samples (first four lanes) immunoprecipitated for cathepsin D, while the absence of this band in the media samples indicated that secretion was not due to cell lysis. In CHO cells the endogenous rodent cathepsin D in the lysosomes remains as a single chain form of ~45kDa and is readily distinguished from the transfected human form, which is efficiently processed to the two-chain form.[8] When the experiment shown in Figure 1 was carried out

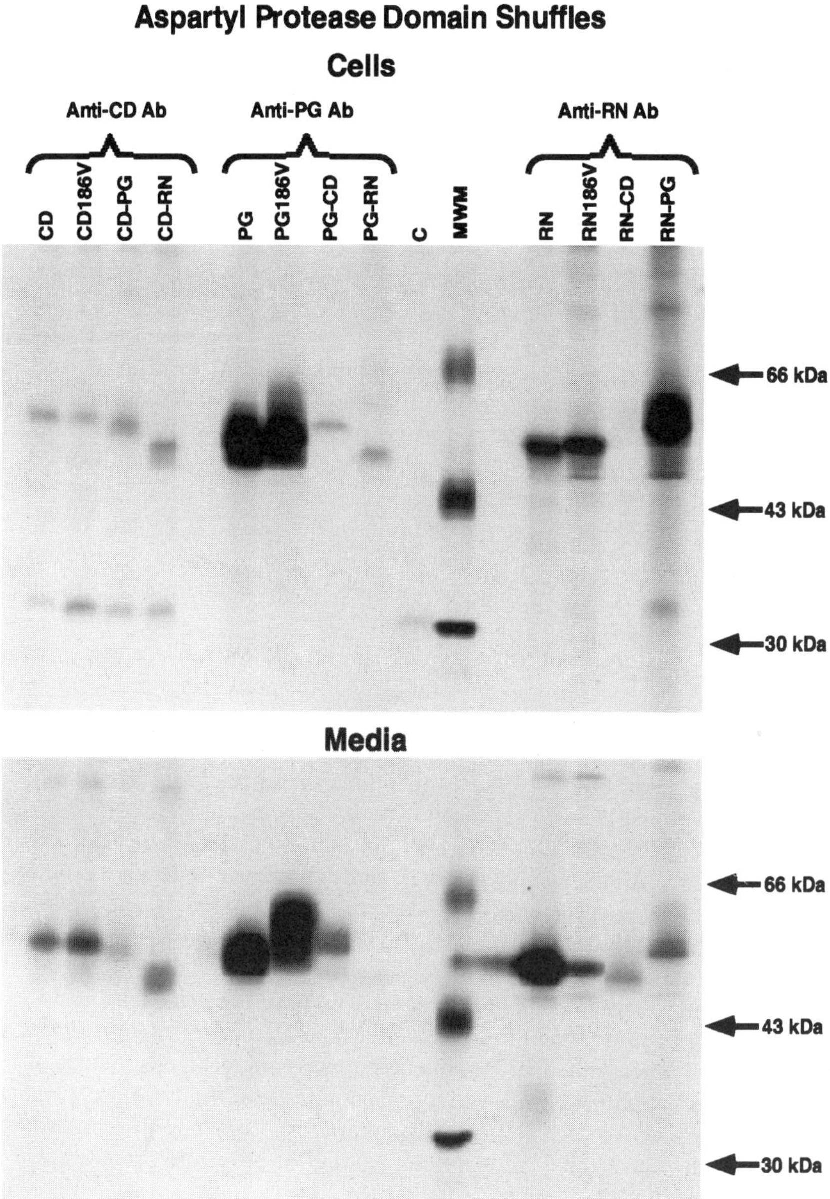

Figure 1. Fluorographic exposure of 12.5% reducing SDS polyacrylamide gel of immunoprecipitates of [^{35}S] labeled proteins of transiently transfected human 293 cells. Upper panel shows the cell lysates, and the lower shows the corresponding conditioned media, with equal loading. Brackets across the top indicate the primary antisera used. Lanes labeled C are a controls immunoprecipitated for endogenous cathepsin D. Arrows along right edge indicate the $m_r s$ of [^{14}C] standard proteins.

with CHO cells, only CD and CD186V showed lysosomal delivery and processing of transfected protein (not shown).

Unlabeled samples, prepared in parallel to those analyzed by radioactive labeling and inmmunoprecipitation, were assayed for cathepsin D activity. Samples of media and lysates were incubated at pH 4.5 and 37°C with radioactive hemoglobin. Under these conditions, purified placental cathepsin D and procathepsin D (recombinant or purified from overexpressing breast cancer cells such as MCF7) are able to digest much of the labeled substrate to peptides soluble in 3% trichloroacetic acid. Under the pH 4.5 assay conditions we found that only the procathepsin D controls (CD and CD186V), both conditioned media and cell lysates, had substantial enzymatic activity. All other samples had less than 5% of the positive control levels, when equivalent volumes of all samples were assayed (data not shown).

DISCUSSION

All six of the pairwise N:C domain chimeras between procathepsin D, pepsinogen, and prorenin were detectably expressed in mammalian cells. Most of these were also secreted from the cells with at least moderate efficiency. Notable is the much higher level of basal expression of glycopepsinogen compared to procathepsin D. Expression in transfected cells showed a pattern pepsinogen > prorenin > procathepsin D. Substitution of the C domain of pepsinogen with sequences from the other two proteases dramatically reduced the level of expression, while the addition of the pepsin C domain to prorenin increased the expression of the chimera. The pepsin C domain is known to fold autonomously *in vitro*.[23] It may be that the presence of this domain facilitates the passage of aspartic proteinases through the ER and Golgi.

When we assayed the unpurified, expressed proteins for low-specificity proteolytic activity at pH 4.5, only nonchimeric procathepsin D was active, either in cells or after secretion, although many of the other aspartic proteinases were present at higher levels. The substrate specificity of renin is absolute, but it was possible that chimeras between cathepsin D and pepsin could be active. The propeptides may not unfold from the active site clefts of pepsinogen and the two chimeras between pepsin and cathepsin D. This would require the stability of the folded pro forms at low pH to be increased over that seen for procathepsin D itself. Another explanation is that the precise molecular geometry is not achieved by the chimeras necessary for productive substrate binding and subsequent proteolytic cleavage by the two active site aspartyl side chains, since these are contributed by heterologous N and C lobes.

Our experiments presented here and previously[8] have not detected any ability for the C-terminal domain of cathepsin D to cause delivery of chimeric proteins to the lysosome, although these and other cathepsin D sequences clearly increase the amount of mannose 6-phosphate modification made to a variety of chimeric constructs with pepsinogen.[1,15,16] Our results are consistent with a major determinant of lysosomal delivery in 293 and CHO cells residing in the cathepsin D propeptide. Binding of propeptide sequences to a recycling receptor, Vps10p, can deliver proteinase A and carboxypeptidase Y to the yeast vacuole. It was found that proteinase A 2propeptide fusions could target invertase to the vacuole,[10] but attempts at mutagenesis of propeptide sequences within intact proproteinase A have been unsuccessful in identifying a specific Vps10p-binding motif, which is the tetrapeptide QRPL in carboxypeptidase Y.[24] Propeptide mutations can result in retention in the ER, where they are degraded, perhaps by a proteasome-mediated mechanism, without reaching the vacuole. Mutations could destabilize the propeptide's binding within the active site cleft, resulting in unfolding of the propeptide prior to exit from the ER. This pro-unfolded form would ressemble the δ form of pepsinogen postulated twenty years ago by Tang,[25] corresponding to the mature conformation with the pro-sequence still uncleaved at the N-terminus. If such a form occured in the ER, the pro-sequences might be recognized as malfolded and retained for degradation. Such a mechanism is compatible with the ER-retention of the imperfect propeptide deletion of cathepsin D described by Conner[26] and the series of propeptide mutations of prorenin reported by Mercure et al.[7]

The observation that rat procathepsin D reaches the vacuole in *Saccharomyces*, while the pro-deleted protein does not,[14] suggests that vertebrate lysosomal targeting may retain a Vps10p:propeptide-like system, in addition to the already redundant mannose 6-phosphate receptor system. We are presently analyzing a series of procathepsin D chimeras carrying short (25aa or less) homologous segments of pepsinogen from the region of the pro:mature junction (Schultz and Chirgwin, unpublished). Cathepsin D sequences within ±10 residues of this junction are necessary for delivery to the lysosome, although they do not confer targeting to glycopepsinogen chimeras. The function of this region in procathepsin D does not appear to be dependent on proteolytic processing at the pro:mature junction, as is the case for prorenin sorting to regulated granules.[9] Richo and Conner found that mutations at the junction within procathepsin D slowed processing, without blocking delivery to the lysosome.[27] Similarly, we have made incorporated up to four Ala mutations at this junction, without effect on delivery to lysosomes in transfected cells (Schultz and Chirgwin, unpublished).

ACKNOWLEDGMENTS

The authors' work has been supported by Merit and Career Scientist Awards to JMC from the Research Service of the Veterans Administration, Audie L. Murphy Memorial VA Medical Center, San Antonio.

REFERENCES

1. Baranski, T.J., Faust, P.L., and Kornfeld, S. (1990). Generation of a lysosomal targeting signal in the secretory protein pepsinogen. *Cell* **63**: 281–291.
2. Tang, J., and Wong, R.N.S. (1987). Evolution in the structure and function of aspartic proteases. *J. Cell. Biochem.* **33**: 53–63.
3. Davies, D.R. (1990). The structure and function of the aspartic proteinases. *Annu. Rev. Biophys. Biophys. Chem.* **19**: 189–215.
4. James, M.N.G., and Sielecki, A.R. (1986). Molecular structure of an aspartic proteinase zymogen, porcine pepsinogen, at 1.8 Å resolution. *Nature* **319**: 33–38.
5. Hartsuck, J.A., Koelsch, G., and Remington, S.J. (1992). The high-resolution crystal structure of porcine pepsinogen. *Proteins: Struct. Funct. Genet.* **13**: 1–25.
6. Baker, D., Shiau, A.K., and Agard, D.A. (1993). The role of pro regions in protein folding. *Curr. Opin. Cell Biol.* **5**: 966–970.
7. Mercure, C., Thibault, G., Lussier-Cacan, S., Davignon, J., Schiffrin, E.L., and Reudelhuber, T.L. (1995). Molecular analysis of human prorenin prosegment variants *in vitro* and *in vivo*. *J. Biol. Chem.* **270**:16355–16359.
8. Fortenberry, S.C., and Chirgwin, J.M. (1995). The propeptide is nonessential for the expression of cathepsin D. *J. Biol. Chem.* **270**: 9778–9782.
9. Brechler, V., Chu, W.N., Baxter, J.O., Thibault, G., and Reudelhuber, T.L. (1996). A protease processing site is essential for prorenin sorting to the regulated secretory pathway. *J. Biol. Chem.* **271**: 20636–20640.

10. Klionsky, D., Banta, L., and Emr, S. (1988). Intracellular sorting and processing of a yeast vacuolar hydrolase: proteinase A propeptide contains vacuolar targeting information. *Molec. Cell. Biol.* **8**: 2105–2116.
11. Cooper, A.A., and Stevens, T.H. (1996). Vps10p cycles between the late-Golgi and prevacuolar compartments in its function as the sorting receptor for multiple yeast vacuolar hydrolases. *J. Cell Biol.* **133**: 529–541.
12. McIntyre, G.F., Godbold, G.D., and Erickson, A.H. (1994). The pH-dependent membrane association of procathepsin L is mediated by a 9-residue sequence within the propeptide. *J. Biol. Chem.* **269**: 567–572.
13. Koelsch, G., Mareš, M., Metcalf, P., and Fusek, M. (1994). Multiple functions of pro-parts of aspartic proteinase zymogens. *FEBS Lett.* **343**: 6–10.
14. Nishimura, Y., Takeshima, H., Sakaguchi, M., Mihara, K., Omura, T., Kato, K., and Himeno, M. (1995). Expression of rat cathepsin D cDNA in *Saccharomyces cerevisiae*: implications for intracellular targeting of cathepsin D to vacuoles. *J. Biochem. (Tokyo)* **118**: 168–177.
15. Dustin, M.L. Baranski, T.J., Sampath, D., and Kornfeld, S. (1995). A novel mutagenesis strategy identifies distantly spaced amino acid sequences that are required for phosphorylation of both the oligosaccharides of procathepsin D by *N*-acetyl glucosamine 1-phosphotransferase. *J. Biol. Chem.* **270**: 170–179.
16. Glickman, J.N., and Kornfeld, S. (1993). Mannose 6-phosphate-independent targeting of lysosomal enzymes in I-cell disease B lymphocytes. *J. Cell Biol.* **123**: 99–108.
17. Schorey, J.S., Fortenberry, S.C., and Chirgwin, J.M. (1995). Lysine residues in the C-terminal lobe and lysosomal targeting of procathepsin D. *J. Cell Sci.* **108**: 2007–2015.
18. Barrett, A.J. (1977). Cathepsin D and other related carboxyl proteinases. In *Proteinases in mammalian cells and tissues* (ed. A.J. Barrett), pp. 209–248. Elsevier/North Holland Biomedical Press, Amsterdam.
19. Andreeva, N.S., Zdanov, A.S., Gustchina, A.E., and Fedorov, A.A. (1984). Structure of the ethanol-inhibited porcine pepsin at 2-Å resolution and binding of the methyl ester of phenylalanyl-diiodotyrsoine to the enzyme. *J. Biol. Chem.* **259**: 11353–11365.
20. Šali, A., Veerapandian, B. Cooper, J.B., Moss, D.S., Hofmann, T., and Blundell, T.L. (1992). Domain flexibility in aspartic proteinases. *Proteins: Struct. Funct. Genet.* **12**: 158–170.
21. Sachdev, D., Schorey, J., and Chirgwin, J. (1991). Efficient mutagenesis, expression, and purification of procathepsin D. *Advan. Exp. Med. Biol.* **306**: 335–338.
22. Isidoro, C., Horst, M., Baccino, F., and Hasilik, A. (1991). Differential segregation of human and hamster cathepsin D in transfected baby-hamster kidney cells. *Biochem. J.* **273**: 363–367.
23. Lin, X., Koelsch, G., Loy, A.J., and Tang, J. (1995). Rearranging the domains of pepsinogen. *Protein Sci.* **4**: 159–166.
24. Westphal, V., Marcuson, E.G., Winther, J.R., Emr, S.D., and van den Hazel, H.B. (1996). Multiple pathways for vacuolar sorting of yeast proteinase A. *J. Biol. Chem.* **271**: 11865–11870.
25. Marciniszyn, J., jr., Huang. J.S., Hartsuck, J.A., and Tang, J. (1976). Mechanism of intramolecular activation of pepsinogen. *J. Biol. Chem.* **251**: 7095–7102.
26. Conner, G.E. (1992). The role of the cathepsin D propeptide in sorting to the lysosome. *J. Biol. Chem.* **267**: 21738–21745.
27. Richo, G.R., and Conner, G.E. (1994). Structural requirements of procathepsin D activation and maturation. *J. Biol. Chem.* **269**: 14806–14812.

CATHEPSIN E

The Best Laid Plans of Mice and Men

P. J. Tatnell,[1] S. D. Fowler,[1] D. Bur,[2] W. E. Lees,[1] and J. Kay[1]

[1]School of Molecular and Medical Biosciences
University of Wales
Cardiff, CF1 3US, Wales, United Kingdom
[2]F. Hoffman La Roche AG
CH-4070 Basel, Switzerland

Aspartic proteinases have been isolated and characterised from a wide variety of organisms, varying from vertebrates to plants, fungi, parasites and retroviruses.[1] In mammals, several are known to exist including the secretory enzymes; pepsin, gastricsin, chymosin and renin and the ubiquitous enzyme cathepsin D found in the lysosomes of most cells. The least well-characterised enzyme, cathepsin E also has an intracellular location but is neither secretory nor lysosomal.

The enzymes of vertebrate, fungal and parasite origin are commonly bi-lobal single-chain enzymes with each domain contributing one Asp residue to the catalytic apparatus.[2] In contrast, in the homodimeric retroviral aspartic proteinases, each monomer provides a catalytic Asp residue to the active site.

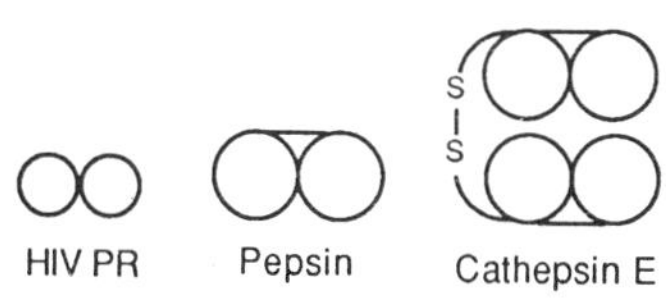

Cathepsin E is exceptional because it has a unique cysteine residue located near to the N-terminus of the mature enzyme. This forms an inter-chain disulphide bond such that the enzyme is a homodimer of ~85kDa. Cathepsin E thus consists of four domains, each containing a catalytic ~Asp–Thr–Gly~ motif arranged in two sets of two (see above).

To date, cDNA sequences encoding procathepsin E from human, rabbit, guinea pig, rat and mouse have been elucidated.[3] The procathepsin E gene, like those of most vertebrate and fungal aspartic proteinase precursors, encodes a signal peptide of ~18 residues, a propart region of ~40 residues fused to ~340 residues of the mature enzyme (Fig. 1). From this alignment, a number of distinguishing features are evident:

Aspartic Proteinases, edited by James.
Plenum Press, New York, 1998.

```
         -18     -10        1P        10P        20P        30P     +1  ◆    10        20
Mouse    MKPLLVLLLLLLL-DLAQAQGALHRVPLRRHQSLRKKLRAQGQLSEFWRSHNLDMTRLSESCNVYSSVNEPLINYLDMEY
Human    **T*--*****V*LE*GE***S*********P**K*****RS******K******IQFT***SMDQ*AK***********
G.pig    **TF--*****V*LE*G**P********S*RE************T*L*K*Q**N*DQ----*STIQ*A***********
Rat      ****F********L*******V***********************D*********IEF******DKGI************
Rabbit   **T*--P******L**G****T*D******QP****************KA*KV**VQYT*T*TMEQ*A************

                 30  ∇    Δ          50        60        70        80        90       100
Mouse    FGTISIGTPPQNFTVIFDTGSSNLWVPSVYCTSPACKAHPVFHPSQSDTYTEVGNHFSIQYGTGSLTGIIGADQVSVEGL
Human    *******S*****************************T*SR*Q****S**SQP*QS**********S*************
G.pig    *******S****************************QT*******L*S**R****S************************
Rat      ***V***S*S*************************************S**M*****************************
Rabbit   *******S***********V****************QM**Q*R****N**S***TP***A*****************Q**

                 110       120       130       140       150       160       170       180
Mouse    TVDGQQFGESVKEPGQTFVNAEFDGILGLGYPSLAAGGVTPVFDNMMAQNLVALPMFSVYLSSDPQGGSGSELTFGGYDP
Human    **V********T*******D***************V****************D*******M**N*E**A****I*****H
G.pig    **V********Q***K***H****************************************M**N*-**************
Rat      **E********************************V********************************************
Rabbit   **V*************************************************S*******M**N*E*************S

                 190       200       210       220  Δ    230       240       250       260
Mouse    SHFSGSLNWIPVTKQAYWQIALDGIQVGDTVMFCSEGCQAIVDTGTSLITGPPDKIKHLQEAIGATPIDGEYAVDCATLD
Human    *********V*************N****G***********************S****Q**N****A*V******E**N*N
G.pig    *********V*******************S***********************G***Q****L***YV*EG*S*Q**N*N
Rat      ***************G*************************************K***Q*********M***********N
Rabbit   *********V*****G*******E****GSP***P*****************S***IQ**A******M******E*EN*N

             ∇   270      ∇280       290       300       310       320       330       340
Mouse    TMPNVTFLINEVSYTLNPTDYILPDLVDGMQFCGSGFQGLDIPPPAGPLWILGDVFIRQFYSVFDRGNNQVGLAPAVP.
Human    V**D***T**G*P***S**A*T*L*F*******S********H**************************R*********
G.pig    M*LD***I**G*P******A*T*L*F*****V*ST**E**E*Q******************A*******R*********
Rat      M*********G*****S**A**********************Q***************K********************
Rabbit   I**D***V**G*P***SA*A*T***F****************Q************************S*R*********
```

Figure 1. Alignment of the amino acid sequences of procathepsin E from several species. The sequence of the mouse protein is given in full. Identical residues in the other sequences are indicated by an asterisk (*). The position of the two active site Asp residues, potential N-linked glycosylation sites and the Cys residue involved in the formation of the intermolecular disulphide bond are indicated by (Δ, ∇ & ♦) respectively.

1. the enzyme from all species is a glycoprotein and has a conserved N-linked oligosaccharide attachment site located at ~Asn34–X–Thr36~ (Fig. 1). This is the only glycosylation motif in the human and rabbit sequences. Cathepsin E from rat and guinea pig each has a second N-glycosylation signal but these are not in the same location; the rat enzyme has this second motif located at Asn266 whereas in the guinea pig sequence, the second site occurs at Asn279. Mouse cathepsin E, in contrast, has all three N-glycosylation signals i.e. at Asn34, Asn266 and Asn279 (Fig. 1). The extent and nature of the glycosylation at the two sites towards the C-terminus of the molecule remains to be characterised.
2. the autoactivation site(s) at which procathepsin E undergoes processing to generate mature cathepsin E are located just upstream from the unique cysteine residue at position 4 (Fig. 2). In the human and mouse proteins, autoactivation has been shown to occur at two neighbouring sites (Fig. 2) so that microheterogeneity is often detected upon N-terminal sequencing of these proteins. Irrespective of its species of origin, cathepsin E is longer at its N-terminus by at least 8 amino acid residues in comparison to other aspartic proteinases such as cathepsin D or pepsin. The prosegment released upon autoactivation of procathepsin E is consequently shorter than its counterparts generated from other zymogens by an equivalent number of residues. The prosegment of many zymogens of aspartic proteinases contains a –Lys–Tyr– motif in which both residues play important roles in the folding and three-dimensional structure of the zymogens.[4,5] This motif is not present in the prosegment of procathepsin E from any of the species (Fig. 1); in its place are the residues which form the autoactivation cleavage sites.
3. the unique cysteine residue at position 4 (Fig. 1) is itself conserved but is located in a region which shows the highest degree of variability in the entire length of procathepsin E sequence. The intermolecular disulphide bond that connects the two chains of cathepsin E (see earlier diagram) is thus positioned on an N-terminal extension to the archetypal aspartic proteinase structure as determined for enzymes of vertebrate/fungal/parasite origin.[2] Whereas it is possible to generate a homology-based model for the remainder of the mature cathepsin E molecule,[6] the region that is unique to cathepsin E (i.e. the N-terminal extension containing the disulphide bond) necessitates structural analyses by crystallisation and X-ray crystallography.

The importance of the intermolecular disulphide bond has been assessed directly by site-directed mutagenesis to change cysteine 4 in the human sequence to alanine or serine.[7,8] The resultant monomeric enzymes, produced in *E. coli* and CHO cells respectively were indistinguishable in terms of their catalytic properties from their dimeric counterparts. However, they were much less stable to pH and temperature changes (Table 1) than the naturally-occurring or recombinant dimeric enzymes. A monomeric form of cathepsin E which is unstable but still able to undergo dimerisation has been detected in gastric can-

```
                                      ♦
Mouse:        ~L D M*T R L*S E S C N V Y S S ~
Human:        ~L D M*I Q F*T E S C S M D Q S ~
Guinea pig:   ~L*N M D Q - - - - C S T I Q S ~
Rat:          ~L D M I E F*S E S C N V D K G ~
Rabbit:       ~V D M*V Q Y T E T C T M E Q S ~
```

Figure 2. Location of sites (*) where procathepsin E undergoes autoactivation to generate mature cathepsin E. The unique cysteine residue is indicated by (♦).

Table 1. Comparisons of the activity and stability of dimeric and Cys4Ala mutant cathepsin E

Enzyme	k_{cat}/K_m ($mM^{-1}sec^{-1}$)	Half-life (min)
Dimeric	2,300	35
Cys4Ala mutant	2,050	<2

Kinetic parameters were determined at pH 3.1 using Lys–Pro–Ile–Glu–Phe*Nph–Arg–Leu as substrate. Stability measurements were made at pH 7.5 and 55°C.

cer cells.[9] Moreover, the major features which determine that cathepsin E is retained within vesicular structures associated with the endoplasmic reticulum/endosome compartments and is not secreted like pepsin/chymosin/renin or trafficked into lysosomes like cathepsin D, appear to be located within the N-terminal 48 residues of the mature enzyme sequence.[10] This region contains both the intermolecular disulphide bond and the one glycosylation site that is conserved in the enzyme from all species. In contrast to the situation in lysosomes or in the lumen of the stomach, these vesicular compartments are not particularly acidic. Nevertheless, the aspartic proteinase cathepsin E is expected not only to be active but also to remain stable (possibly facilitated by ATP[11]) and to be retained within these compartments. It would appear that the region of the molecule which contains the interchain disulphide bridge that is unique to cathepsin E makes a major contribution in directing the protein to its appropriate cytomorphological locations and ensuring that it stays active once it gets there.

In order to facilitate structural investigations, recombinant human cathepsin E has been produced in and purified from CHO cells,[12] *Pichia pastoris*[13] and *E. coli*[14] and the Cys4Ala mutant human enzyme and mouse cathepsin E have also been produced in *E. coli*[7,15]. The enzyme has also been purified to homogeneity from a variety of natural sources such as gastric mucosa, erythrocytes, spleen, neutrophils and epidermis. Isolation of the precursor form, procathepsin E is rather more difficult but has been described.[16,17] Despite considerable attempts, low mol. wt. compounds specific for intracellular inhibition of cathepsin E have proved elusive to develop.[18] In this regard, only the protein inhibitor (M_r~17K) from the parasitic nematode, *Ascaris lumbricoides*, has proven to be effective in permitting distinction between the activities of cathepsin E and cathepsin D inside cells.[19,20] Since only very limited amounts of this protein are available from natural sources, it has been necessary to develop a recombinant strategy to generate sufficient Ascaris protein to use as a selective inhibitor of cathepsin E (see chapter by Zalatoris *et al.*).

Cathepsin E has a limited cell and tissue distribution and this cell-specific expression is not consistent from one vertebrate species to another. For example, it appears to be present in rat and human red blood cells but not in those from pig, goat, cow or guinea pig.[21] Reciprocally, monocytes/neutrophils from rat and guinea pig but not humans give detectable levels of the enzyme and substantially higher levels are found in mouse lymphocytes than those from humans.[16,21,22] This lack of consistency in cell-specific expression implies that different transcription factors may be differentially operative within cell types from various species.

The genomic organisation of the human and mouse procathepsin *E loci* is identical and comparable to those determined for other vertebrate aspartic proteinases, consisting of 9 exons with conservation of the exon/intron boundaries;[23] [PJT, JK, W. Roth, J. Deussing & C. Peters, unpublished]. In contrast to this similarity in the coding regions, the sequences of

5′-regions flanking the mouse and human procathepsin E genes are completely different and distinct motifs for specific transcription factors can be readily identified in each. Analyses of these control elements remain to be performed in detail but transcription of the human (and rat) but not mouse genes results in the generation of alternatively-spliced products[10,24] as well as message for authentic procathepsin E. In the rat, the alternative splicing event would delete exon 9 and thereby remove the C-terminal 30 residues from the protein product. Since this region contains the "thread" strand of the second active site psi loop, its absence would be likely to have serious consequences; current evidence suggests that this alternative transcript is not a major component in rat (spleen) cells.[24] In the human case, two separate events have been observed, both using carcinoma cell lines; in CACO-2 intestinal carcinoma cells, an aberrant event in the splicing out of intron 4 resulted in the retention of an extra 15 nucleotides from the intron connecting exons 4 and 5. This in frame insertion resulted in the introduction of an additional 5 amino acids (–Ala–Phe–Ser–Tyr–Gly) into the sequence of human procathepsin E.[10] This insertion appeared to have little functional consequences on the resultant protein. However, in gastric adenocarcinoma cells, the alternative transcript that was detected was due to the aberrant splicing of the 3′-end of exon 6 to the 5′-end of exon 8 [SDF, PJT, JK - unpublished]. This dictates that the 47 residues encoded by exon 7 are deleted; these include the ~Asp225–Thr226–Gly227~ motif of the second psi loop. In addition, there is a frame-shift so that the predicted amino acid sequence following the deletion is totally different from that of authentic cathepsin E and shows no similarity to any known sequence in the data bases.

Further functional analyses performed on genomic clones of human and mouse origin, together with the generation of knock-out mice deficient in the procathepsin E gene, might be expected to provide additional insights into the limited cellular and tissue distribution of cathepsin E and its putative involvement in such cellular processes as hormone biogenesis,[25] antigen processing for presentation by Class II major histocompatibility (MHC II) antigens,[20] processing of human MHC II invariant chain[26,27] and in neurodegeneration associated with Alzheimer's-type dementia, brain ischaemia and normal ageing.[28–30]

ACKNOWLEDGMENTS

This work was supported by grants from The Wellcome Trust (for PJT), and a ROPA award from the BBSRC (for SDF). We are most grateful to our colleagues in the transgenic Unit led by Prof. Dr. Christoph Peters in the University of Freiburg and to Dr. Ben M. Dunn, University of Florida for their valuable contributions to this work.

REFERENCES

1. Takahashi, K. (ed.) (1995). *Aspartic Proteinases: Structure, Function, Biology and Biomedical Implications,* pp. 1–550, Plenum Press, New York.
2. Silva, A. M., Lee, A. Y., Gulnik, S. V., Majer, P., Collins, J., Bhat, T. N., Collins, P. J., Cachau, R. E., Luker, K. E., Gluzman, I. Y., Francis, S. E., Oksman, A., Goldberg, D. E. & Erickson, J. W. (1996). *Proc. Natl. Acad. Sci.*, **93,** 10034–10039.
3. Kay, J. & Tatnell, P. J. (1997). In, *Handbook of Proteolytic Enzymes* (Barrett, A. J., Rawlings, N. D. & Woessner, J. F., eds), in press, Academic Press, London.
4. Koelsch, G., Mares, M., Metcalf, P. and Fusek, M. (1994). *FEBS Lett.*, **343**, 6–10.
5. James, M. N. G. & Sielecki, A. R. (1986). *Nature*, **319**, 33–38.
6. Rao-Naik, C., Guruprasad, K., Batley, B., Rapundalo, S., Hill, J., Blundell, T.L., Kay, J. & Dunn, B. M. (1995). *Proteins: Struct. Funct. and Genet.* **22**, 168–181.

7. Fowler, S. D., Kay, J., Dunn, B. M. & Tatnell, P. J. (1995). *FEBS Lett.*, **366**, 72–74.
8. Tsukuba, T., Sakai, H., Yamada, M., Maeda, H., Hori, H., Azuma, T., Akamine, A. & Yamamoto, K. (1996). *J. Biochem.* **119**, 126–134.
9. Aoki, T., Takasaki, T., Furukawa, T., Morikawa, J., Yano, T. & Watabe, H. (1995). *Biol. Pharm. Bull.* **18**, 1522–1525.
10. Finley, E. M. & Kornfeld, S. (1994). *J. Biol. Chem.* **269**, 31259–31266.
11. Thomas, D. J., Richards, A. D., Jupp, R. A. , Ueno, E., Yamamoto, K., Samloff, I. M., Dunn, B. M. & Kay, J. (1989). *FEBS Lett.*, **243**, 145–148.
12. Tsukuba, T., Hori, H., Azuma, T., Takahashi, T., Taggart, R. T., Akamine, A., Ezakai, M., Nakanishi, H., Sakai, H. & Yamamoto, K. (1993). *J. Biol. Chem.*, **268**, 7276–7282.
13. Yamada, M., Azuma, T., Matsuba, T., Iida, H., Suzuki, H., Yamamoto, K., Kohli, Y. & Hori, H. (1994). *Biochim. Biophys. Acta*, **1206**, 279–285.
14. Hill, J., Montgomery, D. S. & Kay, J. (1993). *FEBS Lett.*, **326**, 101–104.
15. Tatnell, P. J., Lees, W. E. & Kay, J. (1997). *FEBS Lett.*, in press.
16. Kageyama, T. (1995). *Meth. Enzymol.*, **248**, 120–136.
17. Takeda-Ezaki, M. & Yamamoto, K. (1993). *Arch. Biochem. Biophys.*, **12**, 352–358.
18. Rao, C. M., Scarborough, P. E., Kay, J., Batley, B., Rapundalo, S., Klutchko, S., Taylor, M. D., Lunney, E. A., Humblet, C. C. & Dunn, B. M. (1993). *J. Med. Chem.*, **36**, 2614–2620.
19. Martzen, M. R., McMullen, B. A., Smith, N. E., Fujikawa, K. & Peanasky, R. J. (1990). *Biochemistry*, **29**, 523–526.
20. Bennet, K., Levine, T., Ellis, J. S., Peanasky, R. J., Samloff, I. M., Kay, J. & Chain, B. M. (1992). *Eur. J. Immunol*, **22**, 1519–1524.
21. Yonezawa, S. & Nakamura, K. (1991). *Biochim. Biophys. Acta*, **1073**, 155–160.
22. Sakai, H., Saku, T., Kato, Y. & Yamamoto, K. (1989). *Biochim. Biophys. Acta*, **991**, 367–375.
23. Azuma, T., Liu, W., Vanderlaan, D. J., Bowcock, A. M. & Taggart, R. T. (1992). *J. Biol. Chem.*, **267**, 1609–1614.
24. Okamoto, K., Yu, H., Misumi, Y., Ikehara, Y. & Yamamoto, K. (1995). *Arch. Biochem. Biophys.*, **322**, 103–111.
25. Lees, W. E., Kalinka, S., Meech, J. D., Capper, S. J., Cook, N. D. & Kay, J. (1990). *FEBS Lett.*, **273**, 99–102.
26. Maric, M. A., Taylor, M. D., & Blum, J. S. (1994). *Proc. Natl. Acad. Sci. USA*, **91**, 2171–2175.
27. Kageyama, T., Yonezawa, S., Ichinose, M., Miki, K. & Moriyama, A. (1996). *Biochem. Biophys. Res. Commun.*, **223**, 549–553.
28. Bernstein, H. G. & Wiederanders, B. (1994). *Brain Research*, **667**, 287–290.
29. Nakanishi, H., Tominaga, K., Amano, T., Hirotsu, I., Inoue, T. & Yamamoto, K. (1994). *Exp. Neurology*, **126**, 119–128.
30. Nakanishi, H., Tsukuba, T., Kondou, T., Tanaka, T. & Yamamoto, K. (1993). *Exp. Neurology*, **121**, 215–223.

21

BIOSYNTHESIS AND INTRACELLULAR TARGETING OF THE LYSOSOMAL ASPARTIC PROTEINASE CATHEPSIN D

Gene D. Godbold,[1] Kyujeong Ahn,[1] Susan Yeyeodu,[1] Li-Fen Lee,[2] Jenny P.-Y. Ting,[2] and Ann H. Erickson[1]

[1]Department of Biochemistry and Biophysics
[2]Lineberger Comprehensive Cancer Center
The University of North Carolina
Chapel Hill, North Carolina 27599-7260

Lysosomal proteinases participate in cellular protein turnover and in the degradation of proteins which enter the cell by endocytosis. To serve this physiological role, a lysosomal enzyme must not only be synthesized with a functional catalytic site, it must move along a precise intracellular pathway to reach its site of action. Transport to lysosomes is not a default pathway but rather requires that biosynthetic forms of the lysosomal enzymes be recognized by specific intracellular processing enzymes and receptors in a complex sequence. If a lysosomal enzyme undergoes a mutation that prevents correct cellular targeting, it will either be degraded or secreted from the cell. The resulting absence of a specific enzymatic activity within the lysosomes usually produces a fatal lysosomal storage disorder.

1. BIOSYNTHESIS OF CATHEPSIN D

Like all lysosomal proteinases, the aspartic proteinase cathepsin D undergoes multiple proteolytic processing steps during biosynthesis. The N-terminal 20 amino acid pre sequence or signal peptide mediates transport across the endoplasmic reticulum (ER) membrane[1] through interaction with the signal recognition particle.[2] As this sequence is removed cotranslationally, the preproprotein is not detected in cells. The 44-residue propeptide, an activation peptide, keeps the proteinase inactive during cellular transport.[3] Cleavage of the propeptide, either in an endosome or in lysosomes, generates an active single-chain enzyme of approximately 44 kDa (Fig. 1). In human cells (Fig. 1), but infrequently in rodent cells (Fig. 2), the single-chain enzyme is subsequently cleaved to a two-chain form. This variation in biosynthetic forms is a direct result of species-specific protein sequence differences at the cleavage site.[4] Due to the transient nature of the proenzyme and the single-chain

Aspartic Proteinases, edited by James.
Plenum Press, New York, 1998.

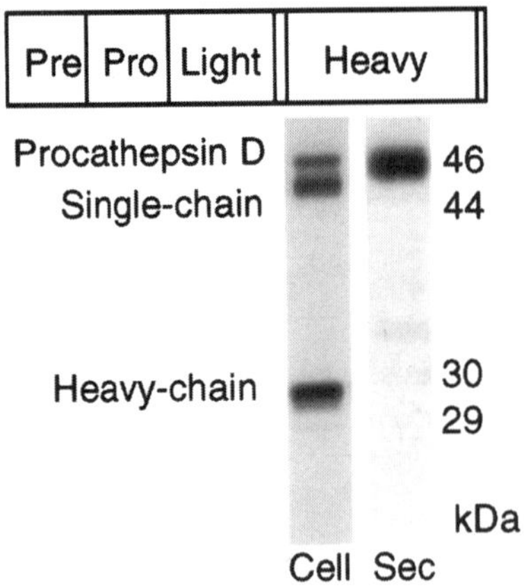

Figure 1. Biosynthetic forms of human cathepsin D. Human BT20 breast tumor cells cultured in a 60-mm dish were labeled continuously with [^{35}S] methionine for 4 h. Cathepsin D, immunoprecipitated from the cells and the cell culture medium as described previously,[10] was resolved on a 10% polyacrylamide gel and visualized by fluorography.

enzyme, the 30-kDa heavy-chain is the predominant form of the proteinase detected in human tissue by western blot analysis. The 7-kDa light chain, derived from the N-terminus, is not detected on polyacrylamide gels designed to resolve proteins of average mass, while the C-terminal heavy chain migrates as a 30-kDa protein (Fig 1). As this processing step generally occurs in lysosomes, several amino acids, varying in number with the species,[4] are lost from the ends of the heavy-chain due to the action of exoproteinases.[5] The resulting 29-kDa form appears as a shadow under the heavy-chain protein band (Fig. 1).

The relationship of the various biosynthetic forms of cathepsin D is revealed by pulse-chase radiolabeling assays. Arrival at the lysosome can be detected between 90 min and 2 h after synthesis in mouse fibroblasts by the appearance of the 44-kDa form generated by proteinase activation (Fig. 2).

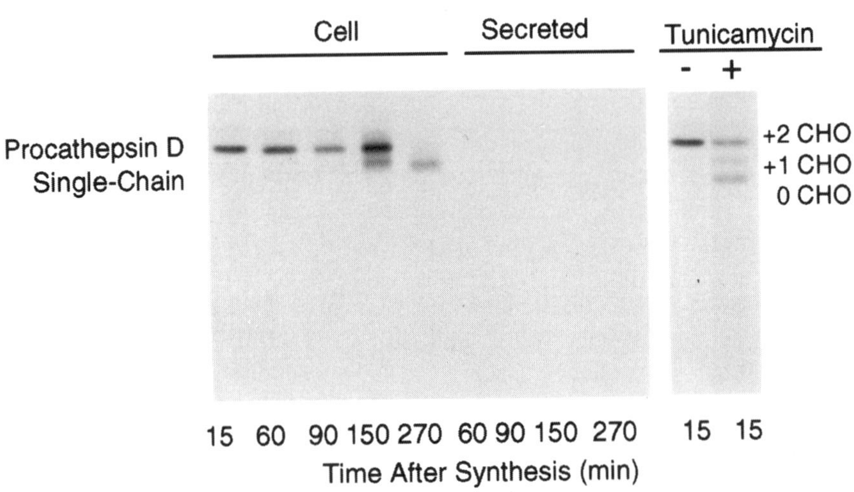

Figure 2. Biosynthetic forms of mouse cathepsin D. KNIH 3T3 mouse fibroblasts cultured in 60-mm dishes were pulsed with [^{35}S]methionine for 15 min and the radiolabel was chased with unlabeled methionine for the times specified. Cathepsin D, immunoprecipitated from the cells (Cell) and the cell culture medium (Secreted),[10] was resolved on a 10% polyacrylamide gel and visualized by fluorography. When cells were pretreated with tunicamycin (1 μg/ml culture medium) for 4 h prior to addition of the radiolabel, oligosaccharide addition was incomplete, resulting in the immunoprecipitation of procathepsin D bearing none (0 CHO), one (+1 CHO) or two (+2 CHO) high-mannose carbohydrate chains.

Procathepsin D acquires two high-mannose oligosaccharide chains cotranslationally, one on the light and one on the heavy chain.[6] Each high-mannose carbohydrate chain contributes approximately 2 kDa to the mass of the protein. The oligosaccharide chains can be visualized by immunoprecipitating procathepsin D from cells cultured in the presence of tunicamycin, a drug which blocks carbohydrate addition with varying degrees of efficiency, resulting in the synthesis of proenzyme possessing two, one or no oligosaccharide chains (Fig. 2). One or both of these high-mannose chains then becomes phosphorylated in the early Golgi by the addition of *N*-acetylglucosamine 1-phosphate.[7] Subsequent removal of the *N*-acetylglucosamine exposes the phosphate, enabling the protein to bind in the trans Golgi network (TGN) to mannose 6-phosphate receptors which mediate targeting to late endosomes.[7]

Lysosomal enzymes are also secreted, despite the fact that they bear phosphorylated mannose residues which should bind to the mannose 6-phosphate lysosomal targeting receptors. The proenzyme is the biosynthetic form found in the culture medium of most cell types; only activated macrophages secrete significant quantities of the mature proteinase (data not shown). The amount of enzyme secreted varies significantly with the cell type. While procathepsin D secretion by fibroblasts is minimal relative to the amount of enzyme in the cells (Fig. 2), breast tumor cells secrete large amounts of the proenzyme (Figs. 1 and 5). A significant proportion of the secreted enzyme acquires complex sugars in the late Golgi, producing a slight increase in mass which causes the modified protein to comigrate with the immunoglobulin heavy-chain released during immunoprecipitation. Although the latter is not radiolabeled, the large amount of protein can cause the secreted cathepsin D protein band to appear slightly concave (Figs. 4 and 6). If the oligosaccharide chains acquire the mannose-phosphate modification, they cannot be modified with complex sugars.[8]

2. MEMBRANE ASSOCIATION OF PROCATHEPSIN D

In addition to binding mannose-phosphate receptors, procathepsin D also becomes associated with membranes by a mannose 6-phosphate-independent mechanism during transport through its biosynthetic pathway. Diment and colleagues initially reported that mouse procathepsin D is membrane-associated by a mannose 6-phosphate-independent mechanism in macrophages.[9] We subsequently found[10] that procathepsin D binds to mouse fibroblast microsomal membranes, while the mature forms of the enzyme were soluble. Strous and co-workers[11,12] and Zhu and Conner[13] both found membrane-association of human procathepsin D in HepG2 cells. Recently, Bergeron and colleagues[14] reported that glucagon is proteolytically cleaved within hepatic endosomes by membrane-associated cathepsin D. The physiological significance of the interaction of procathepsin D with membranes is still not clear.

We found that the membrane association of procathepsin D is pH-dependent when assayed in vitro.[10] The majority of the procathepsin D was solubilized when microsomes were lysed at pH 7 by freezing and thawing in the presence of the detergent saponin, but was membrane-associated if the microsomes were lysed in an acidic buffer (Fig. 3). In vivo, the percentage of the protein which underwent mannose 6-phosphate independent membrane association increased with time after synthesis. Approximately 50% of the newly synthesized proenzyme was membrane-associated as early as 10 min after synthesis, when the protein was presumably still in the ER or cis Golgi network, but the percentage increased to approximately 80% with time (Fig. 3). Similarly, procathepsin D was membrane-associated in cells incubated at 15°C, which blocks protein transport into the

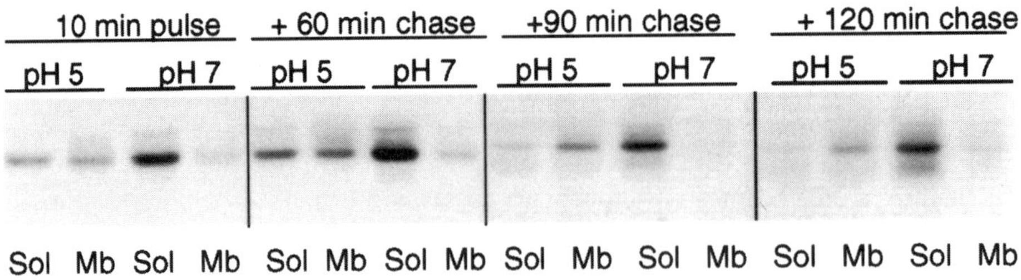

Figure 3. Mannose 6-phosphate-independent membrane association of procathepsin d increases with time after synthesis. KNIH 3T3 mouse fibroblasts cultured in 100-mm dishes were pulsed with [^{35}S]methionine for 10 min and the radiolabel was chased with unlabeled methionine for the times specified. At each time point microsomal membranes were prepared by Dounce homogenization as described.[10] To separate the soluble (luminal) contents of microsomes from the membranes, the microsomal pellet was suspended in a buffer of specified pH containing 50 mM buffer ions (NaOAc for pH 5.0, HEPES for pH 7.0), 1 mM EDTA, 1 mM EGTA, 10 mM mannose 6-phosphate (M6P), 0.1% saponin, 25 μM E-64, and 100 μM DCIC. The membranes were frozen and thawed twice to aid membrane lysis and were recovered by centrifugation at 132,000 × g_{av} for 15 min at 4°C. Cathepsin D immunoprecipitated from the soluble (Sol) and membrane (Mb) fractions was resolved on a 10% polyacrylamide gel, visualized by fluorography, and quantitated by densitometry. Approximately 51% of the procathepsin D was membrane-associated after the 10 min pulse, 54% after a 60 min chase, 82% after a 90 min chase, and 83% after a 120 min chase.

cis Golgi (data not shown). In contrast, the lysosomal cysteine proteinase procathepsin L, immunoprecipitated from the same samples, did not undergo pH-dependent membrane association until the protein reached the trans Golgi (data not shown).

The optimal pH for mannose-phosphate-independent membrane association is somewhat different in vivo, therefore, from that assayed in vitro. There is precedent for such a difference. The receptor for lactotransferrin and the KDEL receptor of the ER salvage pathway have both been shown to have a pH optimum of 5 in vitro,[15,16] yet in vivo these receptors do not function in pH 5 compartments. In contrast, it is well established that the gradual pH drop from the ER to lysosomes regulates the lysosomal biosynthetic pathway by modulating receptor interactions.

It is not clear how the mannose 6-phosphate-independent membrane association of procathepsin D is mediated. The propeptide does not inhibit the pH-dependent membrane association of procathepsin D (data not shown), suggesting that regions of the mature protein are critical for the interaction.

3. ASSOCIATION OF PROCATHEPSIN D WITH PROSAPOSIN

Attempts to cross-link procathepsin D to a specific membrane protein which might mediate mannose 6-phosphate-independent membrane association revealed that in human HepG2 cells procathepsin D associates with prosaposin,[13] the precursor of sphingolipid activator proteins (SAPs) or saposins.[17,18] We have also detected association of procathepsin D with prosaposin in mouse fibroblasts. When saponin-permeabilized wild-type fibroblasts were treated with dithiobis(succinimidylpropionate) (DSP), a 66-kDa protein was detected on immunoprecipitation of procathepsin D from either the soluble (Fig. 4) or membrane fractions (data not shown). The 66-kDa protein was not immunoprecipitated when the cross-linking reagent was omitted prior to immunoprecipitation (Fig. 4). Similarly, the 66-kDa protein was not detected when mouse sap$^-$ fibroblasts, isolated from prosaposin "knockout" mice,[19] were substituted for wild-type fibroblasts (Fig. 4), indicat-

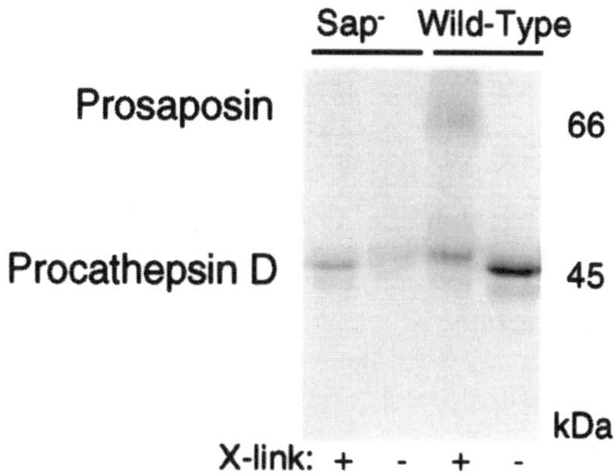

Figure 4. Procathepsin D associates with prosaposin in wild-type primary mouse fibroblasts. Saposin$^-$ (SAP$^-$) and wild-type primary mouse fibroblasts (100-mm culture dish) labeled with [^{35}S]methionine for 15 min and chased for 1 h were permeabilized on ice with phosphate-buffered saline, pH 7.4, containing 0.2% saponin, 5 mM mannose 6-phosphate,. 100 μM 3,4-dichloroisocoumarin and 25 μM E-64. After 1 h, this permeabilization buffer was removed and the solubilized proteins were cross-linked by the addition of dithiobis (succinimidylpropionate) (DSP) to 1 mM, as described by Zhu and Conner.[13] Fresh permeabilization buffer containing 1 mM DSP was added to the cells. Both the cells and the permeabilization buffer containing crosslinker were incubated on ice for 30 min with shaking. To quench the crosslinking reaction, 1 M Tris buffer, pH 7.5, was added to a final concentration of 50 mM and the samples were incubated at room temperature for 30 min. The quenching buffer was discarded and the cells were harvested by scraping into 50 mM Tris buffer, pH 9.0, containing 0.5% SDS, 100 mM NaCl and 2 mM EDTA. Control cells were treated similarly except DSP was omitted. Cathepsin D was immunoprecipitated sequentially from both samples as described[10]. Prior to resolution on an 11% polyacrylamide gel, samples were boiled for 5 min in the presence of 0.1 M DTT to reverse the crosslink. Cathepsin D immunoprecipitated as described previously[10] was resolved on a 10% polyacrylamide gel, visualized by fluorography, and quantitated by densitometry. Procathepsin D isolated from the permeabilization buffer is shown. Approximately 54% of the total procathepsin D was released from the cells by this procedure using a pH 7.4 buffer.

ing, as the molecular mass suggested, that the 66-kDa protein was prosaposin. When prosaposin was immunoprecipitated from cross-linked, permeabilized wild-type fibroblasts, a 46-kDa protein which comigrated with authentic procathepsin D was detected (data not shown). Since prosaposin is a relatively abundant protein, it was important to demonstrate that this interaction was specific. When the cysteine proteinase procathepsin L was immunoprecipitated from the same samples, the 66-kDa protein was not detected (data not shown), suggesting that the interaction was specific for procathepsin D.

The physiological significance of the association of procathepsin D and prosaposin remains unclear. In vivo, prosaposin is cleaved in the lysosome into four molecules which bind to membranes and aid glycosphingolipid degradation by lifting the glycosphingolipids from the plane of the membrane and presenting them for degradation to lysosomal enzymes such as hexosaminidase A.[18,20] Although the membrane association of the saposins appears to require the low pH of the lysosome,[21] prosaposin undergoes mannose 6-phosphate-independent membrane association at neutral pH,[11] as does the procathepsin D--prosaposin complex,[22] raising the possibility that complex formation is required for lysosomal targeting.

Plant aspartic proteinases possess domains which are homologous to mammalian saposins.[23] These inserts have been termed "swaposins" because they have the amino- and carboxy-terminal halves interchanged in sequence order[24] relative to mammalian saposins. The covalent association of the two domains in plant cells suggests that the noncovalent association of procathepsin D and prosaposin detected by cross-linking may serve a similar physiological function in mammalian cells. It has been suggested that the lipid binding of the saposin domain is essential for vacuolar targeting of the plant aspartic proteinases,[23] but at present it is not clear how a lipid-bound peripheral membrane protein would be targeted. Alternatively, it is possible that the association is functionally significant. Precedent for the supramolecular association of enzymes catalyzing related reactions comes from studies of metabolic enzymes[25,26] and from the observation that β-galactosidase and α-neuraminidase, lysosomal enzymes catalyzing the first two steps of ganglioside hydrolysis, form a complex with cathepsin A[27] and *N*-acetylgalactosamine-6-sulfate sulfatase.[28] It appears unlikely that cathepsin D merely activates prosaposin, however; the majority of the cleavages that generate saposins occur after a charged[20], not a hydrophobic, residue.[29]

4. CATHEPSIN D BIOSYNTHESIS IN TRANSFORMED CELLS

Since protein turnover occurs in most cell types, lysosomal enzymes have classically been considered to perform "housekeeping functions" and to be present in lysosomes in all cell types in approximately equal quantities. Recent data indicate, however, that under certain circumstances the biosynthesis of lysosomal enzymes may be regulated and their targeting may be altered.

There have been numerous reports of increased synthesis and secretion of cathepsin D by transformed cells. Human MCF-7 breast cancer cells in culture secrete large amounts of the proenzyme.[30] Cathepsin D activity is elevated in colorectal tissues[31] and in head and neck carcinomas[32] relative to normal tissue. At present, it is not clear how cell transformation increases the synthesis of a specific lysosomal proteinase and alters intracellular targeting to increase its secretion. The hypothesis that activation of secreted lysosomal proteinases which can degrade intact basement membranes[33,34] may correlate with increased metastatic potential of tumors[35–37] remains controversial.[38]

In transformed cells, pH-dependent, mannose 6-phosphate-independent membrane association of procathepsin D is reduced and secretion of the proenzyme is increased. While approximately 80% of the procathepsin D was bound to membranes as assayed at pH 5 in normal human fibroblasts (Fig. 3), 80% of the enzyme was soluble when microsomes prepared from BT20 breast tumor cells were lysed at pH 5 (Fig. 5). When the cells were treated with brefeldin A (BFA), a fungal metabolite which induces fusion of the ER and Golgi membranes,[39] transport to lysosomes was blocked, as seen by the absence of mature forms of cathepsin D, the resulting increased quantity of the proenzyme, and the reduction in secretion of the proenzyme to the cell culture medium (Fig. 3). Yet the percent of procathepsin D which was membrane-associated did not increase (Fig. 3). Preventing the exit of newly synthesized proteins from the fused membrane system should presumably increase the contact of procathepsin D with the protein(s) which mediates membrane association, but increased membrane association of procathepsin D was not detected in microsomes derived from BFA-treated cells.

It is not known whether the reduced membrane association of procathepsin D in transformed cells causes the increased secretion of the proenzyme or is merely the result of other alterations of the cells. In transformed cells either quantities of a receptor which mediates membrane association are limiting due to increased synthesis of the proenzyme, synthesis of the receptor is down-regulated, or the receptor is modified so that it can no longer interact with procathepsin D.

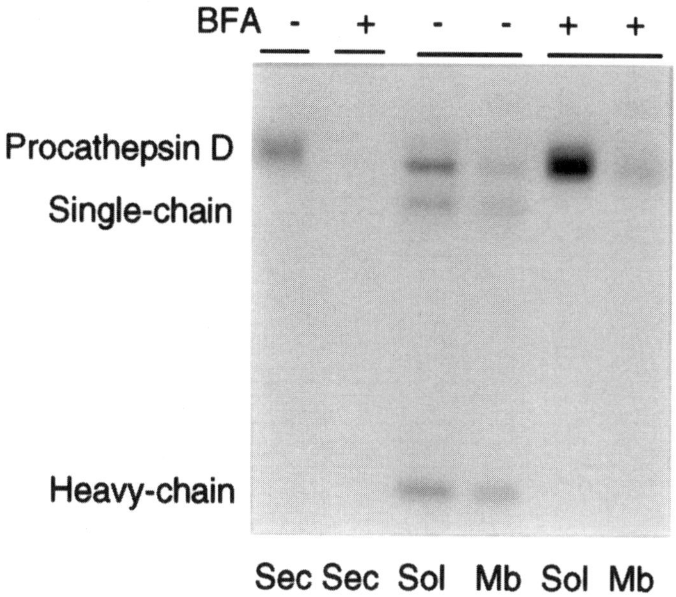

Figure 5. The mannose 6-phosphate-independent membrane association of procathepsin D is reduced in human breast tumor cells. Human BT20 breast tumor cells cultured in a 100-mm dish were labeled continuously with [^{35}S]methionine for 4 h. When specified, brefeldin A (BFA) was added (5 μg/ml culture medium) during the starvation and labeling periods. Microsomes were prepared and lysed at pH 5 as described in Fig. 3. Cathepsin D, immunoprecipitated as described previously[10] from the culture medium (Sec), soluble (Sol), and membrane fractions (Mb) was resolved on a 10% polyacrylamide gel, visualized by fluorography, and quantitated by densitometry.

5. TAXOL INCREASES PROCATHEPSIN D SECRETION

Taxol, a drug which stabilizes microtubules,[40] has been tested clinically as an antitumor agent,[41,42] To determine if the effectiveness of taxol might result from the reduced secretion of proteinases which may aid tumor cell metastasis by degrading extracellular matrix, we treated ovarian tumor cells with taxol. Contrary to our expectations, taxol increased secretion of procathepsin D (Fig. 6). Consistent with this, decreased targeting of cathepsin D to lysosomes was detected, as revealed by the decreased amount of cathepsin D heavy-chain in cells. Similar results were obtained with 4 different ovarian cell lines and when microtubules were depolymerized with colchicine (data not shown). Thus microtubules are apparently required for transport to lysosomes but not for constitutive transport to the plasma membrane. The data suggest that the clinical effectiveness of taxol is related to some event distinct from its modification of procathepsin D targeting. It remains to be determined how, or even if, secreted lysosomal proenzymes are activated to a significant extent extracellularly.

6. ALTERNATIVE TARGETING SITES

There is increasing evidence that in certain cell types lysosomal enzymes are targeted to endosomes in addition to lysosomes and the plasma membrane for secretion, but

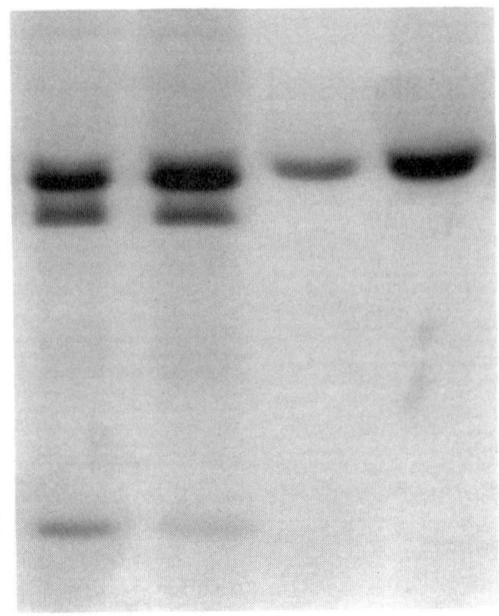

Figure 6. Taxol treatment increases the secretion of procathepsin D. Human ovarian carcinoma cells (OVCA 194), which do not secrete interleukin-8 in response to taxol,[50] were incubated in culture medium with (+) or without (–) 30 μM taxol for 3 h. After the cells were starved for methionine for 30 min in the presence of taxol, 300 μCi of [^{35}S]methionine was added per ml of this starvation medium. After 3 h the cells were harvested in 50 mM Tris, pH 9.0, containing 0.5% SDS, 100 mM NaCl, and 2 mM EDTA and the protein concentration was determined by a Bradford assay.[51] Procathepsin D immunoprecipitated[10] from equal amounts of cellular protein and respective culture medium was resolved on a 12% polyacrylamide gel and visualized by fluorography.

it is not clear whether this is due to a specific targeting mechanism or whether the detected enzymes are merely passing through the endosomes on their way to other sites. Clearly proteolytic activity occurs at cellular sites other than the lysosomes. For example, the transferrin receptor undergoes proteolysis in an endosomal compartment,[43] as does ricin A.[44] Glucagon undergoes proteolysis mediated by membrane-associated cathepsin B and cathepsin D within hepatic endosomes.[14] Proteinases are present in the newly characterized antigen-processing endosomes,[45,46] immature secretory granules,[47] and can reach autophagic vacuoles by an MPR-independent route.[48] They have also been localized to endosomal-like compartments where parasites replicate.[49] In at least two of these cases,[9,14] the enzymes have been shown to be membrane-associated, but the protein(s) mediating this targeting and membrane association have not been identified. Mannose 6-phosphate independent membrane association of procathepsin D may be required for transport of the aspartic proteinase to cellular sites other than the lysosome.

ACKNOWLEDGMENTS

We thank Dr. Kuni Suzuki, The University of North Carolina at Chapel Hill, for providing primary wild-type and sap$^-$ mouse fibroblasts and Dr. Bonnie Sloane, Wayne State University, for sharing MCF10A cells. Dr. Michael Griswold, Washington State University, kindly provided antiserum to rat sertoli cell sulfated glycoprotein 1 which cross-reacts with mouse prosaposin.

REFERENCES

1. Erickson, A. H.; Blobel, G. *J. Biol. Chem.* **1979**, *254*, 11771–11774.
2. Erickson, A. H.; Walter, P.; Blobel, G. *Biochem. Biophys. Res. Commun.* **1983**, *115*, 275–280.
3. Erickson, A. H.; Conner, G. E.; Blobel, G. *J. Biol. Chem.* **1981**, *256*, 11224–11231.
4. Yonezawa, S.; Takahashi, T.; Wang, X.-J.; Wong, N. S.; Hartsuck, J. A.; Tang, J. *J. Biol. Chem.* **1988**, *26*, 16504–16511.
5. Erickson, A. H.; Blobel, G. *Biochemistry* **1983**, *22*, 5201–5205.
6. Takahashi, T.; Schmidt, P. G.; Tang, J. *J. Biol. Chem.* **1983**, *258*, 2819–2830.
7. Kornfeld, S.; Mellman, I. *Ann. Rev. Cell Biol.* **1989**, *5*, 483–525.
8. Varki, A.; Kornfeld, S. *J. Biol. Chem.* **1980**, *255*, 10847–10858.
9. Diment, S.; Leech, M. S.; Stahl, P. D. *J. Biol. Chem.* **1988**, *263*, 6901–6907.
10. McIntyre, G. F.; Erickson, A. H. *J. Biol. Chem.* **1991**, *266*, 15438–15445.
11. Rijnboutt, S.; Aerts, H. M.; Geuze, H. J.; Tager, J. M.; Strous, G. J. *J. Biol. Chem.* **1991**, *266*, 4862–4868.
12. Rijnboutt, S.; Kal, A. J.; Geuze, H. J.; Aerts, H.; Strous, G. J. *J. Biol. Chem.* **1991**, *266*, 23586–23592.
13. Zhu, Y.; Conner, G. E. *J. Biol. Chem.* **1994**, *269*, 3346–3851.
14. Authier, F.; Mort, J. S.; Bell, A. W.; Posner, B. I.; Bergeron, J. J. M. *J. Biol. Chem.* **1995**, *270*, 15798–15807.
15. Hu, W.-L.; Mazurier, J.; Sawatzki, G.; Montreuil, J.; Spik, G. *Biochem. J.* **1988**, *249*, 435–441.
16. Wilson, D. W., M. J. Lewis, and H. R. B. Pelham *J. Biol. Chem.* **1993**, *268*, 7465–7468.
17. O'Brien, J. S.; Kishimoto, Y. *FASEB J.* **1991**, *5*, 301–308.
18. Sandhoff, K.; Klein, A. *FEBS Lett.* **1994**, *346*, 103–107.
19. Fujita, N.; Suzuki, K.; Vanier, M. T.; Popko, B.; Maeda, N.; Klein, A.; Henseler, M.; Sandhoff, K.; Nakayasu, H.; Suzuki, K. *Hum. Mol. Genet.* **1996**, *5*, 711–725.
20. Kishimoto, Y.; Hiraiwa, M.; O'Brien, J. S. *J. Lipid Res.* **1992**, *33*, 1255–1267.
21. Vaccaro, A. M.; Ciaffoni, F.; Tatti, M.; Salvioli, R.; Barca, A.; Tognozzi, D.; Scerch, C. *J. Biol. Chem.* **1995**, *270*, 30576–30580.
22. Zhu, Y. X.; Conner, G. E. *Mol. Biol. Cell* **1994**, *5*, 114a.
23. Guruprasad, K.; Tormokangas, K.; Kervinen, J.; Blundell, T. L. *FEBS Lett.* **1994**, *352*,
24. Ponting, C. P.; Russell, R. B. *Trends Biochem. Sci.* **1995**, *20*, 179–180.
25. Robinson, J.; Srere, P. A. *J. Biol. Chem.* **1985**, *260*, 10800–10805.
26. Srere, P. A. *Annu. Rev. Biochem.* **1987**, *56*, 89–124.
27. Morreau, H.; Galjart, N. J.; Willemsen, R.; Gillemans, N.; Zhou, X. Y.; d'Azzo, A. *J. Biol. Chem.* **1992**, *267*, 17949–17956.
28. Pshezhetsky, A. V.; Potier, M. *J. Biol. Chem.* **1996**, *271*, 28359–28365.
29. van Noort, J. M.; van der Drift, A. C. M. *J. Biol. Chem.* **1989**, *264*, 14159–14164.
30. Rochefort, H.; Capony, F.; Garcia, M.; Cavailles, V.; Freiss, G.; Chambon, M.; Morisset, M.; Vignon, F. *J. Cell. Biochem.* **1987**, *35*, 17–29.
31. Adenis, A.; Huet, G.; Zerimech, F.; Hecquet, B.; Balduyck, M.; Peyrat, J. P. *Cancer Letters* **1995**, *96*, 267–275.
32. Kos, J.; Smid, A.; Krasovec, M.; Svetic, B.; Lenarcic, B.; Vrhovec, I.; Skrk, J.; Turk, V. *Biol. Chem. Hoppe-Seyler* **1995**, *376*, 401–405.
33. Sloane, B. F.; Rozhin, J.; Hatfield, J. S.; Crissman, J. D.; Honn, K. V. *Exp. Cell Biol.* **1987**, *55*, 209–224.
34. Briozzo, P.; Morisset, M.; Capony, F.; Rougeot, C.; Rochefort, H. *Can. Res.* **1988**, *48*, 3688–3692.
35. Rochefort, H.; Capony, F.; Garcia, M. *Cancer Cells* **1990**, *2*, 383–388.
36. Rozhin, J.; Wade, R. L.; Honn, K. V.; Sloane, B. F. *Biochem. Biophys. Res. Commun.* **1989**, *164*, 556–561.
37. Denhardt, D. T.; Greenberg, A. H.; Egan, S. E.; Hamilton, R. T.; Wright, J. A. *Oncogene* **1987**, *2*, 55–59.
38. Johnson, M. D.; Torri, J. A.; Lippman, M. E.; Dickson, R. B. *Cancer Res.* **1993**, *53*, 873–877.
39. Klausner, R. D.; Donaldson, J. G.; Lippincott-Schwartz, J. *J. Cell Biol.* **1992**, *116*, 1071–1080.
40. Schiff, P. B.; Horwitz, S. *Proc. Natl. Acad. Sci. USA* **1980**, *77*, 1561–1565.
41. McGuire, W. P.; Rowinsky, E. K.; Rosenshein, N. B.; Grumbine, F. C.; Ettinger, D. S.; Armstrong, D. K.; Donehower, R. C. *Ann. Intern. Med.* **1989**, *111*, 273–279.
42. Rowinsky, E. K.; Onetto, N.; Canetta, R. M.; Arbuck, S. G. *Sem. Oncol.* **1992**, *19*, 646–662.
43. Rutledge, E. A.; Green, F. A.; Enns, C. A. *J. Biol. Chem.* **1994**, *269*, 31864–31868.

44. Holst, B.; Bruun, A. W.; Kielland-Brandt, M. C.; Winther, J. R. *EMBO J.* **1996**, *15*, 3538–3546.
45. Amigorena, S.; Drake, J. R.; Webster, P.; Mellman, I. *Nature* **1994**, *369*, 113–120.
46. Tulp, A.; Verwoerd, D.; Dobberstein, B.; Ploegh, H. L.; Pieters, J. *Nature* **1994**, *369*, 120–126.
47. Kuliawat, R.; Arvan, P. *J. Cell Biol.* **1994**, *126*, 77–86.
48. Punnonen, E.-L.; Autio, S.; Kaija, H.; Reunanen, H. *Eur. J. Cell Biol.* **1993**, *61*, 54–66.
49. Coulombe, R.; Grochulski, P.; Sivaraman, J.; Menard, R.; Mort, J. S.; Cygler, M. *EMBO J.* **1996**, *15*, 5492–5503.
50. Lee, L.-F.; Schuerer-Maly, C.-C.; Lofquist, A. K.; van Haaften-Day, C.; Ting, J. P.-Y.; White, C. M.; Martin, B. K.; Haskill, J. S. *Cancer Res.* **1996**, *86*, 1303–1308.
51. Bradford, M. M. *Anal. Biochem.* **1976**, *72*, 248–254.

22

AN ASPARTIC PROTEINASE EXPRESSED IN THE EQUINE PLACENTA

Jonathan Green,[2] Sancai Xie,[3] Xinsheng Gan,[3] and R. Michael Roberts[1,2,3]

[1]Department of Animal Sciences
[2]Department of Biochemistry
[3]Department of Veterinary Pathobiology
University of Missouri
Columbia, Missouri 65211

INTRODUCTION

The pregnancy-associated glycoproteins (PAG) are a large and diverse group of proteins produced by the placenta of cattle, sheep, pigs, and presumably most other species within the *Artiodactyla* order.[1,2] PAG are members of the aspartic proteinase gene family having greatest sequence identity (~50%) with pepsinogens.[1] However, analysis of the putative amino acid sequence of several PAG cDNA revealed that most of these molecules are probably not proteinases due to mutations in key residues involved directly or indirectly in the catalytic mechanism.[3] Neither native or recombinant PAG have detectable enzymatic activity in proteinase assays with ^{14}C-hemoglobin as substrate. However, these changes within the active site do not interfere with their ability to bind peptide ligands since both native and recombinant PAG can be isolated by affinity chromatography with pepstatin agarose (JG, SX, RMR; unpublished).

This study was aimed at determining whether the PAG genes are widely distributed in placental mammals, possibly indicative of a fundamental contribution to placental function. To identify PAG-related transcripts in placentas of animals outside of the *Artiodactyla* order (namely in the *Perissodactyla* which are believed to have diverged from the even-toed ungulates ~70 million years ago[4]), an equine d25 placental library was screened for the presence of PAG. A PAG-like transcript (equine PAG or ePAG) was identified that had 50–60% amino acid sequence identity with members of the PAG family. Once a horse PAG-like molecule was identified, the search was extended to the cat, a carnivore whose ancestors diverged from the *Perissodactyla* and *Artiodactyla* ~80 million years ago.[4]

Aspartic Proteinases, edited by James.
Plenum Press, New York, 1998.

MATERIALS AND METHODS

cDNA Cloning and Sequencing

An equine d25 placental cDNA library (a gift from K. McDowell, University of Kentucky) was screened with a probe mixture consisting of ^{32}P-labeled PAG cDNA clones from cattle, sheep and pig. Hybridization and washing of the filters was carried out under standard conditions.[5] Phage positives were *in vivo* excised and sequenced by the dideoxy-nucleotide chain termination method.[5] One of the clones, ep12.1 (Genebank accession # L38511), was sequenced in both directions.

RNA Isolation and Northern Analysis

Early stage equine conceptus tissue (d25-d32) was collected from pregnant mares and the extraembryonic membranes were separated from the embryo. The tissues were then frozen on dry ice and stored at –70°C until RNA isolation. The tissues were extracted by the guanidinium thiocyanate-phenol-chloroform method.[6] Total RNA was separated by electrophoresis on 1.1% formaldehyde-agarose gels and transferred onto nylon membranes. The ePAG mRNA was detected by hybridization with ^{32}P-labeled full length ep12.1 cDNA under standard conditions (42°C in 50% formamide, 5x SSC, 0.5% SDS, 5X Denhardts, 0.1mg/ml herring sperm DNA).

RT-PCR Cloning from Cat Placenta

Total RNA isolated from a d40 cat placenta was reverse transcribed with AMV reverse transcriptase (Promega) and the resulting cDNA was used as template for PCR with oligonucleotides directed towards conserved regions within various PAG cDNA. The resulting PCR product was subcloned into the TA vector (Invitrogen) and sequenced. The 3′ end of the cDNA was obtained by RT-PCR by using an oligo dT primer in conjunction with an internal primer specific to the cat transcript. The 5′ portion was obtained with an internal cat PAG-specific oligonucleotide and an oligonucleotide directed toward a highly conserved portion of the 5′ untranslated region of most PAG cDNA.

Genomic Southern Blotting

Bovine genomic DNA (5 μg) was digested with various restriction endonucleases and resolved on a 0.8% (w/v) agarose gel by electrophoresis and blotted to nylon membrane. A ~350 bp genomic fragment encompassing exon7, intron 7, and exon 8 of the equine PAG gene was labeled with [^{32}P-α]dATP for use as probe. A second probe from the bovine PAG1 gene (portion of intron 6, exon 7, intron 7, exon 8, and a portion of intron 8) was generated for probing the same blot. Hybridization was carried out under standard conditions.[5]

RESULTS AND DISCUSSION

ePAG cDNA

The cDNA for ep12.1 was 1352 bp and contained an open reading frame of 388 codons encoding a polypeptide of predicted M_r 42,897. From the inferred amino acid sequence, ePAG is a secretory protein and has a single putative N-linked glycosylation site

Table 1. Sequence identity between ePAG and other aspartic proteinases

	% Amino acid identity
Pepsinogen F (rabbit)	68.5
PAG 2 (pig)	57.2
Pepsinogen A (pig)	54.9
PAG 2 (cattle)	54.1
PAG 1 (cattle)	51.5
Cathepsin E (human)	46.5
Cathepsin D (human)	42.2
Renin (mouse)	35.0

(Asn–Asn–Ser). Equine PAG has greatest identity (68.5% at the amino acid level) with rabbit pepsinogen F, a gastric proteinase expressed for a short time in the neonatal rabbit.[7] The sequence identity between ePAG and other members of the pepsinogen family is around 55%, whereas its identity with members of the PAG family ranges from 52–57% (Table 1). Sequence conservation is greatest within the signal sequence and in the regions immediately adjacent to the catalytic aspartic acids. Equine PAG possesses the Asp–Thr–Gly–Ser sequence in the NH2-terminal domain and Asp–Thr–Gly–Thr sequence in the COOH-terminal domain, characteristic of most mammalian aspartic proteinases.[8]

As stated in the introduction, most of the PAG cloned so far have multiple mutations within each molecule that likely render them incapable of functioning as proteinases. In contrast, analysis of the sequence for ePAG revealed that this protein is likely to have proteolytic activity.[3] This was confirmed by using ePAG in proteinase assays with denatured ^{14}C-hemoglobin as substrate (data not shown). EPAG has an acidic pH optimum (pH 4.5) although activity can be detected under neutral conditions provided the pro-peptide is removed prior to the start of the assay. The enzymatic activity of ePAG can be inhibited with pepstatin A.

Expression of ePAG mRNA

To screen the cDNA library, ten thousand plaques were plated. After hybridization and washing of the filters it was calculated that 0.5% of the plaques hybridized to the probe, indicating that the ePAG transcript is expressed at high levels in the early stage placenta. Northern blot analysis revealed that the 1.7 kb ePAG transcript is highly expressed in pre-implantation conceptus tissue and that the expression is restricted to the extraembryonic membranes. No signal could be detected in RNA extracted from the embryo (Fig. 1).

ePAG-like Genes within Other Species

Recently, a clone similar to ePAG was obtained by RT-PCR upon RNA isolated from a d40 cat placenta. The full length transcript contained an open reading frame of 389 codons encoding a protein of predicted Mr 43,035. This cat PAG cDNA shared 78.5% nucleotide sequence identity (70.5% amino acid identity) with ePAG, and like ePAG, it appears to be a proteinase (XG, RMR; unpublished).

An ePAG-like gene also appears to be present within the bovine genome. When Southern blotting was performed with a short ePAG genomic fragment as probe, only a

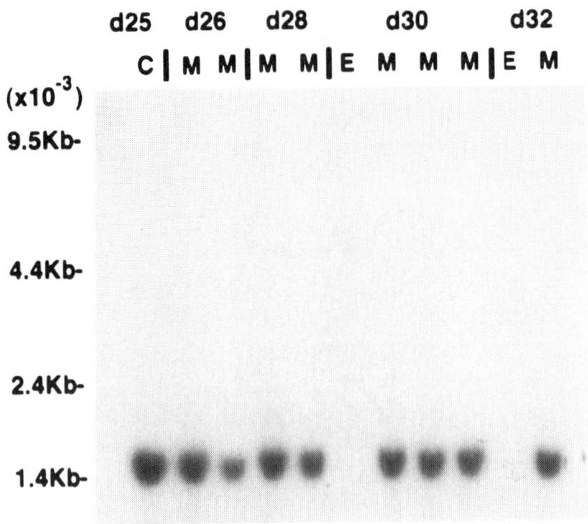

Figure 1. Northern blot of RNA isolated from early stage equine conceptuses. C: whole conceptus (embryo + extraembryonic membranes), E: embryo, M: extraembryonic membranes. Total RNA was extracted from the indicated tissues and resolved on formaldehyde-agarose gels, transferred to nylon membranes, and probed with the full length [^{32}P]ep12.1 cDNA. The ePAG transcript is highly expressed in extraembryonic membranes but is not detectable in early stage embryos.

single band hybridized strongly to the probe, indicating that the bovine genome has a gene similar to ePAG (Fig. 2). However, no cDNA with high sequence identity to ePAG has been cloned from bovine placenta, possibly indicating that the gene is expressed at very low levels compared to the PAG. Alternatively, the bovine homolog of ePAG may be a psuedogene. For comparison, also shown in Figure 2 is the same blot probed with a bovine PAG-1 genomic fragment. The multiple banding pattern is indicative of the many PAG genes present in the bovine genome.

SUMMARY

This manuscript describes the cloning of a novel aspartic proteinase expressed in the placenta of the horse (order *Perrisodactyla*). Evidence for similar genes in the cat

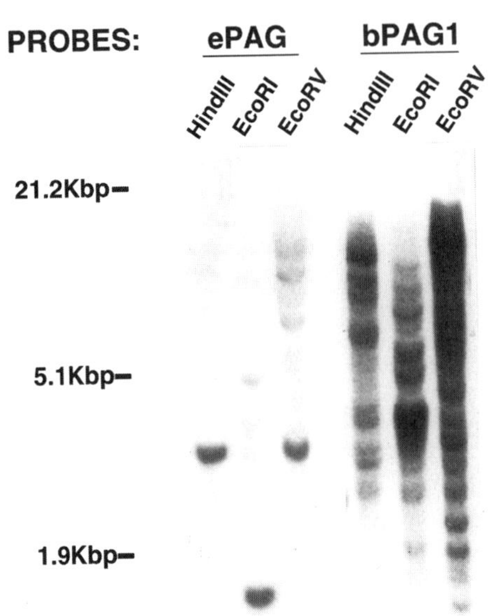

Figure 2. Bovine genomic Southern blot. Bovine genomic DNA (5 µg) was digested with the indicated enzymes, resolved on a 0.8% agarose gel, and blotted to nylon membrane. The blot was then probed sequentially, under the same hybridization conditions, with short genomic fragments (ePAG and bovPAG1) as described in the material and methods. Probing with the ePAG genomic fragment revealed the presence of a single strongly hybridizing band. Probing with the bovPAG1 fragment produced several hybridizing bands indicative of the many PAG1-related genes present within the bovine genome.

(*Carnivora*) and ruminants (*Artiodactyla*), indicates that these molecules have been conserved within widely divergent species with distinct types of placentation.

Since ePAG is produced by the outer cell layer (trophoblast) of the placenta, it can tentatively be grouped with the pregnancy-associated glycoproteins (PAG) of cattle, sheep, and pig.[1,2] The high sequence identity that ePAG shares with pepsinogens as well as the PAG, indicates that ePAG may be the evolutionary bridge that links these two groups of aspartic proteinases.

ACKNOWLEDGMENTS

This work is supported by NIH grant HD29483 to R.M.R. The authors thank Adam Newman for assistance with sequencing the equine cDNA clones.

REFERENCES

1. Xie S, Low BG, Kramer KK, Nagel FJ, Anthony RV, Zoli AP, Beckers J-F, Roberts RM Identification of the major pregnancy-specific antigens of cattle and sheep as inactive members of the aspartic proteinase family. Proc Natl Acad Sci USA 1991; 88:10247–51.
2. Szafranska B, Xie S, Green J, Roberts RM Porcine pregnancy-associated glycoproteins: New members of the aspartic proteinase gene family expressed in trophectoderm. Biol Reprod 1995; 53:21–28.
3. Guruprasad K, Blundell TL, Xie S, Green J, Szafranska B, Nagel RJ, McDowell K, Baker CB, Roberts RM. Comparative modeling and analysis of amino acid substitutions suggests that the family of pregnancy-associated glycoproteins includes both active and inactive aspartic proteinases. Protein Engineering 1996; 9:849–856.
4. Novacek MJ. Mammalian phylogeny: shaking the tree. Nature 1992; 356:121–125.
5. Sambrook J, Fritsch EF, Maniatis T (Eds). Molecular Cloning: A laboratory manual. 2nd Edition, Cold Spring Harbor Press. 1989.
6. Chomczynski P, Sacchi N. Single-step method of RNA isolation by acid guanidinium thiocyanate-phenol-chloroform extraction. Anal Biochem 1987; 162:156–159.
7. Kageyama T, Tanabe K, Koiwai O. Structure and development of rabbit pepsinogens. J Biol Chem 1990; 265:17031–17038.
8. Szecsi PB. The aspartic proteinases. Scand J Clin Lab Invest 1992; 52(Suppl 210):5–22.

23

PROTEIN ENGINEERING ASPARTIC PROTEINASES

Site-Directed Mutagenesis, Biochemical Characterisation, and X-Ray Analysis of Chymosins with Substituted Single Amino Acid Substitutions and Loop Replacements

Armando Albert,[1,2] Tom L. Blundell,[1,2] V. Dhanaraj,[1,2,*]
Luis Enrique Donate,[1,2] Matthew Groves,[1] Kunchur Guruprasad,[1,2]
Phil G. Nugent,[1] P. Orprayoon,[1] Jim E. Pitts,[1] Stephen Rufino,[1]
N. Srinivasan,[1,2] Mark Williams,[1,2] and Julie Wilsher[1]

[1]Department of Crystallography
Birkbeck College, University of London
Malet Street, London WC1E 7HX, United Kingdom
[2]Department of Biochemistry
University of Cambridge
Tennis Court Road, Cambridge CB2 1QW, United Kingdom

1. INTRODUCTION

Aspartic proteinases (1) are presently used in commercial milk, soya and cocoa processing, but there may be other applications in the longer term for designed biocatalysts for the food industry if the thermal stability, pH dependence, size and specificity can be engineered. With this objective in mind we have carried out protein engineering programmes to understand the roles of both individual amino acids and those of loops in aspartic proteinases. We have used chymosin as our model system.

The aspartic proteinases are an excellent choice for a protein engineering study as they have been well characterised in terms of three-dimensional structure, catalytic activity and specificity. They are of particular interest as they have evolved to carry out a wide range of functions at different pHs and with different specificities in fungi, plants and animals. Recombinant chymosin is used commercially for cheese processing. It has already been the subject of site-directed mutagenesis (2,3). Here we describe the design, expres-

* To whom correspondence should be addressed.

Aspartic Proteinases, edited by James.
Plenum Press, New York, 1998.

sion, purification, crystallisation, kinetic and structural analysis of several such amino acid and loop-exchange mutants.

2. DESIGN

Efficient design requires knowledge of the three-dimensional structures defined by X-ray analysis or NMR at high resolution. For the aspartic proteinases we are fortunate to have three-dimensional structures not only for chymosin (4,5) but also for other mammalian, fungal and retroviral aspartic proteinases. A significant deficiency at the start of this study was the absence of a crystal structure for a chymosin inhibitor complex. This was rectified by the determination at 2.3Å resolution of the structure of chymosin complexed with a reduced bond inhibitor (6). In the crystal structure of uncomplexed native chymosin a β-hairpin at the active site, known as "the flap", is rearranged when compared with other aspartic proteinases. This conformation would prevent the normal mode of binding of substrates/inhibitors. Our analysis of the complex showed that the inhibitor adopts the usual extended β-sheet conformation within the active site and the flap region closes over the active site cleft in a way that closely resembles other aspartic proteinase-inhibitor complexes. We conclude that the usual position and conformation of the flap are available to native chymosin.

For our design work we have developed an approach to the prediction of mutant stability using knowledge of amino acid replacements that are tolerated within families of homologous proteins of known three-dimensional structure (7). Amino acid variations in families of homologous proteins were converted to propensity and substitution tables in order to provide quantitative information about the existence of an amino acid in a structural environment and the probability of replacement by any other amino acid. These were used to estimate the difference in pseudo-free energy between a mutant and its wild type of known three-dimensional structure. The method was tested using the high resolution structures for T4 lysozyme and 159 site-specific mutants, where the calculated free energy differences are correlated with experimentally observed differences in free energy and melting temperature. Blind tests on a combined set of 83 staphylococcal nuclease and 68 barnase mutants showed a correlation of 0.80 in the predicted stability changes with experimental thermodynamic data. Approximately 86% of the predictions were correctly classified as destabilising or stabilising. We applied this approach to estimating all possible single-site mutations in chymosin.

Using this and more subjective approaches, we selected mutants in the vicinity of S1 and S3, for example I120H and A115T. These were synthesised by the Prague group of Dr J. Sedlacek to extend our earlier work with them on substitutions at residue 111 (3). Mutants at the periphery of the active site cleft such as G243D and K220E were synthesised by ourselves. We have also collaborated on mutants in the flap, such as chymosin Y75F and Y75N for synthesis by Prof. T. Beppu of the Univeristy of Tokyo (8). Although these were originally introduced into chymosin, equivalent mutants for mucorpepsin proved more amenable to production of large quantities of material that crystallised (V. Dhanaraj unpublished results).

For the loop mutants the basis of the design was provided by the three-dimensional structures of many aspartic proteinases. These provided information about the ways in which nature has varied loop size while retaining function. Beyond this the retroviral proteinases provide a precedent for size reduction; nature has designed a small dimeric enzyme with subunits each of 99 amino acids (albeit with weak catalytic activity) comparable to lobes of pepsin (~164 amino acids). The challenge is to understand the steps that should be taken to make such radical redesigns in the laboratory.

We began by identifying the most variable loops in the structures of single-chain, pepsin-like aspartic proteinases, by comparing hydrogen-bonding diagrams, superposing their structures using MNYFIT (9) and producing structure-based alignments using COMPARER (10) (computer programs written in our research group). Eight variable loops were identified: D57-K60, M155-E162, I199-V202, Q239-E244, I262-M266, Q277-F281, E289-W299, C206-C210 (pepsin numbering). We also required procedures that would enable us to select a loop connecting any two regions.

For this we first modified our procedures, encoded in COMPOSER (11), so that loops could be selected not only on the geometry of the "framework" but also on sequence patterns. This was achieved by using propensity and substitution tables to generate a template or profile for each loop in the database (12). Fragments, which have been selected on a geometric basis, were ranked according to the sequence to be modelled. We have also developed a loop prediction procedure, SLoop, to identify the conformational class of the loop, rather than to select a specific loop from a set of fragments extracted from known structures or generated *ab initio*. To achieve this, we analysed all irregular regions connecting secondary structures in a database of 2024 loops of length between one and eight residues, and developed algorithms for recognising the appropriate conformations (13,14). Loops were grouped according to their length and the type of their bounding secondary structures. From the pairwise superposition and clustering of the loops in each group, 161 loop conformational classes corresponding to 63% of all loops, were identified.

We selected loops using three procedures: initially from homologues, where sequence and conformation were available; then from the procedure of Topham et al. (12) where we proposed sequences on the basis of loops selected geometrically; and later using SLoop (14), where loop conformational classes and sequences compatible with the framework secondary structures were selected from the database.

All loops were modelled using COMPOSER, energy minimised and checked using computer graphics. Often the sequences needed modification on the basis of the model, and most loop changes required redesign in adjacent regions. It is clearly easier to redesign a protein by growing it rather than reducing its size! We focused on the region 155–162; this has fewer interactions with the main body of the protein than other loops, it influences the substrate only outside the catalytic cleft and it shows considerable variation amongst the aspartic proteinases.

3. MUTAGENESIS, EXPRESSION, AND CHARACTERISATION OF MUTANTS

The expression of recombinant chymosin was initially investigated using *E. coli* transformed with the chymosin expression vector pCT70, which contains the coding sequence for bovine prochymosin B under the control of the Trp promoter/operator (kindly provided by CellTech Ltd; ref. 15). Prochymosin was expressed as insoluble inclusion bodies, and needed solubilisation, refolding and activation to yield active enzyme before functional and structural studies could be instigated. Although the yield of correctly folded and active chymosin mutants by this route was always low, several point mutant chymosins have been constructed and characterised (2,16,17).

The low yield of mutant chymosin isolated using this system led us to investigate the use of filamentous fungi for subsequent expression work. Initial expression work with

Trichoderma reesei employed strain RutC 30 and the expression vector pAMH104 (both kindly provided by Dr. M. Penttila, Biotechnical Laboratory, VTT, Espoo, Finland; ref. 18) which utilises the strong cellobiohydrolase I promoter/operator and signal sequence to drive the expression of soluble bovine chymosin A and its secretion into the culture medium. Removal of the cbh1 signal sequence on secreted prochymosin occurs co-translationally, while removal of the prochymosin pro-peptide occurs post-translationally, as the enzyme accumulates in the culture medium and the culture medium pH falls to 3.5 during fermentation. The amdS sequence, which codes for the enzyme acetamidase, allows for the selection of recombinant *Trichoderma* protoplasts transformed with the modified chymosin gene, by growth in the presence of acetamide as the sole nitrogen source. The prochymosin A coding sequence in this vector was substituted with the coding sequence for the B isoenzyme carrying the mutation of interest, prior to the generation of recombinant *Trichoderma* protoplasts. Mutations in the prochymosin B gene were introduced by site-directed mutagenesis of the prochymosin B sequence cloned into M13, using the Oligonucleotide-Directed Mutagenesis System. *Trichoderma* strain RutC 30 does not secrete Trichoderma proteinase under the standardised fermentation conditions employed by us.

We have used *Trichoderma reesei* to express a number of chymosin loop mutants, where the 155–162 loop region has been replaced with the corresponding loop from other aspartic proteinases (19). Attempts were made to improve the yield, firstly, by employing a modified expression medium including spent grain, and secondly, by employing the fusion vector pCBH104 (VTT, Finland). Unfortunately, neither approach was successful in generating increased yields of chymosin mutants. Although the inclusion of spent grain in the medium did indeed slightly increase the amount of chymosin enzyme secreted during fermentation, it also induced the expression and secretion of trichodermapepsin.

In order to overcome these problems, the alternative filamentous fungus *Aspergillus niger* was investigated. We employed the strain AB 1–13 (kindly provided by TNO Medical Biology Laboratory, The Netherlands), which is deficient in a number of important proteolytic enzymes, including aspergillopepsin, and has been used for the heterologous expression of a number of different proteins, including chymosin. The vector employed, pUCpyrGRG3 (kindly supplied by Genecor, USA), utilises the strong glucoamylase A (glaA) promoter/operator and the prochymosin signal sequence to drive the expression of soluble bovine chymosin B and its secretion into the culture medium. As with expression in *Trichoderma,* the removal of the secretion signal sequence occurs co-translationally, while removal of the prochymosin pro-peptide occurs post-translationally in the culture medium during fermentation. The plasmid also contains the pyr4 sequence which codes for the enzyme orotidine-5′-phosphate decarboxylase, and allows for the selection of recombinant pyrG-*Aspergillus* protoplasts by growth in the absence of uridine. Mutations in the preprochymosin B gene were introduced by site-directed mutagenesis of this sequence cloned into M13, using the Oligonucleotide-Directed Mutagenesis System (Amersham). We have generated recombinant wild type chymosin B, and mutants at the 155–162 loop, where some have loops replaced by the equivalent regions of other aspartic proteinases, for example of endothiapepsin, penicillopepsin and trichodermapepsin, and another is a loop with sequence DDG. However, enzymes with severely truncated loops rarely expressed well and often not at all, presumably due to their decreased stability.

Kinetic values for chymosin mutant enzymes and wild type enzyme were determined using two synthetic peptides, the first based on residues 98–112 of κ-casein and the second a shorter peptide Leu-Ser-Phe(NO_2)-Nle-Ala-Leu-OMe. Kinetic measurements were performed in 0.1M acetate buffer, pH5.6, 1mM $CaCl_2$, 30°C and the change of absorbance at 305nm was followed. Results for the point mutants are shown in Figures 1 and 2.

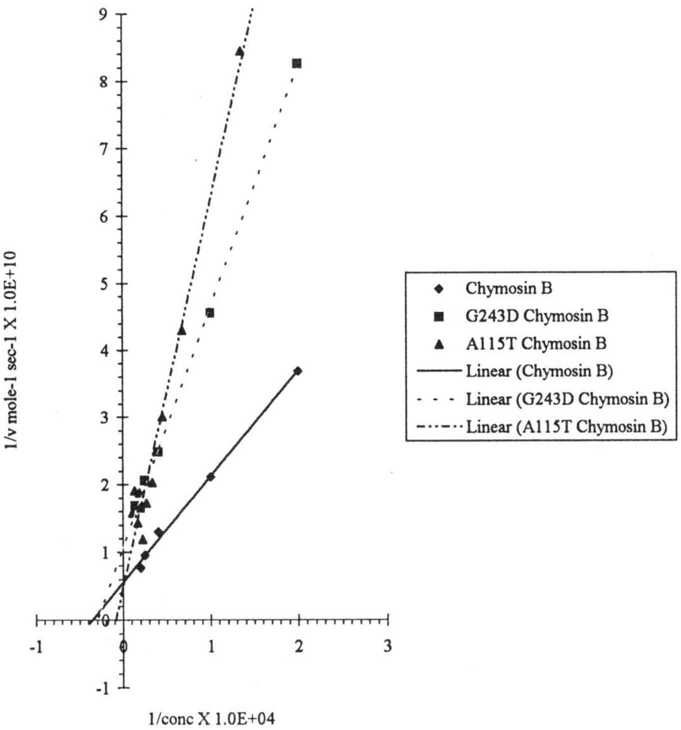

Figure 1. Kinetics of point mutants A115T and G243D with a hexapeptide substrate.

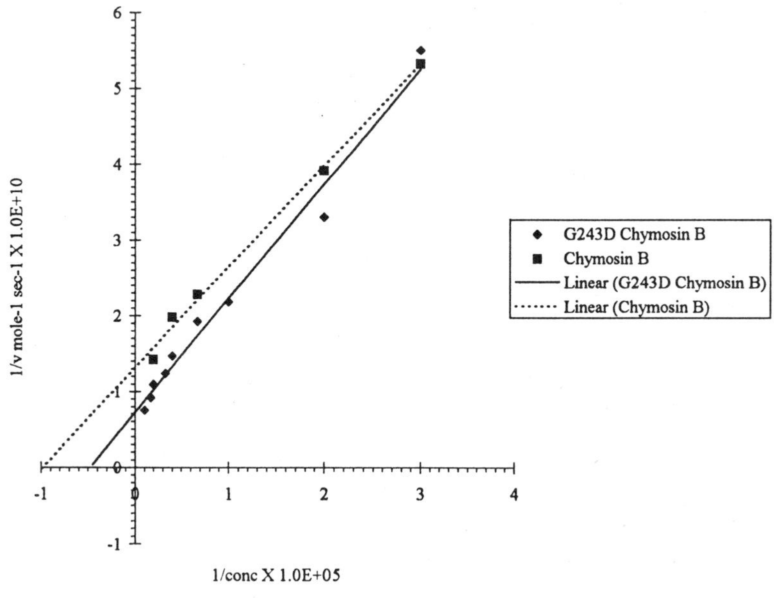

Figure 2. Kinetics of point mutant G243D with a 15 residue κ-casein peptide.

With the shorter peptide the chymosin A115T mutant showed a reduction of K_m with respect to the native, possibly caused by the changes in the vicinity of S1 and S3. Although the kinetic anlysis of the G243D mutant using the short peptide showed a reduced catalytic activity, use of the longer substrate revealed an increase of k_{cat} compared to chymosin B, probably as a consequence of the introduced charge that may interact with the substrate between P4 and P8. Results for a loop mutant are reported in Nugent et al. (19).

The effect of mutation on the stability of the mutants was assessed by heat, urea and guanidine HCl denaturation studies. The degree to which chymosin secondary structure was altered to accommodate each mutation was examined by CD spectroscopy.

4. CRYSTAL STRUCTURES

A115T chymosin crystallises in the orthorhombic space group I222, with cell dimensions similar to those of native chymosin. The structure was refined to a final crystallographic R value of 0.18 for all data from 17.0 to 2.3 Å. The mutant is altered only by the addition of a methyl group and a hydroxyl to a single side-chain and it is not surprising that the structure is largely unchanged compared to the native structure. The Thr115 side-chain is accommodated with relatively slight local movements and forms a new side-chain to main-chain hydrogen bond (Figure 3). Residue 115 is at the end of an irregular helix; it appears that the helix shifts by about 0.6Å extending ten or more residues back along the chain with the loss of two hydrogen bonds and a slight tilting of the 'flap'. These small shifts are likely to affect the specificity pockets at S1 and S3.

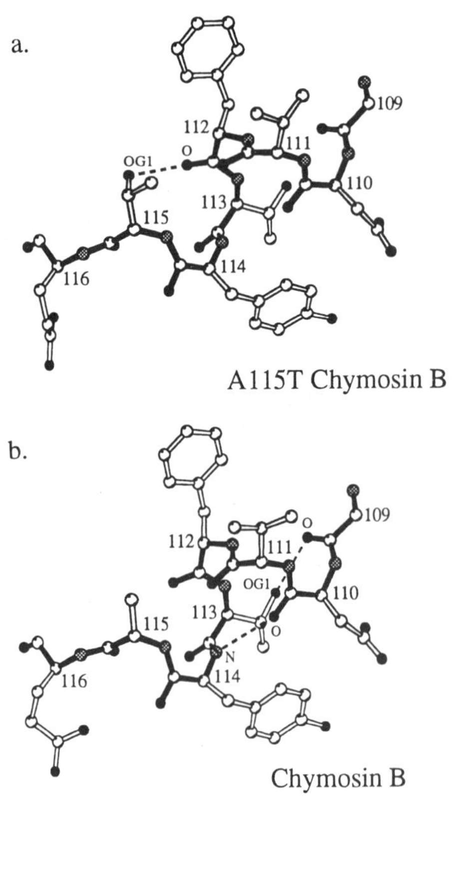

Figure 3. The conformation of (a) A115T chymosin in the region of the Thr115 compared to Ala115 and (b) the wild type, demonstrating that the side-chain is accommodated with relatively slight local movements and forms a new side-chain to main-chain hydrogen bond.

G243D chymosin (also known as chymosin A) also crystallised isomorphously with the native enzyme and was refined to a final R of 0.19. The structure of chymosin G243D is very similar to that of the native. The main differences between chymosin G243D and native chymosin occur in the loop 278–281 and unsurprisingly, in loop 240–244, where the only sequence difference in the protein arises. The main chain has a slightly different conformation and the carboxylate oxygens of Asp243 (Figure 4) form intramolecular hydrogen bonds with the side-chain amide Nitrogen of Gln239 and a water molecule.

It has been postulated that the increased activity of chymosin G243D (chymosin A) may be the result of an enhanced binding affinity for its natural substrate, κ-casein, due to His102 at P4 of the substrate and Asp243 of the enzyme. Our studies indicate that Asp243 is more likely to interact at P8His or residue 98 of the casein sequence.

Loop exchange mutant chymosin 155–164 rhizopuspepsin, expressed in *Trichoderma reesei*, was crystallised and found to be isomorphous with the native crystals. The final model has an R-factor of 0.20 between 8 and 2.5Å resolution (0.22 between 8 and 2.2 Å) and provided the definition of previously ill-defined loops and side chains. The conformation of the mutated loop, shown in Figure 5, involves hydrogen bonding of a water molecule to five loop main-chain atoms. Comparisons of the 155–164 hairpin in the mutated protein with the equivalent ones in native chymosin and in rhizopuspesin are shown in Figure 6. The SLoop database shows that the equivalent loop in rhizopuspepsin is the most similar to the mutated loop in terms of the main chain atoms RMS deviations. Interestingly, the second closest loop to the mutated one, in terms of RMS deviation, is

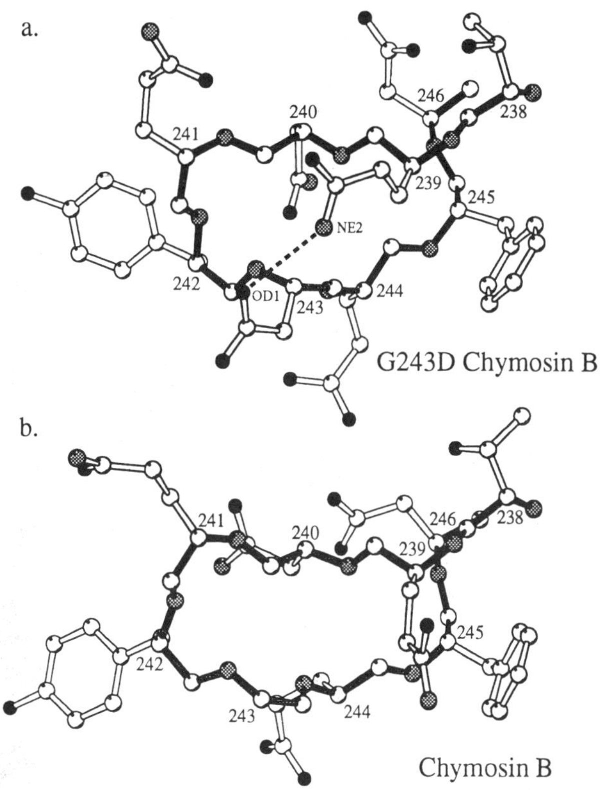

Figure 4. The conformation of the mutated loop in (a) G243D chymosin compared to (b) the wild type demonstrates that there is a slightly different conformation and that the carboxylate oxygens of Asp243 form intramolecular hydrogen bonds with the sidechain amide of Gln239 side-chain and a water molecule.

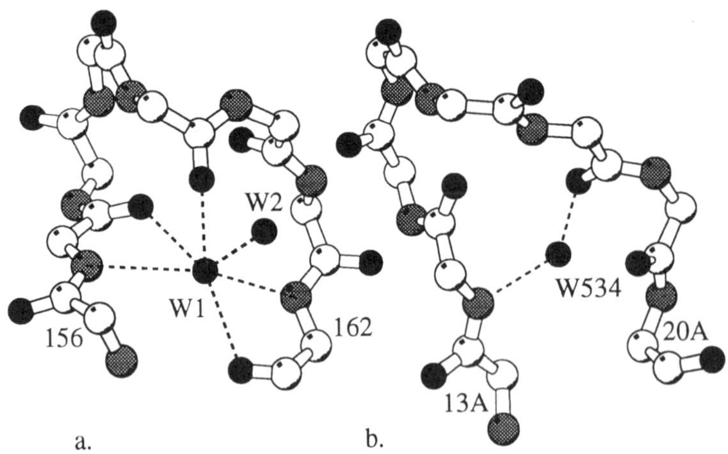

Figure 5. The conformation of (a) the mutated loop in chymosin 155–164 rhizopuspepsin, showing the hydrogen bonding pattern of a water molecule, which interacts with five loop main-chain atoms, and with another water molecule, compared to (b) loop 13A-20A of histocompatibility antigen.

from the histocompatibility antigen HLA-B*2705, which also contains a water molecule in the middle of the loop, albeit in a different position.

In order to assess the effect of the loop mutation on the proteolysis of the 15 residue substrate, the chymosin substrate transition state complex was modelled on the basis of the structure of the chymosin inhibitor complex and the mouse renin inhibitor complex. It was evident from these studies that, in some conformations the substrate should be able to interact with the mutated 155–164 loop. The changed enzyme activity of the mutated protein towards the 15 residue κ-casein peptide and κ-casein in whole milk could result from either steric interference or electrostatic repulsion.

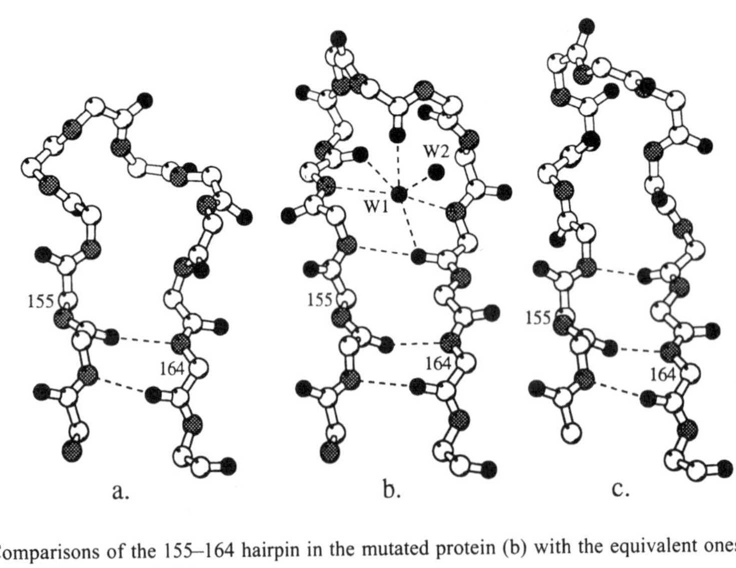

Figure 6. Comparisons of the 155–164 hairpin in the mutated protein (b) with the equivalent ones in native chymosin (a) and in rhizopuspepsin (c).

5. CONCLUSIONS

These experiments have demonstrated that it is possible to engineer enzymes with modified loop regions that are still functional. They have shown that a good knowledge of the protein structure, combined with careful choice of replacement loops, can lead to stable enzymes. However, they have emphasised the danger of assuming that loops, even from homologous proteins, will have indentical conformations to those of the donor enzyme. They have also underlined the sensitivity of the whole enzyme structure to substantial reductions in loop size, unless complementary changes are made in the core structure, a conclusion also reached by antibody engineers.

Aspartic proteinases such as chymosin, pepsin and the microbial enzymes from *Mucor pusillus* (mucor pepsin) and *Endothia parasitica* (endothiapepsin) have been widely used in the cheese industry and also for proteolysis in the processing of soya and other foods. Their value resides not only in the limited proteolysis of specific bonds (such as Phe105 - Met106 of κ-casein by chymosin) but also in the production of small peptides that give flavour. These characteristics are dependent on the specificities, pH optima and temperature profiles of the various aspartic proteinases. We have shown that we can express recombinant prochymosin and we can use site-directed mutagenesis to produce enzymes of different characteristics which could provide the basis of design of enzymes for other specific applications in food processing and the production of new flavours.

REFERENCES

1. Kostka, V. (1985) Aspartic Proteinases and Their Inhibitors. P 613, Walter de Gruyer, Berlin.
2. Mantafounis, D. and Pitts, J.E. (1989). Biochem.Soc.Trans., 17, 763–764.
3. Strop, P., Sedlacek, J., Stys, J., Kaderabkova, Z., Blaha, I., Pavlickova, L., Pohl, J., Fabry, M., Kostka, V., Newman, M., Frazao, C., Shearer, A., Tickle, I.J. and Blundell, T.L. (1990) Biochemistry 29, 9863–9871.
4. Gilliland, G.L., Winbourne, E.L., Nachman, J. and Wlodawer, A. (1990). Proteins, 8, 82–101.
5. Newman, M., Safro, M., Frazao, C., Khan, G., Zdanov, A., Tickle, I.J., Blundell, T.L. and Andreeva, N. (1991). J.Mol.Biol., 221, 1295–1309.
6. Groves, M., Dhanaraj, V., Badasso, M., Pitts, J.E., Hoover, D.J. and Blundell, T.L. (1997). Prot.Engg., Submitted.
7. Topham, C.M., Srinivasan, N. and Blundell, T.L. (1997) Prot.Engg. 10, 7–21.
8. Park, Y.-N., Aikawa, J.-i., Nishiyama, M., Horinouchi, S. and Beppu, T. (1996). Prot.Engg. 9, 869–875.
9. Sutcliffe, M.J., Hannef, I. and Carney, D. and Blundell, T.L. (1987) Prot.Engg. 1, 377–384.
10. Sali, A. and Blundell, T.L. (1990) J.Mol.Biol. 212, 403–428.
11. Carney, D., Gardner, S., Hayes, F., Howlin, B., Hubbard, T., Overington, J., Singh, D., Sibanda, B.L. and Sutcliffe, M. (1988) Eur.J Biochem. 173, 513–520.
12. Topham, C.M., McLeod, A., Eisenmenger, F., Overington, J.P., Johnson, M.S. and Blundell, T.L. (1993) J.Mol.Biol. 229, 194–220.
13. Donate, L.E., Rufino, S., Canard, L. and Blundell, T.L. (1996) Prot.Sci. 5, 2600–2616.
14. Rufino, S., Donate, L.E. and Blundell, T.L. (1997) J.Mol.Biol. 267. In press.
15. Emtage, J.S., Angal, S., Doel, M.T., Harris, T.J.R., Jenkins, B., Lilley, G. and Lowe, P.A. (1983). Proc.Nat.Acad.Sci., 80, 3671–3675.
16. Pitts, J.E., Dhanaraj, V., Dealwis, C., Mantafounis, D., Nugent, P.G., Orprayoon, P., Cooper, J.B. Newman, M. and Blundell, T.L. (1992) Scand.J.Clin.Lab.Invest. 52, 39–50.
17. Pitts, J.E., Uusitalo, J.M., Mantafounis, D., Nugent, P.G., Quinn, D.D., Orprayoon, P. and Pentilla, M. (1993) J.Biotechnology. 28, 69–83.
18. Harkki, A., Uusitalo, J., Bailey, M., Penttilla, M. and Knowles, J.K.C. (1989) Bio-Technology, 7, 596–602.
19. Nugent, P.G., Albert, A., Orprayoon, P., Wilsher, J., Pitts, J.E., Blundell, T.L. and Dhanaraj, V. (1996). Prot.Engg. 9, 885–893.

24

POST X-RAY CRYSTALLOGRAPHIC STUDIES OF CHYMOSIN SPECIFICITY

The Role of Histidine-Proline Cluster of κ-Casein in Catalytic Reactions

E. A. Gustchina,[1,3] P. Majer,[3] L. D. Rumsh,[1] L. M. Ginodman,[1] and N. S. Andreeva[2]

[1]M. Shemyakin and Y. Ovchinnikov Institute of Bioorganic Chemistry
Russian Academy of Sciences
Moscow, Russia
[2]W. Engelhardt Institute of Molecular Biology
Russian Academy of Sciences
Moscow, Russia
[3]Structural Biochemistry Program
SAIC, NCI-FCRDC
Frederick, Maryland 21702

1. INTRODUCTION

X-ray crystallographic studies of bovine chymosin (Gilliland et al. 1990; Newman et al. 1991) have shown that the conserved Tyr75 residue (pepsin numbering is used), located in the so-called 'flap', takes up an unusual position in molecules of this enzyme. In almost all native aspartic proteinase structures and in all their complexes with inhibitors, it forms the wall of the S1 binding pocket, and the hydroxyl group of the Tyr75 is bound to Ser35 hydroxyl through a water molecule, which is also highly conserved. In contrast, the tip of the flap in the orthorhombic crystals of chymosin has an unusual conformation, and the phenolic ring of the Tyr75 residue is embedded in the S1 binding pocket, and partly in the S3 binding pocket (Fig. 1).

There were two explanations of the observed results:

1. The position of this ring is the result of the flap distortion, and such a distortion is a consequence of tight crystal packing in orthorhombic crystals (1).

Aspartic Proteinases, edited by James.
Plenum Press, New York, 1998.

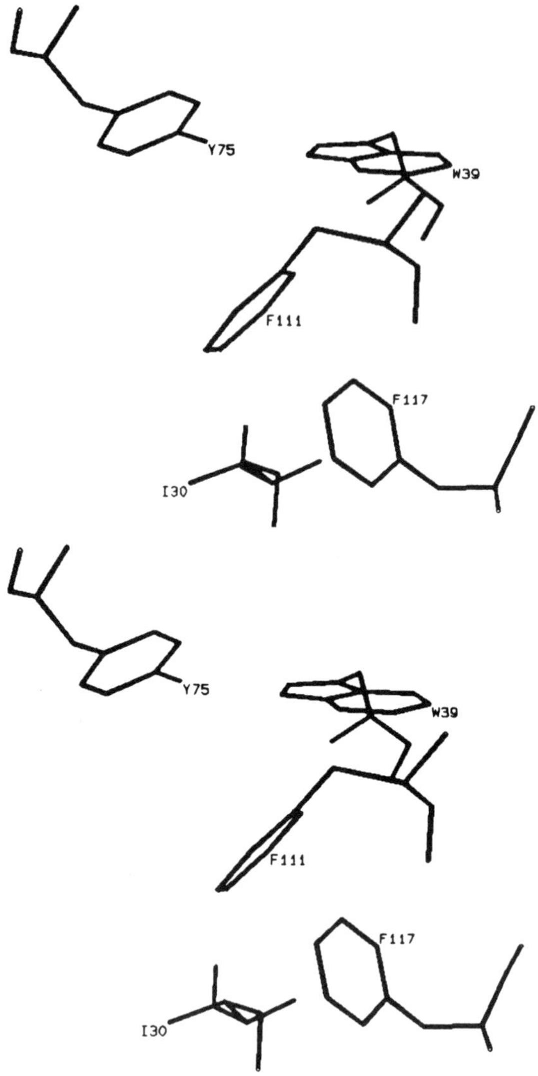

Figure 1a. The S1 binding pocket of porcine pepsin (8).

2. The detailed analysis of intermolecular interactions in the region of the flap does not confirm the presence of close intermolecular contacts and stereochemical parameters of the flap do not correspond to any distorted state. At the same time, the flexibility of the flap is a well-known phenomenon. Therefore it was concluded that Tyr75 ring of chymosin can occlude S1/S3 binding pockets, and such ability is probably an intrinsic property of the enzyme (2).

The second explanation made it possible to formulate a new hypothesis on the structural grounds of the enzyme's specificity (3).

Chymosin has very low general proteolytic activity and it is highly specific towards the peptide bond Phe105–Met106 of its natural substrate, κ-casein. The structural state of the enzyme when S1 and partly S3 binding pockets are occupied by its own residue Tyr75 (Tyr77 in (1)) has been called a self-inhibited state. It was proposed that in

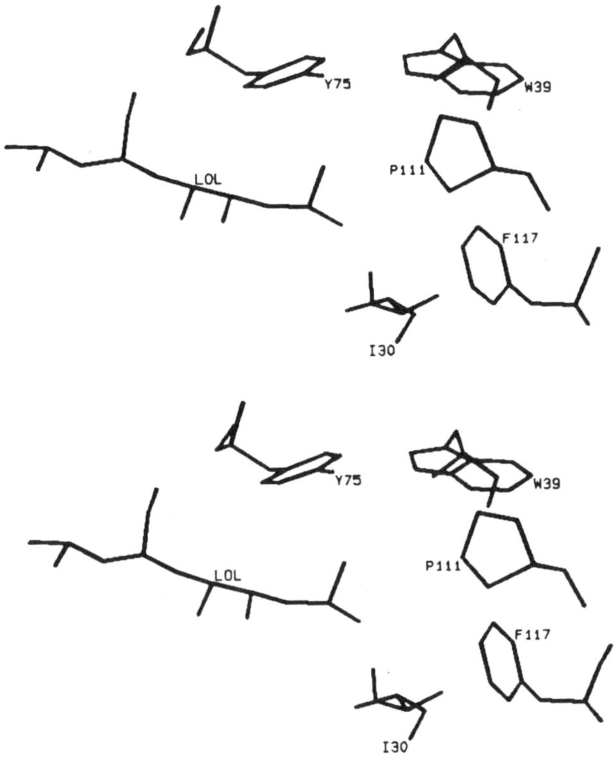

Figure 1b. The position of a substrate in this pocket (complex of mouse renin with CH-66 inhibitor) (7).

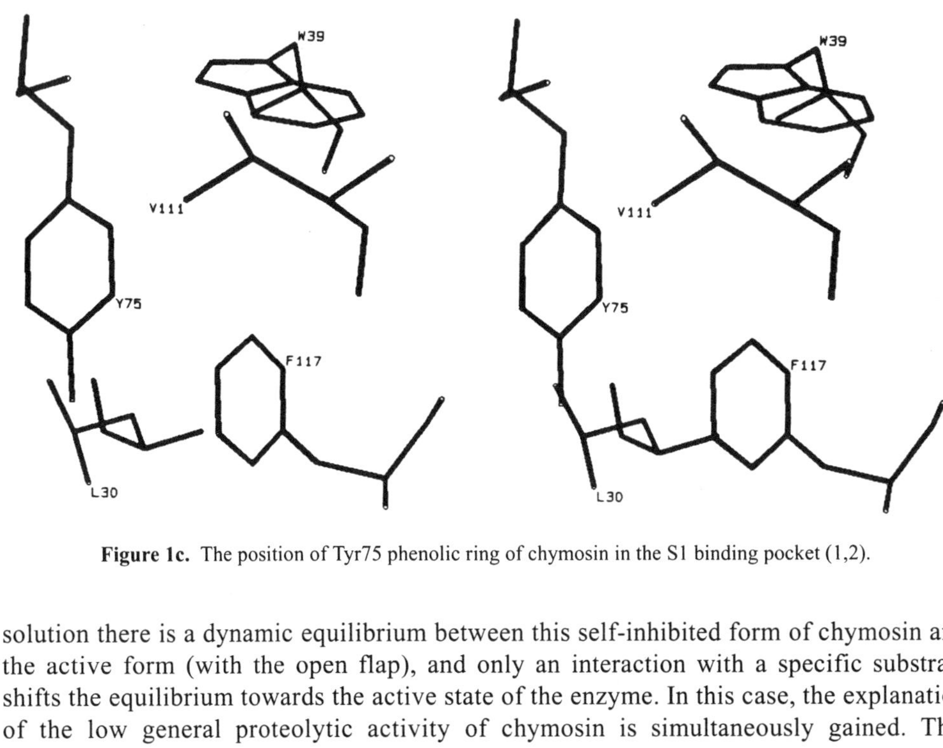

Figure 1c. The position of Tyr75 phenolic ring of chymosin in the S1 binding pocket (1,2).

solution there is a dynamic equilibrium between this self-inhibited form of chymosin and the active form (with the open flap), and only an interaction with a specific substrate shifts the equilibrium towards the active state of the enzyme. In this case, the explanation of the low general proteolytic activity of chymosin is simultaneously gained. This hypothesis implies that there is a special fragment in the structure of κ-casein molecule which activates chymosin.

This work represents the results of experimental studies of chymosin interaction with various fragments of bovine κ-casein. It shows a special role of the histidine-proline cluster of κ-casein in hydrolytic reaction catalyzed by the enzyme.

2. MATERIALS AND METHODS

The following fragments of κ-casein molecule have been synthesized (4):

1. NS1-peptide: 98–109 fragment of κ-casein containing the scissile bond Phe105–Met106: His–Pro–His–Pro–His–Leu–Ser–Phe–Met–Ala–Ile–Pro–NH_2
2. NS1A-peptide: 103–109 fragment of κ-casein: Leu–Ser–Phe–Met–Ala–Ile–Pro–NH_2
3. NS2-peptide: 98–102 fragment of κ-casein (the histidine-proline cluster): His–Pro–His–Pro–His–NH_2

Several preparations of chymosin have been used in this work:

1. Commercial preparation from Sigma, St. Louis, MO.
2. Crystalline samples of chymosin B kindly provided by Prof. B. Foltmann, University of Copenhagen, Denmark.
3. Preparations of recombinant chymosin A kindly provided by Dr. Marianne Harboe, Hansen Laboratories, Denmark.

The experiments described here included the comparison of the hydrolysis rates of the peptides NS1 and NS1A by recombinant chymosin A in the absence and in the presence of the excess of the histidine-proline cluster (the NS2 peptide). Mixtures containing 50 mM sodium acetate buffer, pH 5.0, enzyme (about 5 nM, according to the protein concentration) and histidine-proline cluster (5 mM) were preincubated for 4 min at 37°C prior to initiation of reaction by addition of the substrates (500 μM). Reactions were stopped at various times by freezing in liquid nitrogen. The cleavage of peptides was monitored by the reverse phase HPLC at λ=222 nm using VYDAC C-18 column (4).

3. RESULTS AND DISCUSSION

A comparison of the hydrolysis rates of the NS1 and the NS1A peptides is in agreement with the result obtained by Visser and his colleagues several years ago (5). The rate of hydrolysis of the short peptide, NS1A, is very low compared with that for the long peptide, NS1 (Fig. 2). The short peptide contains the following groups: P3, P2, P1, P1′, P2′, P3′ and P4′. The the long NS1 peptide contains the additional groups at the N-terminus P8, P7, P6, P5. P4. A considerable increase of the catalytic efficiency of the enzyme following the enlargement of the the substrate by the addition of P8...P4 groups at the N-terminus is a remarkable feature of chymosin. The P8...P4 groups correspond to the histidine-proline cluster (residues 98–102) and the possible role of this cluster has been already discussed in previous publications (3,6).

One can think that the cluster assists in correct positioning of the peptide to be cleaved in the long substrate binding cleft of chymosin (6). At the same time, if the two structural forms of the enzyme really exist in solution, the cluster can play a role of the activator promoting the conversion from the self-inhibited to the active state (3). The aim of this work was to show which one of the two possible explanations corresponds to the real function of the cluster.

In case that the first explanation is correct, the histidine-proline cluster facilitates the enzyme action only if a substrate contains this cluster as its own part. The binding of the isolated histidine-proline cluster to the enzyme should not affect the rate of hydrolysis either of the short NS1A peptide or of the long NS1 peptide. If the second explanation is valid, then the cluster plays the role of an activator, and the binding of the isolated peptide NS2 to the enzyme promotes a more efficient enzyme action on the short peptide NS1A and has little influence on the rate of hydrolysis of the long peptide NS1, which contains the cluster as its own part.

The two experiments performed one after another have shown that the preliminary incubation of chymosin A with the NS2 peptide decreases the rate of hydrolysis of the long peptide NS1, while the short peptide, NS1A, has been cleaved faster than in experiments without the cluster (Fig. 2). These data can be explained if the NS2 peptide binds in S8...S4 pockets taking the position of a part of the real substrate. Thus the NS2 peptide acts as an activator for the hydrolysis of the substrates containing P3, P2...P4′ side chains, and it is an inhibitor for the long substrates containing P8, P7...P4′ side chains. Although the variations in the rates of hydrolysis of the substrate by the chymosins obtained from different sources were observed in the experiments published previously (4) and described here, the general trend of the effect of the histidine-proline cluster is still present. Therefore, the second possible explanation of the role of the histidine-proline cluster, discussed above, looks more plausible.

The results obtained here support the hypothesis that chymosin undergoes structural transformation in the process of binding of the histidine-proline cluster, which activates the enzyme. Since the cluster appears to occupy the S8...S4 pockets which are on the periphery of the active site, a concerted movement of several atomic groups must be involved in the activation process.

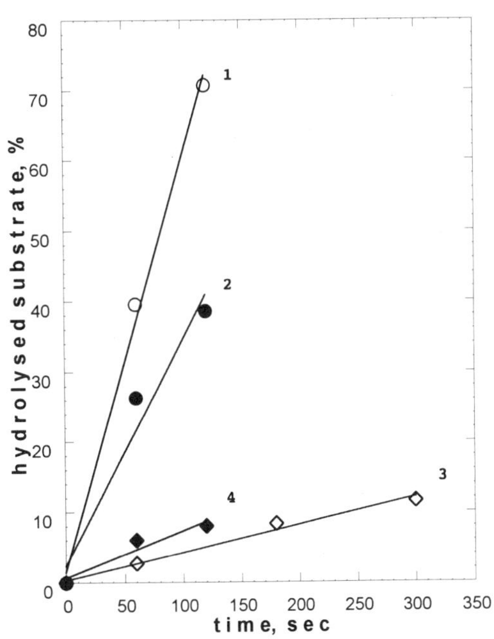

Figure 2. Hydrolysis of the substrates by chymosin A (percentage of hydrolysed substrate as a function of time). (1) NS1 peptide; (2) NS1 peptide + histidine–proline cluster; (3) NS1A peptide; (4) NS1A peptide + histidine–proline cluster.

The structural grounds of the activation of chymosin are an interesting problems to be elaborated. The preliminary model building studies have shown that in case of a completely extended conformation of the whole bound peptide 98–111 (P8...P4′), there would be difficult to explain the observed phenomenon. At the same time, the cluster cannot form any direct contacts with the Tyr75 residue. These results call for the development of X-ray crystallographic studies of chymosin complexes with inhibitors containing the histidine-proline cluster, and such studies are under way.

All non-viral aspartic proteinases which do not have a bulky residue like phenylalanine at position 111 (pepsin numbering) look as virtually capable to adopt the self-inhibited form. These enzymes include renin, cathepsin E, and some others.

REFERENCES

1. Gilliland, G.L., Winborn, E.L., Nachman, J. & Wlodawer A, (1990) The three-dimensional structure of recombinant bovine chymosin at 2.3 Å resolution. Proteins, 8, 82–101.
2. Newman, M., Safro, M., Frazao, C., Khan, G., Zdanov, A., Tickle, I., Blundell, T., Andreeva, N. (1991) X-ray analyses of aspartic proteinases. IV. Structure and refinement at 2.2 Å resolution of bovine chymosin. J.M.B., 221, 1295–1309.
3. Andreeva, N., Dill, J., Gilliland G. (1992) Can enzymes adopt a self-inhibited form. BBRC, 184, 1081–1084.
4. Gustchina, E., Rumsh, L., Ginodman, L. Mayer, P. Andreeva, N. (1996) Post X-ray crystallographic studies of chymosin: the existence of two structural forms and the regulation of activity with the histidine-proline cluster of κ-casein. FEBS Lettr.,379, 60–62.
5. Visser, S., Slangen, J. and van Rooijen P. (1987) Biochem. J. 244, 553–558.
6. Safro, M.G., Andreeva, N.S. (1990) On the role of peripheral interactions in the specificity of chymosin. Biochem. Int. 20, 555–561.
7. Dhanaraj v., Dealwis C.G., Frazao C., Badasso M., Sibanda B.L., Tickle I.J., Cooper J.B., Driessen H.P.C., Newman M., Aquilar C., Wood S.P., Blundell T.L., Hobart P.M., Geoghegan K.F., Amirati M.J., Danley D.E., O'Connor B.A., Hoover D.J. (1992). X-ray analysis of peptide-inhibitor complexes define the structural basis of specificity for human and mouse renin. Nature, 357, 466–472.
8. Sielecki A., Fedorov A., Boodhoo A., Andreeva N., James M. (1990) Molecular and crystal structure of monoclinic porcine pepsin refined at 1.8 Å resolution. J.Mol.Biol. 214, 143–170.

25

EPITOPE MAPPING OF RECOMBINANT HUMAN PROCATHEPSIN D

Deepali Sachdev,[1,2] Yasu Ohsaki,[1,2] G. David Roodman,[1,2] and John M. Chirgwin[1,2]

[1]Audie L. Murphy Veterans Administration Research Service
[2]Departments of Medicine and Biochemistry
The University of Texas Health Science Center at San Antonio
San Antonio, Texas 78284-7877

1. INTRODUCTION

Procathepsin D is the proform of the aspartic proteinase, cathepsin D. The mature form, cathepsin D, is a low specificity proteinase found in the lysosome where the intracellular protein catabolism carried out by it is well understood. The proform, procathepsin D is also expressed in secretory and membrane-associated forms.[1,2] These non-lysosomal forms of procathepsin D are associated with a wide variety of physiological and pathophysiological functions including metastasis of breast cancer,[3] tissue remodeling, antigen presentation in the MHC class II pathway,[4] and general role in the immune system.[5] The structural bases whereby this proteinase can perform these varied and specific functions are unknown. Several possibilities for membrane-association have been proposed, none has been conclusively demonstrated. The propeptide has no hydrophobic regions or C-terminal membrane anchor sequences that may be responsible for membrane association. Membrane association may be due to surface exposed residues of the propeptide. It seems likely that most of the non-lysosomal functions involve interactions of the cellular machinery with the surface features of the folded propeptide region such as propeptide specific receptors.[6] Therefore there is interest in the conformation of procathepsin D. A demonstration of a difference in the conformations of the various forms of this proteinase will add to the limited knowledge about the structural bases of the functions of the non-lysosomal forms of this proteinase.

The crystallographic structure of mature cathepsin D is known.[7,8] Like other members of the family, cathepsin D has a bilobed structure with an amino terminal domain and a carboxy terminal domain. Both the domains have predominantly β-sheets. They lie on either side of a 30Å long active site cleft and each contributes an aspartate to the active site. The two domains are held together by a central connecting region which has a six stranded β-

Aspartic Proteinases, edited by James.
Plenum Press, New York, 1998.

pleated sheet contributed by the amino and carboxy terminal domains. The crystallographic structure of the precursor form, procathepsin D, is not known. However computer generated models of procathepsin D[9] have been described based on the coordinates of pepsinogen,[10,11] another member of the aspartic proteinase family. Procathepsin D shows substantial structural similarity to pepsinogen. In pepsinogen, the 44 amino acid propeptide (residues 1p–44p) covers the active site cleft and the first 10 amino acids of the propeptide, 1p–10p, form a β-sheet in the connecting region between the two domains. In mature pepsin the propeptide is removed, and in pepsin relative to pepsinogen, the first amino acid of the mature amino terminus (residue 1) moves about 40Å from its position in the precursor form[11] and residues 1–10 of the mature form replace the strand contributed by residues 1p–10p in pepsinogen. Procathepsin D can be predicted to have the following 4 surface exposed antigenic portions—the amino terminal domain surface, carboxy terminal domain surface, folded propeptide region, and the connecting region. The connecting region in procathepsin D will therefore have one strand (contributed by residues 1p–10p) which is different from that in the mature form. Thus, substantial differences are predicted in the surface regions of the two forms particularly in the region around the amino terminus. Polyclonal antibodies against procathepsin D are available but monoclonal antibodies (Mabs) against it are not readily prepared due to the difficulty in obtaining large quantities of pure procathepsin D.

We hypothesized that since procathepsin D and cathepsin D differ in surface regions of the protein structure and in size, it is possible to raise MAbs against secreted procathepsin D that recognize native features of the protein. The MAbs can be used to probe the surface differences between procathepsin D and cathepsin D and characterize the surface of the proform. MAbs are highly specific and therefore are powerful tools for conformational analysis, and can provide useful information on the conformation of the proform. We have purified recombinant human procathepsin D and developed and characterized six non-competing groups of MAbs specific for native secreted procathepsin D.

2. MATERIALS AND METHODS

2.1. Production of Monoclonal Antibodies

The expression vector, pCMVPCDMf1, encoding human procathepsin D with its C-terminus bearing an epitope tag of 13 amino acids derived from *c–myc*[12] was transiently transfected into human embryonal kidney cells (293) by a modified calcium phosphate procedure.[13] Procathepsin D–*myc* (pCDM) secreted into the conditioned medium of transfected cells was partially purified by concanavalin A affinity chromatography. The concanavalin A purified material was then applied to superose 12 column and subjected to gel filtration with fast phase liquid chromatography (FPLC) and used as antigen. Splenocytes from female Balb/c mouse with positive anti-pCDM antibody titer were fused with myeloma cells, X63Ag8.563, using 50% PEG / 10% DMSO. Hybridomas were screened by enzyme-linked immunosorbent assay (ELISA) with antibody capture. Hybridomas secreting antibodies against pCDM were single cell-cloned, expanded, and obtained as ascitic fluids. IgG from ascitic fluids was purified by successive precipitations with caprylic acid and ammonium sulfate.

2.2. Antibody Competition Binding ELISA

To sort the MAbs into groups based on the recognition of similar or distinct epitopes on pCDM, competition binding assays were performed by ELISA using antibody capture.

MAbs were added in pairs to wells coated with pCDM. One MAb in the pair was labeled with biotin using NHS–LC biotin (Pierce Chem. Co., Rockford, IL) and the other was unlabeled. The efficiency of the binding of the biotinylated antibody (Ab) was monitored in the presence of the unlabeled Ab. Streptavidin–alkaline phosphatase conjugate followed by phosphatase substrate was used for detection and absorbance at 405 nm determined.

3. RESULTS AND DISCUSSION

3.1. Production of Monoclonal Antibodies

2688 wells with hybridoma supernatants were screened for antibodies against procathepsin D. 11 positive hybridoma clones were identified. Of these, 9 were single cell-cloned, collected as ascitic fluids, purified, and sorted into epitopic groups.

3.2. Sorting into Epitopic Groups by Antibody Competition ELISAs

Competition ELISAs were done to sort the nine MAbs into groups based on the epitopes recognized. In a competition ELISA if the labeled MAb and the unlabeled MAb recognize distinct epitopes, the labeled MAb will bind to the same extent in the absence or presence of the unlabeled MAb and the two Abs are non-competitors. If the two MAbs recognize the same epitope, then the binding of the labeled Ab is decreased in the presence of the other MAb and they are competitors. If the unlabeled Ab partially decreased the binding of the labeled Ab, the two were classified as semi-competitors and recognize overlapping epitopes. Competition ELISAs were performed with all possible combinations of the 9 MAbs assayed in pairs. Figure 1 shows the results of some of the competition ELISAs of biotinylated MAb 3v–G7 and three other MAbs. Binding of biotinylated MAb 3v–G7–F4 is shown by the line with filled circles. Binding of labeled 3v–G7–F4 is competed off by unlabeled 3v–G7–F4 (self-competitor) as expected and is shown by the line with open circles. MAbs 5–F8–G8 and 14–A6–G11 are shown by lines with filled tri-

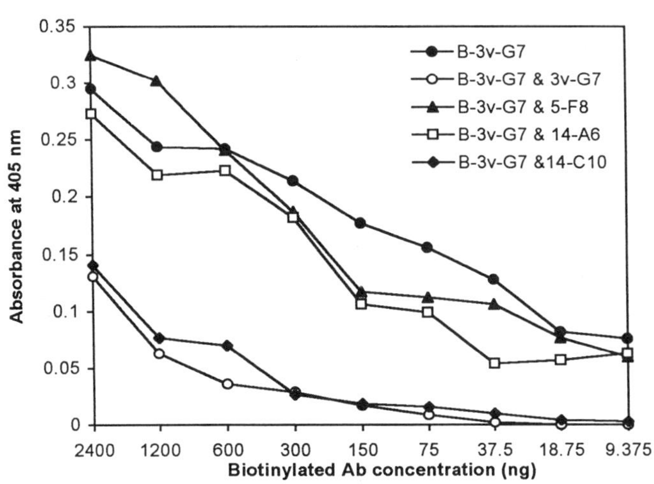

Figure 1. Antibody competition ELISA showing that binding of biotinylated MAb 3v–G7 to procathepsin D is competed by MAb 14–C10 but not by MAbs 5–F8 and 14–A6.

angles and open squares respectively. They have no effect on the binding of biotinylated 3v–G7–F4 and were classified as non-competitors and hence recognize different epitopes on procathepsin D. MAb 14–C10–F12 shown by the line with filled diamonds decreased the binding of biotinylated 3v–G7–F4 in a manner similar to the self-competitor and thus competed with 3v–G7–F4.

When the assay was done with biotinylated 14–C10–F12 and unlabeled 3v–G7–F4 competition was also observed as shown in Table 1. Therefore MAbs 3v–G7–F4 and 14–C10–F12 recognize the same epitope. Table 1 below shows the results of all the Ab competition ELISAs. In all cases, the results obtained for any given pair of MAbs was the same irrespective of which Ab in the pair was biotinylated. In the table competition is indicated by a '+', no competition by '–', and semi-competition by '±'. Based on all these competition ELISAs the Abs were sorted into epitopic groups. As shown in table 1, MAbs 14–A6–G11, 14–A7–C11, and 14–C6–E8 competed with each other for binding to procathepsin D and were therefore sorted into the same epitopic group. MAbs 5–F8–G8, 7–D4–B2, 18–B3–D3, and 18–C5–C3 did not compete with any of the MAbs and hence recognize epitopes distinct from each other. MAbs 3v–G7–F4 and 14–C10–F12 as described above recognize the same epitope. Thus the nine monoclonal antibodies, although of low affinity, sorted into six non-competing epitopic groups, with each group binding to a nearly nonoverlapping epitope on the surface of procathepsin D distinct from the surface regions bound by the remaining 5 groups.

The 3-dimensional structures of several protein antigens such as hen egg white lysozyme,[14,15] neuraminidase of influenza virus,[16] and histidine phosphocarrier protein, (HPr) from *E. coli*[17] with the Fab fragments of the monoclonal antibodies raised against them have been determined. These structures have revealed that the surface area comprising an epitope on the protein antigen recognized by the different monoclonal antibodies varies between 680–916 Å^2.[17,18] In the case of lysozyme, 3 different non-competing MAbs cover more than 40% of the total surface area,[18] and for HPr 50% of the total surface area was monitored using 3 different MAbs.[19] Based on these calculations it is deduced that for a globular protein of the size of procathepsin D, six distinct epitopic regions will cover nearly all of its surface-exposed areas. Our results show that all of the surface exposed

Table 1. Results of the antibody competition ELISAs to sort the MAbs into epitopic groups[a]

	Labeled MAb								
Competitor	3v–G7	5–F8	7–D4	14–A6	14–A7	14–C6	14–C10	18–B3	18–C5
3v–G7–F4	+	–	–	–	–	–	+	–	–
5–F8–G8	–	+	nt	–	–	–	–	–	–
7–D4–B2	–	–	+	–	–	–	–	–	–
14–A6–G11	–	–	–	+	+	+	–	–	–
14–A7–C11	–	–	nt	+	+	+	–	–	–
14–C6–E8	–	–	–	+	+	+	–	–	–
14–C10–F12	+	–	nt	±	±	±	+	–	–
18–B3–D3	–	–	–	±	±	±	–	+	–
18–C5–C3	–	–	nt	–	–	–	–	–	+

+ = competitor ± = semi-competitor (overlapping epitope)
– = non-competitor nt = not tested

[a]The competition ELISAs were performed with the MAbs added in pairs to antigen coated wells. One MAb in the pair was biotinylated and the efficiency of its binding in the presence of the unlabeled MAb was monitored to determine if the two Abs were competitors. All possible combinations of the 9 MAbs were assayed and the final results of the assays are tabulated here.

areas of recombinant human procathepsin D are antigenic. Thus monoclonal antibodies can be generated against all of the surface areas of procathepsin D. Antibodies so generated can be used to characterize further the surface areas of procathepsin D and their interactions with cellular machinery to understand the structural bases by which the proenzyme performs its diverse roles.

REFERENCES

1. Rijnboutt, S., Aerts, H.M.F.G., Geuze, H.J., Tager, J.M., and Strous, G.J. (1991). Mannose 6-phosphate-independent membrane association of cathepsin D, glucocerebrosidase, and sphingolipid-activating protein in HepG2 cells. J. Biol. Chem. **266**: 4862–4868.
2. McIntyre, G.F., and Erickson, A.H. (1991). Procathepsins L and D are membrane-bound in acidic microsomal vesicles. J. Biol. Chem. **266**: 15438–15455.
3. Capony, F., Rougeot, C., Montcourrier, P., Cavailles, V., Salazar, G., and Rochefort, H. (1989). Increased secretion and altered processing and glycosylation of procathepsin D in mammary cancer cells. Cancer Res. **49**: 3904–3909.
4. van Noort, J.M., and van der Drift, A.C.M. (1989). The selectivity of cathepsin D suggests an involvement of the enzyme in the generation of T-cell epitopes. J. Biol. Chem. **264**: 14159–14164.
5. Saftig, P., Hetman, M., Schmahl, W., Weber, K., Heine, L., Mossmann, H., Koster, A., Hess, B., Evers, M., and von Figure, K. (1995). Mice deficient for the lysosomal proteinase cathepsin D exhibit progressive atrophy of the intestinal mucosa and profound destruction of lymphoid cells. EMBO J. **14**: 3599–3608.
6. McIntyre, G.F., Gobold, G.O., and Erickson, A.H. (1994). The pH-dependent membrane association of procathepsin L is mediated by a 9-residue sequence within the propeptide. J. Biol. Chem. **269**: 567–572.
7. Baldwin, E.T., Bhat, T.N., Gulnik, S., Hosur, M.V., Sowder, R.C., II, Cachau, R.E., Collins, J., Silva, A.M., and Erickson, J.W. (1993). Crystal structures of native and inhibited forms of human cathepsin D: implications for lysosomal targeting and drug design. Proc. Natl. Acad. Sci. USA. **90**: 6796–6800.
8. Metcalf, P., and Fusek, M. (1993). Two crystal structures for cathepsin D: the lysosomal targeting signal and active site. EMBO J. **12**: 1293–1302.
9. Koelsch, G., Metcalf, P., Vetvicka, V., and Fusek, M. (1995). Human procathepsin D: three-dimensional model and isolation. Advan. Exp. Med. Biol. **362**: 273–278.
10. James, M.N.G., and Sielecki, A.R. (1986). Molecular structure of an aspartic proteinase zymogen, porcine pepsinogen, at 1.8 Å resolution. Nature **319**: 33–38.
11. Hartsuck, J.A., Koelsch, G., and Remington, S.J. (1992). The high-resolution crystal structure of porcine pepsinogen. Proteins: Struct. Funct. Genet. **13**: 1–25.
12. Sachdev, D., Schorey, J., and Chirgwin, J. M. (1991). Efficient mutagenesis, expression, and purification of procathepsin D. Advan. Exp. Med. Biol. **306**: 335–338.
13. Gorman, C., Gies, D., and McCray, G. (1990). Transient production of proteins using an adenovirus transformed cell line. DNA Prot. Engineer. Tech. **2**: 3–10.
14. Fischman, T.O., Bentley, G.A., Bhat, T.N., Boulot, G., Mariuzza, R.A., Phillips, S.E.V., Tello, D., and Poljak, R.J. (1991). Crystallographic refinement of the three-dimensional structure of the Fab1.3-lysosome complex at 2.5Å resolution. J. Biol. Chem. **266**: 12915–12920
15. Padlan, E.A., Silverton, E.W., Sheriff, S., Cohen, G.H., Smith-Gill, S.J., and Davies, D.R. (1989). Structure of an antibody-antigen complex: Crystal structure of the HyHEL-10 Fab-lysozyme complex. Proc. Natl. Acad. Sci. USA. **86**: 5938–5942.
16. Colman, P.M., Laver, W.G., Varghese, J.N., Baker, A.T., Tulloch, P.A., Air, G.M., and Webster, R.G. (1987). Three-dimensional structure of a complex of antibody with influenza virus neuraminidase. Nature **326**: 358–363. [17] Prasad, L., Sharma, S., Vandonselaar, M., Quail, J.W., Lee, J.S., Waygood, E.B., Wilson, K.S., Dauter, Z., and Delbaere, L.T.J. (1993). Evaluation of mutagenesis for epitope mapping. J. Biol. Chem. **268**: 10705–10708.
18. Davies, D.R., Padlan, E.A., and Sheriff, S. (1990). Antibody-antigen complexes. Annu. Rev. Biochem. **59**: 439–473.
19. Sharma, S., Georges, F., Delbaere, L.T.J., Lee, J.S., Klevit, R.E., and Waygood, E.B. (1991). Epitope mapping by mutagenesis distinguishes between the two tertiary structures of the histidine containing protein, Hpr. Proc. Natl. Acad. Sci. USA. **88**: 4877–4881.

26

CONSTRUCTION OF CHIMERIC ENZYMES TO PROBE SUBSITE CONTRIBUTIONS TO CATALYTIC SPECIFICITY

Deepa Bhatt and Ben Dunn

Department of Biochemistry and Molecular Biology
University of Florida
Gainesville, Florida 32610

1. INTRODUCTION

Aspartic proteinases are characterized by a high degree of sequence as well as structural homology.[1] The subtle differences in specificity of these enzymes arise from interactions between the side chains of the ligand and the solvent exposed amino acids in the subsites that line the active site cleft. Studies using site-directed mutagenesis have suggested that although specific residues may contribute towards enzymatic efficiency, long range electrostatic interactions may also be involved.[2,3] These interactions may arise from the interdependency between subsites that are present on the two domains. The subsites present within the N-terminal domain include S_5, S_3, S_1, and the S_2' while those contributed by the C-terminal domain include S_4, S_2, S_1' and S_3'.

It has been demonstrated that the two domains of pepsinogen expressed from separate genes can independently fold and recombine to form a viable two-chain zymogen.[4] Also, active enzyme was obtained following the removal of the propeptide from a rearranged pepsinogen polypeptide which contained the C-terminus followed by the pro peptide and the N-terminal lobe.[5] These studies indicate the innate property of the domains of pepsinogen to independently fold.

Since both the domains of aspartic proteinases are involved in substrate binding and catalysis, the study of chimeras constructed by exchanging domains between the different enzymes of this family can be therefore be utilized to provide information on the structure–function relationship of the parent proteins.

2. CHIMERIC GENE CONSTRUCTION

Amplification of the amino and carboxyl terminal domains of porcine pepsinogen and rhizopuspepsinogen was carried out separately using the polymerase chain reaction.

Aspartic Proteinases, edited by James.
Plenum Press, New York, 1998.

In each case, one of the oligo primers used in the reaction had a unique SpeI restriction site. After purification and restriction digest, the PCR products were ligated to yield a chimeric gene that had the N-terminus of pepsinogen fused to the C-terminus of rhizopuspepsinogen. This was followed by a double digest with XbaI and PvuI to create compatible ends for ligation into the vector pET3d.

The vector with the chimeric gene was transformed into the HMS174 strain of *E. coli* . Successful transformants were grown overnight in LB media containing ampicillin at 37°C. Plasmids were isolated using the Promega 'Magic' Miniprep DNA purification system. Positive clones were screened by restriction analysis using SpeI and further verified by DNA sequencing (Sequenase Version 2.0. Kit USB).

3. EXPRESSION, REFOLDING, AND PURIFICATION OF CHIMERIC PROTEIN

The pET3d vector with the chimeric gene was transformed into an expression cell line *E. coli* BL21(DE3). The transformed cells were grown in LB media containing ampicillin to an $O.D._{600}$ of 0.5 at 37°C. IPTG (0.5 mM) was added to induce expression of the chimeric gene. After 3 hours of induction, the cells were harvested. Inclusion bodies containing the chimeric protein was recovered from lysed cells.

The inclusion bodies were solubilized in 8 M urea to a final concentration of 1 mg/ml. After refolding by dialysis, the chimeric protein was concentrated using an Amicon apparatus. The refolded protein was purified by gel filtration using a Superdex G75 column with 20 mM MOPS and 150 mM NaCl pH 6.0 as the eluting buffer. The fractions were analyzed by SDS-PAGE and protein concentrations determined by A_{280}. The activity of the fractions was determined by measuring the cleavage of a chromogenic octapeptide using a spectrophotometric assay as described below. Homogenous fractions that had similar activity were pooled and used for kinetic studies.

4. ACTIVITY STUDIES

Kinetic studies were performed on a Hewlett–Packard diode array Spectrophotometer according to the assay described by Dunn et al.,[6] using chromogenic octapeptide substrates. The enzyme-catalyzed hydrolysis of the substrate was monitored by the average decrease in absorbance over a range of 284–324 nm after incubating the enzyme for 3 mins at 37°C in 0.1 M sodium formate at pH 3.0. The kinetic parameters are given in Table 1.

5. RESULTS

Table 1 lists the kinetic parameters obtained with substrates of the form K–P–A–K–F*Nph–R–L having substitutions in the P_2 and P_3 positions. The substrates with hydrophobic residues in the P_2 position are slightly more favored. In the case of the P_3 substituents, the chimera accommodates small residues such as alanine and serine. Kinetic data obtained with P_2 substituents of the form K–P–I–E–F*Nph–R–L are also shown. Peptides with glutamate and alanine at this position were readily cleaved.

Table 1. Comparison of the kinetic parameters of the chimera with porcine pepsin[2] and rhizopuspepsin,[7] obtained with the chromogenic octapeptides with systematic substitutions in P_2 and P_3 positions

								Pepsin			Chimera			Rhizopuspepsin		
P_5	P_4	P_3	P_2	P_1	$*P_1'$	P_1'	P_2'	k_{cat} (s^{-1})	K_m (μM)	k_{cat}/K_m ($\mu M^{-1}s^{-1}$)	k_{cat} (s^{-1})	K_m (μM)	k_{cat}/K_m ($mM^{-1}s^{-1}$)	k_{cat} (s^{-1})	K_m (μM)	k_{cat}/K_m ($\mu M^{-1}s^{-1}$)
K	P	A	K	F	*Nph	R	L	65	83	0.8	0.49	68	7.2	12	11	1.13
			A					31	40	0.8	2.25	107	21.0	20	8	2.39
			L					19	28	0.7	1.05	61	17.2	9	4	2.31
			R					17	35	0.5	0.77	69	11.2	9	4	2.17
			S					30	58	0.5	1.30	132	9.9	9	9	1.03
			D					43	84	0.5	1.11	120	9.3	7	8	0.83
		S						10	57	0.2	0.62	343	1.8	9	16	0.6
K	P	I	E	F	*Nph	R	L	72	70	1.02	1.13	29	38.9	22	6	3.84
			A					30	25	1.20	0.49	30	16.3	8	5	1.56

Nph: p-nitrophenylalanine

Pepsin and rhizopuspepsin: 0.1 M Sodium formate buffer, pH 3.5

Chimera: 0.1 M Sodium formate buffer, pH 3.0

6. DISCUSSION

Aspartic proteinases have subtle differences in specificity in spite of sharing a high degree of primary and tertiary structural homology. The construction and kinetic characterization of chimeric enzymes, created by exchanging domains between the various aspartic proteinases, is a novel approach to explore these differences.

A chimeric protein with the N-terminal domain of porcine pepsinogen and the C-terminal domain of rhizopuspepsinogen has been expressed in a bacterial system. After solubilization in urea, the protein was refolded by dialysis. The chimeric zymogen was purified by gel filtration. The majority of the protein obtained was polymeric, however a small amount of monomeric material was also obtained. Autoactivation of the monomeric zymogen to the mature enzyme at low pH has been observed by SDS-PAGE and verified by N-terminal sequencing.

Kinetic studies were done with homogeneous purified protein. Since the C-terminal domain of the chimera is that of rhizopuspepsin, the S_2 subsite of the chimera would be similarly hydrophobic, and peptides with hydrophobic amino acids in the P_2 position would be expected to be readily cleaved. This effect, which is similar to rhizopuspepsin, is seen with the various P_2 substitutions of the octapeptides, based on the K–P–A–K–F*Nph–R–L sequence. When substrates with P_2 variations of the form K–P–I–E–F*Nph–R–L were examined, glutamate was preferred over alanine, in the case of both the chimera as well as rhizopuspepsin as expected. In the P_3 position pepsin is known to favor the hydrophobic alanine more than serine. As the S_3 subsite of the chimera, which is present in the N-terminal domain, is that from pepsin, a similar preference is noted for the chimeric enzyme when substrates with P_3 variations of the parental peptide K–P–A–K–F*Nph–R–L were studied.

Future efforts will be directed towards increasing the yield of monomeric purified protein. More extensive kinetic analysis, as well as pH dependence studies, will be done to investigate the contributions of the subsites towards catalytic specificity.

REFERENCES

1. Davies, D. R. (1990) *Ann. Rev. Biophys. Chem. 19,* 189–215.
2. Rao, C. M. (1994) Ph.D. Thesis, Univ. of Florida, Gainesville, FL.
3. Lowther, T., Majer, P. and Dunn, B. M. (1995) *Protein Science 4*, 689–702.
4. Lin, X., Lin, Y., Koelsch, G., Gustchina, A., Wlodawer, A. and Tang, J. (1992) *J Biol Chem 267*, 17257–17263.
5. Lin, X., Koelsch, G., Loy, J. and Tang, J. (1995) *Protein Science 4*, 159–166.
6. Dunn, B. M., Scarborough, P. E., Davenport R. and Swietnicki W. (1994) Peptide Analysis and Protocols ed. Pennington M. and Dunn B. M. *Methods in Molecular Biology*, Humana Press 36, 225–243.
7. Lowther, T., (1994) Ph.D. Thesis, Univ. of Florida, Gainesville, FL.

27

EXPRESSION OF MULTIPLE GENES FOR PREGNANCY-ASSOCIATED GLYCOPROTEINS IN THE SHEEP PLACENTA

Sancai Xie, Jonathan Green, and R. Michael Roberts

Departments of Animal Science, Biochemistry and Pathobiology
University of Missouri-Columbia
Columbia, Missouri 65211

INTRODUCTION

Bovine pregnancy-associated glycoprotein 1(boPAG1) (1), also known as pregnancy-specific protein B (2), was initially identified as a unique placental antigen in cattle by raising antisera against total placental proteins in rabbits. Serologically similar antigens have also been found in sheep and other ruminant ungulates (3). Subsequently, molecular cloning showed that bo- and ovine (ov) PAG1 belong to the superfamily of aspartic proteinases (4). Unlike other known aspartic proteinases, however, both bo- and ovPAG1 appear to be enzymatically inactive due to key mutations around the catalytic site regions (4,5). For instance, the catalytic aspartic acid residue in the C-terminal lobe of ovPAG1 is mutated to glycine. A similar mutation in the N-terminal lobe of porcine pepsin completely inactivates the molecule (6). By analogy, it seems unlikely that ovPAG1 could function as a proteinase. BoPAG1, despite retaining both active site aspartic acid residues, has a mutation within the N-terminal D–T–G triad, in which the highly conserved glycine is replaced by an alanine. It is hypothesized that the methyl group of the alanine would displace the catalytic water molecule from its normal position between the two catalytic aspartic acid residues (4,7), thereby impairing the activity of the molecule. In addition, other critical alterations around the substrate binding pockets are likely to disfavor the potential of both PAG1 as active proteinases (7).

Immunocytochemistry and in situ hybridization experiments showed that bo- and ovPAG1 are expressed exclusively in the outer epithelial cell layer of the placenta and are localized specifically to a group of binucleate trophoblast cells (4,5,8). The latter are the invasive component of the trophoblast and can fuse with the adjacent uterine epithelium to form a multinucleate syncytium (9). As a result, binucleate trophoblast cells can deliver their products directly to maternal blood. Consequently, PAG1 is found in the maternal circulation and can be used as a biochemical marker for pregnancy (10).

Aspartic Proteinases, edited by James.
Plenum Press, New York, 1998.

Soon after the initial cloning of ov- and boPAG1, several other PAG were identified (5,8,11). Here, we attempt to estimate the number of PAG-related genes and their products in the ovine placenta.

MATERIALS AND METHODS

Southern Blotting

Ovine and bovine genomic DNA (5 μg) was digested to completion with a series of restriction endonucleases. The DNA fragments were resolved on 0.8% (w/v) agarose gel by electrophoresis and blotted onto nylon membrane. An 803 bp DNA probe from the boPAG1 gene (a part of intron 6, exon 7, intron 7, exon 8 and a part of intron 8) was labeled by using the polymerase chain reactions (PCR) procedure in the presence of $[^{32}P–\alpha]$dATP (11). Hybridization was performed under standard conditions (11).

Cloning of ovPAG Gene Fragments

DNA isolated from sheep white blood cells was used as a template to amplify PAG genes at regions corresponding to part of exon 3, intron 3 and part of exon 4 by PCR procedure with two sets of primers. The first set was derived from regions well conserved in both ovPAG1 and boPAG1 (bp314e3f, 5′TGGGTAACATCACCATTGGAA3′; bp314e4r, 5′TTG TTGTTCATGACACAGTTC3′). The second set had sequences identical to ovPAG–2 in corresponding regions (bp314e3f, 5′TGGGTAACATCACCATTGGAA3′; op13e4r, 5′CTT GCCTATGACATGTTGG3′). The amplified ovPAG gene fragments were then cloned into TA cloning vectors (Invitrogen, CA, USA) and partially sequenced. If a clone represented a distinct ovPAG gene, it was fully sequenced.

Cloning of ovPAG Transcripts from Early Conceptuses and Late Placenta

Two procedures were used to clone ovine PAG transcripts from day 13 conceptuses and day 100 placenta. First, a day 13 ovine conceptus cDNA library was screened with a mixture of boPAG1 and ovPAG1 cDNA labeled with ^{32}P by using random primer reactions. The positive clones were isolated and partially sequenced. Novel PAG transcripts from day 100 ovine placenta were cloned by reverse transcription (RT) and PCR procedures. In brief, cellular RNA extracted from a day 100 placenta was first reverse transcribed into cDNA and amplified by PCR with a pair of well conserved primers (ovPAG forward, 5′AGGAAG-AAGCATGAAGTGGCT3′; ovPAG reverse, 5′TAAATGCTTGGACTTGTTCAGGAA3′). The RT–PCR products were cloned into TA cloning vectors (Invitrogen, CA, USA). All the novel ovPAG cDNA were fully sequenced in both directions.

RESULTS AND DISCUSSION

There Are Multiple PAG Genes in Both the Ovine and Bovine Genomes

Southern blotting was performed to determine the number of genes related to boPAG1 and ovPAG1 in the bovine and ovine genomes. Multiple strong hybridization bands of varying intensities were detected in each lane in the bovine genomic blot (Fig. 1A), suggesting that several genes closely related to boPAG1 gene were probably present. In addition, several bands that hybridized much more weakly with the probe were also detected. The latter likely represent genes that belong to the PAG family, but share lower sequence identity with the probe. Similarly, graded hybridization signals could be detected in digests of ovine DNA (Fig. 1B). Since the probe used spanned only a small part of the PAG–1 gene that lacked sites for the restriction enzymes employed for DNA digestion (11), the data indicate that there are many genes with close identity to PAG–1 in cattle and in sheep.

To confirm the existence of multiple PAG genes in the ovine genome, fragments of ovPAG genes, corresponding to exon 3, intron 3 and exon 4, were amplified from ovine genomic DNA with two sets of primers, and cloned into TA cloning vectors. Fourteen

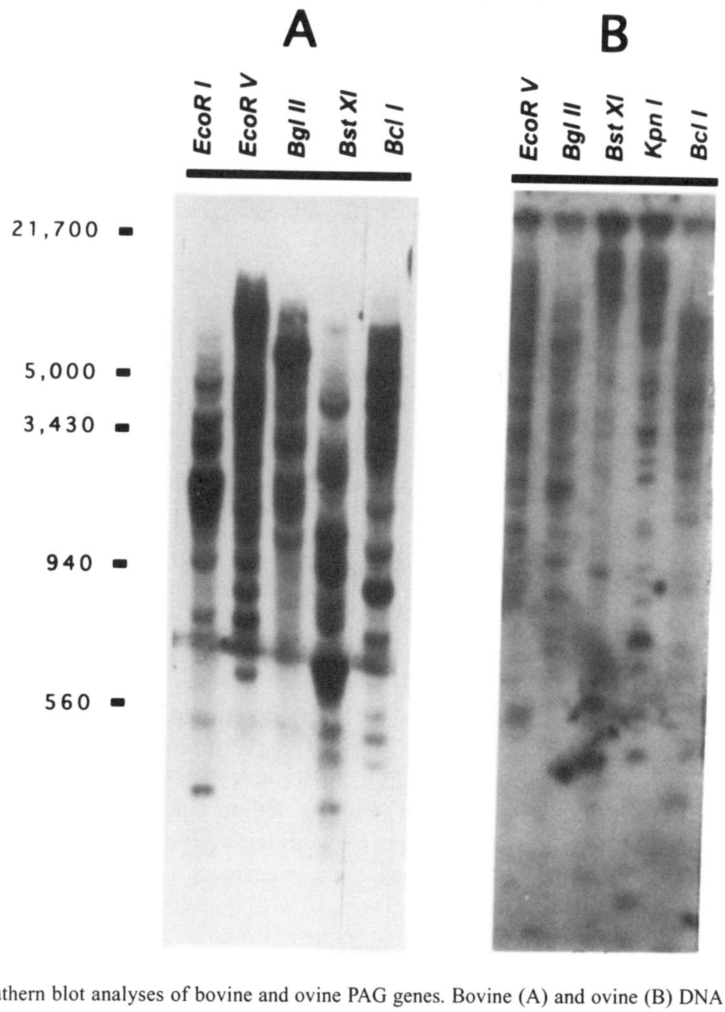

Figure 1. Southern blot analyses of bovine and ovine PAG genes. Bovine (A) and ovine (B) DNA (5 μg) was digested to completion with the restriction endonucleases indicated above the gel. The DNA fragments were resolved on 0.8% (w/v) agarose gel by electrophoresis and blotted to nylon membrane. An 803 bp DNA probe from boPAG1 gene (a part of intron 6, exon 7, intron 7, exon 8 and a part of intron 8) was labeled by PCR. Hybridization was performed under standard conditions (11).

```
Consensus  TGGGTAACAT CACCATTGGA ACACCYCYTS AGGAATTCCA GGTTVTCTTY GACACAGSYT CAKCTKWHTT GTKGGTGSCS TCYRWCWWTT   90
OPGB17     .......... .......... .....C.C.C .......... ....G....T .......GC. ..T..GAC.. ..T....C.C ..CAT.AA..   90
OPGB26     .......... .......... .....C.C.G .......... ....G....T .......GC. ..T..GAT.. ..G....C.C ..TGA.TT..   90
OPGB 6     .......... .......... .....C.C.C .......... ....G....T .......GC. ..G..GTT.. ..G....C.C ..CGT.TT..   90
OPGB16     .......... .......... .....T.C.C .......... ....G....T .......CT. ..T..GAC.. ..G....C.C ..CAT.TT..   90
OPGB14     .......... .......... .....C.C.C .......... ....G....T .......GC. ..T..TTT.. ..G....C.C ..TGT.TT..   90
OPGB25     .......... .......... .....C.C.C .......... ....C....T .......GC. ..T..TTT.. ..G....C.C ..TGT.TT..   90
OPGB21     .......... .......... .....C.C.C .......... ....A....T .......GC. ..T..GAC.. ..G....C.C ..TGT.TT..   90
OPGB22     .......... .......... .....C.C.C .......... ....A....T .......GC. ..T..GAC.. ..G....C.C ..CGA.TT..   90
OPGB23     .......... .......... .....C.C.C .......... ....G....C .......GC. ..T..GAC.. ..G....C.C ..CAA.TT..   90
OPGB 1     .......... .......... .....T.T.C .......... ....G....T .......GC. ..T..GAA.. ..G....C.C ..CAT.TT..   90
OPGB24     .......... .......... .....C.C.C .......... ....G....T .......GC. ..T..GAA.. ..G....G.G ..CGA.TT..   90

Consensus  GHVHBAKYBM ARCCAAGAGA CYCTGKDGTS AGTAYRAGAY AAMMYNVTAC CBRRABBRTC YTTCACYKRC CYTGYYRCYY TRTTYHSRHC  180
OPGB17     .CCTC.GCCC .A........ .C...TA..G ....-G...C -.CCCTG... .CAG.CTA.. C.....TGG. .T..CCA.CC .G..TACGA.  178
OPGB26     .CACC.GTCC .G-------- -C...TT..G ....-G...- ..CCCTC... .TAG.CCG.. C.....TTG. .C..CCA.CC .A..TCCAC.  169
OPGB 6     .CAAC.GCTC .A-------- -C...TT..G ....CA-..C -.CCCCA... .CGA.TGA.. T.....TTG. .C..CCA.CC .G..TTCAC.  169
OPGB16     .CAAC.GCTC .A-------- -C...TT..G ....CA-..C -.CATCC... .CGG.CCA.. T.....TTA. .C..CCA.CC .G..TCCAC.  169
OPGB14     .CAAG.GCTC .A-------- -C...TT..G ....TG-..C -.CCCCC... .GGG.CCA.. T.....CTG. .C..CCA.CC .G..CCCAC.  169
OPGB25     .CAAG.GCTC .A-------- -C...TT..G ....TG-..C -.CCCCC... .GGG.CCA.. T.....CTG. .C..CCA.CC .G..CCCAC.  169
OPGB21     .AAAC.GCTC .A-------- -C...TT..C ....CA-..- -.CCCCC... .CAG.GCA.. C.....TTG. .C..CTG.CC .G..TCGAC.  168
OPGB22     .CACC.GTCC .G-------- -C...TT..G ....CA-..C -.ACCCA... .CAG.GCA.. C.....TTG. .C..CTG.CC .G..TCGAC.  169
OPGB23     .TGCC.TTGA .G-------- -C...TT..G ....CA-..T -.CCTTG... .CGG.CCA.. C.....TTG. .C..CCA.TC .G..TCCAT.  169
OPGB 1     .CAAC.GCTC .G-------- -C...GT..G ....CA...C ..CCCAG... .CAG.CCA.. T.....TTG. .C..CCA.CT .A..TCCAC.  171
OPGB24     .CACT.GTCA .G-------- -T...TG..G ....CA-.-- -.CCCGA... .CGG.CCA.. C.....TTG. .C..TCA.CC .G..TTCAT.  167

Consensus  CKTGRCAMMT GRTGACAYYC AYCWCTTKTD TCTVCMRSYA MACRNRHTAR GTTYARAYAH CDYSWKTCKT CYAYCYTCYG GHYTRYMVRW  270
OPGB17     .T..G..CC. .A.....CT. .T.T...G.T ...G.AGGT. A..AAGA..A ...C.A.C.T .ACCAG..T. .C.C.T..C. .TT.ACCAAT  268
OPGB26     .G..G..CA. .A.....TT. .T.T...G.G ...A.AGCT. C..GTGT..G ...C.G.C.A .ATCAG..T. .C.C.T..C. .TC.ACCCGA  259
OPGB 6     .T..G..CC. .A.....CT. .T.T...G.G ...G.AGCT. C..ACGT..G ...C.G.C.C .TTGAG..T. .C.T.T..C. .CC.ACCAAT  259
OPGB16     .T..G..CC. .A.....CC. .T.T...G.G ...G.AGCT. C..GTGT..G ...C.G.C.T .GCCAG..T. .C.C.T..C. .CT.GCCAAT  259
OPGB14     .T..G..CC. .A.....CT. .T.T...G.G ...A.AGCT. A..GCAT..G ...C.G.C.T .TTCAG..T. .C.C.C..C. .CT.ACAGAT  259
OPGB25     .T..G..CC. .A.....CT. .T.T...G.G ...A.AGCT. A..GCAT..G ...C.G.C.T .TTCAG..T. .C.C.C..C. .CT.ACAGAT  259
OPGB21     .T..G..AC. .G.....CT. .T.T...G.G ...G.CACC. C..AGGT..G ...C.A.C.T .TTCAT..G. .C.C.T..T. .CT.ATCAAT  258
OPGB22     .T..G..AC. .G.....CT. .T.T...G.G ...G.AGCT. C..AGGC..G ...C.G.T.T .GTCAG..T. .C.C.T..C. .CT.ACCAAT  259
OPGB23     .T..G..CC. .A.....CT. .T.T...G.G ...G.AGCT. C..AGGC..G ...C.G.T.T .GTCAG..T. .C.C.T..C. .CT.ACCAAT  259
OPGB 1     .T..A..CC. .A.....CT. .C.T...T.A ...G.AGCT. C..ATGT..G ...T.G.C.T .TTCAG..T. .C.C.T..T. .CC.ACCAAT  261
OPGB24     .T..G..CC. .A.....CT. .T.A...G.G ...C.AGCT. C..ACAA..G ...C.G.T.T .GTCTG..T. .T.C.T..C. .AT.ACCAAT  257

Consensus  RRVAYMTKCD GNWTMDHSTW YRRWKYYRGR RBMMTKRRAR RADTTGTTGT TCATGACACA GTTC  334
OPGB17     GAC.CC.T.A .AA.CTAC.T TGGTTCTG.A ACAA.GAG.A G.T....... .......... ....  332
OPGB26     AAG.CC.T.A .CA.CAAC.A TGGATCTG.G ACAA.GAG.G A.G....... .......... ....  323
OPGB 6     AAG.CA.T.T .GA.CAAC.A TGGAGCTG.G AGAA.TGA.G G.G....... .......... ....  323
OPGB16     AAG.CA.T.G .GA.CATG.A TGGAGTTG.G ATAA.TGA.G G.A....... .......... ....  323
OPGB14     AAG.TC.G.A .GA.AACC.A CGGATCCA.A AGAA.TAA.G G.A....... .......... ....  323
OPGB25     AAG.TC.G.A .GA.AACC.A CGGATCCA.G AGAA.TAA.G G.A....... .......... ....  323
OPGB21     AGA.CC.T.T .TA.CACC.A TGGATCTG.G GGCC.GAA.G G.G....... .......... ....  322
OPGB22     AGA.CC.T.T .TA.CACC.A TGGATCCG.G AGAA.TAA.G G.G....... .......... ....  323
OPGB23     AGA.CC.T.T .TA.CACC.A TAGATCTG.G GGCC.GAA.G G.G....... .......... ....  323
OPGB 1     AAG.CC.T.A .CA.CACC.A TGGATCCG.G AGAA.TAA.G G.G....... .......... ....  325
OPGB24     AAG.CC.T.A .CT.CGCC.A CGAATCTG.G AGAA.GAA.G G.G....... .......... ....  321
```

Figure 2. Alignment of the nucleotide sequences of cloned ovPAG gene fragments. Sheep genomic DNA was amplified by PCR with primers that were identical to ovPAG1 cDNA at the regions corresponding to exon 3 and 4 of PAG genes. Eleven of amplified gene fragments were cloned, sequenced and aligned using GeneWorks 2.3. The first and the last 21 nucleotides were the sequence of the primers that had been used to amplify the ovPAG gene fragments. Sequences of exon 3, encoding for the N-terminal active site region, were more conserved (about 90%) than those of exon 4 (about 80%). The nucleotide sequences at the 5′ donor and 3′ acceptor sites of all intron 3, except opgb21 (GT–CA), conform to the GT–AG rule. Note that the sequences of intron 3 of those cloned ovPAG gene fragments are as conserved as those in exon 4. Gaps were introduced to produce maximal alignment. The primer sequences that were used to amplify the inserts were boxed and shaded. The sequence of intron 3 was boxed.

distinct gene fragments were identified with the first set of primers and eleven of these were fully sequenced. As shown in Fig. 2, exons 3 and 4 of the eleven ovPAG genes were well conserved in nucleotide sequence and identical in length, except opgB17, which had a three-codon insertion in exon 3. Intron 3, whose length ranged from 96 to 100 bp was also well conserved. An additional group of five ovPAG gene fragments were amplified with the second set of primers. Again, their exons were very similar in length, but, in this instance, intron sizes differed markedly, ranging from 70 bp (opgA12) to 144 bp (ogpA5) (data not shown). The extent of nucleotide identities among all the cloned ovPAG gene fragments varied from 62 to 94%. Those amplified with the first set of primers shared approximately 80 to 90% identity while the ones primed by the second set were much more variable (62 to 80%).

This data confirmed what had been suspected from southern blotting, namely that there were multiple genes for ovPAG. A total of 19 distinct PAG genes were identified,

```
          -1            -15
Consensus MKWLVLLGLVAFSECIVKIPLRRVKTMRKTLSGKNMLNNFLKEHPYRLSQISFR.SNLTIHPLRN..D..YVGNITIGTPPQEFQVVFDT   90
ovPAG4    ................F.....................DV........P.....G...I......IR.TF................L...   90
ovPAG7    ................F.V....L.........N.....V...Y....P.....D..V..L....MK.IF...........P.......S   90
ovPAG3    ......F...........V....L.........N.I...V...Y....P.....D..V..L....MK.IF....................   90
ovPAG6    ..........SI....................E...........A.........G..........TK.LV.L...............S..   90
ovPAG1    ............................N.....K...S.....A.........A..........IM.ML....................   90
ovPAG5    ....SV....RV.....E...MK.NV.QE..NE.SL......KY....A.N.T.N.KMAFV.M..FL.LA..P.M...RGGEQ.R.....   90
ovPAG9    ..................M..........NP.....M.......A.........GL...-...S.YI.ML.T..........K...I...   89
ovPAG2    .......W...L.....IM..TKT....EI.RE..L.....E.QAN.M.DD.ASDPK.ST.....AL.MA....V......K..R.....   90
ovPAG8    .......A........I..........SN.A...........K...........G....T...M.IW.LL.L..............L..R   90

Consensus GSSDLWVPS.FC.S.T---CS.H.RFRH.QSSTFR.TNKTF.I.YG.G.MKGVVAHDTVRIGDLVSTDQPFGLS..E.GF...PFDGVLG  180
ovPAG4    ...V.....VL.N.S......I.V....L......T.....W.T..A.T.................I.......MA.Y..HGRR......  177
ovPAG7    T........I.WN.S......TLV.K..R......T.....W.T..A.T.................I.......MA.Y..MDRR...I..  177
ovPAG3    A........I..N.S......TRV....R......LA....G.M..A.K.....V...................VV.S..EHRQ......  177
ovPAG6    .........D..AIEA.....L.T....L......P..R..S.T..C.TV....V...................TA.HVSRCT.......  177
ovPAG1    .....L...IN.L.P.KRP..KQDK.K.H......F..D..R.YF.S.T.R.F....................IFL.SWLD-I....I..  179
ovPAG5    .........N..T.PA....YS.IT.KYWE...Y.H.T.P.E.A..S.RI..HL.Y..IQ..N...........LV.Y..NGL....I..  177
ovPAG9    .........N..T.PA.....TQA...YR......L..R..C.T..S.GL..I.........N..........TLK.Y..ENI.......  176
ovPAG2    .........IK.I.PA....YT.IT.D.HK.....L.RRP.R.L..S.M.N..L.Y......N...........LQQF..DNA.......  177
ovPAG8    .........LL.N.S.....AK.VM...RL...Y.L.....M.F.RV.KIE...VR..........A..T....IA.T..ENTTL..I..  177

Consensus LNYP..S..G.IPIFDKLKNEGAISEPVFAFYLSKD.QEGSVVMFGGVDH.YYKGELNWVPL..AGDW.I.VDRISM.R.VIACS.GC.A  270
ovPAG4    ....RQ.CCRPT........Q...............E.............R...........VK.D..T.Q......R.E.....D..D.  267
ovPAG7    ....RQ.CSKTKW......SQ...............E.............T.......SL..VK.D..S.H......R.E.....D..D.  267
ovPAG3    ....NL.FSKT.........................E.............S....K......IK....SVR..S.T.K.E.....D..R.  267
ovPAG6    ....SI.FWST.........................G.............R...........IP..N.MVHM...YIE.N.....A..K.  267
ovPAG1    ....KI.FS.A............F.........N..K.............R...........IHP.E.S.PL.....R.K.....G..E.  269
ovPAG5    ....N-.IL.A.....N..KQ......I.......GTVN...L.L....KA........I..IRV...R.R..H...KGKL.G..G..E.  266
ovPAG9    ....NI.FS.AV............L.......F...E.............R..E........TK....IVRL....IG.K.....GD.E.  266
ovPAG2    .S..SLAVP.T........QQ......I......TNKEN...L.L.....S.H..K...I.VSQTKS.L.T......NGK..G.EH..E.  267
ovPAG8    .S..NT.CF.A.......................N----------------------------------.TMK.-----E.....D..R.  228

Consensus LVDTG.S.I.GP.RLVDNIQKLIGA.P.G.KHYVSCSAVNTLPSIIFTINGINYPVPAQAYILK.S.G.CYT.F...RVR.STE.WVLGD  360
ovPAG4    .L...A.F.H..G..I.D......SE.RDL...I..............................G.T.H...A.RAK...T...S.....  357
ovPAG7    .L...A.F.H..G..I.D......SEQRDF...I..............................G.T.H...A.RAK...T...S.....  357
ovPAG3    .....S.H.Q..G..I..V.....TM.Q.SM...P.............................GFT.H...A.RAK...T...S.....  357
ovPAG6    V....AAF.E..KSQ...M..F...R.R.S.Y..P..V.........R..S......GR.....NHR.R...T.KENQWSP...I.I...  357
ovPAG1    ..G..T.L.L..RTV.E....H...TQQCFEYF......YA....V...........P...LV.D.R.Q..SP.QVN.ANP.A.N.I...  359
ovPAG5    .....P.L.N..T...T...R....M.L.P.......TI.I.............TL........T.Q.G.F.T.KGA..FKHR.T.....  356
ovPAG9    V....TAF.G..RK..KK.......RRR.N.---------------------------------DDA.R.S.T.QEN...L...S.L..N  323
ovPAG2    .....T.L.H..AGP.T....F.H.V.YDSEYM...PVISI..PV.....S.D.S...E...Q.I.NSL.LSS.HGDN----.DK.I...  353
ovPAG8    .....S.H.Q..G.....V..H..TM.Q.SM...P......................GR.....DFR.R...N.KKKIA.TF..Y....E  318

Consensus VFLRLYFSVFDRGNDRIGLAPAV  383
ovPAG4    ......................M  380
ovPAG7    ......................M  380
ovPAG3    .............H......R.G  380
ovPAG6    .............H......R..  380
ovPAG1    ....R...............R..  382
ovPAG5    I........Y..EHN..A..Q.R  379
ovPAG9    ....V.L................  346
ovPAG2    .........Y....N........  376
ovPAG8    P......................  341
```

Figure 3. Alignment of the inferred amino acid sequences from ovPAG cDNA. The inferred amino acid sequences were aligned using GeneWorks 2.3. OvPAG1 and ovPAG2 were isolated from a day 100 placental cDNA library and a day 13 conceptus cDNA library, respectively, by homologous hybridization. Other ovPAG, ovPAG3 to 9, were isolated from a day 100 placenta using an RT–PCR procedure. Two alternatively-spliced transcripts of ovPAG4 were identified, only the full length one was shown here. No full length clones for ovPAG8 and 9 were isolated and their sequences shown here are the alternatively-spliced forms, that lacked one exon. Gaps were introduced to produce maximal alignment. The conserved active site regions were boxed.

among which 16 were fully sequenced. However, only three of those corresponded with the individual ovPAG cDNA shown in Fig. 3 (opgB17, ovPAG–1; opgA12, ovPAG–2; opgA5, ovPAG–5). Clearly there are many more PAG genes than those represented in Fig. 2.

Expression of Multiple ovPAG Genes in the Placenta

By day 13 of pregnancy the ovine conceptus has begun to elongate from its previous spherical form to a longer thread tubular structure which is still not attached to the uterine wall. At this stage the embryo proper is very small while the elongating trophoblast (preplacenta) constitutes >95% of the total mass of tissue. To gain an estimate of how many PAG genes are transcribed in the early conceptus at this stage, a day 13 ovine conceptus cDNA library was screened with a mixture of ^{32}P-labeled boPAG–1 and ovPAG–1 cDNA

(4). Twenty-seven individual clones were isolated and partially sequenced. All represented an identical transcript, named ovPAG2. OvPAG transcripts in later stage tissue were isolated by employing an RT–PCR procedure on RNA from day 100 placenta. A total of 39 clones was partially sequenced, from which seven novel transcripts were identified. They were named from ovPAG3 to ovPAG9 according to the order in which they were discovered. In Fig. 3, the amino acid sequences interpreted from the cDNA are compared with that of the already known ovPAG–1.

Of the nine PAG shown, ovPAG–4 most resembles that of the consensus. Individual sequences differed from each other within the range of 10 to 40%, and three groupings could be recognized. The first group consists of those that resemble ovPAG1, including ovPAG1, 3, 4, 6, 7, 8, and 9 (Fig. 3), which differ by no more than 30% from each other in amino acid sequence. OvPAG4 and 7, in particular, share more than 90% identity. OvPAG2 and 5, however, have retained low resemblance, either to each other (about 60%) or to the remaining PAG (about 60%). Interestingly, a PAG (boPAG–11) recently cloned from a day 25 bovine placental cDNA library closely resembles ovPAG–2 in sequence (94% identity) (Xie S. and Roberts R.M., unpublished data) and is probably the bovine homolog of ovPAG2.

SUMMARY

Together these experiments have showed that there are many PAG genes in the ovine genome and many of these are expressed in the placenta. These PAG genes vary as much as 40% in nucleotide sequence identity. Thus, a wide range of ovPAG molecules are secreted at the placental-uterine surface during pregnancy.

ACKNOWLEDGMENTS

This work is supported by NIH grant HD29483 to R.M.R. The authors thank Kelli E. Valdez and Livija Hakami for sequencing cDNA clones and Jim Bixby and Gail Foristal for aligning DNA sequences.

REFERENCES

1. Zoli AP, Beckers JF, Woutters-Ballman P, Closset J, Falmagne P, Ectors F. Biol Reprod 1991; 45:1–10.
2. Butler JE, Hamilton WC, Sasser RG, Ruder CA, Hass GM, Williams RJ. Biol Reprod 1982; 26:925–933.
3. Willard JM, White DR, Wesson CAR, Stellflug J, Sasser RG. J Anim Sci 1995; 73:960–966.
4. Xie S, Low BG, Kramer KK, Nagel RJ, Anthony RV, Zoli AP, Beckers JF, Roberts RM. Proc Natl Acad Sci USA 1991; 88:10247–10251.
5. Roberts RM, Xie S, Nagel RJ, Low BG, Green J, Beckers JF. Glycoproteins of the aspartic proteinase gene family secreted by the developing placenta. Adv Exp Med Biol 1995; 362:231–240.
6. Lin XL, Wong RNS, Tang J.. J Biol Chem 1989; 264:4482–4489.
7. Guruprasad K, Blundell TL, Xie S, Green J, Szafranska B, Nagel RJ, Mcdowell K, Baker CB, Roberts RM.. Protein Engineering. 1996; 9:849–856.
8. Xie S, Low BG, Nagel RJ, Beckers J-F, Roberts RM. Biol Reprod 1994; 51:1145–1153.
9. Wooding FBP. Placenta 1992; 13:101–113.
10. Sasser RG, Ruder CA, Ivani KA, Butler JE, Hamilton WC. Biol Reprod 1986; 35:936–942.
11. Xie S, Green J, Beckers JF, Roberts RM. Gene 1995; 159:193–197.

28

MASS SPECTROMETRY AS A TOOL FOR STUDYING THE ACTION OF HUMAN ASPARTIC PROTEASES ON PEPTIDES AND PROTEINS

N. I. Tarasova,[1] G. Czerwinski,[1] N. B. Roberts,[2] J. Lubkowski,[1] T. L. Sumpter,[1] B. N. Green,[3] and C. J. Michejda[1]

[1]ABL-Basic Research Program
NCI-FCRDC
Frederick, Maryland 21702, USA
[2]Royal Liverpool Hospital
University of Liverpool
United Kingdom
[3]Micromass UK Ltd.
Altrincham, United Kingdom

INTRODUCTION

In the late eighties, two novel mass spectrometric methods with the sensitivity in the femtomole range were introduced, electrospray mass spectrometry (ESMS) and matrix assisted laser desorption ionization time-of-flight mass spectrometry (MALDI TOF). These methods have been used for a variety of applications in protein chemistry (reviewed in [1] and [2]). In MALDI TOF the analyte is mixed with UV light-absorbing substance (matrix). The matrix and the compound to be analyzed are allowed to cocrystallize on the probe surface of the mass spectrometer. The sample is irradiated with UV light from a laser. The light is absorbed by the matrix causing the sample to be vaporized and ionized. The generated ions are accelerated in an electric field and are separated and detected by a time-of-flight detector. MALDI TOF was used for analysis of proteins up to 10^5 Da. ESMS is performed by evaporating droplets formed by the electrospraying of the protein solution followed by the transfer of the ions into the high vacuum. The evaporation generated highly charged ions are analyzed by the quadrupole mass spectrometer. We have applied the two techniques to the study of action of proteases on natural peptide and protein substrates. The present paper describes the advantages and limitations of the methods.

Aspartic Proteinases, edited by James.
Plenum Press, New York, 1998.

METHODS

Digestion of Peptides

Synthetic human gastrin and sulfated cholecystokinin octapeptide were purchased from Research Plus Inc. (Eugene, OR). Human liver cathepsin D was from Calbiochem. Human stomach gastricsin and the major isoform of pepsin A (3b) were purified as described in [3].

1mM solutions of peptides in 0.1 M Na–acetate buffer pH 4.5 were incubated with the enzymes at 1:4000 molar ratio at 37°C for 1 hour, 3 hours and overnight. The mixtures were frozen in liquid nitrogen and kept at –70°C before analysis.

Matrix-Assisted Laser Desorbtion Mass Spectrometry

The mass spectra were generated on Bruker Reflex II time of flight mass spectrometer. The instrument was operated with an acceleration voltage of 28.5kV. Samples were desorbed/ionized with an neodymium: YAG laser (l=355 nm). α–Cyano–4–hydroxycinnamic acid was used as the matrix. Samples were mixed 1:1 with a saturated solution of a–cyano–4–hydroxycinnamic acid in 30:70 acetonitrile / 0.1% TFA in water and allowed to dry on a sample holder.

A computer program was written for the analysis of the data. The program finds the matches for the detected molecular masses within the inputted sequence taking in consideration possible salt formation with sodium or potassium ions in the defined mass error interval. The program was written for PC and can be obtained by writing to Dr. Jacek Lubkowski.

Analysis of Hemoglobin Hydrolysis by Electrospray Mass Spectrometry

Hemoglobin (HbA) (3 mg/ml) was digested with pig pepsin (0.06 mg/ml) at 37°C and pH = 2, 3 and 4 (adjusted with HCl). After various incubation times, 20ml of the digest was diluted with 80 ml of water and 100 ml of acetonitrile and analyzed by ESMS on Quattro II (Micromass UK, Ltd, Altrincham, UK). Data were acquired for 2.5 min over the m/z range 150–1650 from 20 ml of the diluted digest introduced into the electrospray source at 5 ml/min.

RESULTS AND DISCUSSION

Cleavage of Peptide Hormones

Peptide hormones have short lifetimes in the circulation. The half life of gastric hormone gastrin, for example, is approximately 6 min [4]. Hormone deactivation and clearance are important physiological processes but the mechanisms are largely unknown. The most common pathway of peptide hormone deactivation takes place after interaction with specific receptors. We have demonstrated that gastrin undergoes endocytosis after interaction with the receptor and ends up in lysosomes, where it gets degraded by lysosomal proteases [5]. Primary clearance sites for gastrin are the major sites of gastrin receptor expression, the kidney and stomach [6,7]. Besides acting through the blood stream, gastrin is also secreted into the lumen of gastric glands where it can be either degraded by proteases or can interact with the receptor on parietal and chief cells. We have chosen gastrin

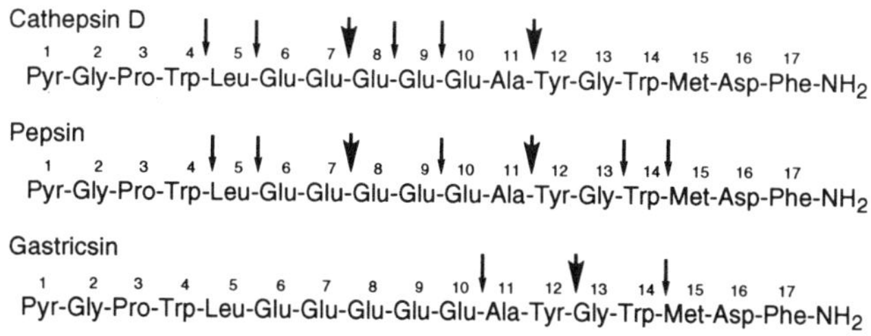

Figure 1. Cleavage of human gastrin with human aspartic proteases.

cleavage by aspartic proteases of gastric juice (pepsin and gastricsin) and lysosomal protease cathepsin D as a system to evaluate the applicability of MALDI–TOF spectrometry for studying the enzymatic cleavage of peptides and to characterize the degradation pathways of an important hormone.

Gastrin shares the same carboxyamidated C-terminus with cholecystokinin (CCK). The sequence –Trp–Met–Asp–Phe–NH_2 constitutes the active site of both hormones. Gastrin and CCK have many overlapping activities, so it was interesting to compare the action of the proteases on the structurally and physiologically related peptides.

All three proteases readily cleaved gastrin (Figure 1). Cathepsin D and pepsin primarily cleaved the hormone at Glu7–Glu8 and Ala11–Tyr12 bonds. Both enzymes also hydrolyzed at Trp4–Leu5, Leu5–Glu6 and Glu9–Glu10 sites. Cathepsin D also cleaved Glu8–Glu9 bond that was not hydrolyzed by pepsin and pepsin cleaved Gly13–Trp14 and Trp14–Met15 sites that were not touched by cathepsin D. Gastricsin hydrolyzed mainly Tyr12–Gly13 bond, which was not cleaved by pepsin and cathepsin D. Cleavage of Glu10–Ala11 and Trp14–Met15 was also detected.

Cholecystokinin remained almost intact even after overnight incubation with the enzymes at 37°C and only minor cleavage with pepsin and gastricsin could be detected.

The reaction mixtures were analyzed by MALDI–TOF directly after incubation (Figure 2). The results from the direct analysis of the hydrolysates were later verified by conventional separation of reaction mixtures by HPLC and MS and/or amino acid analysis of the fragments. The mass spectral data alone allowed for the correct identification of most of the fragments. The gastrin sequence has a characteristic cluster of five glutamic acid residues that makes the fragments difficult to analyze by any method. In the case of MS numerous carboxylic groups form salts and the majority of the ions appear on the spectra in multiple bands with different numbers of attached sodium ions. Our computer program takes care of possible salt formations. However broadening of the peaks can result in the overlaps and make the spectra difficult to analyze. In some cases the assignment of molecular masses gave more than one possible cleavage sites. For instance, hydrolysis at Glu7–Glu8 and Ala11–Tyr12 bonds results in the fragments that have overlapping molecular weights, although with different numbers of sodium atoms. HPLC separation and MS analysis of fractions provided the evidence that indeed both sites were cleaved by pepsin and cathepsin D.

Hemoglobin Cleavage with Porcine Pepsin

Electrospray mass spectrometry was used to study both the time and pH dependent cleavage of human hemoglobin. The intact alpha and beta chains were completely lost after 2 min incubation at pH 2.0 and 3.0 with pig pepsin. However, at pH 4.0 and under

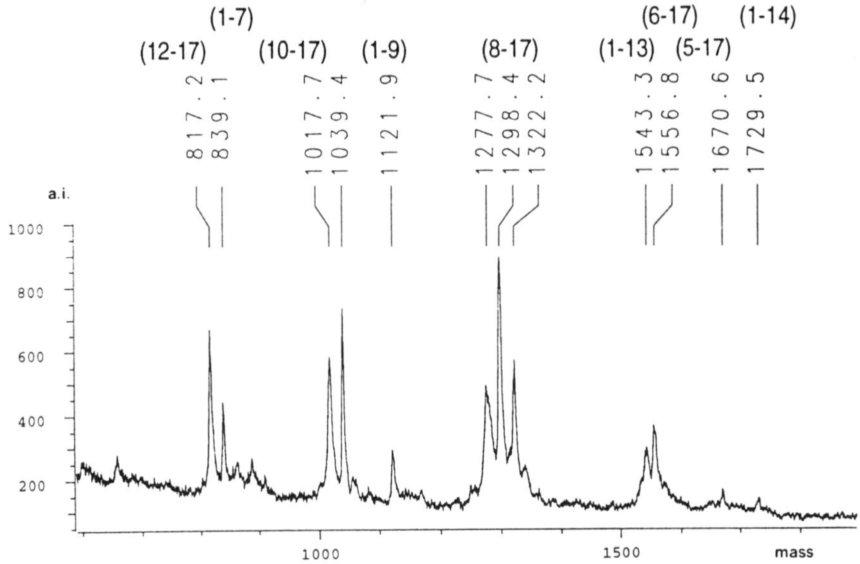

Figure 2. MALDI–TOF spectrum of human gastrin hydrolysed with human pepsin for 1 hour at pH 4.5 and 37°C.

comparable conditions, the process was significantly slower with intact chains remaining for up to 120 mins. Using the less complex peptide fragment profile observed at pH 4.0, it was possible to identify mass fragments resulting from cleavage of residues Met 32–Phe 33 and Leu 106–Val 107.

The mass/charge (M/Z) profiles of human hemoglobin after digestion with pig pepsin at 37°C and pH 4.0 are shown in Fig. 3A.

The pattern at time zero shows the typical profile of intact hemoglobin indicating the presence of both alpha (A) and beta (B) chains. This profile is lost completely after 2 min digestion at pHs 2.0 and 3.0 with an increasing number of low mass fragments seen and no evidence of the intact chains (data not shown). The pattern (fingerprint) changes relatively little for up to 10 mins but after 60 mins digestion indicates further cleavage with loss of fragments 749.1, 1248.1 and appearance of new fragments at 600.3 and 728.7. Unfortunately none of these fragments have yet been identified.

Incubation at pH4.0 indicates a much slower rate of cleavage (Fig. 3) and accordingly we were able to identify four fragments (Fig. 3B) from the alpha chain after 2 min digestion, namely residues 1–32, 33–141, 1–1–6 and 1–7–141. These correspond to cleavage at bonds Met 32–Phe 33, Leu 106–Val 107 residue 141 being the C-terminal of the alpha chain. Free amino acids were not detected in any of the digests.

CONCLUSIONS

Advantages of Application of Mass Spectrometry for the Analysis of the Specificity of Proteases

1. The techniques are fast and direct. They do not require separation of the products, allow for the detection of even minor components that can be lost during purification.

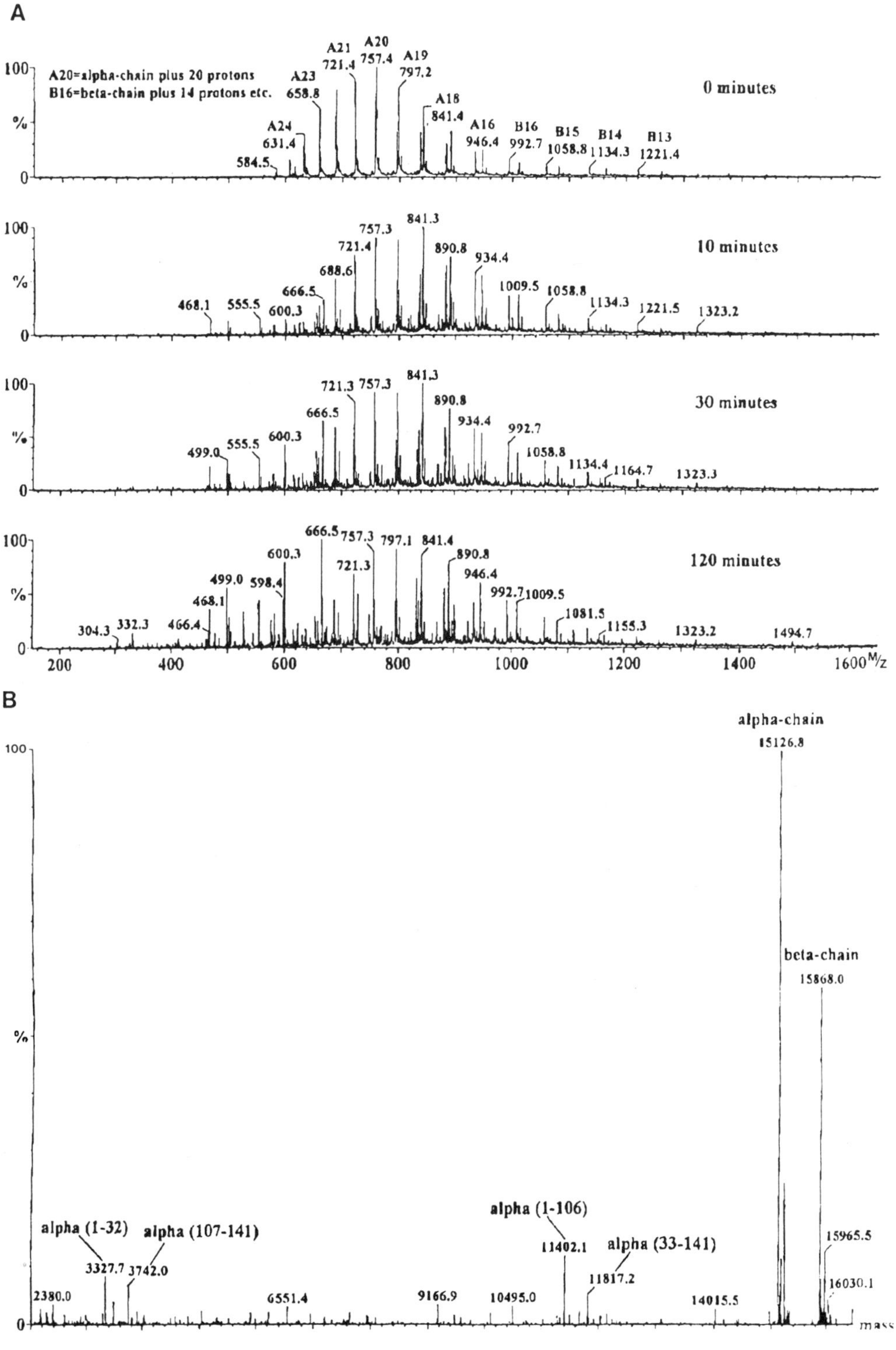

Figure 3. **A**, ES mass spectra of human hemoglobin after incubation with porcine pepsin at pH 4.0 for different time periods. **B**, ES mass spectrum of human hemoglobin cleaved with porcine pipsin at pH 4.0 and 37°C for 2 min.

2. High sensitivity (femtomolar) permits to conduct the studies on the proteins, peptides and enzymes that are expensive or difficult to obtain.
3. The methods allow for the study on *any* natural or synthetic substrate.

Limitations

1. The methods are largely qualitative, not quantitative.
2. Detection of fragments smaller than four amino acids residues long can be a problem because of the background from the matrixes.
3. Although the cases of several possible fragments of identical molecular masses within one sequence are not very frequent, they may occur. Additional analysis may be necessary in such cases.

The study also shows that gastrin can be hydrolyzed rapidly in the lumen of gastric glands by pepsin and gastricsin and in lysosomes by cathepsin D. The hydrolysis however only slowly destroys the active site of the hormone. Closely related peptide hormone, cholecystokinin is very resistant to degradation by human aspartic proteases.

ACKNOWLEDGMENT

Research sponsired in part by the National Cancer Institute, DHHS under contract with ABL.

REFERENCES

1. Roepstorff P 1992 Analyst **117**:299–303.
2. Zaluzec EJ; Gage DA; Watson JT, 1995 Protein Expr Purif **6:**109–123.
3. Tarasova NI, Denslow ND, Parten BF, Tran N, Nhuyen HP, Jones A., Roberts NB, Dunn BM. 1995, in "Aspartic Proteases", K. Takahashi Ed., Plenum Press, NY, pp. 77–81.
4. Walsh JH, Isenberg JI, Ansfield J, Maxwell V, 1976 J Clin Invest **57**:1125–31.
5. Tarasova NI, Wank SA, Hudson EA, Romanov VI, Czerwinski G, Resau JH, Michejda CJ 1997 Cell Tissue Res, in press.
6. Tarasova NI, Romanov VI, Da Silva PP, Michejda CJ 1996 Cell Tissue Res **283**:1–6.
7. Pisegna JR, Tarasova NI, Kopp JA, Asico LD, Jose P, Farnsworth DW, Michejda CJ, Wank SA 1996 Gastroenterology **110**:A1106.

29

SKIN SURFACE PROTEOLYTIC ACTIVITY

Partial Characterization and Identification

Uri Wormser,[1*] Berta Brodsky,[1] Eldad Victor Moor,[3] Arie Eldad,[4] Rivka Gal,[5] Abraham Nyska,[6] and Ron Kohen[2]

[1]Department of Pharmacology
[2]Department of Pharmacy
School of Pharmacy
Faculty of Medicine, The Hebrew University
[3]Plastic Surgery Department
[4]Burn Unit
Hadassah University Hospital
School of Medicine, The Hebrew University
[5]Department of Pathology
Rabin Medical Center
Campus Golda, Petah Tiqva
[6]Department of Pathology
Kimron Veterinary Institute
Beit Dagan, The Hebrew University, Israel

SUMMARY

Skin surface proteolytic activity in the living animal was determined by a sensitive, non-invasive methodology developed in our laboratory. A non-leaky well was constructed on the shaved back of an anesthetized guinea pig. The well contained the reaction mixture including the substrate ^{125}I–S–carboxymethylated insulin B-chain (ICMI). The proteolytic activity was shown to be time-dependent. The activity was strongly inhibited by pepstatin A, indicating the involvement of aspartic proteinase(s) such as cathepsin D and/or E. Pre-treatment of the skin with propylene glycol blocked the proteolytic activity. The present study demonstrates the presence of proteolytic activity located on skin surface using a unique, non-invasive method for *in situ* proteinase determination in the living animal.

* Corresponding author: Uri Wormser, Department of Pharmacology, School of Pharmacy, Faculty of Medicine, The Hebrew University, POB 12065, 91120 Jerusalem, Israel. Fax. 972-2-6431094; Tel. 972-2-6757515; E-mail: wormser@yam-suff.cc.huji.ac.il

Aspartic Proteinases, edited by James.
Plenum Press, New York, 1998.

INTRODUCTION

The increasing skin exposure to irritating and ulcerative factors such as ultraviolet irradiation and hazardous chemicals has stimulated an intensive research for understanding the physiological and pathological processes occurring in the dermis and particularly in the epidermal layer of the skin. Among a variety of cellular functions, proteolytic activity may play a crucial role in controlling the pathophysiology of the skin and especially of the epidermis. In recent years a number of studies have characterized several epidermal proteolytic enzymes possessing unique properties. Brysk et al. (1) has reported that desquamine, a cell adhesion molecule with lectin-like properties localized in the lipid envelopes of the stratum corneum, was a heat stable trypsin-like serine proteinase. Additional human stratum corneum serine proteolytic activities but of chymotryptic type have been characterized (2), cloned and expressed (3). A different type of proteolytic enzyme, cathepsin D, which belongs to the aspartic proteinase group, has been characterized in rabbit (4) and rat (5) epidermis. Immunohistochemical localization of cathepsin D in rabbit skin showed its presence in the basal layer of the epidermis, hair follicles, sebaceous glands, fibroblasts and endothelial cells (6). Recent studies (7) have shown that cathepsin E, additional aspartic proteinase, was distributed diffusely in the cytoplasm of almost all epidermal cells, though its concentration increased above suprabasal cells, whereas cathepsin D appeared in particulate form only in cells of the granular layer (7).

Although skin proteinases has been characterized and localized, *in situ* skin surface proteolytic activity in the living animal has never been demonstrated. Moreover, the literature dealing with epidermal proteinases describes the activity and/or presence of several proteolytic enzymes in tissues that were isolated from the body and were appropriately processed for enzymatic or immunohistochemical analyses. These *in vitro* procedures required animal sacrifice or were accompanied by human discomfort due to procedures for obtaining epidermal layer (2).

In the present study we have developed a simple, sensitive and non-invasive procedure to determine, qualitatively and quantitatively, *in situ* proteolytic activity on the outer layer of the skin of a living animal. This procedure is based on the cleavage of ^{125}I–S–carboxymethylated insulin B-chain (ICMI) located on the guinea pig skin surface. The assay enabled us to test the effect of specific proteinase inhibitors and cosmetics such as propylene glycol on skin surface proteolytic activity.

MATERIALS AND METHODS

Animals and Well Construction

A guinea (500–800 g) pig was shaved (with electrical shaver) twenty-four hours prior the experiment. Small wells were constructed on the back of the anesthetized (induced by 30 mg/kg pentobarbital ip) by the following procedure. The basis of a plastic tube cap was cut and the edge of the cap was applied with a thin layer of commercial silicon sealing ointment. The cup was attached to the guinea pig back so that no liquid inside the well could leak out. The skin area covered by the well was 0.5 cm^2. Six wells were constructed on each animal back (Fig. 1). Maintenance of anesthesia along the experiment was induced by 15 mg/kg pentobarbital ip whenever needed.

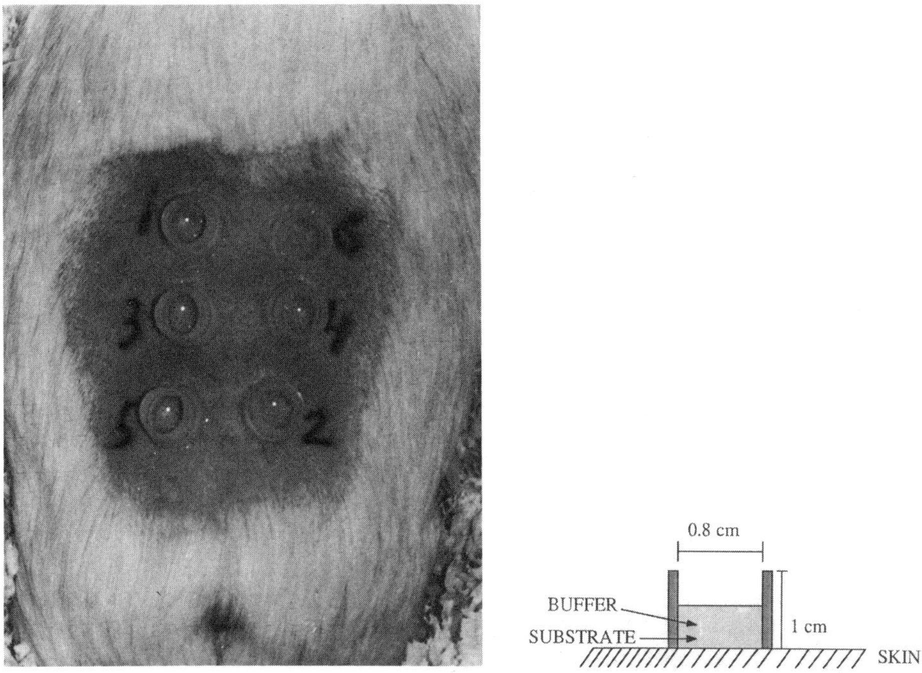

Figure 1. The non-invasive proteolytic activity apparatus on the guinea pig back.

The Proteolytic Activity Assay

Substrate Preparation. ICMI was prepared according to our previous publications (8,9). Briefly, bovine insulin (Sigma) was separated to chain A and B in the presence of 8M urea and mercaptoethanol followed by carboxymethylation of the cysteine residues by iodoacetic acid. S–carboxymethylated insulin B-chain was separated from A chain by ion exchange chromatography using Amberlite CG-50 resin (Mallinkrodt). Radioiodination of the tyrosine residues was carried out with ^{125}ICl (radiolabled Na^{125}I obtained from Amersham) and separation of the radiolabled peptide was performed with ion exchanger Amberlite IRA-400 (Mallinkrodt).

Assay System. In most experiments each well contained total volume of 350 μl of which 175 μl consisted the radioactive substrate (80 μg equivalent to 200,000 cpm) and 175 μl tris-HCl buffer 10mM pH 7.0. When proteinase inhibitors were tested they were added to the well together with the buffer and substrate was added after 5 minutes. An aliquots of 10 μl (in triplicate) was removed every 30 minutes. The aliquots were put into Eppendorf tubes placed in ice water until the end of the experiment. Ninety microliters of buffer were added to each tube followed by 100 μl 2% casein, after mixing, 200 μl 25% trichloroacetic acid were added and the tubes were mixed and centrifuged for 3 min at 14,000 RPM. Radioactivity of the supernatant (300 μl) was counted and the background (all assay components without exposure to the skin) was subtracted from each value. Percent of degradation was obtained by dividing each value by total counts (all assay components but trichloroacetic acid was replaced by water). Amount of degraded substrate was calculated by multiplying percent of degradation by the amount of substrate present in the well.

Effect of Propylene Glycol. Propylene glycol (10μl) was applied on the skin of preshaved guinea pigs. The assay system was carried out at certain time intervals following application.

RESULTS

There was a linear relationship between the degree of proteolytic activity and the reaction time (Fig. 2). It can also be shown that the proteolytic activity remained attached to the skin surface and was not released into the reaction medium (Fig. 2).

An interesting phenomenon was observed when propylene glycol (propane 1,2 diol) was applied on the skin prior to the proteolytic assay. As shown in Fig. 2, skin surface proteinase activity was strongly inhibited following propylene glycol application. This phenomenon occurred even when the time interval between propylene glycol exposure and proteolytic assay was as short as 2 min. Longer interval such as 60 min intensified the blocking effect of propylene glycol. It should be noted that no propylene glycol was left 2 h after application whereas its inhibitory activity lasted for at least 24 h following treatment (data not shown). In some experiments, particularly of 24 hr interval, the adjacent control wells showed reduced activity (data not shown) which may be due to penetration of propylene glycol into the skin, resulting in inhibition in areas not applied with propylene glycol.

The effect of certain proteinase inhibitors on skin surface proteolytic activity is shown in Table 1. The cathepsin D and E inhibitor, pepstatin A demonstrated a powerful inhibitory effect, namely, at a low concentration of 1 nM the activity was blocked by 98%. Additional inhibitory effect was demonstrated by the metal chelator ethylenediaminetetraacetic acid (EDTA) whereas alkylators such as N–ethyl maleimide and phenylmethylsulfonylfluoride did not influence the skin surface proteolytic activity.

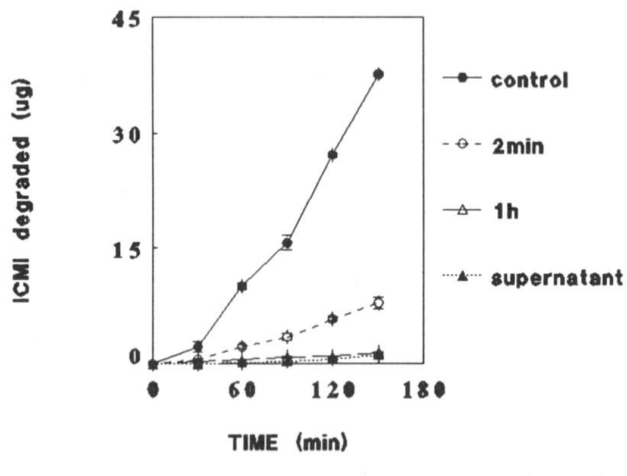

Figure 2. Time course, effect of propylene glycol and testing spontaneous release of proteinase activity into reaction medium. Wells were constructed on the guinea pigs' back and the assay was performed as described in Materials and Methods. Each value (in triplicate) is the mean of 5 experiments. In order to test the effect of propylene glycol, neat liquid (10 μl) was applied on a square of 0.7 × 0.7 cm (0.5 cm^2) at the indicated time intervals prior to reaction. To examine spontaneous release of the proteolytic activity from the skin, buffer (without substrate) termed 'supernatant' was incubated in the well and after the indicated time intervals an aliquot of 80 μl was removed and incubated with 20 μl substrate (in an Eppendorf tube) for 60 minutes followed by casein and trichloroacetic acid addition as described in Materials and Methods.

Table 1. Effect of proteinase inhibitors on skin surface proteolytic activity[a]

Inhibitor	Concentration	Degradation (% of control)
Pepstatin A	1 nM	2
EDTA	1 mM	3
PMSF	1 mM	100
N–ethylmaleimide	5 mM	100

[a]The stock solution of pepstatin A (10 mM in dimethylsulfoxide (DMSO)) was kept in −18°C and for every experiment the substance was appropriately diluted with the reaction buffer. Phenylmethylsulfonylfluoride (PMSF) was dissolved in DMSO and the other inhibitors were dissolved in water. The inhibitors were added to the well 5 minutes prior to substrate addition. Each value is a mean of triplicate (SD up to 15%).

DISCUSSION

The present study demonstrates the presence of proteolytic activity on skin surface of mammalian body. This *in situ* activity is measured on a living animal using a simple, sensitive and non-invasive procedure for qualitative and quantitative determination of proteolytic enzymes located on the outer layer of the skin. The powerful inhibitory effect of the small peptide pepstatin A, indicates that aspartic proteinases i.e. cathepsin D and/or E (7) compose the measured skin surface proteolytic activity. Nevertheless, the blocking activity of propylene glycol may be a result of some conformational alterations in the skin which affect proteases and perhaps other skin enzymatic activities. Such explanation may also be applicable for the inhibitory effect of EDTA, however, the involvement of a metal in the catalytic activity can not be ruled out as well.

The unique assay system may provide better understanding of the physiological reactions occurring on skin surface. Moreover, the present methodology may serve as a powerful tool for investigating the pathophysiology of skin disorders and the *in situ* proteolytic processes accompanying these illnesses. Being a non invasive procedure convey our assay a big advantage, not only from the ethical-experimental point of view of animal protection and prevention of animal suffering, but also because of its powerful ability to test proteolytic activity on human skin *in situ*. It can be used as an important tool in understanding molecular mechanisms accompanying a variety of skin disorders such as skin allergy, psoriasis, pemphigus, herpes, various kinds of dermatitis and malignancies. Future development of our assay may employ a non radioactive substrate which can be safely applied on human skin.

The role of skin proteolytic activity is not yet understood. A possible explanation may be related to a unique protecting mechanism of the skin against biological damage. Ginsburg and Cohen (10) have shown that oxidative damage to fibroblasts in culture may occur as a synergistic activity between proteinases and oxidants. Skin cooperation between proteolytic enzymes and oxidants released from the skin such as $O_2^{\bullet -}$ (induced by polymorphnuclears in inflamed skin) may play a role in inducing biological damage to outside invaders such as bacteria and viruses. The role of this surface enzymatic activity in skin aging and in pathophysiological skin disorders is under investigation.

REFERENCES

1. Brysk, M. M., Bell, T., Brysk, H., Selvanayagam, P., and Rajaraman, S. (1994) Exp. Cell Res. **214,** 22–26
2. Lundstrom, A., and Egelrud, T. (1991) Acta Derm. Venereol. **71,** 471–474
3. Hansson, L., Stromqvist, M., Backman, A., Wallbrandt, P., Carlstein, A., and Egelrud, T.(1994) J. Biol. Chem. **269,** 19420–19426.
4. Jasani, B., Jasani, M. K., and Talbot, M. D. (1978) Biochem. J. **169,** 287–295
5. Heikkinen, J. E., Jarvinen, M., and Jansen, C. R. (1975) J. Invest. Dermatol. **65,** 272–278
6. Lazarus, G. S., and Poole, A. R. (1975) Arch. Dermatol. **111,** 1150–1153
7. Hara, K., Fukuyama, K., Sakai, H., Yamamoto, K., and Epstein, W. L. (1993) J. Invest. Dermatol. **100,** 394–399
8. Wormser, U., and Zbinden, G. (1986) Arch. Toxicol. **59,** 111–114
9. Wormser, U., and Zbinden, G. (1985) Biochem. Biophys. Res. Commun. **127,** 191–197
10. Ginsburg, I., and Cohen, R. (1995) Free Rad. Res. **22,** 489–517

30

CATHEPSIN E IN THE CENTRAL NERVOUS SYSTEM

Hiroshi Nakanishi and Kenji Yamamoto

Department of Pharmacology
Kyushu University Faculty of Dentistry
Fukuoka 812, Japan

1. INTRODUCTION

There is growing evidence that altered intracellular proteolytic system is closely associated with different neuropathological conditions in the central nervous system. Irregular proteolysis of physiologically important neuronal proteins is thought to have severe consequence for the integrity of neuronal structure and function and the disorder or breakdown of the endosomal/lysosomal proteolytic system is suggested to be of pathological significance in neurodegenerative diseases such as Alzheimer's disease. Recent studies have also suggested that the neuronal cell death observed in certain neurodegenerative processes, as well as in the early neuronal development, proceeds in the apoptotic pathway, in which the activation of intracellular proteolytic systems plays an essential role. Furthermore, proteinases produced by activated microglia at pathological sites are considered to be involved in neuronal damage, removal of neuronal debris, and tissue remodeling during regeneration via degradation of extracellular matrix compartments. Therefore, alterations in proteolytic systems in both neurons and microglias plays an essential role in cellular events that lead to neuropathological conditions.

Cathepsin E (CE), a non-lysosomal aspartic proteinase that has a limited tissue and cell-specific distribution in the peripheral tissues such as the thymus, stomach and spleen, appears to exist as an enzymatically inactive proform in the storage pool of these tissues, such as the endoplasmic reticulum and Golgi, and to be mobilized to the endosomal acidic compartments in response to certain stimuli, where it is converted to the catalytically active mature form in an autocatalytic manner. In contrast to the peripheral tissues, CE appears to be highly repressed for its expression in the central nervous system and to be increasingly expressed in both neurons and microglias during neuropathological changes associated with senescence,[1,8,10,11] ischemia,[8,9] and epilepsy[12] in the rat. Recently, it has been shown that CE is highly expressed in degenerating neurons and that it is also localized in the core of the senile plaques of Alzheimer's disease brain.[3] It should be noted that

Aspartic Proteinases, edited by James.
Plenum Press, New York, 1998.

increased CE in senescent neurons and activated microglias is proteolytically processed to the mature form. These observations strongly suggest that the newly synthesized pro-CE on the membrane-bound ribosomes in aged neurons and activated microglias is transported to the intracellular acidic compartments through the Golgi complex and converted into the mature form. However, there is little understanding of the biosynthesis and processing of CE in the central nervous system. Thus the present study was attempted to determine accumulation sites of CE in neurons of the aged rat brain by subcellular fractionation and immunocytochemistry at both light and electron microscopic levels. Furthermore, the pulse-chase experiments were carried out in the primary cultured rat brain microglia. Possible pathological significance of CE upregulated in the aged neuron and activated microglia will be discussed.

2. AGE-DEPENDENT INCREASE IN CATHEPSIN E IN NEURON OF THE RAT BRAIN

Immunoblot analyses were performed in the cerebral cortex and the brain stem of Fischer 344 rats with different four ages (embryo, 2, 17 and 30 months). Nonlysosomal CE was barely detectable in the rat brain at the embryonic stages, whereas the relatively high expression of lysosomal cathepsin D (CD) was observed in embryonic tissues. After birth, CE was increasingly expressed in both brain tissues with aging to attain the maximal levels at 30 months of age. Unlike peripheral tissues, CE was present as the mature enzyme (42 kDa in the reduced condition) in the cerebral cortex and the brainstem at 2 and 17 months of age, whereas it was present as not only the mature enzyme but also the proenzyme (46 kDa) at 30 months of age (Table 1). The predominant existence of the mature form of CE in neurons was confirmed in primary cultured rat cortical neurons. These results suggest that the newly synthesized pro-CE in these brain tissues is transported to the intracellular processing sites connected with its activation. On the other

Table 1. Cathepsin E upregulated in neurons and microglias during senescence, neurodegeneration and cellular activation

Cell types	Induced by	System	Molecular forms	References
Neuron	senescence	rat, in vivo	42 kDa (major) 46 kDa (minor)	8, 10, 11
	ischemia	rat, in vivo	n.d.	8, 9
	kainate	rat, in vivo	54 kDa (major) 42 kDa (minor)	12
	glutamate	rat, primary culture	54 kDa (major) 42 kDa (minor)	unpublished data
Microglia	senescence	rat, in vivo	n.d.	8, 10, 11
	ischemia	rat, in vivo	n.d.	8, 9
	kainate	rat, in vivo	n.d.	12
	spongiform degeneration	SAMP8, P10 in vivo	42 kDa	1
	LPS + IFN-γ	rat, primary culture	42 kDa	unpublished data

Molecular forms were determined in the reduced condition. n.d.: not determined, SAM: senescence-accerelated mouse, LPS: *Escherichia coli* lipopolysaccharide, IFN-γ: interferon-γ.

hand, CD was mainly present in the mature form throughout development, although its level was also significantly increased with aging. The CE-positive cortical and brainstem neurons of the aged rat corresponded well with cells emitting autofluorescence for lipopigments. The electron microscopy clearly revealed that CE in brainstem neurons of the aged rat was colocalized with CD in the lipofuscin-containing lysosomes. Lipofuscin is a major type of residual body observed in various types of neurons during the aging process and is considered to be formed initially by autophagocytosis. Taken together, the results indicate that the age-related accumulation of CE in lipofuscin-containing lysosomes is associated with enhancement of both autophagy and endosomal/lysosomal system in the neuronal aging process. The increased levels of CE as well as CD may contribute to neuronal apoptotic process during senescence since CD has been recently reported to mediate cytokine-mediated apoptosis.[4] Subcellular fractionation showed that the amounts of CE and CD in the soluble fraction of the aged rat brain tissues was markedly increased as compared with those of the young tissues. The increase of CE and CD in the soluble fraction may be related to the age-dependent increase in the fragility of the endosome/lysosome system. The leakage of CE as well as CD into the cytoplasm of neurons may be associated with age-related pathological changes since both enzymes have been shown to be capable of degrading cytoskeletal proteins and bioactive peptides.

Immunohistochemically, CE was also colocalized with the amyloid precursor protein (APP) reacted with anti-$APP_{643-695}$ antibody in cortical and brain stem neurons of the aged rat. The anti-$APP_{643-695}$ antibody intensely stained only in neurons of the aged rat but not those of the young rat. On the other hand, anti-APP_{60-100} antibody intensely stained cortical and brain stem neurons of both young and aged rats, where there was no obvious difference in staining patterns between both groups. The age-related accumulation of the C-terminal fragment of APP was further confirmed by immunoblot analysis of the brainstem extracts of both young and aged rats with anti-$APP_{643-695}$ antibody. The results indicated that two immunoreactive bands of 19 and 22 kDa corresponding to the C-terminal fragment of APP were markedly increased in the brain extracts of aged rats. The size of these immunoreactive bands strongly suggests that they might contain an intact β/A4 domain and therefore be amyloidogenic. As APP has been shown to accumulated in lipofuscin-containing lysosomes in neurons from normal and Alzheimer's disease subjects and as CE was predominantly found in lipofuscin-containing lysosomes of aged rat neurons, CE may be involved in altered APP catabolisms to produce amyloidogenic C-terminal APP fragments. Recently it has been reported that CD as well as CE efficiently cleaved not only the wild-type APP to yield amyloidogenic fragments but also a Swedish mutant form when synthetic peptide was used as substrates.[6] Further study will be needed to determine whether CE can process full-length APP into small amyloidogenic fragments.

3. BIOSYNTHESIS AND PROCESSING OF CATHEPSIN E IN THE RAT BRAIN MICROGLIA

Microglias were obtained from the mixed primary cultures of the cerebral cortex from 3-day old Wistar rats. Most of isolated microglias had ameboid and rod-shaped cell bodies expressing CR3 complement receptor and MHC class II. These cells also showed phagocytic ability via Fc and CR3 complement receptors. This cellular phenotype resembles activated microglias at sites of pathology. In these primary cultured microglias, CE was found to exist predominantly as the mature form (42 kDa). Indirect immunofluorescence microscopy revealed its endosome-like distribution. When *Escherichia coli* lipopolysaccharide (LPS, 1 μg/ml) and rat recombinant interferon-γ (IFN-γ, 100 U/ml) were added in the culture medium for 48–72 hr, microglias became circular in shape with ruffled edges. In these cells, the amount of CE was further increased (Table 1). On the other hand, CE was not found in astrocytes even after treatment with c-AMP.

The intracellular processing kinetics of CE was analyzed by pulse-labeling experiment with ^{35}S-methionine in primary cultured microglias. CE was initially synthesized as the proenzyme (46 kDa) and then converted into the mature form (42 kDa) via the intermediate form (44 kDa) after 4 hr-chase. This conversion was completely blocked by bafilomycin A_1, a vacuolar type H^+-ATPase inhibitor, indicating that its maturation occurs in the intracellular acidic compartment. It was also noted that CE was secreted into the culture medium as the proform throughout the chase period and the secreted enzyme was markedly increased in the presence of bafilomycin A_1. The processing of CE in microglias was further analyzed in the presence of brefeldin A, which is known to have a potent inhibitory effect on the intracellular protein transport from the endoplasmic reticulum to the Golgi apparatus. In the presence of brefeldin A, both the processing of pro-CE was also completely inhibited. In this case, however, the amount of pro-CE released into the medium was markedly reduced. In order to examine whether CE in these cells undergoes the carbohydrate modification during the biosynthesis, the nature of oligosaccharide chains of this protein was determined by endoglycosidase H (endo H) which cleaves high-mannnose glycans from glycoprotein. Within 2 hr chase, CE was clearly sensitive to endo H, whereas after 4 hr chase an approximately half amount of the enzyme was converted to an endo H-resistant type and the other half was still the endo H-sensitive high-mannose type.

These observations strongly suggest that the newly synthesized pro-CE in the microglia is transported its final processing sites connected with its activation, through which its carbohydrate moieties is converted from the high-mannose type to the complex-type. On the other hand, CD was sensitive to endo H during entire chase period. It has also been shown that CE is localized as the mature form in endosome-like structures of other monocyte phagocytic cells such as macrophages. More recently, CE has been suggested to be responsible for processing of the antigen [2] and the invariant chain.[5,7] Taken together with the present observations, it is likely considered that CE is activated and localized mainly in endosomal vesicles of these cells having the common role of phagocytosis and/or antigen presentation.

REFERENCES

1. Amano, T., Nakanishi, H., Oka, M. and K. Yamamoto : Increased expression of cathepsins E and D in reactive microglial cells associated with spongiform degeneration in the brain stem of senescence-accelerated mouse, *Exp. Neurol.* 136–182; 171 (1995).
2. Bennett, K., Levine, T., Ellis, J. S., Peanasky, R. J., Samloff, I. M., Kay, J. and B. M. Chain : Antigen processing for presentation by class II major histocompatibility complex requires cleavage by cathepsin E, *Eur. J. Immunol.* 22: 1519–1524 (1992).
3. Bernstein H. -G. and B. Wiederanders : An immunohistochemical study of cathepsin E in Alzheimer-type dementia brains, *Brain Res.* 667: 287–290 (1994).
4. Deiss, L. P., Galinka, H., Berissi, H., Cohen, O. and A. Kimchi : Cathepsin D protease mediates programmed cell death induced by interferon-g, Fas/APO-1 and TNF-a, *EMBP J.* 15: 3861–3870 (1996).
5. Kageyama, T., Yonezawa, S., Ichinose, M., Miki, K. and A. Moriyama : Potential sites for processing of the human invariant chain by cathepsins D and E, *Biochem. Biophys. Res. Commun.* 223: 549–553 (1996).
6. Lador, U. S., Snyder, S. W., Wang, G. T., Holzman, T. F. and G. A. Krafft : Cleavage at the amino and carboxyl termini of Alzheimer's amyloid-β by cathepsin D, *J. Biol. Chem.* 269: 18422–18428 (1994).

7. Maric, M. A., Taylor, M. D. and J. S. Blum : Endosomal aspartic proteinases are required for invariant-chain processing, *Proc. Natl. Acad. Sci. USA* 91: 2171–2175 (1994).
8. Nakanishi, H., Tominaga, K. and K. Yamamoto : Increased expression and specific localization of cathepsins E and D in vulnerable brain regions of aged and postischemic rats, *Adv. Exp. Med. Biol.* 362: 335–339 (1995).
9. Nakanishi, H., Tsukuba, T., Kondou, T., Tanaka, T. and K. Yamamoto : Transient forebrain ischemia increased expression and specific localization of cathepsins E and D in rat hippocampus and neostriatum, *Exp. Neurol.* 121: 215–223 (1993).
10. Nakanishi, H., Tominaga, K., Amano, T., Hirotsu, I., Inoue, T. and K. Yamamoto : Age-related changes in activities and localizations of cathepsins D, E, B and L in the rat brain tissues, *Exp. Neurol.* 126: 119–128 (1994).
11. Nakanishi, H., Amano, T., Sastradipura, D. F., Yoshimine, Y., Tsukuba, T., Tanabe, K., Hirotsu, I., Ohono, T. and K. Yamamoto : Increased expression of cathepsins E and D in neurons of the aged rat brain and their colocalization with lipofuscin and carboxy-terminal fragments of Alzheimer amyloid precursor protein, *J. Neurochem.* in press.
12. Tominaga, K., Nakanishi, H., Yajima, M. and K. Yamamoto : Characterization of cathepsins E and D accumulated at early stages of neuronal damage in hippocampal neurons of rats, *Adv. Exp. Med. Biol.* 362: 341–343 (1995).

31

DETAILED ANALYSIS OF HUMAN CATHEPSIN E PRIME REGION SPECIFICITY

Marina Bukhtiyarova,[1] Chetana Rao-Naik,[1] Peter J. Tatnell,[2] Paul C. White,[2] John Kay,[2] and Ben M. Dunn[1]

[1]Department of Biochemistry and Molecular Biology
University of Florida
Gainesville, Florida 32610
[2]Department of Biochemistry
University of Wales
College of Cardiff
Cardiff, Wales, United Kingdom

1. INTRODUCTION

Aspartic proteinases are produced in the human body by a variety of cells. Some of these proteins, such as pepsin, gastricsin and renin, are secreted and exert their effects in the extracellular environment. Cathepsin D and cathepsin E, on the other hand, are intracellular enzymes. Three-dimensional structures have been determined for four of these enzymes by X-ray crystallography with the exception of cathepsin E.

The functional role of cathepsin E is not well established. It has been suggested that cathepsin E is involved in the processing of antigens[1] and prohormones[2] and in the infection of human cells by HIV.[3] In addition, it has been reported by Azuma *et al.* that resected human pancreatic ductal adenocarcinomas express cathepsin E at levels higher than seen in normal tissues, suggesting that this enzyme may be a useful tumor marker.[4]

In this study we attempt to understand the specificity in the prime region of human cathepsin E. A peptide series based on Lys–Pro–Ile–Glu–Phe*Nph–Arg–Leu was generated to elucidate the specificity in the individual binding pockets. The systematic amino acid substitutions were introduced in the P_2' and P_3' positions with respect to charge, hydrophobicity and hydrogen bonding. The kinetic parameters (k_{cat} and K_m) were determined for both sets of peptides. The results were correlated to the known rule-based structural model of human cathepsin E.[5]

Aspartic Proteinases, edited by James.
Plenum Press, New York, 1998.

2. MATERIALS AND METHODS

2.1. Isolation of Procathepsin E and Cathepsin E

Procathepsin E was purified by modification of the method previously described.[5] All chemicals were obtained from Sigma Chemical Co., St. Louis, Mo, USA, and the chromatography resins and accessories were from Pharmacia LKB Inc., Piscataway, NJ, USA.

2.2. Spectrophotometric Analyses

All the chromogenic substrates under study were synthesized by the solid phase method using an Applied Biosystems Model 430A (Protein Core Facility at the University of Florida). Stock peptide solutions were made in distilled water and quantified by amino acid analysis. The purity of each peptide (= 90%) was verified by HPLC analysis.

The hydrolysis of each substrate was monitored at 37°C, pH 4.5, in a 0.1 M sodium formate buffer by averaging the decrease in absorbance over a range of 284–324 nm using a Hewlett Packard Diode Array Spectrophotometer.[6] The K_m and V_{max} values were determined from the initial rates using the equation:

$$v = \frac{V_{\max} \times S}{K_m + [S]}$$

The cleavage of the peptides was also monitored by HPLC followed by sequence determination of the respective peaks to confirm that each substrate had been cleaved at the ~Phe*Nph~ bond.

The amount of active enzyme used in the assay was determined by competitive inhibition of human cathepsin E by a potent inhibitor, N-(4-morpholinylsulfonyl)-L-phenylalanyl-N-[1(S)-(cyclohexylmethyl)-3,3-di-fluoro-4-[(2-methylbutyl)amino]-2,4-dioxobutyl]-4,5-didehydro-L-norvalinamide (K_i = 0.5 ± 0.07 nM). The resulting curve was fitted with the Henderson equation on the Enzfitter program.[7]

The k_{cat} values (sec^{-1}) were determined by the equation:

$$k_{cat} = \frac{\Delta_{\max}(AU/\sec)}{\Delta A/[S] \times [E]}$$

where, ΔA = Total change in absorbance observed with 50μM substrate [S], [E] = total enzyme concentration (Molar).

2.3. Modelling

A three-dimensional model of cathepsin E minus its 6 N-terminal residues was constructed using the rule-based comparative modelling approach encoded in the COMPOSER suite of computer programs.[5] The model was refined using energy minimization techniques in SYBYL (Tripos Associates, USA).

3. RESULTS

Previous studies by Rao-Naik *et al.* have shown that the peptide Lys–Pro–Ile–Glu–Phe*Nph–Arg–Leu is an extraordinarily good substrate for cathepsin E.[5] To analyze the

Table 1. Kinetic parameters of peptide substrates of the form K–P–I–E–F*X–Xaa–L with substitutions in P_2'

P_5	P4	P3	P2	P1	P1′	P_2'	P_3'	Human cathepsin E k_{cat} (sec^{-1})	 K_m (μM)	 k_{cat}/K_m ($\mu M^{-1}s^{-1}$)
K	P	I	E	F	X	R	L	105	46	2.3
K	P	I	E	F	X	**E**	L	252	25	10.1
K	P	I	E	F	X	**L**	L	98	16	6.2
K	P	I	E	F	X	**A**	L	234	44	5.4
K	P	I	E	F	X	**S**	L	158	59	2.6
K	P	I	E	F	X	**I**	L	113	44	2.5
K	P	I	E	F	X	**K**	L	45	21	2.1
K	P	I	E	F	X	**V**	L	69	51	1.4
K	P	I	E	F	X	**D**	L	62	109	0.6

specificity in the prime region of cathepsin E, we introduce the systematic substitutions in the P_2' and P_3' positions. The results from these studies are listed in Tables 1 and 2. The most notable contributions to prime region specificity come from S_2'–P_2' interactions; however, the nature of the P_3' residue occupying the S_3' subsite seems to be of secondary importance.

S_2' subsite preferences: The peptide containing Glu in P_2' was found to be cleaved with the highest rate (k_{cat}/K_m=10.1 $\mu M^{-1}sec^{-1}$, Table 1). Hydrophobic amino acids such as Leu, Ile, Ala as well as charged Arg were all acceptable in P_2' with 2- to 5-fold variation in k_{cat}/K_m values. However, the aspartate side chain was unacceptable in this position under standard assay conditions. The S_2' subsite of most aspartic proteinases has been shown to

Table 2. Kinetic parameters of peptide substrates of the form K–P–I–E–F*X–R–Yaa for P_3' substitutions

P_5	P_4	P_3	P_2	P_1	P_1'	P_2'	P_3'	Human cathepsin E k_{cat} (sec^{-1})	 K_m (μM)	 k_{cat}/K_m ($\mu M^{-1}s^{-1}$)
K	P	I	E	F	X	R	L	105	46	2.3
K	P	I	E	F	X	R	**D**	256	80	3.2
K	P	I	E	F	X	R	**V**	65	21	3.1
K	P	I	E	F	X	R	**Y**	128	58	2.3
K	P	I	E	F	X	R	**Nle**	158	70	2.2
K	P	I	E	F	X	R	**A**	39	20	1.9
K	P	I	E	F	X	R	**I**	83	49	1.7
K	P	I	E	F	X	R	**F**	188	121	1.6
K	P	I	E	F	X	R	**R**	48	31	1.5
K	P	I	E	F	X	R	**S**	NC[a]	NC	NC

NC[a] = not cleaved under standard assay conditions.

be largely hydrophobic in nature, and cathepsin E is no exception. A Glu or an Arg side chain in P_2' may be able to interact with the hydrophobic S_2' subsite through its methylene groups if the charged groups can be extended into solvent. However, in the case of Asp in the P_2' position, the negatively charged side-chain carboxyl group is unlikely to be able reach the solvent due to the shorter length of the side chain. This arrangement would place the negative charge near Tyr 75, a structurally and catalytically essential residue of all aspartic proteinases, which would cause disruption of the active site and could easily lead to the drastic reduction in cleavage observed for both peptides.

S_3' subsite preferences: The charged residues as well as small and hydrophobic amino acids were all acceptable in the S_3' subsite, but it rejects a Ser (Table 2). Otherwise, no significant variations of catalytic efficiency were noted in the results derived from changes in the P_3' residue, though Asp was the best substituent by a factor of 2.

ACKNOWLEDGMENTS

We would like to thank the Protein Chemistry Core at the University of Florida for the peptide synthesis and for N-terminal sequence analyses. This work was supported by NIH grant DK 18865 to BMD.

REFERENCES

1. Bennett, K. et. al (1992) *Eur. J. Immunol.* 22, 1519–1524.
2. Lees, W.E. et. al (1990) *FEBS Lett.* 273, 99–102.
3. Clements, G.J. et. al (1991) *Aids Research and Human Retroviruses* 7, 3–16.
4. Azuma, T. et. al (1995) *"Aspartic Proteinases: Structure, Function, Biology and Biomedical Implications."* (ed. Takahashi, K.) Plenum press, New York, 363–366.
5. Rao-Naik, C. et. al (1995) *Proteins: Structure, Function and Genetics* 22, 168–181.
6. Dunn, B.M. et. al (1994) *"Peptide Analysis Protocols"*, Humana Press, Clifton, N.J., 225–243.
7. Henderson, P. (1972) *Biochem. J.* 127, 321–333.

ABSTRACTS

THE ROLE OF ASPARTIC PROTEINASES IN ANTIGEN PROCESSING

B. M. Chain, J. Kay, P. Medd, D. O'Neil, L. Sealy, and P. J. Tatnell

Department of Immunology, University College London Medical School, Cleveland St., London W1P 6DB, and School of Molecular and Medical Biosciences, University of Wales, Cardiff, CF1 3US, United Kingdom

The presentation of antigen to CD4 T cells, which is an essential step in the majority of immune responses, requires proteolytic degradation of most protein antigens. The molecular details of this processing remain obscure, however. We and others have previously reported that processing of several proteins can be blocked by the aspartic proteinase inhibitor pepstatin, and have also shown that in at least one system, this processing can be attributed to the non-lysosomal enzyme cathepsin E. In this presentation, we will discuss three more recent lines of evidence which lend support to the hypothesis that cathepsin E plays a role in the antigen processing pathway. Firstly, we have developed a cathepsin E-specific monoclonal antibody; using this antibody we have shown that both the intracellular and tissue distribution of cathepsin E is consistent with, albeit indirect evidence for, a role for this enzyme in antigen processing. Secondly, the expression of cathepsin E, as opposed to the lysosomal enzyme cathepsin D, is tightly regulated during both B cell and monocyte differentiation. Finally, in vitro digestion of the model enzyme, hen egg lysozyme by cathepsin E suggest the intriguing possibility that susceptibility to aspartic proteinase digestion may be a key factor in determining the epitope hierarchy associated with responses to any complex protein antigen.

X-RAY STRUCTURE OF CHYMOSIN COMPLEXED WITH A REDUCED-BOND INHIBITOR

V. Dhanaraj,[1] M. Groves,[1] M. Badasso,[1] J. E. Pitts,[1] D. J. Hoover,[2] and T. L. Blundell[1]

[1]ICRF Unit of Structural Molecular Biology, Department of Crystallography, Birkbeck College, Malet Street, London WC1E 7HX, United Kingdom
[2]Department of Medicinal Chemistry, Central Research Division, Pfizer Inc., Groton, Connecticut 06340

Calf chymosin, the enzyme used for many centuries in the manufacture of cheese, is a bilobal neonatal gastric proteinase secreted as an inactive precursor (prochymosin) which is activated at acidic pH by the proteolytic cleavage of the N-terminal propeptide. The main physiological function of chymosin is to cleave the milk protein k-casein, which acts as a stabiliser of the micelle, at an exposed Phe–Met bond.

In order to engineer new specificities for other substrates that may have commercial applications in the food and feed processing industries, the crystal structures of the native and the inhibitor-bound forms are essential pre-requisites. However, the active site of the native crystal structures of chymosin differed radically from those of mammalian aspartic proteinases in the conformation of the hairpin loop above the active site, known as the flap. In the absence of an inhibitor, chymosin crystallises in the orthorhombic space group I222. In this form, the flap is involved in intermolecular lattice contacts and its observed conformation would preclude the binding of substrates or inhibitors in the S1 pocket. However, when complexed with a reduced-bond peptidomimetic inhibitor, CP-113972, chymosin crystallises in a novel crystal form of space group R32.

The structure, solved by the molecular replacement technique and refined by least-squares to a crystallographic R-factor of 0.19, shows that the flap adopts the conformation normally observed in other inhibitor complexes. The inhibitor is bound in an extended conformation, occupying the specificity pockets S4–S1′, in a manner similar to that of other aspartic proteinase inhibitors. This suggests that the normal position and conformation of the flap are available to chymosin and that the conformation observed in the native structures may be attributable to intermolecular interactions between symmetry-related molecules.

DESIGN OF SENSITIVE FLUOROGENIC SUBSTRATES FOR CATHEPSIN D

S. V. Gulnik,[1] L. I. Suvorov,[1] P. Majer,[1] J. Collins,[1] G. Krafft,[2] and J. W. Erickson[1]

[1]Structural Biochemistry Program, SAIC, NCI-FCRDC, Frederick, Maryland 21702
[2]Department of Molecular Pharmacology, Northwestern University Medical School, Chicago, Illinois 60611

Cathepsin D is a lysosomal aspartic proteinase that has been implicated in several pathological processes such as breast cancer and Alzheimer's disease which makes this enzyme an attractive therapeutic target. In order to investigate the enzymatic properties of Cathepsin D and evaluate inhibitor binding it is necessary to have a reliable, sensitive, and continuous assay method for monitoring enzyme activity. We designed and synthesized a number of quenched fluorogenic substrates in the series AcEE(EDANS)KPIXFFRLGK (DABCYL)E-NH_2, X = cysteine, methylcysteine, ethylcysteine, tert-butylcysteine, carboxymethylcysteine, methionine, valine or isoleucine). The EDANS (5-[2-aminoethyl)amino]naphthalene-1sulfonic acid) and DABCYL (4-(4-dimethylaminophenylazo) benzoic acid) groups were used as fluorophor and quencher, respectively. The peptide portion of the substrate was derived from the sequence K–P–I–L–F–F*–R–L (F* is p-nitrophenylalanine) as described by Scarborough et al. for a cathepsin D substrate (K_m = 28 μM, k_{cat} = 45 s^{-1}). Several charged amino acids were added to the N- and C-termini of the peptide to facilitate attachment of donor and acceptor molecules and to increase substrate solubility. F* was replaced by phenylalanine in P1′ since the nitro group is not necessary for substrate turnover. Introduction of the hydrophobic fluorophor and quencher groups combined with an increased length of the substrate resulted in a 5–6 fold reduction in K_m with no change in catalytic rate constant. The presence of cysteine, modified cysteine or methionine in P2 slightly improved binding (K_m = 3.5 μM), with the exception of ethylcysteine. Interestingly the later substrate had the highest k_{cat} — around 200 s^{-1} — ever described for a cathepsin D substrate. Most of the fluorogenic substrates exhibited greater kcat/Km ratios than the best cathepsin D substrates described so far. The increase in assay sensitivity afforded by our fluorogenic substrates allowed us to decrease the enzyme concentration used for the assay to 0.2–0.3 nM.

Differences in kinetic constants might make certain substrates useful for particular applications. For example, the substrate with X = ethylcysteine (K_m = 27.5 μM, k_{cat} = 196 s^{-1}) might be convenient for active site titrations since it permits the use of substrate concentrations far below the K_m. The substrate with X — carboxymethylcysteine, which has a low Km with a slow kcat (K_m = 3.9 μM, k_{cat} = 5.8 s^{-1}), could be used in an assay format that requires the reaction velocity to remain constant during a prolonged period, such as the cycling time of a fluorescent plate reader.

REFINING THE RECOMBINANT MODEL OF HUMAN CATHEPSIN D

Brian M. Beyer and Ben M. Dunn

Department of Biochemistry and Molecular Biology, University of Florida, JHMHC, Gainesville, Florida 32610-0245

Cathepsin A overexpression and activity have been correlated with breast cancer aggressiveness and plaque formation in Alzheimer's disease. To facilitate inhibitor design, we are working on the subtle differences in ligand/enzyme interactions that exist between cathepsin D and other human aspartic proteinases. In correlation with kinetic studies using native enzyme and several series of chromogenic octapeptides, computer modeling of the tertiary structure has been utilized to suggest specific interactions between substrate side-chains and subsite residues within the active site cleft.

To accurately confirm these suspected interactions, we are now utilizing an improved recombinant model of cathepsin D that has kinetic properties identical to native material (1). A cleavable junction, suitable for self-processing, was engineered into the pro-segment and effectively removes twelve of the eighteen otherwise retained amino acids present in pseudocathepsin D (including Lys31p and the Lys36p–Tyr37p dipeptide). This recombinant species, which reproduces the specific catalytic activity of native human liver cathepsin D, has been employed to assess the contributions of specific S_2' residues (I128 and V130) to enzymatic specificity. Furthermore, a unique structural component of cathepsin D, the only-proline loop, has been changed to the corresponding sequence found in porcine pepsin to determine the role of this region in refolding and specificity.

Supported by NIH grant DK18865 to BMD.

REFERENCE

1. Beyer, B.M. and Dunn, B.M. (1996) *J. Biol. Chem.* **271**, 15590–55596.

PURIFICATION OF AN ASPARTIC PROTEINASE FROM THE LIVER OF THE ANTARCTIC ICEFISH *Chionodraco hamatus*

E. Parisi,[1] C. Capasso,[1] A. Capasso,[1] and J. Kay[2]

[1]CNR Institute of Protein Biochemistry and Enzymology, Naples, Italy
[2]Department of Biochemistry, University of Wales, Cardiff, United Kingdom

The coastal waters of Antarctica are characterized by cold temperatures (–1.9 to +2°C). Under these extreme conditions poikilotermic Antarctic fishes have evolved compensatory adaptations to maintain normal metabolic functions (1). This implies that enzymes must possess features which make them capable to work at these extreme temperatures.

We have initiated studies on the proteolytic enzymes present in the liver of Antarctic fishes as these enzymes may have an important role not only in metabolism, but also in the processing of specific proteins. We identified a proteolytic activity in liver extracts obtained from the icefish *Chionodraco hamatus.* The most peculiar characteristic of icefishes is that they lack erythrocytes and the respiratory transport pigment hemoglobin (2,3). The proteolytic activity has been purified by a combination of anion-exchange chromatography and affinity chromatography. The extract (120 mg protein) was loaded on a 1.5 × 20 cm DEAE-Sephacel column which was eluted with a linear gradient from 0 to 0.5 M KCl in 50 mM Tris-HCl pH 7.5. The column was monitored for absorbance at 280 nm and enzyme activity determined using denatured hemoglobin as substrate. A single peak of proteolytic activity was eluted at an ionic strength of 10 mMHO. This peak was pooled and loaded on a 0.6 × 6 cm Conc A-agarose column which was washed extensively with 0.2 M KCl in 50 mM Tris-HCl pH 7.5 and eluted with 0.5 M methylmannopyranoside in the same buffer. After this step, the final purification was 150-fold with a recovery of 70% (see Table 1).

Table 1.

Step	Volume (ml)	Units*	Protein (mg)	Units/ mg protein	Yield (%)
100,000g sup	20	2,660	115	23	100
DEAE-cellulose	25	2,500	4.8	53	94
ConcA	27	1,840	0.56	3,300	69

*The assay was carried out at 37°C in the absence and in the presence of pepstatin. One unit = amount of enzyme for which $OD_{280} - OD_{280}$ + papstatin = 0.1 in 30 min.

The purified enzyme was concentrated by ultrafiltration in Centricon tubes and subjected to SDS-PAGE. The results of such an analysis showed the presence of a 40 kDa band. The enzyme displayed optimal activity at pH 2.0, and was fully inhibited by pepstatin A. Although the highest level of activity was obtained at 37°C, the enzyme retained 25% of activity when assayed at a temperature of 5°C. These results indicate that the protease activity isolated from icefish liver has the properties of a cathepsin-like aspartic proteinase.

We thank Professor G. di Prisco for generously providing icefish liver. This work was supported by the Programma Nazionale per le Ricerche in Antartide (P.N.R.A.).

REFERENCES

1. Ruud JT (1954) Nature (London) 173: 858–850.
2. Hureau JC, Petit D and Marneux M (1977) in Adaptation Within Antarctic Ecosystems, ed. Llano GA (Smithsonian Inst. Washington, DC), pp 459–477.
3. Barber DL, Mills Westermann JE and White MG (1981) J. Fish Biol. 19: 11–28.

NOVEL FORMS OF CATHEPSIN E

S. D. Fowler,[1] P. J. Tatnell,[1] D. Bur,[2] and J. Kay[1]

[1]School of Molecular and Medical Biosciences, University of Wales College of Cardiff, Cardiff CF1 3US, Wales, United Kingdom
[2]F. Hoffmann La Roche AG, CH-4002 Basel, Switzerland

Aspartic proteinases of vertebrate and fungal origin are commonly bi-lobal, single-chain enzymes. Each domain contributes one Asp residue to the catalytic apparatus. In the dimeric retroviral aspartic proteinases, each monomer provides a catalytic Asp residue.

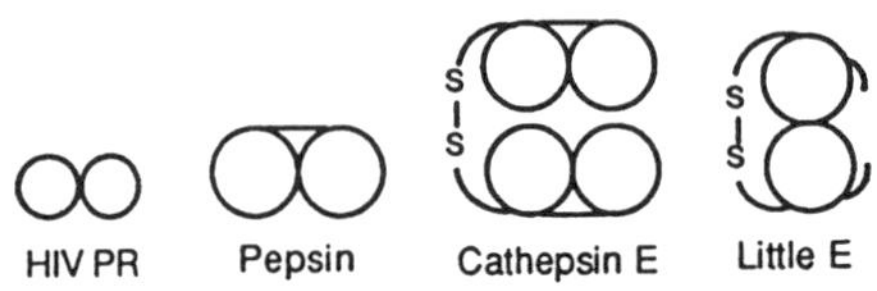

Cathepsin E is exceptional because it has an extra cysteine residue near its N-terminus. This forms an inter-chain disulphide bond such that the enzyme is a homodimer of ~ 85 kDa. Cathepsin E thus consists of *four* domains, each containing a catalytic –Asp–Thr–Gly– motif arranged in two sets of two (see Figure). This inter-chain disulphide bond has been shown to be important for correct intracellular trafficking and localisation of Cathepsin E. Its significance in terms of enzyme stability and activity was investigated by mutating the cysteine residue to alanine. The activity of the resultant monomeric form of human cathepsin E was not altered significantly compared to its dimeric counterpart and its susceptibility to inhibition (e.g. by pepstatin) was not affected either. However, the stability of the monomeric enzyme to temperature and to alkaline pH was markedly reduced.

During our ongoing investigations into the organisation and transcription of genes encoding (pro)cathepsin E, an alternatively spliced variant of the procathepsin E mRNA was found to be present in gastric adenocarcinoma cell lines of human origin e.g. AGS and Kato III cells. The sequence of the first 732 bp at the 5′-end of this variant was identical to that of authentic procathepsin E. The next 142 bp were deleted due to the aberrant splicing of the 3′-end of exon 6 to the 5′-end of exon 8. This resulted in (i) the loss of the 47 residues encoded by exon 7 which include the second –Asp–Thr–Gly– motif and (ii) a frame-shift so that the predicted amino acid sequence following the deletion is totally different to that of authentic cathepsin E. The unique cysteine residue, however, is conserved in the putative gene product, suggesting that two N-terminal domains may be covalently cross-linked by an –S–S-bond to form a bi-lobal structure akin to pepsin and the retroviral proteinases (see Figure). This putative gene product has been termed little E.

A truncated form of the little E cDNA was generated and expressed in *E. coli* in order to determine whether two N-terminal domains combined in this way would display enzymatic activity, in a manner similar to that demonstrated for propep domains by Tang and colleagues. The S-S linked little E arrangement contains two flaps and may be a reflection of the ancestral dimer of this enzyme superfamily.

32

PROTEOLYTIC AND NON-PROTEOLYTIC ACTIVATION OF PRORENIN

Timothy L. Reudelhuber, Véronique Brechler, Isabelle Jutras, Chantal Mercure, and Danielle Methot

Laboratory of Molecular Biochemistry of Hypertension and MRC Multidisciplinary Research Group on Hypertension
Clinical Research Institute of Montreal
Montreal, Quebec H2W 1R7, Canada

1. INTRODUCTION

Largely because of its importance in the pathophysiology of cardiovascular diseases, renin has been intensely studied as a target for drug inhibition. Like most other aspartyl proteases, mammalian renins are synthesized *in vivo* as inactive precursors called prorenins. Although the crystal structure of prorenin has not yet been reported, it seems likely by analogy with the known structure of the closely related porcine pepsinogen (1) that the amino terminal prosegment of prorenin blocks access to the substrate-binding cleft of the enzyme. *In vivo*, activation of prorenin is associated with the proteolytic cleavage of the prosegment. Unlike many other aspartyl proteases, however, prorenin does not undergo an autocatalytic processing, but rather requires the action of one or more distinct proteases to remove its prosegment.

In vitro, purified prorenins can undergo significant transient and reversible activation by acidification (2) which presumably leads to a protonation of charged residues responsible for holding the prosegment in the substrate-binding cleft and causes a charge repulsion between the body of renin and its prosegment. Prorenin may also be reversibly activated by prolonged exposure to cold (3) which can destabilize hydrophobic interactions. Comparison of the structures of human renin and porcine pepsinogen has led to the suggestion that activation of prorenin would also require a significant conformational change in which the polypeptide corresponding to the amino terminus of the mature renin would bind at the position on the renin molecule previously occupied by the amino-terminus of the prosegment (4). Several groups have obtained evidence that these macromolecular rearrangements can also occur at physiological pH and temperatures as demonstrated by the ability of active site-directed inhibitors to quantitatively "trap" recombinant purified prorenin over time *in vitro* (5,6). These results have led to the suggestion that prorenin, which is secreted by

Aspartic Proteinases, edited by James.
Plenum Press, New York, 1998.

numerous tissues and is present in 5–10 times the concentration of proteolytically active renin in the circulation of mammals (7), is in an equilibrium between a "closed" (inactive) and "open" (active) conformation and might, therefore, contribute to overall renin activity *in vivo*.

In the current manuscript, we review experiments that identify the region of the prosegment responsible for its binding to the body of renin and review evidence for the "breathability" of the prosegment during its transition in the secretory pathway and after its secretion from cells.

2. CULTURED CELL MODELS IN THE STUDY OF PRORENIN BIOSYNTHESIS AND SECRETION

2.1. Intracellular Fates of Prorenin

The juxtaglomerular (JG) cells of the kidney are the primary source of active renin in humans. However, removal of the kidneys results not only in the virtual disappearance of active renin from the circulation, but also in a dramatic reduction in prorenin (8–10), suggesting that JG cells are capable of secreting both the active enzyme and its zymogen precursor. Biochemical studies suggest that the two forms are secreted with different kinetics and from different intracellular compartments (11): While the bulk of the prorenin is secreted rapidly and constitutively by low density vesicles budding off from the trans-Golgi network (TGN), a portion of the prorenin in the TGN is packaged into secretory granules where it is converted to active renin by proteolytic cleavage (12). Active renin is stored in these dense cytoplasmic secretory granules until the cell receives a signal triggering its release. Thus, in vivo, prorenin is largely secreted into the circulation by the constitutive secretory pathway while active renin is released from JG cells by the regulated secretory pathway (see Fig. 1 and ref. (13) for a general review of constitutive and regulated secretion).

Because JG cells have proven difficult to culture in the differentiated state (14), alternative cell culture models have been developed to study the biosynthesis and intracellular sorting of prorenin. Chinese Hamster Ovary (CHO) cells, have been used by several groups to produce recombinant prorenin (15–17) because they constitutively secrete large

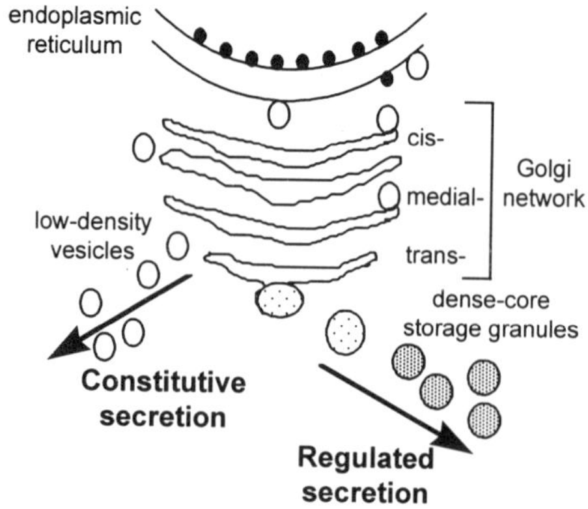

Figure 1. Schematic representation of the constitutive and regulated secretory pathways. See text for details.

quantities of the uncleaved zymogen when transfected with the appropriate recombinant expression vector. In contrast, AtT-20 cells, derived from an ACTH-producing mouse corticotrophic tumor, not only secrete recombinant prorenin constitutively, but they also sort a portion of the expressed prorenin to secretory granules where it is proteolytically cleaved to, and stored as, active renin (18). This active renin can be released by stimulation of the cells with depolarizing agents or secretagogues (18), much like the JG cells of the kidney. Thus, these two cell lines can be used to analyze two distinct stages in the biosynthesis and secretion of prorenin; its folding in the early stages of the constitutive secretory pathway and the characteristics of its proteolytic cleavage in the regulated secretory pathway. Each of these stages reveals complementary information on the way in which prorenin is folded into its native conformation.

2.2. CHO Cells and the Initial Steps in Prorenin Biosynthesis

In an effort to identify the peptide region which provides the "glue" in holding the prosegment in the active site cleft of renin (thereby maintaining the inactive state of prorenin), we introduced single amino acid mutations in the prosegment of recombinant human prorenin (19). Expression vectors for the native and mutated prorenins were transfected into CHO cells and the degree to which each mutation caused an enzymatic activation of prorenin was determined (Fig. 2). It should be noted that under the experimental conditions used, CHO cells do not proteolytically activate human prorenin (18). Therefore, activation of the prorenin caused by the point mutations in its prosegment is presumed to be due to a repulsion of the prosegment from the body of renin and an exposure of the active site of the enzyme. Our results demonstrated that while numerous point mutations led to activation of prorenin, they did so to different extents (19). Of particular note is that regardless of whether the mutation consisted of a charge reversal (positively charged amino acids changed to negative and vice-versa) or an alanine substitution, the most dramatic effect on activation of prorenin came from mutation of amino acids 10–15 relative to the amino terminus (Fig. 3). Of the residues in this region, by far the most important for maintaining prorenin in its inactive state was leucine 13, where even the relatively conservative substitution of alanine resulted in the secretion of a prorenin molecule which was 95% active. These results suggest that a fairly short peptide region is responsible for holding the human prorenin prosegment in the substrate-binding cleft and are consistent with a model in which the interaction of leucine in a hydrophobic pocket is stabilized by flanking salt bridges between basic residues in the prosegment and carboxylate groups on renin. A similar proposal was made by Heinrikson et al. (5) based on the ability of prorenin molecules containing various lengths of prosegment to reversibly activate in the presence of acid.

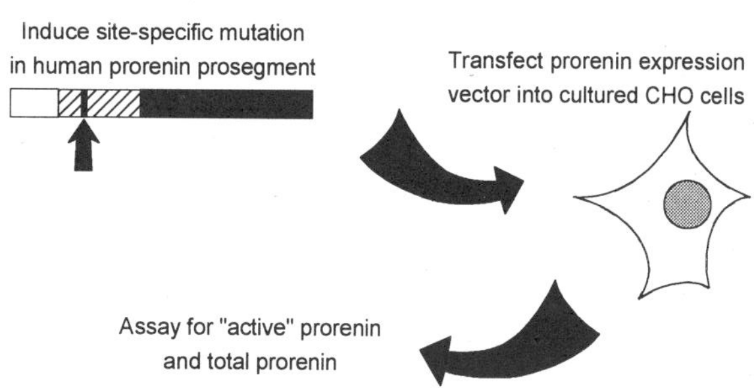

Figure 2. Schematic representation of the approach used to identify the prosegment amino acids involved in maintaining the enzymatically inactive state of prorenin.

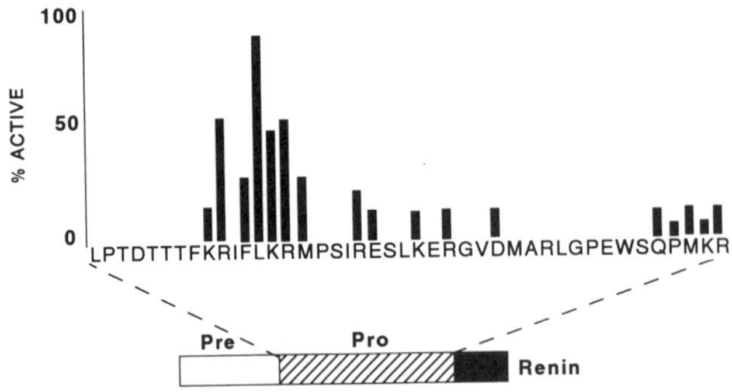

Figure 3. Schematic summary of the effect of mutating prosegment amino acids on activation of prorenin. A bar graph depicting the maximal activation obtained at each position is shown above the single amino acid code of the human prorenin prosegment. Data were compiled from ref. 19.

In the course of these studies, we also noticed that there existed an inverse relationship between the ability of a given mutation to activate the enzyme and its effect on the level of expression of the mutated protein (Fig. 4). In other words, while a prorenin molecule which displays 5–7% renin activity is expressed at levels comparable to the nonmutated protein, one which is 95% enzymatically active is only expressed at 5% of the level of the native protein. Pulse-chase experiments confirmed that this was due to a decrease in secretion of the mutated prorenins and was not due simply to the secretion of normal amounts of prorenin, various portions of which were catalytically inactive (19). One possible explanation for this finding is that weakening the interaction of the prosegment with the body of renin is detrimental to the biosynthesis of the prorenin molecule, thereby leading to its incorrect folding within the secretory pathway and its resulting degradation prior to its secretion from the cell. Indeed, the role of prosegments as folding "chaperones" has been demonstrated in the case of other proteases (20) and our data provided the first demonstration that the prosegment of prorenin (and perhaps that of other aspartyl proteases) plays a similar role.

What was perhaps most surprising about the role of the prosegment as a folding "chaperone," however, was the fact that the effect was not stochastic. In other words, there was not an "all-or-none" effect of the activation mutations, where beyond a certain percentage of activation, the protein could no longer be correctly folded and secreted. Rather, the effect of the mutations followed a smooth continuum and we were able to derive a mathematical relationship between the percentage activation by a mutation and the level of expression of the resulting protein (Fig. 4). These results suggest that individual muta-

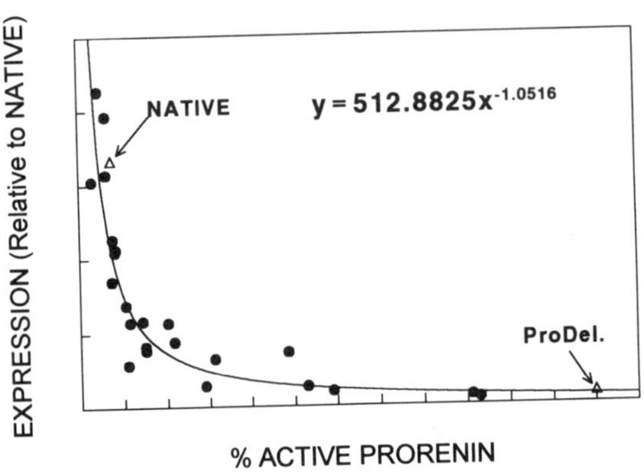

Figure 4. Relationship between the enzymatic activity of the various prorenins and their level of expression in transfected CHO cells. The open triangles depict the position of native prorenin (NATIVE) and prorenin in which the entire prosegment was deleted by mutagenesis (ProDel.) Data are reproduced from ref. 19.

tions cause a shift in an equilibrium in the renin-bound and -unbound states of the prosegment and that the same interactions which are necessary to hold the prosegment in the substrate-binding cleft are necessary for the "chaperone" effect of the prosegment in the initial folding of prorenin.

2.3. Secretory Granule Sorting and Proteolytic Activation of Prorenin

AtT-20 cells have been used by numerous investigators to analyze the molecular signals responsible for the sorting of proteins to dense core secretory granules and the regulated secretory pathway (21). Indeed, some years ago, we demonstrated that a portion of the human prorenin expressed by transfection in AtT-20 cells was secreted as active renin and with all of the characteristics of a protein stored in secretory granules (18). Furthermore, AtT-20 cells appear to cleave human prorenin only within the granules of the regulated secretory pathway and at a unique site which is identical to that reported for *bona fide* human kidney renin (18,22). For these reasons, we have used AtT-20 cells as a model to understand the targeting of human prorenin to the regulated secretory pathway (23).

Of particular interest with regard to prorenin folding and the mechanism of its activation is the selection of the cleavage site by the endogenous protease in AtT-20 cells. While this cleavage takes place after a pair of basic amino acids (Lys^{42}, Arg^{43}), there are 6 other pairs of basic amino acids in prorenin, including two additional pairs in the prosegment proximal to the natural cleavage site (Fig. 5A). Somewhat surprisingly, when the natural cleavage site is mutated (Fig. 5A), there is no other proteolytic cleavage of prorenin and AtT-20 cells secrete the full-length zymogen (Fig. 5B). Is this simply because the proximal dibasic pairs in the prosegment are poor substrates for the activating protease? This is apparently not the case, because if a portion of the prosegment containing the two proximal paired basic residues is engineered into a heterologous fusion protein (Fig. 6A), they are cleaved in transfected AtT-20 cells (Fig. 6B). These results suggest that the two proximal cleavage sites in the prosegment would be good substrates for the AtT-20 protease if they were exposed during the transit through the secretory pathway, but that they must be inaccessible. These results argue against a model in which the prosegment "breathes", even in the relatively acid environment (pH 5.5) of the secretory granule.

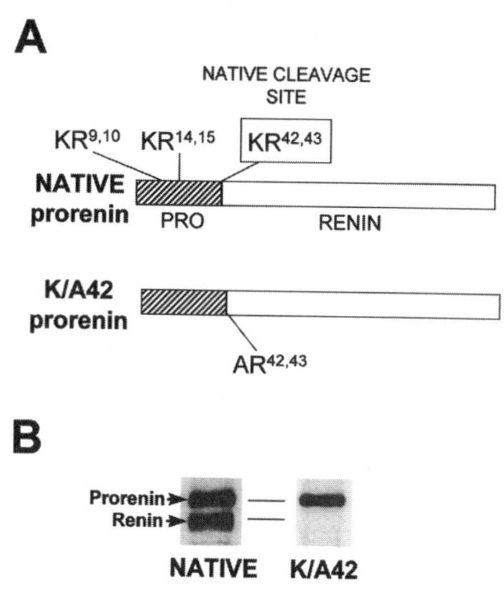

Figure 5. Effect of mutating the native cleavage site on prorenin expressed in transfected AtT-20 cells. **A:** shown are the three dibasic (Lys,Arg) residues in the prosegment of native human prorenin as well as the point mutation of the lysine residue at the native cleavage site (K/A42). **B:** Immunoprecipitation of radiolabeled native and K/A42 prorenin from transfected AtT-20 cells. Note that mutation of lysine 42 results in secretion of full-length prorenin. Data are derived from ref. 23.

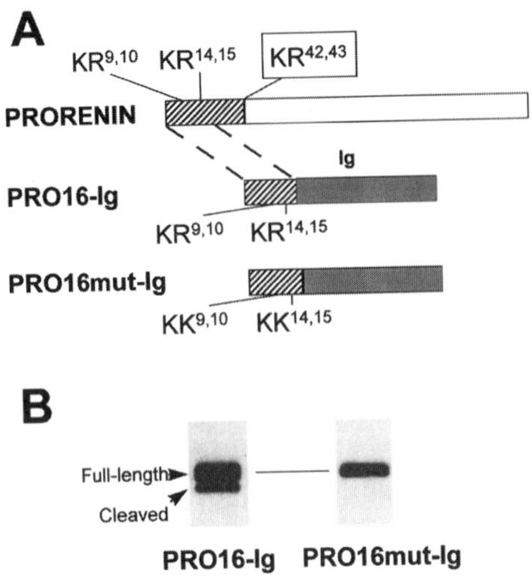

Figure 6. Cleavage of the proximal dibasic pairs in the prorenin prosegment in a heterologous fusion protein expressed in AtT-20 cells. **A:** Amino acids 1–16 of the prosegment were fused to a fragment of mouse IgG heavy chain and expressed in AtT-20 cells. The prorenin prosegment is either in its native form (PRO16-Ig) or the two Lys–Arg residues were mutated to Lys–Lys (PRO16mut-Ig). **B:** Immunoprecipitation of radiolabeled fusion proteins expressed in AtT-20 cells. Note that while the Pro16-Ig protein can be cleaved by the cells, mutation of the Lys–Arg pairs prevents its processing (PRO16mut-Ig) suggesting that cleavage of PRO16-Ig is occurring at these residues. Data are derived from ref. 23.

These findings have two implications regarding the evolution of prorenin structure during its biosynthesis: First, they suggest that once the prorenin is folded, the prosegment remains tightly bound to the body of renin for the remainder of its intracellular transition. Second, they suggest that proteolytic cleavage of prorenin occurs in the absence of any substantial conformational change since molecules in which the natural cleavage site was mutated did not apparently "open" the prosegment to expose the two other proximal dibasic sites to cleavage.

3. PRORENIN ACTIVATION *IN VIVO*

While our data suggest that prorenin goes from a relatively dynamic to a more constrained conformation during its transition through the cell secretory pathway, what happens to it once it exits the cell? Could there be conditions in the circulation or in tissues which would allow the prosegment to unfold, thereby rendering prorenin transiently active? While it is very difficult to test all possible conditions to which prorenin might be exposed *in vivo*, some information relevant to this question can be obtained from experiments in which prorenin was introduced by either infusion or by transgenic expression into whole animals.

3.1. Injection of Recombinant Prorenin

Because both prorenin and renin are present in the circulation, it is difficult to test for the activation of the endogenous precursor after its secretion. However, Lenz et al. (24,25) took advantage of the fact that recombinant human renin is active in cynomolgous monkeys (common marmoset) to ask whether injection of recombinant human *prorenin* led to any of the known *renin* responses (thereby testing whether prorenin had been activated in the circulation). Monkeys were infused with either moderate amounts of human prorenin over a period of 40 minutes, resulting in a 4-fold increase of circulating prorenin over endogenous levels (24), or were given a bolus of human prorenin over five minutes which resulted in over a 100-fold increase in circulating prorenin (25). In neither case did the investigators detect any evidence that the prorenin had been converted to renin as they did not detect any increase in blood pressure, heart rate, glomerular filtration rate, renal plasma flow or urinary sodium and potassium (known physiological effects of renin). These results strongly suggest that prorenin is neither proteolytically cleaved to active renin, nor "opened" to generate active prorenin once it is in the circulation.

3.2. Over-Expression of Prorenin in Transgenic Animals

Several groups have used the approach of over-expressing components of the renin-angiotensin system in transgenic animals with the goal of generating hypertensive animals (26–28). One consistent finding is that expression of either human renin or its substrate (angiotensinogen) alone in either mice or rats does not result in elevations in blood pressure, presumably due to the weak affinities of human renin for the rodent substrate and vice-versa (29,30). However, when an animal expressing human renin is mated with one expressing human angiotensinogen, blood pressure goes up since the components of the renin-angiotensin system downstream of renin (angiotensin converting enzyme, angiotensin peptides and angiotensin receptors) are interchangeable in rodents and humans (26–28). We have taken advantage of this natural species-specificity to test for the activity of human prorenin within tissues of transgenic mice (31). By over-expressing human active renin specifically within the growth hormone-producing cells of the anterior pituitary and subsequently mating these animals with mice in which human angiotensinogen is expressed in the same cells, we generate large quantities of angiotensin I within the pituitaries of the double-transgenic mice (Fig. 7). Because of the small size of the pituitary gland (<20μl in volume), it is virtually certain that the angiotensin I being measured in

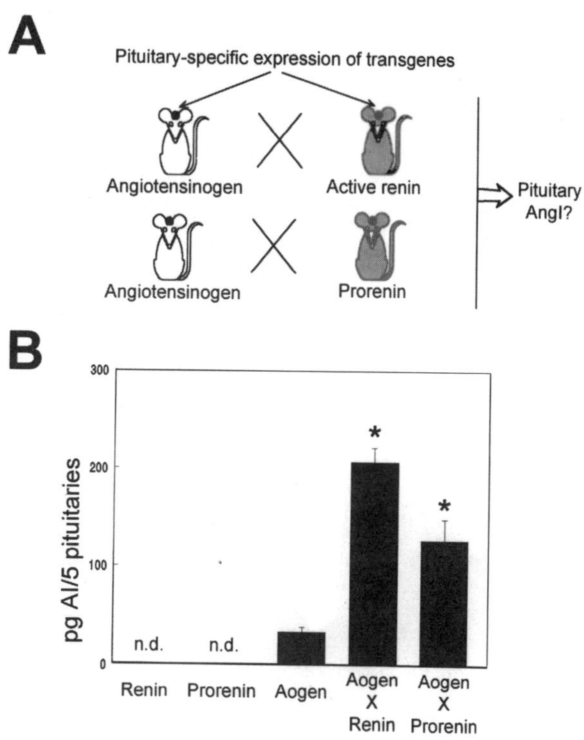

Figure 7. A transgenic mouse model to test for the enzymatic activity of human prorenin within tissues. **A:** Strategy for the study. Human active renin or human prorenin were expressed specifically in the pituitary of transgenic mice. These were individually mated to mice expressing human angiotensinogen within the same tissue and male offspring were tested for pituitary angiotensin I (AngI) content (the product of renin cleavage of angiotensinogen). **B:** Pituitary content of AngI in transgenic mice. Pituitaries from 5 animals of each line were pooled and tested for angiotensin I (AngI) content by HPLC and RIA. n.d., not detectable; Aogen., angiotensinogen; *$P < 0.01$ as compared to animals transgenic for angiotensinogen alone as determined by Student's t-test. Data are derived from ref. 31.

these experiments is synthesized within the tissue and is not derived from the circulation. This system therefore provides us with an extremely sensitive assay for the activity of the transgenic human renin. What happens when prorenin is expressed instead of renin? In this case, pituitary angiotensin I levels are also clearly elevated as compared to animals expressing angiotensinogen alone (Fig. 7B). As the prorenin expressed in this case is not cleaved to renin (data not shown), these results strongly suggest that human prorenin can display enzymatic activity within expressing tissues, although this activity is not maintained once the prorenin is released in the circulation.

4. CONCLUSIONS

Taken together, our results suggest that the prosegment of prorenin serves two main functions: It provides a folding "chaperone" for the zymogen within the cell and it conditionally maintains the enzyme in the inactive state. Thus, prorenin appears to undergo multiple, alternating states of flexibility during its biosynthesis: The prosegment is in a binding equilibrium with the body of the nascent protein during its initial folding in the endoplasmic reticulum. It then stabilizes for the remainder of the intracellular transition of the folded protein but is once again amenable to unfolding after secretion, but while still within the tissue. Once prorenin is released in the circulation, the prosegment remains firmly bound to the body of renin, maintaining it in the inactive state. Why is the prosegment able to unfold in tissues and not within the circulation? One possible explanation (among many) is the potential for acidification of interstitial spaces in tissues which might favor prosegment unfolding and resulting prorenin activation, while the relatively neutral pH and high protein concentrations in the circulation would favor the maintenance of the inactive (prosegment attached) state of the zymogen.

Prorenin can also be activated by proteolysis prior to secretion from cells. Indeed, several enzymes have been proposed as prorenin processing enzymes (PPEs) in mammals (32–35) and more may yet be discovered. The distribution of these enzymes suggests that different PPEs might activate prorenin in different tissues (35,36). The physiological implication of this finding is that the activity of the renin-angiotensin system in any given tissue could be determined both by the activity of an accompanying PPE or by modulation of conditions (interstitial pH, etc.) which would cause the prosegment to unfold. Variations in either of these factors could provide control for activity of tissue renin-angiotensin systems at different stages in development or in different physiological or pathophysiological conditions.

Does this mean that circulating prorenin is physiologically inactive? While we still don't have the definitive answer to this question, the recent report of membrane binding sites for prorenin (37,38) and the discovery that life-long hyper-proreninemia might lead to multiple cardiovascular lesions (39) suggest that even in the absence of catalytic activity, circulating prorenin may not be physiologically inert.

ACKNOWLEDGMENTS

This work was supported by a grant from the Medical Research Council of Canada (MRCC), a post-doctoral fellowship from the MRCC (V.B.), a studentship from the Heart and Stroke Foundation of Canada (D.M.) and the Merck-Frosst Chair in Pharmacology (T.L.R.).

REFERENCES

1. Sielecki, A. R., Fujinaga, M., Read, R. J., and James, M. N. (1991) *J. Mol. Biol.* **219**, 671–692
2. Derkx, F. H., Schalekamp, M. P., and Schalekamp, M. A. (1987) *J. Biol. Chem.* **262**, 2472–2477
3. Pitarresi, T. M., Rubattu, S., Heinrikson, R., and Sealey, J. E. (1992) *J. Biol. Chem.* **267**, 11753–11759
4. Sielecki, A. R., Hayakawa, K., Fujinaga, M., Murphy, M. E., Fraser, M., Muir, A. K., Carilli, C. T., Lewicki, J. A., Baxter, J. D., and James, M. N. (1989) *Science* **243**, 1346–1351
5. Heinrikson, R. L., Hui, J., Zürcher-Neely, H., and Poorman, R. A. (1989) *Am. J. Hypertens.* **2**, 367–380
6. Derkx, F. H., Deinum, J., Lipovski, M., Verhaar, M., Fischli, W., and Schalekamp, M. A. (1992) *J. Biol. Chem.* **267**, 22837–22842
7. Hsueh, W. A. and Baxter, J. D. (1991) *Hypertension* **17**, 469–479
8. Sealey, J. E. and Rubattu, S. (1989) *Am. J. Hypertens.* **2**, 358–366
9. Campbell, D. J., Kladis, A., Skinner, S. L., and Whitworth, J. A. (1991) *J. Hypertens.* **9**, 265–274
10. Hosoi, M., Kim, S., Tabata, T., Nishitani, H., Nishizawa, Y., Morii, H., Murakami, K., and Yamamoto, K. (1992) *J. Clin. Endocrinol. Metab.* **74**, 680–684
11. Pratt, R. E., Carleton, J. E., Richie, J. P., Heusser, C., and Dzau, V. J. (1987) *Proc. Natl. Acad. Sci. USA* **84**, 7837–7840
12. Taugner, R., Kim, S. J., Murakami, K., and Waldherr, R. (1987) *Histochemistry* **86**, 249–253
13. Kelly, R. B. (1991) *Curr. Opin. Cell Biol.* **3**, 654–660
14. Pinet, F., Corvol, M. T., Dench, F., Bourguignon, J., Feunteun, J., Menard, J., and Corvol, P. (1985) *Proc. Natl. Acad. Sci. USA* **82**, 8503–8507
15. Ishizuka, Y., Saito, M., Hori, H., Poorman, R. A., Yoshida, S., and Murakami, K. (1988) *Biochem. Biophys. Res. Commun.* **152**, 849–856
16. Poorman, R. A., Palermo, D. P., Post, L. E., Murakami, K., Kinner, J. H., Smith, C. W., Reardon, I., and Heinrikson, R. L. (1986) *Proteins* **1**, 139–145
17. Carilli, C. T., Vigne, J., Wallace, L. C., Smmith, L. M., Wong, M. A., Lewicki, J. A., and Baxter, J. D. (1988) *Hypertension* **11**, 713–716
18. Fritz, L. C., Haidar, M. A., Arfsten, A. E., Schilling, J. W., Carilli, C., Shine, J., Baxter, J. D., and Reudelhuber, T. L. (1987) *J. Biol. Chem.* **262**, 12409–12412
19. Mercure, C., Thibault, G., Lussier-Cacan, S., Davignon, J., Schiffrin, E. L., and Reudelhuber, T. L. (1995) *J. Biol. Chem.* **270**, 16355–16359
20. Baker, D., Shiau, A. K., and Agard, D. A. (1993) *Curr. Opin. Cell Biol.* **5**, 966–970
21. Miller, S. G. and Moore, H. (1990) *Curr. Opin. Cell Biol.* **2**, 642–647
22. Do, Y. S., Shinagawa, T., Tam, H., Inagami, T., and Hsueh, W. A. (1987) *J. Biol. Chem.* **262**, 1037–1043
23. Brechler, V., Chu, W. N., Baxter, J. D., Thibault, G., and Reudelhuber, T. L. (1996) *J. Biol. Chem.* **271**, 20636–20640
24. Lenz, T., Sealey, J. E., Lappe, R. W., Carilli, C., Oshiro, G. T., Baxter, J. D., and Laragh, J. H. (1990) *Am. J. Hypertens.* **3**, 257–261
25. Lenz, T., Sealey, J. E., Maack, T., James, G. D., Heinrikson, R. L., Marion, D., and Laragh, J. H. (1991) *Am. J. Physiol.* **260**, R804-R810
26. Ganten, D., Wagner, J., Zeh, K., Bader, M., Michel, J. B., Paul, M., Zimmermann, F., Ruf, P., Hilgenfeldt, U., Ganten, U., and et al (1992) *Proc. Natl. Acad. Sci. USA* **89**, 7806–7810
27. Fukamizu, A., Sugimura, K., Takimoto, E., Sugiyama, F., Seo, M. S., Takahashi, S., Hatae, T., Kajiwara, N., Yagami, K., and Murakami, K. (1993) *J. Biol. Chem.* **268**, 11617–11621
28. Merrill, D. C., Thompson, M. W., Carney, C. L., Granwehr, B. P., Schlager, G., Robillard, J. E., and Sigmund, C. D. (1996) *J. Clin. Invest.* **97**, 1047–1055
29. Cumin, F., Le-Nguyen, D., Castro, B., Menard, J., and Corvol, P. (1987) *Biochim. Biophys. Acta* **913**, 10–19
30. Takahashi, S., Fukamizu, A., Hatae, T., Yamada, Y., Sugiyama, F., Kajiwara, N., Yagami, K., and Murakami, K. (1992) *J. Vet. Med. Sci.* **54**, 1191–1193
31. Methot, D., Silversides, D., Mercure, C., and Reudelhuber, T. L. (submitted, 1997) *Catalytic activity of human prorenin in transgenic mouse tissues.*
32. Kim, W. S., Nakayama, K., and Murakami, K. (1991) *FEBS Lett.* **293**, 142–144
33. Wang, P. H., Do, Y. S., Macaulay, L., Shinagawa, T., Anderson, P. W., Baxter, J. D., and Hsueh, W. A. (1991) *J. Biol. Chem.* **266**, 12633–12638
34. Benjannet, S., Reudelhuber, T., Mercure, C., Rondeau, N., Chretien, M., and Seidah, N. G. (1992) *J. Biol. Chem.* **267**, 11417–11423
35. Mercure, C., Jutras, I., Day, R., Seidah, N. G., and Reudelhuber, T. L. (1996) *Hypertension* **28**, 840–846

36. Reudelhuber, T. L., Ramla, D., Chiu, L., Mercure, C., and Seidah, N. G. (1994) *Kidney Int.* **46**, 1522–1524
37. Campbell, D. J. and Valentijn, A. J. (1994) *J. Hypertens.* **12**, 879–890
38. Sealey, J. E., Catanzaro, D. F., Lavin, T. N., Gahnem, F., Pitarresi, T., Hu, L. F., and Laragh, J. H. (1996) *Am. J. Hypertens.* **9**, 491–502
39. Véniant, M., Ménard, J., Bruneval, P., Morley, S., Gonzales, M., and Mullins, J. (1997) *J. Clin. Invest.* **98**, 1966–1970

33

THE ROLES OF THE BASIC RESIDUES IN THE PROSEGMENT OF ASPERGILLOPEPSINOGEN I

H. Inoue,[1] X.-P. Huang,[1,2] T. Hayashi,[3] S. B. P. Athauda,[4] H. Yamagata,[1] S. Udaka,[5] and K. Takahashi[1]

[1]School of Life Science
Tokyo University of Pharmacy and Life Science, Japan
[2]Graduate School of Agriculture
[3]Graduate School of Science
University of Tokyo, Japan
[4]Department of Biochemistry
Faculty of Medicine
University of Peradeniya, Sri Lanka
[5]Department of Brewing and Fermentation
Faculty of Agriculture
Tokyo University of Agriculture, Japan

1. INTRODUCTION

Aspartic proteinases are normally synthesized as inactive precursors. In many cases, the precursor is activated by proteolytic removal of the propart under acidic conditions. The propart is considered to participate not only in inactivation of the proenzyme by masking the active site, but also in control of the activation.[1–3] Additionally, the prosequence is considered to play an essential role in folding.[4,5]

Generally, the propart of aspartic proteinase is abundant in basic residues while the enzyme moiety is abundant in acidic residues. Some of the basic residues are supposed to be important for the roles of the prosequence in electrostatic interaction with acidic residues in the enzyme moiety. The most noteworthy interaction is between a lysyl residue in the propart and the aspartyl residues in the active site of the enzyme moiety, as observed in pepsinogen and progastricsin.[5–8] Through this interaction, the lysyl residue is thought to mask the active site of the proform at neutral pH. The lysyl residue is conserved as a Lys–Tyr motif or Lys–Phe motif in many aspartic proteinase zymogens such as pepsinogen, progastricsin, procathepsin D, and fungal pepsinogen, whose mature forms normally function under acidic conditions (Fig. 1). The corresponding residue in aspergillopepsinogen I from *Aspergillus niger* var. *macrosporus*, which is also called proctase B or acid

Aspartic Proteinases, edited by James.
Plenum Press, New York, 1998.

	21 30 40
Aspergillopepsinogen I	APAPTRKGFTINQIARPANKTRTINL
Rhizopuspepsinogen	VEAAPNGKKINI--PLAKNNSYKPSAK
Human pepsinogen A	IMYKVPLIRKKSLRRTLSERGLLKDFLK
Human progastricsin	AVVKVPLKKFKSIRETMKEKGLLGEFLR
Bovine prochymosin	AEITRIPLYKGKSLRKALKEHGLLEDFLQ
Human procathepsin D	LVRIPLHKFTSIRRTMSEVGGSVEDLI
Human procathepsin E	QGSLHRVPLRRHPSLKKKLRARSQLSEFWK
Human prorenin	TFGLPTDTTTFKRIFLKRMPSIRESLK

	50 ↓ 60 69
Aspergillopepsinogen I	PGMYARSLAKFGGTVPQSVKEAA
Rhizopuspepsinogen	NALN-KALAKYNRRKVGSGGITTE
Human pepsinogen A	KHNL-NPARKYFPQWEAPTL
Human progastricsin	THKY-DPAWKYRFGDL
Bovine prochymosin	KQQYG-ISSKYSGF
Human procathepsin D	AKG---PVSKYSQAVPAVTE
Human procathepsin E	SHNLDMIQFTES
Human prorenin	ERGVDMARLGPEWSQPMKR

Figure 1. Comparison of the prosequences of aspartic proteinases.

proteinase B, is Lys56 (numbering is based on the preproform).[9] Except for this lysyl residue, the roles of the other basic residues in the prosegment are not clear. The present study was performed to shed further light on the roles of their basic residues in the prosegments of aspergillopepsinogen I using site-directed mutagenesis.

2. EXPERIMENTAL PROCEDURE

Materials and methods used in this study are described elsewhere.[10–12]

3. RESULTS AND DISCUSSION

3.1. Heterologous Expression of Aspergillopepsinogen I[10–12]

3.1.1. Expression by Escherichia coli. In order to investigate the roles of the basic residues in the propart by using site-directed mutagenesis, we constructed a heterologous expression system of aspergillopepsinogen I. First, we constructed an expression system using *E. coli* BL21(DE3) and a vector containing T7 promoter (Fig. 2A). In this expression system, most part of the expressed proenzyme formed inclusion bodies so that *in vitro* refolding was required to obtain an enough amount of the potentially active proform. Previously, we prepared aspergillopepsinogen I in the potentially active form by using this system.[11] However, the mutant proenzyme K56N, in which Lys56 was replaced with asparagine, could not be obtained in the correctly folded, potentially active form. This result suggests the importance of Lys56 for folding, but the possibility cannot be ruled out

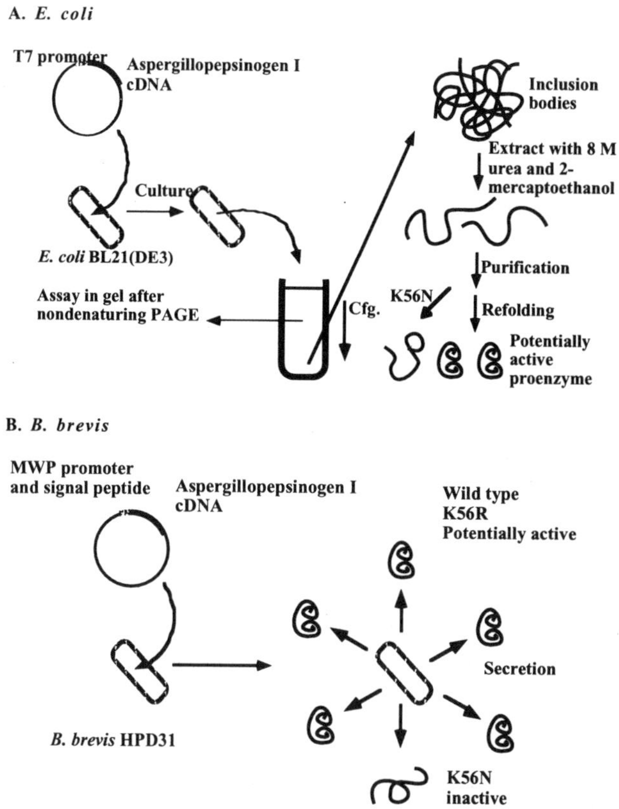

Figure 2. Heterologous expression by *E. coli* (pabel A) and *B. brevis* (panel B).

that the artificial refolding conditions were not adequate for the mutant proenzyme. Therefore the expression system using *E. coli* did not seem to be good enough to study the roles of the basic residues by site-directed mutagenesis, and another expression system was desirable in which folding proceeds under physiological conditions.

3.1.2. Expression by Bacillus brevis[12]. Therefore, we constructed an expression system using *B. brevis* (Fig. 2B). In this system, aspergillopepsinogen I was secreted into the medium as a potentially active form, and was indistinguishable from that obtained by the *E. coli* system in all the properties analyzed. The amount of the expressed proenzyme was estimated to be about 30–50 mg per liter of culture. The replacement of Lys56 with Asn using this expression system also resulted in loss of the potential activity. On the other hand, the replacement of Lys56 with Arg gave a potentially active proenzyme. These facts suggest that the lysyl residue is important for folding through electrostatic interaction with the catalytic aspartyl residues. By using this expression system, the roles of the basic residues other than Lys56 were also investigated.

3.2. Activation of Aspergillopepsinogen I

Aspergillopepsinogen I is activated at pH < 6. The conversion into the mature form was observed on SDS-PAGE, and was completely inhibited by pepstatin A. The K56N mutant, which is not correctly folded and is inactive, did not undergo proteolysis at all. These results suggest an autoproteolytic activation mechanism. During the processing, two intermediates were transiently observed. Only the larger intermediate appeared first, then the smaller one and the mature form appeared successively. The N-terminus of the mature processed form was Glu67, which is inconsistent with the N-terminal Ser70 of the native enzyme (Fig. 3). A similar difference between the recombinant and the native processed form was observed also in the case of aspergillopepsin II.[14] Some other proteinase(s) is considered to participate in the maturation of the native enzymes. One of the intermediate forms, possibly the larger one, was obtained in the presence of pepstatin A. The N-terminus of this intermediate was found to be Ser53.

3.3. Site-Directed Mutagenesis of the Basic Residues at the Processing Sites

3.3.1. Mutation of the Putative Initial Processing Site. Arg52 was replaced with glutamine by site-directed mutagenesis. The resulting mutant, R52Q, was expressed by *B. brevis*. The R52Q mutant thus obtained was inactive even at acidic pH, and no processing was observed on SDS-PAGE after incubation at acidic pH. However, addition of a small amount of native aspergillopepsin I converted the mutant proform completely into the processed form that was indistinguishable from the wild-type mature form on SDS-PAGE and was active toward acid denatured hemoglobin. These facts suggest that the R52Q mutant was correctly folded but could not be activated by autolysis, and that the processing required the already activated enzyme.

3.3.2. Mutation of the Processing Site for Maturation. Lys66 was replaced with glutamine. The resulting mutant K66Q was potentially active toward hemoglobin, but the relative molecular size of the mature form observed on SDS-PAGE was larger than that of the wild type. The N-terminus of the mature form was shown to be Ala55, indicating the proteolysis between Leu54 and Ala55 (Fig. 3). It is plausible that this form is the smaller intermediate.

3.4. The Roles of the Basic Residues in the Propart of Aspergillopepsinogen I

The model of activation mechanism is shown in Fig. 4. At neutral pH, Lys56 interacts with the catalytic aspartyl residues and the prosequence masks the active site. Lys56

```
21
A-P-A-P-T-R-K-G-F-T-I-N-Q-I-A-R-P-A-N-

                                K66Q mutant
                 Intermediate-1 (Intermediate-2?)
                           ↓53   ↓55
K-T-R-T-I-N-L-P-G-M-Y-A-R-S-L-A-K-F-G-

    Recombinant mature   Native enzyme
                   ↓67    ↓70
G-T-V-P-Q-S-V-K-E-A-A-S-K-G-S-A-
```

Figure 3. Processing sites.

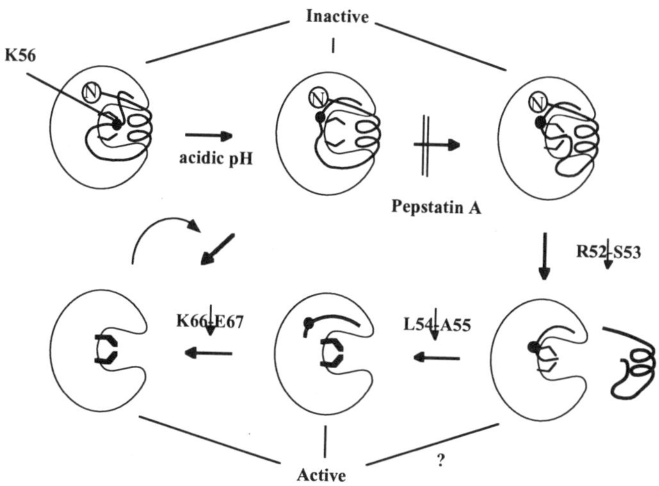

Figure 4. Activation mechanism of aspergillopepsinogen I.

is also considered to be important for folding of the proform. At acidic pH, there is no interaction between Lys56 and the aspartyl residues, and the bond between residues 52 and 53 is cleaved. Under these acidic conditions, pepstatin A, if present, can enter the active site to inhibit this initial processing. Arg52 is essential for the autolysis at this position. Unless the bond between residues 52 and 53 is cleaved, the prosequence continues masking the active site. The cleavage of this peptide bond allows the prosegment to be removed, and the active site to be unmasked. Then the bonds between Leu54 and Ala55, and between Lys66 and Glu67 are successively cleaved. Lys66 is essential for processing between Lys66 and Glu67, but the processing at this position is not necessary for the activation of aspergillopepsinogen I. Thus the activation of aspergillopepsinogen I is thought to proceed by a stepwise mechanism.

In the propart of aspergillopepsinogen I, there are five basic residues other than Arg52, Lys56 and Lys66. The individual replacement of the five residues with glutamine did not significantly affect the enzyme activity and the molecular size of the mature processed form observed on SDS-PAGE. However, the contributions of these residues to control of the activation cannot be ruled out.

REFERENCES

1. Tang, J., and Wong, R. N. S. (1987) J. Cell. Biochem. 33, 53–63.
2. Foltmann, B. (1988) Biol. Chem. Hoppe-Seyler 369 Suppl., 311–314.
3. Koelsh, G., Mares, M., Metcalf., and Fusek, M. (1994) FEBS Lett. 343, 6–10.
4. Van den Hazel, H. B., Kielland-Brandt, M. L., and Winther, J. R. (1993) J. Biol. Chem. 268, 18002–18007.
5. Fukuda, R., Horiuchi, H., Ohta, A., and Takagi, M. (1994) J. Biol. Chem. 269, 9556–9561.
6. James, M. N. G., and Sielecki, A. R. (1986) Nature 319, 33–38.
7. Sielecki, A. R., Fujinaga, M., Read, R. J., and James, M. N. G. (1991) J. Mol. Biol. 219, 671–692.
8. Hartsuck, J. A., Koelsh, G., and Remington, S. J. (1992) Proteins 13, 1–25.
9. Moore, S.A., Sielecki, A. R., Chernaia, M. M., Tarasova, N. I., and James, M. N. G. (1995) J. Mol. Biol. 247, 466–485
10. Lu, J.-F., Inoue, H., Kimura, T., Makabe, O., and Takahashi, K. (1995) Biosci. Biotech. Biochem. 59, 954–955.

11. Inoue, H., Hayashi, T., Huang, X.-P., Lu, J.-F., Athauda, S. B. P., Kong, K.-H., Yamagata, H., Udaka, S., and Takahashi, K. (1996) Eur. J. Biochem. 237, 719–725.
12. Inoue, H., Lu, J.-F., Athauda, S. B. P., Kong, K.-H., Hayashi, T., Kimura, T., Makabe, O., and Takahashi, K. (1995) in: Aspartic Proteinases: Structure, Function, Biology, and Biomedical Implications, K. Takahashi, ed., pp. 581–587, Plenum Press, New York.
13. Hayashi, T., Inoue, H., Kato, M., Udaka, S., and Takahashi, K. (1995) in: Aspartic Proteinases: Structure, Function, Biology, and Biomedical Implications, K. Takahashi, ed., pp. 589–596, Plenum Press, New York.
14. Kagami, N., Inoue, H., Kimura, H., Makabe, O., and Takahashi, K. (1995) in: Aspartic Proteinases: Structure, Function, Biology, and Biomedical Implications, K. Takahashi, ed., pp. 597–603, Plenum Press, New York.

34

ACTIVATION MECHANISM OF PEPSINOGEN AS COMPARED TO THE PROCESSING OF HIV PROTEASE *gag–pol* PRECURSOR PROTEIN

Gerald Koelsch, Jeff Loy, Xinli Lin, and Jordan Tang

Protein Studies Program
Oklahoma Medical Research Foundation
University of Oklahoma Health Sciences Center
Oklahoma City, Oklahoma 73104

INTRODUCTION

The structural evidence suggests that the eukaryotic aspartic proteases and the retroviral proteases are related in three-dimensional structures and in their evolution. The internal two-fold structural homology between the two lobes of the eukaryotic aspartic proteases (Tang *et al.*, 1978) is similar to the homodimeric structure of HIV (Wlodawer *et al.*, 1989) and other retroviral proteases. Also, the folding topology of individual lobes in eukaryotic aspartic proteases is related to that of the retroviral enzyme monomers (Rao *et al.*, 1991). It is now well established that these two groups of proteases share active-site structure and catalytic mechanism.

Eukaryotic and retroviral aspartic proteases are both synthesized as larger precursors. In the former, each single-chain zymogen contains, upstream from the mature protease, a *pro* region which is removed upon activation. On the other hand, the precursor of the human immunodeficiency virus (HIV) protease, the *gag–pol* polyprotein, contains in a single chain several protein units: p17–p24–p7–protease–reverse transcriptase–integrase. Not only is the protease monomer linked at both termini, the *gag–pol* protein is also assembled with *gag* protein as a tightly packed shell inside the immature HIV virion (Nermut *et al.*, 1994). Therefore, the activation of HIV protease precursor must be more complicated than that of the eukaryotic aspartic protease zymogens and the relationship between the two mechanisms is not immediately obvious.

In the following, we will discuss the current understanding and compare the activation mechanisms of both groups using pepsinogen and HIV *gag–pol* polyprotein as models.

Aspartic Proteinases, edited by James.
Plenum Press, New York, 1998.

ACTIVATION MECHANISM OF PEPSINOGEN

When pepsinogen is converted to pepsin in an acid solution, two kinetically distinguishable mechanisms, the intramolecular and bimolecular activations, may take place. The intramolecularly catalyzed pepsinogen activation was demonstrated by the acid activation of pepsinogen covalently attached to Sepharose (Bustin and Conway-Jacobs, 1971) and by the first-order activation kinetic of pepsinogen in solution (Al-Janabi *et al.*, 1972; McPhie, 1972). The current understanding of the intramolecular mechanism is schematically summarized in Fig. 1, A. The three-dimensional structure of porcine pepsinogen shows that the central part of the 44-residue *pro* region is in helical conformation covering over the active-site cleft of the pepsin moiety (James and Sielecki, 1986; Hartsuck *et al.*, 1992). The spectral evidence has shown that the initial event after the acidification of pepsinogen solution is a very rapid change of pepsinogen conformation (Glick *et al.*, 1986). Evidence supports the hypothesis that this spectral change is associated with the selective denaturation and 'conformational dissociation' of the *pro* region resulting in the exposure of the active-site cleft (Fig. 1, A, step (1)). This activation intermediate can be trapped on a pepstatin-Sepharose column and alkaline eluted as pepsinogen. Also, hemoglobin, a pepsin substrate, competes with the pro peptide for the binding to the active site of this intermediate and inhibits the further steps in the intramolecular activation of pepsinogen (Marciniszyn, Jr., *et al.*, 1976). The binding of the denatured *pro* region as an intramolecular substrate to the active-site results in the first cleavage (Fig. 1, A, step (2)), between Leu^{16p}–Ile^{17p} (Dyke and Kay, 1976). The dissociation of *pro* fragment 1p–16p, step (3), the conformational dissociation of peptide 17p–44p region, step (4), and the bimolecular cleavage of the bond between residues 44p–1, step (5), yield pepsin. The rate-limiting step in this mechanism is at the first bond cleavage, step (2). Since the down-stream steps are faster, the rate for the generation of pepsin activity during the intramolecular activation is the same as that for the appearance of the pepsin N-terminus, step (5) (Sanny *et al.*, 1975).

The importance of step (3) (Fig. 1), the dissociation of *pro* peptide 1p–16p from the rest of pepsinogen, in maintaining the rate of the intramolecular activation has been recently demonstrated. The native pepsinogen single chain is arranged in the order of *pro–pep–sin*, where *pep* and *sin* are the N- and C-terminal lobes respectively. By protein engineering, pepsinogen domains were rearranged into a recombinant protein, *sin–pro–pep* (Lin *et al.*, 1995), in which the chain between the two lobes, *pep* and *sin*, was cut and the N-terminus of *pro* is linked through a 10-residue linker to the C-terminus of *sin* (Fig. 2). The complete activation of *sin–pro–pep* produced a two-chain pepsin which had been shown previously to be active (Lin *et al.*, 1992). The intermolecular activation mechanism of *sin–pro–pep* is shown in Fig. 1, B. The steps (1), the conformational change, and (2), the first cleavage, are fast and are apparently similar to those in the activation of native pepsinogen (Lin *et al.*, 1995). Since the N-terminus of the *pro* region is linked and more difficult to dissociate, the fast initial steps suggest that the acid-triggered conformational change in the native pepsinogen may involve only the central helical part of the *pro* region, an idea previously proposed by Kageyama *et al.* (1989). The major difference in the intramolecular activation of *sin–pro–pep* is the slowness in generating pepsin activity. While the activation of native pepsinogen in pH 2, 20°C, has a $t_{1/2}$ about 16 s (Al-Janabi *et al.*, 1972), the activation of *sin–pro–pep* under a similar conditions produced a $t_{1/2}$ of bout 45 min (Lin *et al.*, 1992). This is apparently due to the ineffectiveness in the 'conformational dissociation' of the linked *pro* peptide 1p-16p from the pepsin moiety (Fig. 1, B, step (4)). The cleaving of this peptide, step (6), then becomes rate-limiting.

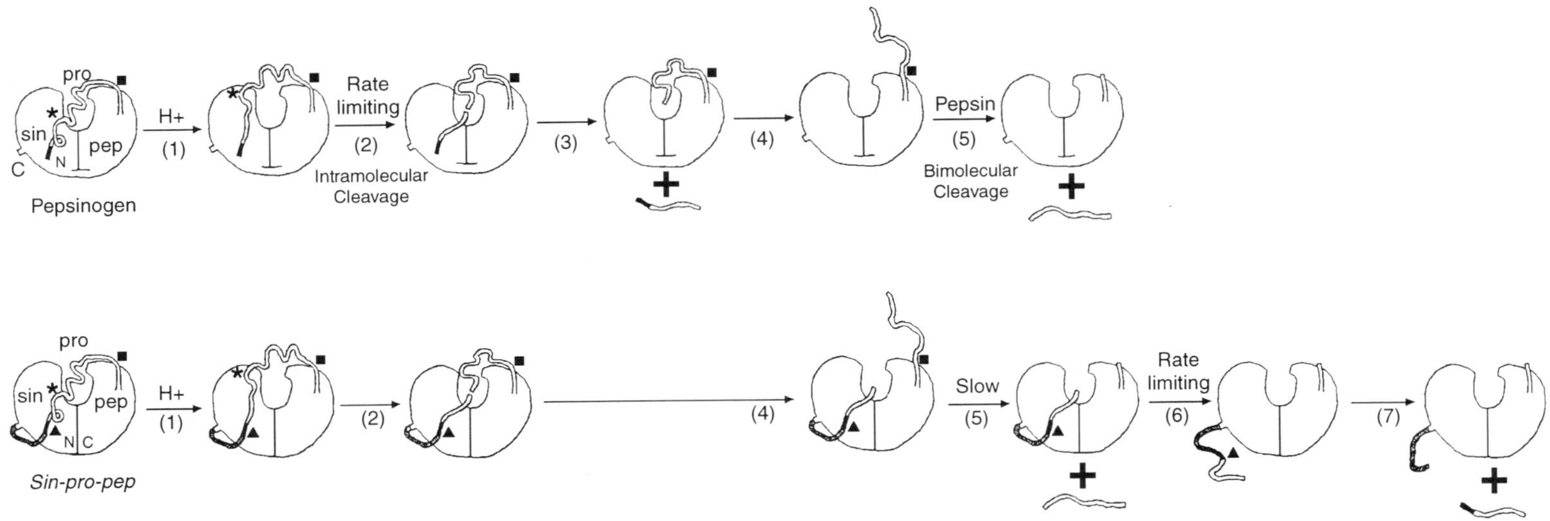

Figure 1. Schematic presentation of activation mechanism of pepsinogen (upper portion) and domain-rearranged pepsinogen, *sin–pro–pep* (lower portion). Zymogens are marked to denote their N- and C-termini, and connectivity of domains *pro, pep,* and *sin*. Cleavage sites 1 (★), 2(■), and 3 (▲) are marked where appropriate. The N-terminal residues 1P–9P of pepsinogen or *sin–pro–pep* are marked solid within the *pro* domain coil. The connecting strand between *sin* and *pro* (lower portion) is shaded. Mechanistic details are described in the text.

NdeI	*Sin*	Connecting Strand	*pro*	*pep*	*NdeI*
CAT ATG	TCT-----------GCC	GGG CCC $(GGT)_6$ CCC GGG	CTC--	----------TCT TGA	CAT ATG AT
Met	Ser^{173}-----------Ala^{326}	Gly Pro $(Gly)_6$ Pro Gly	Leu^{1P}-Val^{2P}----Leu^{16P}-Ile^{17P}----Leu^{44P}-Ile^{1}	------Ser^{172} Stop	

↑ Site 3 (▲) ↑ Site 1 (★) ↑ Site 2 (■)

Figure 2. The rearrangement of pepsinogen domains *pro–pep–sin* to *sin–pro–pep*. The domains of pepsinogen are depicted (not to scale) with their respective coding sequence and peptide sequence beneath. Amino acids relevant to the scheme are numbered. Activation cleavage sites are designated as in Figure 1.

Pepsinogen activation involving a bimolecular digestion was originally observed by Harriott (1938). The mechanism of this activation route is less clear. Kageyama and Takahashi (1983) observed the release of the entire 44-residue *pro* peptide from pepsinogen activation. There are good reasons to suggest that the release of entire *pro* peptide is the result of the bimolecular activation. Firstly, in the experiment of Kageyama and Takahashi (1983), this peptide accounted for about 20% of all *pro* peptides, which is near the fraction of bimolecular activation (16%) calculated by Al-Janabi *et al.* (1972) for a 1 mg/ml of pepsinogen at pH 2. Also, there was a clear concentration dependence of the activation rate, suggesting the presence of bimolecular activation. Secondly, the peptide strand near Leu^{44p}–Ile^{1} is well exposed on the tertiary structure of pepsinogen and should be accessible for pepsin hydrolysis. It is also known that an equivalent site in prorenin is cleaved bimolecularly by another protease. Finally, judging from the crystal structure of pepsinogen, it is difficult to see how the peptide region near Leu^{44p}–Ile^{1} can be placed in the active site without seriously disrupting the tertiary structure. The cleavage may be followed by the dissociation from the pepsin moiety either the entire or part of the *pro* peptide, which is then further hydrolyzed. The second-order kinetic constant for this mechanism has a maximum at pH 3, in contrast to that for the intramolecular first-order constant which is highest near pH 1 (Al-Janabi *et al.*, 1972). Under the physiological conditions, the pepsinogen is activated predominantly by the intramolecular mechanism.

PROCESSING OF HIV PROTEASE PRECURSOR

Since the HIV protease moiety is the only source of proteolytic activity present in the *gag–pol* polyprotein, it is generally assumed that this protease unit catalyzes all the cleavages in the immature HIV virions. Since *gag–pol* polyproteins are tightly co-packaged as a '*gag* shell' in a ratio of about 1 to 20 (for active protease homodimer: 1:40) inside the immature virions, it is reasonable to assume that dimerization of the protease precedes the packaging of the virions. The dimerization may also be aided by the dimerization of reverse transcriptase and integrase monomers present in the *gag–pol* polyprotein. Since the lateral motion of *gag–pol* inside of the '*gag* shell' is very much limited by the tight packing, it is also reasonable to assume the cleavages at the termini of HIV protease moiety and the release of the protease to be an early event in the processing. It is therefore logical to study the processing of an abbreviated precursor of HIV protease.

We have developed an *E. coli* expression system for an HIV-1 protease mini-precursor, which has a 25 residue extension from the N-terminus of protease (ten residues next to protease are native *gag* residues). Most heterologously-expressed HIV protease precursors spontaneously process into protease. However, the mutation of Ala^{28} to Ser resulted in the reduction of k_{cat}/K_m about 200 fold as compared to that of the wild-type enzyme and thus enabled the accumulation and isolation of this mini-precursor (Ido *et al.*, 1991). The conversion of the isolated 14-kDa mini-precursor to the 11-kDa mature A28S protease was observed in the pH range from 3.5 to 6.8. The highest conversion rate occurred near pH 5, the optimal pH for the matured protease. During the course of conversion from mini-precursor to A28S protease, the specific activity remained the same as measured by using a chromogenic substrate (Co *et al.*, 1994). This observation suggests that the mini-precursor has a proteolytic activity which is nearly as efficient as that for the mature A28S-protease. The rate of the conversion was measured as follows. The ^{14}C-labeled mini-precursor was incubated in pH 4.5 to convert it to an 11-kDa mature A28S-HIV protease. Samples from different time of incubation were separated on SDS-polyacrylamide electrophoresis and the 14-kDa and 11-kDa bands were excised for the determination of radioactivity.

The time-conversion curves could be fit with nine first-order kinetic models to obtain the apparent first-order rate constants, k_{app}. A plot of mini-precursor concentration against k_{app} is shown in Fig. 3. This plot clearly shows that the apparent first-order constant is concentration dependent, implying the presence of both the first-order (intra-homodimer) processing and the second-order (inter-homodimer) processing. The first-order rate constant, k_1, and the second-order rate constant, k_2, were calculated from the intercept and slope respectively (Co *et al.*, 1994). These constants from 5 independent experiments are shown in Table 1. The average k_1 is 0.003 ± 0.001 min^{-1} and the average k_2 is 0.03 ± 0.01 $min^{-1}{\cdot}mg^{-1}{\cdot}mL$. These results imply that the mini-precursor is proteolytically processed by both the inter-homodimer and the intra-homodimer mechanisms (Fig. 4). Louis *et al.* (1994), using a similar kinetic approach with a different construct of mini-precursor, also observed the intra-homodimer processing.

The intra-homodimer processing is well suited for the proteolytic processing of *gag–pol* polyprotein in the maturation of HIV virions. In the immature HIV virions, the *gag* and *gag–pol* polyproteins form a '*gag* shell', which, based on the electron microscopic evidence, constitutes the tight packing of cylinder-shaped *gag* in a two-dimensional icosahedral array just inside of the lipid membrane (Nermut *et al.*, 1994). As illustrated in Fig. 5, the N-terminal ends of the *gag* and *gag–pol* cylinders are anchored in the lipid membrane by their myristoyl groups. Because of the tight packing, the individual *gag* and *gag–pol* in the '*gag* shell' would not be able to engage in lateral diffusion. In a *gag* to *gag–pol* ratio of 1 to 20, the random packing of *gag–pol* monomers is unlikely to form

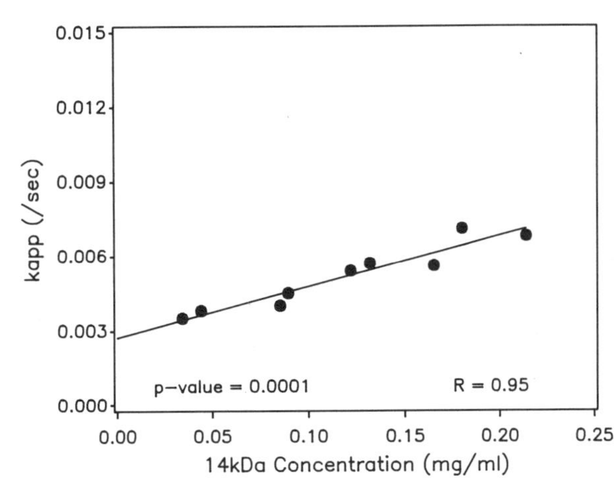

Figure 3. Dependence of apparent first-order rate of activation (k_{app}) on the concentration of A28S HIV-1 protease mini-precursor. Separate k_{app} were obtained from profiles of activation at various 14 kDa mini-precursor concentrations. The correlation coefficient significance of the non-zero Y-intercept (p-value labeled in the figure) is conclusive for first-order activation.

Table 1. Conversion rate constants for the processing of 14-kDa miniprecursor to 11-kDa mature HIV A28S protease

Prep	Concentration (mg/mL)	N	k_1 (min^{-1})	k_2 [min^{-1}–$(mg/mL)^{-1}$]
A	2.50	8	0.003 (0.001)[a]	0.045 (0.006)
B	0.65	8	0.005 (0.002)	0.03 (0.01)
C	0.353	8	0.0027 (0.0003)	0.021 (0.003)
D	0.30	5	0.0009 (0.0001)	0.029 (0.005)
E	NA[b]	6	0.004 (0.001)	NA
Average			0.003 (0.001)[c,d]	0.03 (0.01)[d]

[a]Standard deviation in parentheses.
[b]NA, not available.
[c]The k_1 was averaged from five experiments as noted under Methods. The k_1 is significant (p-value = 0.007 for HO: mean = 0, Student's T test).
[d]Reported constants are averages weighted by the product of the correlation coefficient and the inverse of standard error from linear regression.

adjacent homodimers with one another. It is assumed that the dimerization of *gag–pol* is accomplished prior to its insertion into the icosahedral packed '*gag* shell'. This may not be difficult to occur considering that many *gag–pol* monomers should emerge from the same mRNA during the polysome synthesis in the cytosol. Another important point derived from the icosahedrally packed *gag–pol* dimers is that, at the ratio of 1:40 for the *gag–pol* dimer to *gag*, it is even more unlikely that a *gag–pol* dimer is able to reach another *gag–pol* dimer in order to catalyze the release of HIV protease from its polyprotein. It is more likely that each *gag–pol* dimer is first cleaved by an intra-dimer cleavage at the N-terminal position of the HIV protease, as demonstrated above in the cleavage of mini-precursor of HIV. This first cleavage would result in the release of a protease-reverse transcriptase-integrase (PR–RT–IN) dimer (Fig. 4), which should then be free to engage in further processing shown in Fig. 4. It is also interesting to note that the PR–RT–IN dimer is free to diffuse so the intra-dimer processing at the C-terminal position of the protease is

DEMONSTRATED

$(–P)_2 \rightarrow P_2 + 2–$

$(P–)_2 + (P–)_2 \rightarrow P_2 + 2– + (P–)_2$

IMPLIED

$–(17–24–7–P–R–I)_2 \rightarrow –(17–24–7)_2 + (P–R–I)_2$

$2\,(P–R–I)_2 \rightarrow P_2 + (–R–I)_2 + (P–R–I)_2$

OTHER POSSIBILITIES

$(P–R–I)_2 \rightarrow P_2 + (–R–I)_2$

$2\,(P–R–I)_2 \rightarrow (P–R–)_2 + I_2$

$(P–R–I)_2$, $(P–R–)_2$ and P_2 digestion of other sites

Figure 4. Reactions in the activation of HIV-1 protease precursors. In the upper portion of the schematic, the symbol –P indicates experimental HIV-1 protease precursor construct with an amino-terminal peptide extension; P– indicates that with a carboxyl-terminal extension. The middle portion of the schematic depicts the implied *in vivo* activiation mechanism first at the amino terminus by intra-dimeric processing, followed by inter-dimeric processing of the carboxy terminus. The p160 *gag–pol* precursor is represented by symbols indicating p17, p24, p7, protease (P), reverse transcriptase (R), and integrase (I). The subscripted entity indicates a dimeric species at least with respect to the protease moiety. The lower portion indicates the possibility of intra-dimer processing, and digestion of other sites by active precursor.

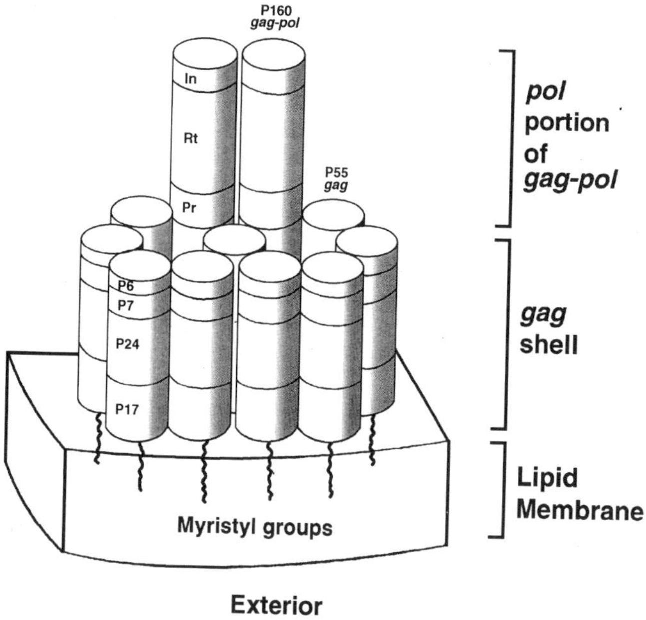

Figure 5. Packing of p55 *gag* and p160 *gag–pol* within the immature HIV virion. Each p55 or p160 is anchored in the membrane by myristoyl groups. The two p160 cylinders depict the protease precursor *gag–pol* dimer. The unprocessed protease dimer, in a ratio of 1:40 within the virion, is laterally inaccessible for processing by a second p160 *gag–pol* protease precursor dimer. The packing dimensions are derived from Nermut *et al.* (1994).

not required. Indeed, Wondrak *et al.* (1996) have observed that the cleavage of this position is by inter-homodimer digestion. The remainder of the cleavages of the eight processing sites in *gag* and *gag–pol* is also likely to be ordered due to the packing of the '*gag* shell' (Tang and Hartsuck, 1995).

POSSIBLE EVOLUTIONARY RELATIONSHIP BETWEEN THE ACTIVATION MECHANISMS OF THE PEPSINOGEN FAMILY AND RETROVIRAL POLYPROTEIN PRECURSORS

The mechanism of the intra-homodimer processing of the HIV protease precursor dimer is obviously similar to that for the intramolecular activation of pepsinogen. Likewise, the inter-homodimer activation of *gag–pol* at the protease N-terminus and the intermolecular activation of pepsinogen are similar. It is tempting to suggest that these mechanisms may have been preserved in the evolutionary process of the aspartic proteases. Especially in the light that HIV protease and pepsin are likely to be related in evolution by the gene duplication and fusion (Tang *et al.*, 1978). However, in this case, there is an absence of structural evidence to support the relationships in functional mechanism of the precursor processings. It would be interesting to compare the three-dimensional structure of the regions immediately adjacent to the HIV protease in *gag–pol* with the conformation of the *pro* region in pepsinogen. In the absence of the three-dimensional structure of *gag–pol*, it seems equally conceivable that the relatedness of these activation mechanisms may evolve from a convergent evolutionary process as demanded by the need of particular processing situations.

ACKNOWLEDGMENT

The authors wish to thank Dr. J. Sedlacek for pointing out the possible role of polysome synthesis in *gag–pol* dimerization.

REFERENCES

Al-Janabi, J., Hartsuck, J.A. and Tang, J. (1972) "Kinetics and Mechanism of Pepsinogen Activation" *J. Biol. Chem.* **247:** 4628–4632.

Bustin, M. and Conway-Jacobs, A. (1971) "Intramolecular Activation of Porcine Pepsinogen" *J. Biol. Chem.* **246:** 615–620.

Co, E., Koelsch, G., Lin, Y., Ido, E., Hartsuck, J.A. and Tang, J. (1994) "Proteolytic Processing Mechanisms of a Miniprecursor of the Aspartic Protease of Human Immunodeficiency Virus Type 1" *Biochemistry* **33:** 1248–1254.

Dyke, C.W. and Kay, J. (1976) "Conversion of Pepsinogen into Pepsin is not a One-Step Process" *Biochem. J.* **153:** 141–144.

Glick, D.M., Auer,H.E., Rich, D.H., Kawai, M. and Kamath, A. (1986) "Pepsinogen Activation: Genesis of the Binding Site" *Biochemistry* **25:** 18158–18164.

Hartsuck, J.A., Koelsch, G. and Remington, S.J. (1992) "The High Resolution Crystal Structure of Porcine Pepsinogen" *Proteins* **13:** 1–25.

Ido, E., Han, H.-P., Kezdy, F.J. and Tang, J. (1991) "Kinetic Studies of Human Immunodeficiency Virus Type 1 Protease and its Active-Site Hydrogen Bond Mutant A28S" *J. Biol. Chem.* **266:** 24359–24366.

James, M.N.G. and Sielecki, A.R. (1986) "Molecular Structure of an Aspartic Proteinase Zymogen, Porcine Pepsinogen, at 1.8 Å Resolution" *Nature* **319:** 33–38.

Kageyama, T., Ichinose, M., Miki, K., Athauda, S.B., Tanji, M. and Takahashi, K. (1989) "Difference of Activation Processes and Structure of Activation Peptides in Human Pepsinogen A and Progastricsin" *J. Biochem. (Tokyo)* **105:** 15–22.

Kageyama, T. and Takahashi, K. (1983) "Occurrence of Two Different Pathways in the Activation of Porcine Pepsinogen to Pepsin" *J. Biochem.* **93:** 743–754.

Lin, X.L., Koelsch, G., Loy, J.A. and Tang, J. (1995) "Rearranging the Domains of Pepsinogen" *Protein Science* **4:** 159–166.

Lin, X.L., Lin, Y.Z., Koelsch, G., Gustchina, A., Wlodawer, A. and Tang, J. (1992) "Enzymic Activities of Two-Chain Pepsinogen, Two-Chain Pepsin, and the Amino-Terminal Lobe of Pepsinogen" *J. Biol. Chem.* **267:** 17257–17263.

Louis, J.M., Nashed, N.T., Parris, K.D., Kimmel, A.R. and Jerina, D.M. (1994) "Kinetics and Mechanism of Auto-Processing of Human Immunodeficiency Virus Type 1 Protease from an Analog of the Gag-pol Polyprotein" *Proc. Natl. Acad. Sci. USA* **91:** 7970–7974.

Marciniszyn, J., Jr., Huang, J.S., Hartsuck, J.A. and Tang, J. (1976) "Mechanism of Intramolecular Activation of Pepsinogen" *J. Biol. Chem.* **251:** 7095–7102.

McPhie, P. (1972) "Pepsinogen: Activation by a Unimolecular Mechanism" *Biochem. Biophys. Res. Commun.* **56:** 789–792.

Nermut, M.V., Hockley, D.J., Jewett, J.B.M., Jones, I.M., Garreua, M. and Thomas, D. (1994) "Fullerene-like Organization of HIV gag-protein Shell in Virus-like Particles Produced by Recombinant Baculovirus" *Virology* **198:** 288–296.

Rao, J.K.M., Erickson, J.W. and Wlodawer, A. (1991) "Structural and Evolutionary Relationships Between Retroviral and Eucaryotic Aspartic Proteinases" *Biochemistry* **30:** 4663–4671.

Sanny, C.G., Hartsuck, J.A. and Tang, J. (1975) "Conversion of Pepsinogen to Pepsin. Further Evidence for Intramolecular and Pepsin-Catalyzed Activation" *J. Biol. Chem.* **250:** 2635–2639.

Tang, J. and Hartsuck, J.A. (1995) "A Kinetic Model for Comparing Proteolytic Processing Activity and Inhibitor Resistance Potential of Mutant HIV-1 Protease" *FEBS Letters* **367:** 112–116.

Tang, J., James, M.N.G., Hsu, I.N., Jenkins, J.A. and Blundell, T.L. (1978) "Structural Evidence for Gene Duplication in the Evolution of the Acid Protease" *Nature* **271:** 618–621.

Wlodawer, A., Miller, M., Jaskolski, M., Sathyanarayana, B.K., Baldwin,E., Weber, I.T., Selk, L.M., Clawson, L., Schneider, J. and Kent, S.B.H. (1989) "Conserved Folding in Retroviral Proteases: Crystal Structure of a Synthetic HIV-1 Protease" *Science* **245:** 616–621.

Wondrak, E.M., Nashed, N.T., Baber, M.T., Jerina, D.M. and Louis, J.M. (1996) "A Transient Precursor for the HIV-1 Protease, Isolation, Characterization and Kinetics of Maturation" *J. Biol. Chem.* **271:** 4477–4481.

CARDOSINOGEN A

The Precursor Form of the Major Aspartic Proteinase from Cardoon

Miguel Ramalho-Santos,[1] José Pissarra,[2] Euclides Pires,[1] and Carlos Faro[1]

[1]Departamento de Bioquímica
Universidade de Coimbra
Apartado 3126, 3000 Coimbra, Portugal
[2]Centro de Citologia Experimental
Universidade do Porto
4100 Porto, Portugal

1. INTRODUCTION

A feature common to many proteases, including aspartic proteinases, is that they are synthesised as inactive zymogens, subsequently undergoing proteolytic processing to yield the active enzyme. This is a way of assuring the correct folding of the proteinase, regulating its activity during biosynthesis and avoiding unwanted proteolysis. Most aspartic proteinases have a conserved N-terminal pro-segment, which is later removed (1). In pepsinogen, the pro-segment is located over the active site cleft, stabilized by salt bridges that are disrupted at low pH (2). In cathepsin D, apart from the N-terminal pro-segment and a C-terminal dipeptide there is another sequence which is also removed, located within the protein (3). Removal of this 2–7 amino acid long sequence gives rise to the two chain active cathepsin D.

Aspartic proteinases have also been purified and cloned from several plant species (4). Although they bear a high sequence and structure homology to the animal and fungal representatives of the class, they all have an extra Plant Specific Insert (PSI) of about 102 amino acids in a precise position of the protein with no homology whatsoever to the other aspartic proteinases (5). The function of this unexpected portion of the plant proteinases remains unknown.

Recently, we isolated and characterized two APs from fresh flowers of the Asteraceae (Compositae) *Cynara cardunculus* L., that belongs to the same genus as the artichoke (6). The enzymes, which are derived from two different, though related, genes, were named cardosin A and cardosin B (7). They are both two chain enzymes, and cardosin A's chains have

Aspartic Proteinases, edited by James.
Plenum Press, New York, 1998.

apparent molecular masses of 31 and 15 kD (6). We have shown that cardosin A is accumulated in protein storage vacuoles of the stigmatic papillae, the female pollen-receptive surface (8). Here we identify cardosinogen A, the precursor form of the major aspartic proteinase from cardoon. This precursor still contains the PSI, which is entirely and precisely removed during processing. Localization studies indicate that removal of the PSI occurs in the vacuole. In view of these results, the vacuolar targeting and the processing of the enzyme is discussed, as well as the physiological relevance of the PSI.

2. IDENTIFICATION OF CARDOSINOGEN A

Our previous results suggested that the PSI in cardosin A was at least in part removed during proteolytic processing (6). In order to identify a precursor form of cardosin A that still contained the PSI an antibody was produced in rabbit against the synthetic peptide KNNVKSSGGIHDE. This sequence corresponds to a region in cardosin A's PSI (Fig. 1A) predicted to be exposed and antigenic. The peptide was conjugated to the carrier thyroglobulin and produced by standard procedures (9). On immunoblots of SDS-polyacrylamide gels the anti-PSI antibody showed strong reactivity with a polypeptide with an apparent molecular mass of 64 kD mainly present in extracts of immature pistils (female part of the flowers), and complete absence of reactivity with other pistil proteins (Fig. 1B, lane 1). This 64 kD protein is also detected by a polyclonal antibody

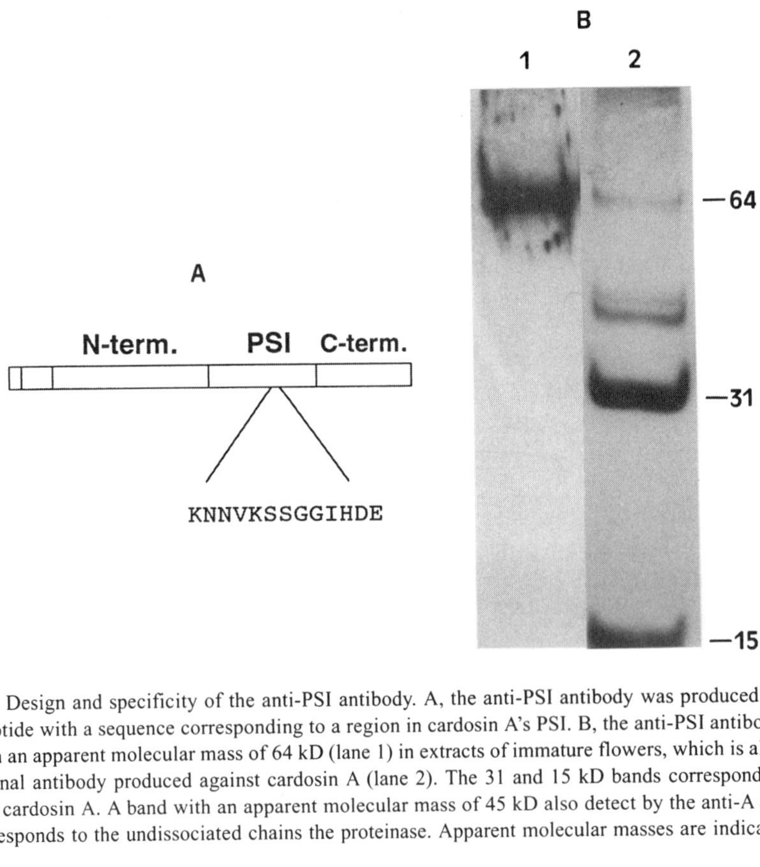

Figure 1. Design and specificity of the anti-PSI antibody. A, the anti-PSI antibody was produced against a synthetic peptide with a sequence corresponding to a region in cardosin A's PSI. B, the anti-PSI antibody detects one band with an apparent molecular mass of 64 kD (lane 1) in extracts of immature flowers, which is also detected by a polyclonal antibody produced against cardosin A (lane 2). The 31 and 15 kD bands correspond to the mature chains of cardosin A. A band with an apparent molecular mass of 45 kD also detect by the anti-A antibody probably corresponds to the undissociated chains the proteinase. Apparent molecular masses are indicated in kilodaltons.

produced against cardosin A (Fig. 1B, lane 2) and binds to pepstatin A–Agarose, but does not display proteolytic activity (data not shown). Following the descriptive nomenclature used for other zymogens the 64 kD cardosin A precursor has been named cardosinogen A. The absence of reactivity of the anti-PSI antibody with any of the chains of mature cardosin A indicates that the PSI is at least in part removed during proteolytic processing.

3. PROTEOLYTIC CLEAVAGE OF CARDOSINOGEN A

N-terminal sequencing (6) and analysis of the crystal structure of cardosin A (manuscript in preparation) allowed the identification of the sites of cleavage that occurs during processing of the PSI (Table 1). These results indicate that during proteolytic processing the PSI is entirely and quite precisely removed, since the cleavage sites are located just at the borders of 31 kD chain/PSI and PSI/15 kD chain. Based on these results, a model for the proteolytic processing of cardosin A is proposed (Fig. 2). The faint reaction of anti-PSI and anti-15 kDa with a band with an apparent molecular mass of 30 kDa (data not shown) suggests the existence of an intermediate form. Thus, removal of the PSI appears to happen in a sequential manner, cleavage occurring first between the 31 kD chain and the PSI and then between the PSI and the 15 kD chain.

Cardosin A itself probably does not process the cleavage sites Pro/31 kDa (R/D) and PSI/15 kDa (S/T), in view of its primary specificity (10,11). In the case of the 31 kDa/PSI cleavage site, however, a very susceptible bond (V–M) is located just two residues away from the C-terminal of the 31 kDa chain, as determined by 3D structure analysis. It is possible that this is the bond that is actually cleaved, the additional two residues not being detected by crystallography. Enzymatic analysis of the C-terminal sequence of the 31 kDa chain is currently underway to test this hypothesis. Interestingly, preliminary results on the activation of cardosinogen A indicate that the conversion is partially inhibited by pepstatin

Table 1. N- and C-terminal sequences of the 31 and 15 kD chains of cardosin A as determined by Edman degradation (6) and 3D structure analysis

	N-terminal	C-terminal
31 kD chain	DSGSA...	...AIGAN
15 kD chain	TSSEE...	...VGFAEAA

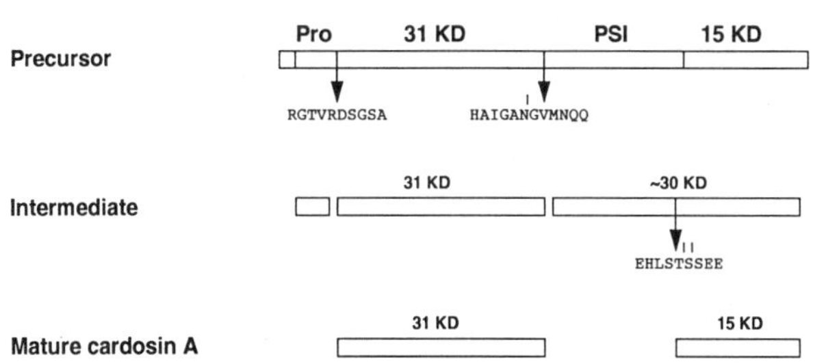

Figure 2. Proposed model for the proteolytic processing of cardosin A. The precise Pro/31 kD chain, 31 kD chain/PSI and PSI/15 kD chain cleavage sites are indicated by arrows. Removal of the PSI appears to occur in the sequential order indicated, with the existence of an intermediate (see text for detailed discussion). Vertical traces in the intermediate stage indicate secondary processing sites.

A (data not shown). This is in agreement with the existence of the intermediate (Fig. 2) and the proposed sequential processing of the PSI. The excised PSI was not detected in immunoblots of protein extracts from different stages of development of the pistils, indicating that it is degraded after being removed from cardosinogen A.

4. LOCALIZATION OF CARDOSINOGEN A

Cardosin A accumulates in protein storage vacuoles of the epidermic papillae of the stigma, i.e., the pollen-receptive surface (Fig. 3a, and ref. 8). Expression of cardosinogen A was studied by immuno electron microscopy using the anti-PSI antibody, as described (12). Cardosinogen A in also localized in the vacuolar protein masses of the papillae (Fig. 3b), indicating that proteolytic processing leading to the activation of the enzyme occurs in the vacuole. The slightly acidified pH of the vacuole and its proteinase content may trigger the removal of the PSI.

5. POSSIBLE TARGETING OF CARDOSINOGEN A AND FUNCTION OF THE PSI

The PSI has been suggested to be involved in the vacuolar targeting of plant aspartic proteinases, in particular barley aspartic proteinase (13). This proposal was based on the circa 33% identity that the PSI shares with saposins, animal proteins activators of glucosidases that remove sugar residues from glycosphingolipids (14). Saposins have been suggested to be involved in a mannose 6-phosphate independent pathway of the lysosomal targeting of cathepsin D (15). In plants, multiple vacuolar targeting sequences have been identified (for review see ref. 16). In the case of cardosin A the amino acid sequence

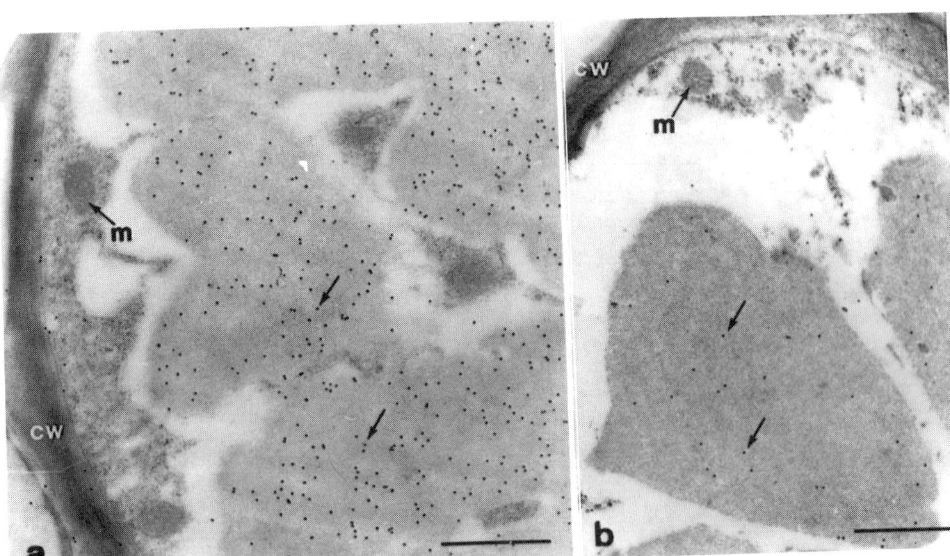

Figure 3. Immunocytochemical localization of cardosin A (a) and cardosinogen A (b) in epidermic papillae of the stigma. a, probed with anti-15 kD. b, probed with anti-PSI. Labelling (arrows) is largely confined to the protein storage vacuoles. No labelling was detected in control sections (data not shown). cw, cell wall; m, mitochondria. Scale bar, 1 μm.

Table 2. Comparison of C-terminal sequences of plant aspartic proteinases with the vacuolar targeting determinant of barley lectin

Barley lectin	V FAEAIAA[a]
Cardosin A	VGFAE AA
Barley aspartic proteinase	IGFAK AA
Oryzasin	VGFAK SA
Cyprosin	VGFAE AA

VGFAEAA was identified at the C-terminal (Table 1). This sequence has a striking similarity to a region in the barley lectin C-terminal propeptide, known to be a vacuolar protein sorting determinant (17), namely VFAEAIAA. In particular, the segment FAEAI was shown to be sufficient for probarley lectin targeting to the vacuole (18). Homologous C-terminal sequences are also present in other plant aspartic proteinases (Table 2), but of these only the barley aspartic proteinase has been shown to be a vacuolar protein (19). The high similarity between cardosin A's C-terminal sequence and a vacuolar targeting signal from barley lectin suggests that the former may direct the vacuolar targeting of cardosin A.

As concerns the physiological relevance of the PSI, other possibilities can be considered. The PSI is obviously not required for proteolytic activity, since it must be removed to render the active enzyme. The PSI may block the active site of cardosinogen A, and thus serve as a way of avoiding unwanted proteolysis. It may also be necessary for the correct folding of the proteinase, acting as an intramolecular chaperone. In addition, the homology with saposins may be relevant, but not necessarily for targeting. Saposins activate glucosidases by binding their substrates or the enzymes directly. They have also recently been reported to have a neurotrophic effect (20). It is possible that during biosynthesis or during execution of the physiological role of the proteinase there is need for the PSI to bind glycosphingolipids or glucosidases, or to participate in a signal transduction pathway, in a way similar to saposins.

6. CONCLUDING REMARKS

Cardosinogen A, the precursor form of the major aspartic proteinase from cardoon, is targeted to the vacuole. Proteolytic processing occurs in the vacuole, most likely with the involvement of other proteinases. It is possible that cardosin A processes one of the cleavage sites (31 kDa/PSI). The PSI is entirely and quite precisely removed. Processing appears to occur first at the 31 kDa/PSI site, and after at the PSI/15 kDa site. The PSI is most likely degraded after being removed from cardosin A.

Here a possible signal for the targeting of vacuolar plant aspartic proteinases is proposed, based on the high homology of a sequence at the C-terminal of cardosinogen A and other plant aspartic proteinases with a vacuolar sorting determinant identified in barley lectin.

The PSI may be necessary for correct folding or to avoid unwanted proteolysis during biosynthesis. It is also possible that the PSI plays a molecular role similar to that of saposins, in view of the homology shared between them. These hypothesis are currently being investigated in our laboratory.

This research was supported by JNICT, Portugal. Miguel Ramalho-Santos is the recipient of a fellowship from the PRAXIS XXI program (JNICT). The skilful photographic assistance of Mrs Andrea Costa is gratefully acknowledged.

REFERENCES

1. G. Koelsch, M. Mares, P. Metcalf, M. Fusek, Multiple functions of pro-parts of aspartic proteinase zymogens. FEBS Lett. 343: 6 (1994).
2. M.N.G. James, A.R. Sielecki, Molecular structure of an aspartic proteinase zymogen, porcine pepsinogen, at 1.8 Å resolution. Nature 319: 33 (1986).
3. S. Yonezawa, T. Takahashi, X.J. Wang, R.N. Wong, J.A. Hartsuck, J. Tang, Structures at the proteolytic processing region of cathepsin D. J. Biol. Chem. 263: 16504 (1988).
4. J. Kervinen, K. Törmäkangas, P. Runeberg-Roos, K. Guruprasad, T. Blundell, T.H. Teeri, Structure and possible function of aspartic proteinases in barley and other plants. In "Aspartic Proteinases: Structure, Function, Biology and Biomedical Implications", K. Takahashi, ed., p 241, Plenum Press, New York (1995).
5. C. Faro, M. Ramalho-Santos, P. Veríssimo, J. Pissarra, C. Frazão, J. Costa, X.-l. Lin, J. Tang, E. Pires, Structural and functional aspects of cardosins. In "Structure and Function of Aspartic Proteinases: Retroviral and Cellular Enzymes", M.N.G. James, ed., p. , Plenum Press, New York (1997).
6. P. Veríssimo, C. Faro, A.J.G. Moir, Y. Lin, J. Tang, E. Pires, Purification, characterization and partial amino acid sequencing of two novel aspartic proteinases from fresh flowers of *Cynara cardunculus*. Eur. J. Biochem. 235: 762 (1996).
7. C. Faro, P. Veríssimo, A.J.G. Moir, Y. Lin, J. Tang, E. Pires, Cardosin A and B, aspartic proteases from the flowers of cardoon. In "Aspartic Proteinases: Structure, Function, Biology and Biomedical Implications", K. Takahashi, ed., p 241, Plenum Press, New York (1995).
8. M. Ramalho-Santos, J. Pissarra, P. Veríssimo, S. Pereira, R. Salema, E. Pires, C.J. Faro, Cardosin A, an abundant aspartic proteinase, accumulates in protein storage vacuoles of the stigmatic papillae of *Cynara cardunculus* L. Planta (accepted for publication) (1997).
9. J. Sambrook, E.F. Fritsch, T. Maniatis, Molecular cloning: a laboratory manual, Second Ed., Cold Spring Harbour Laboratory Press, New York (1989).
10. P. Veríssimo, C. Esteves, C.J. Faro, E.V. Pires, The vegetable rennet of *Cynara cardunculus* contains two proteinases with chymosin and pepsin-like specificities. Biotech. Lett. 17: 621 (1995).
11. M. Ramalho-Santos, P. Veríssimo, C. Faro, E. Pires, Action on bovine α_{s1}-casein of cardosins A and B, aspartic proteinases from the flowers of the cardoon *Cynara cardunculus* L.. Biochim. Biophys. Acta 1297: 83 (1996).
12. S. Pereira, H. Carvalho, C. Sunkel, R. Salema, Immunocytolocalization of glutamine synthetase in mesophyll and phloem of leaves of *Solanum tuberosum* L. Protoplasma 167: 66 (1992).
13. K. Guruprasad, K. Törmäkangas, J. Kervinen, T. Blundell, Comparative modelling of barley-grain aspartic proteinase: a structural rationale for observed hydrolytic specificity. FEBS Lett. 352: 131 (1994).
14. J.S. O'Brien, Y. Kishimoto, Saposin proteins: structure, function, and role in human lysosomal storage disorders. FASEB J. 5: 301 (1991).
15. Y. Zhu, G.E. Conner, Intermolecular association of lysosomal protein precursors during biosynthesis. J. Biol. Chem. 269: 3846 (1994).
16. K. Nakamura, K. Matsuoka, Protein targeting to the vacuole in plant cells. Plant Physiol. 101: 1 (1993).
17. S.Y. Bednarek, N.V. Raikhel, The barley lectin carboxyl-terminal propeptide is a vacuolar protein sorting determinant in plants. Plant Cell 3: 1195 (1991).
18. J.E. Dombrowski, M.R. Schroeder, S.Y. Bednarek, N.V. Raikhel, Determination of the functional elements within the vacuolar targeting signal of barley lectin. Plant Cell 5: 587 (1993).
19. P. Runeberg-Roos, J. Kervinen, V. Kovaleva, N.V. Raikhel, S. Gal, The aspartic proteinase of barley is a vacuolar enzyme that processes probarley lectin in vitro. Plant Physiol. 105: 321 (1994).
20. J.S. O'Brien, G.S. Carson, H.-C. Seo, M. Hiraiwa, Y. Kishimoto, Identification of prosaposin as a neurotrophic factor. Proc. Natl. Acad. Sci. USA 91: 9593 (1994).

36

CRYSTAL STRUCTURE OF HUMAN PEPSINOGEN A

Katherine S. Bateman,[1] Maia M. Chernaia,[1] Nadya I. Tarasova,[2] and Michael N. G. James[1]

[1]MRC Group in Protein Structure and Function
Department of Biochemistry
University of Alberta
Edmonton, Alberta, Canada T6G 2H7
[2]Molecular Aspects of Drug Design Section
ABL-Basic Research Program
National Cancer Institute
FCDRC, P. O. Box B, Frederick, Maryland 21702

1. INTRODUCTION

Human pepsinogen A is the inactive protein precursor of pepsin, an aspartic proteinase found in the stomach. Pepsinogen is synthesized in the chief cells and secreted into the gastric lumen. After pepsinogen has been exposed to the acidic pH of the stomach, a 47 amino acid, N-terminal prosegment is removed by autolytic cleavage to form the active enzyme.

Pepsin has a huge excess of negatively charged residues over positively charged ones. Out of 326 amino acids, thirty-five are acidic and only four are basic. In contrast, the pepsinogen prosegment contains 13 basic residues. The salt bridges formed between the prosegment and pepsin (Lys4p–Asp171, Arg14p–Asp11, His30p–Glu7, Lys37p–Asp32, Lys37p–Asp215) (p denotes residues belonging to the prosegment) will be broken in the acidic conditions of the stomach as the aspartate and glutamate residues become protonated. This facilitates the rearrangement of the prosegment, such that the substrate binding site becomes accessible and the enzyme becomes active.

The three-dimensional structure of the human pepsin precursor has been solved by the method of X-ray crystallography to a resolution of 1.8 Å. Human pepsinogen A shares the same overall conformation of other aspartic proteinase zymogens previously determined such as porcine pepsinogen[1] and human progastricsin.[2]

Aspartic Proteinases, edited by James.
Plenum Press, New York, 1998.

2. STRUCTURE DETERMINATION

Human pepsinogen A, purified from stomach mucosa, was crystallized in space group $P2_12_12$ from 1.7M K_2HPO_4/NaH_2PO_4 buffer at pH 7.3. X-ray diffraction data were collected at the Photon Factory in Tsukuba, Japan to 1.8 Å resolution. The data were processed with the program Weiss[3] and the three-dimensional structure was solved by molecular replacement using the program AMORE.[4] Porcine pepsinogen[1] was used as the search model. Refinement was initially carried out with X-PLOR[5] and then completed with TNT.[6] Both programs were used with Maximum Likelihood targets as implemented by Pannu and Read.[7] Refinement statistics are found in Table 1. In between the rounds of refinement, the model was inspected and fitted to the electron density using the programs FRODO[8] and O.[9] The current R-factor is 20.7%.

3. RESULTS AND DISCUSSION

Human pepsinogen A is made up of a central, six-stranded, anti-parallel β-sheet and N- and C-terminal lobes, also consisting mostly of strands of β-sheet (Figure 1). The first strand of the central β-sheet is provided by the N-terminal portion of the prosegment (Tyr3p to Leu7p). Following this strand are two α-helices, a 3_{10} helix, and a coil region. The junction between the prosegment and the beginning of the mature enzyme is located in the coil region.

A substrate binding cleft is formed where the N- and C-terminal lobes meet and each of the lobes contributes one of the catalytic aspartic acid residues (Asp32 and Asp215). Pepsinogen is inactive because the prosegment fills the active site cleft thereby precluding substrate binding. In addition, catalytic residues Asp32 and Asp215 are electrostatically stabilized by residues Lys37p, Tyr38p and Tyr9 (Figure 2).

Table 1. Crystallographic data for human pepsinogen A

Crystal system	Orthorhombic
Space group	$P2_12_12$
Unit cell dimensions (Å)	a=91.6, b=105.2, c=40.2, α=β=γ=90º
Total number of reflections	143166
Total number of unique reflections	33491
Merging R-factor[a] (%)	7.36
Completeness	80.2% to 1.8 Å resolution
Resolution range used in refinement	20Å to 1.8Å
Refinement progress at present	
R-factor[b] (%)	20.7
R-free[c] (for 10% of reflections) (%)	25.8
Total number of protein atoms	2852
Total number of water molecules	305
RMS deviations from ideal stereochemistry	
Bond lengths (Å)	0.019
Bond angles (°)	2.050
Deviation of planar groups (°)	0.018

[a]Merging R-factor = $\Sigma_{hkl}\Sigma_i |I_{i,hkl} - \langle I\rangle_{hkl}| / \Sigma_{hkl}\Sigma_i\langle I\rangle_{hkl}$ where $\Sigma_i I_{i,hkl}$ indicates summation over i measurements of a particular hkl intensity, and $\langle I\rangle_{hkl}$ is the mean value of that intensity.

[b]R-factor = $\Sigma||F_o|-|F_c||/\Sigma|F_o|$ where $|F_o|$ and $|F_c|$ are observed and calculated structure factor amplitudes respectively.

[c]R-free = An R-factor calculated from 10% of the data which have not been included in refinement.

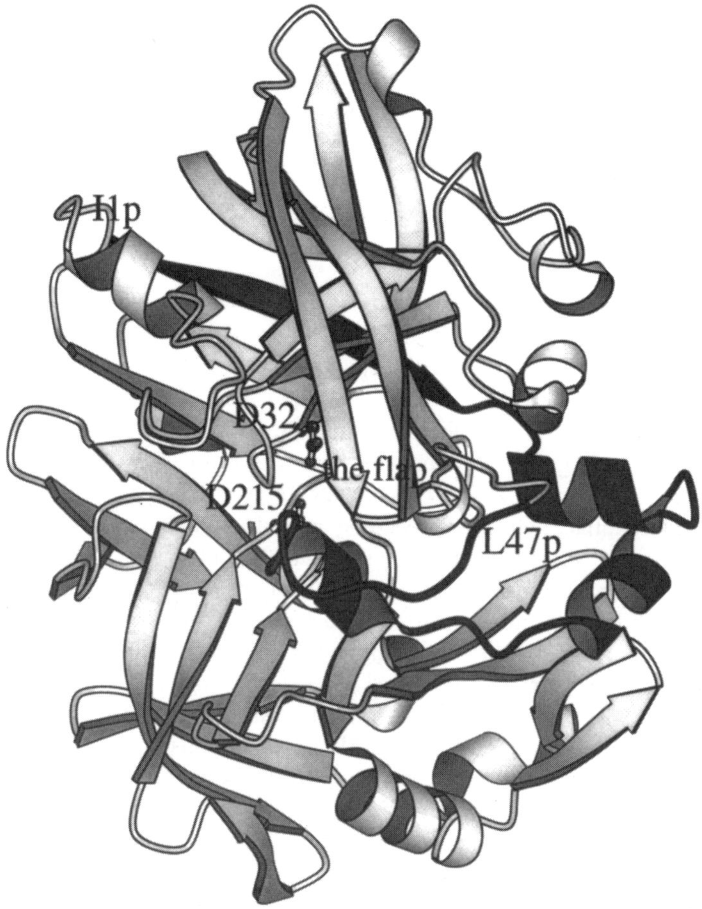

Figure 1. Human pepsinogen A. The prosegment of human pepsinogen A is drawn in dark gray. The remaining pepsin part of the molecule is drawn in light gray with the two catalytic aspartic acid residues depicted in ball-and-stick. This figure was made with Bobscript.[10]

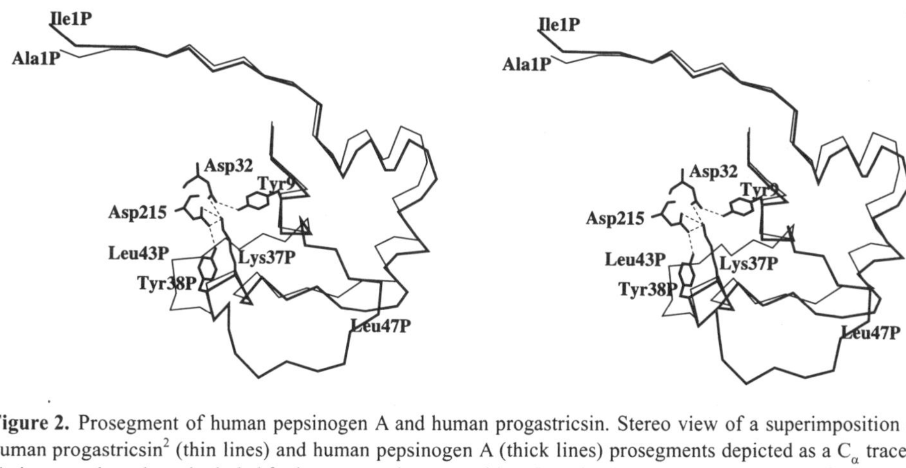

Figure 2. Prosegment of human pepsinogen A and human progastricsin. Stereo view of a superimposition of the human progastricsin[2] (thin lines) and human pepsinogen A (thick lines) prosegments depicted as a C_α trace. Side chain atoms have been included for human pepsinogen residues Lys37p, Tyr38p and Tyr9 and also for the catalytic aspartic acid residues Asp32 and Asp215. Hydrogen bonds to the catalytic residues are drawn in dashed lines. This figure is rotated approximately 90° towards the reader with respect to Figure 1.

Table 2. Superimposition of two aspartic acid proteinase zymogens and human pepsin onto human pepsinogen A

Protein	RMS deviation (Å)	Number of C_α atoms used in calculation
Human progastricsin	2.52	336
Porcine pepsinogen	0.91	362
Human pepsin	1.74	314

Table 2 indicates the root-mean-squared (r.m.s.) deviations (Å) between human pepsinogen A and the structures of porcine pepsinogen,[1] human progastricsin,[2] and human pepsin.[11] The low r.m.s. deviations between the pairs of these structures indicate that they share similar overall conformations. The similarity also extends to the catalytic aspartic acid residues at the active site and the stabilizing residues Lys37p, Tyr38p and Tyr9 of the zymogens. However, the prosegment of human pepsinogen A follows a different path than that of human progastricsin[2] (Figure 2). The zymogens diverge at residue 39p (human pepsinogen A numbering) with human pepsinogen A going around 'the flap' (residues 68–86). The flap can be seen in Figure 1. It is made up of the two long β-strands, at the front and center of the figure. In contrast to human pepsinogen A, the human progastricsin prosegment passes behind the flap. In order to accommodate the prosegment in this arrangement, another loop from residues 125–135 must turn away with respect to the same loop in human pepsinogen A. The paths of the prosegments are reunited at residue 6 (Figure 2).

The human pepsinogen A prosegment is more similar to that of the porcine pepsinogen structure[1] than to human progastricsin (Figure 3) and this is reflected in the r.m.s. deviation calculations (Table 2). Human pepsinogen A has an extra N-terminal residue and two additional prosegment residues with respect to porcine pepsinogen but the overall structural differences are small and are most evident in the region from 39p to 3. However, this is an area with weak density and is probably a region of considerable flexibility.

The first thirteen pepsin residues are located in a different position in the zymogen than they are in the active enzyme. Otherwise, the pepsin portion of the zymogen resembles that of mature, human pepsin[11] (Table 2). Upon enzyme activation, pepsin residues

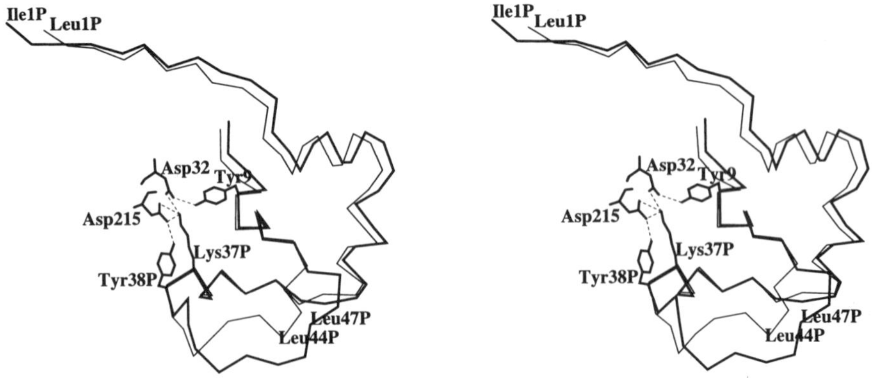

Figure 3. Prosegment of human pepsinogen A and porcine pepsinogen. Stereo view of a superimposition of the porcine pepsinogen[1] (thin lines) and human pepsinogen A (thick lines) prosegments depicted as a C_α trace. Side chain atoms have been included for human pepsinogen residues Lys37p, Tyr38p and Tyr9 and also for the catalytic aspartic acid residues Asp32 and Asp215. Hydrogen bonds to the catalytic residues are drawn in dashed lines.

1–13 move from their position in the precursor where they help to block the active site cleft, to replace the first residues of pepsinogen and complete the central six-stranded β-sheet. Pro6p and Leu7p are replaced by Pro5 and Leu6. These residues probably contribute towards the stability of the β-strand, the analogous leucine side chains fitting into a hydrophobic pocket. The remaining residues of this strand are quite different between zymogen and mature enzyme. Pepsin 1–11 (V D E Q P L E N Y L D), contains two aspartic acid residues and two glutamic acid residues. In contrast, the first 12 pepsinogen residues, (I M Y K V P L I R K K S), contain three lysine residues, one arginine and no negatively charged residues. At neutral pH values, the positive charges on these residues are balanced by the negatively charged carboxylates of the aspartate and glutamate residues on the mature enzyme portion of the zymogen. At the low pH of the stomach, the negative charges are lost due to protonation and the positively charged residues of the prosegment would, therefore, become destabilized.

Additional factors must contribute towards the conformational change which is triggered by the drop in pH. One possibility is that some amino acids will adopt more favourable conformations once the rearrangement of the prosegment has begun. For example, Phe111 has a χ^2 angle of 70° in the structure of mature human pepsin.[11] In human pepsinogen, Phe111 is packed into a hydrophobic pocket with a χ^2 angle of 0°, a very unfavourable angle for aromatic side chains. Phe111 is held in this position by Gln4 and several hydrophobic residues from the first two α-helices of the prosegment. These helices are held to pepsin on either side by salt bridges between Arg14p and Asp11 and between His30p and Glu7. After the initial disruption of these salt bridges, Phe111 may try to adopt a more favourable χ^2 angle, which would further disrupt the hydrophobic pocket of the zymogen and release the prosegment from packing against the pepsin portion of the molecule.

REFERENCES

1. Sielecki, A.R., Fujinaga, M., Read, R.J., & James, M.N.G. 1991. Refined structure of porcine pepsinogen at 1.8 Å resolution. *J. Mol. Biol.* 219:671–692.
2. Moore, S.A., Sielecki, A.R., Chernaia, M.M., Tarasova, N.I., & James, M.N.G. 1995. Crystal and molecular structures of human progastricsin at 1.62 Å resolution. *J. Mol. Biol.* 247:466–485.
3. Higashi, T. 1989. The processing of diffraction data taken on a screenless Weissenberg camera for macromolecular crystallography. *J. Appl. Cryst.* 22:9–18.
4. Navaza, J. 1994. AMoRe: an automated package for molecular replacement. *Acta Cryst.* A50:157–163.
5. Brünger, A.T. 1993. X-PLOR: A System for X-ray Crystallography and NMR. Yale University Press, New Haven.
6. Tronrud, D.E. 1992. Conjugate-direction minimization: an improved method for the refinement of macromolecules. *Acta Cryst.* A48:912–916.
7. Pannu, N.S., & Read, R.J. 1996. Improved structure refinement through maximum likelihood. *Acta Cryst.* A52:659.
8. Jones, T.A. 1985. Interactive Computer Graphics: FRODO. Meth. Enzymol. 115:157–171.
9. Jones, T.A., Zou, J.-Y., & Cowan, S.W. 1991. Improved methods for building protein models in electron density maps and the location of errors in these models. *Acta Cryst.* A47:110–119.
10. Esnouf, R. 1996. Bobscript version 2.0 based on Kraulis, P.J. 1991. MOLSCRIPT: a program to produce both detailed and schematic plots of protein structures *J. Appl. Cryst.* 24:946–950.
11. Sielecki, A.R., Fedorov, A.A., Boodhoo, A., Andreeva, N.S., & James, M.N.G. 1990. Molecular and crystal structures of monoclinic porcine pepsin refined at 1.8 Å resolution. *J. Mol. Biol.* 214:143–170.

37

CRYSTALLOGRAPHIC STUDIES OF AN ACTIVATION INTERMEDIATE OF HUMAN GASTRICSIN

Amir R. Khan,[1] Maia M. Chernaia,[1] Nadezhda I. Tarasova,[2] and Michael N. G. James[1]

[1]MRC Group in Protein Structure and Function
Department of Biochemistry
University of Alberta
Edmonton, Alberta T6G 2H7, Canada
[2]Molecular Aspects of Drug Design Section
ABL-Basic Research Program
National Cancer Institute
FCRDC, P.O. Box B, Frederick, Maryland 21702

INTRODUCTION

The human digestive enzyme progastricsin (hPGC) is an aspartic proteinase zymogen that is synthesized as an inactive precursor, having a positively charged and inhibitory N-terminal prosegment of 43 residues (Ala1p to Leu43p; the "p" suffix refers to the prosegment). Conversion of progastricsin to mature gastricsin occurs upon the lowering of pH, and involves conformational changes that uncover the active site.[1] The initial hydrolytic event is the uni-molecular, auto-catalytic cleavage of the peptide bond between Phe26p and Leu27p in the prosegment.[1,3] Further proteolytic processing is then initiated, ultimately resulting in the removal of the entire prosegment and formation of the mature enzyme. Our lab has previously determined the crystal structure of hPGC.[2] The structure of mature gastricsin is currently undetermined, although it is expected to resemble human pepsin and the other aspartic proteinases.

During the stepwise activation of hPGC, a transient intermediate (intermediate II[1]) is formed in which the first 26 residues of the prosegment are non-covalently associated with the enzyme, while the remainder of the prosegment is removed.[1,3] Intermediate II of gastricsin can be stabilized upon transfer of the protein to neutral pH, thereby preventing further processing events while maintaining the intermediate in a proteolytically inactive form. When activation intermediate II is subjected to mass spectrometry, a peak corresponding to the molecular weight of mature gastricsin is observed (N. Tarasova, unpub-

Aspartic Proteinases, edited by James.
Plenum Press, New York, 1998.

lished results), confirming that hydrolysis at the site of the pro/mature junction of the zymogen has taken place. However, it has been unclear how the first 26 residues of the prosegment are non-covalently associated with mature gastricsin in the intermediate.

Here we present the structure of activation intermediate II of human gastricsin (hGSI; residues Ala1p–Phe26p of the prosegment, and residues Ser1–Ala329 of the mature segment). We discuss the implications for the nature of the activation process.

METHODS

Purification and Crystallization

The intermediate was prepared from human progastricsin by auto-activation at pH 2.5, essentially as described by Bent Foltmann[1], and is termed "Intermediate II" in his paper. We will abbreviate this intermediate "hGSI" for the remainder of our paper. hGSI was purified by FPLC, and the purity was checked by SDS-polyacrylamide gel electrophoresis and electro-spray mass spectrometry. The protein was crystallized in 4M sodium formate, 100 mM Bis–Tris–propane (BTP), pH 7.8. Data were collected at room temperature at the Photon Factory in Tsukuba, Japan.

Crystallographic Refinement

The crystals of hGSI belong to the monoclinic space group (C2), with cell dimensions a=160.6Å, b=51.7Å, c=131.0Å, β=118.7°. The structure was solved by AMoRe[4] using the previously determined structure of hPGC as the search model.[2] Residues Ala1p–Leu23p in the prosegment of hPGC, and residues Met7–Ala329 of the mature segment of hPGC were used in the rotation and translation search. Two solutions were obtained, and were subjected to rigid body refinement in AMoRe. The R-factor was initially 45% (R_{free}=46%). The starting model, with two molecules in the asymmetric unit, was refined using a maximum likelihood target function,[5] implemented in XPLOR,[6] using data between 20Å and 2.5Å. Non-crystallographic symmetry restraints (in XPLOR) and density averaging (CCP4 package[7]) were used during model building and refinement. Manual intervention, display and map interpretation were performed using the program O.[8] The structure is presently refined to an R-value of 23.2% (R-free = 28.5%) for all data between 20Å and 2.5Å. It is important to point out, however, that a significant portion of the data is absent in the higher resolution shells. This is due, in part, to anisotropic scattering by the crystal. The overall completeness between 20 and 2.5Å is 65%, and only 43% in the highest resolution bin (2.6–2.5Å). However, we have used these data during the course of refinement because there was improvement in the electron density maps. Nevertheless, we judge the overall resolution of the model to be about 2.8Å, due to the relative incompleteness of the data set (Table 1).

RESULTS AND CONCLUSIONS

Molecular Structure of hGSI. The structure of hGSI reveals that the first 22 residues of the prosegment are held in the same position and conformation as observed in the structure of the proenzyme (hPGC). This prosegment consists of a β-strand (Val3p to Lys8p) which makes antiparallel β-sheet hydrogen bonds with the mature gastricsin moiety, and a subsequent α-helix (Ile13p to Lys20p, Fig. 1a/b). Asp11 of mature gastricsin forms a salt bridge with Arg14p, located in the helical prosegment of hGSI, as it does in hPGC. In

Table 1. Crystallographic refinement statistics for gastricsin intermediate

Resolution range (Å)	20.0 - 2.50
R-factor (R-free)	0.23 (0.28)
Number of unique reflections	21572
R-merge	8.5%
Number of protein atoms	2654
Number of molecules in the asymmetric unit	2
Number of water molecules	55
Completeness of data	
20.0 - 2.50Å	65%
20.0 - 3.10Å	80%
Root-mean-square deviations from ideal stereochemistry	
bond distances	0.01Å
bond angles	1.6°
Ramachandran plot	
residues in most favoured regions	89.4%
number of residues in disallowed regions	0

R-Factor = $\Sigma||F_o|-|F_c||/\Sigma|F_o|$ where $|F_o|$ and $|F_c|$ are observed and calculated structure factor amplitudes, respectively.

addition, Lys11p forms a salt bridge with Asp19p, providing additional stabilization of the helix in this region (Fig. 2). In the mature segment of gastricsin, Tyr9 occupies the S1 pocket in the enzyme and makes a hydrogen bond with the carboxylate of Asp32 at the active site. (Fig. 2). This interaction is identical to the one observed in hPGC. However, a water molecule now resides between the catalytic Asp residues, Asp32 and Asp217, in place of the εNH_3^+ side chain of Lys37p. Lys37p is a component of the peptide fragment from Leu28p–Leu43p that has been cleaved and removed during the formation of hGSI. In addition, there are significant conformational changes in the mature segment of hGSI, as discussed below.

There is a major conformational change in the loop between residues Tyr125–Thr134. In hPGC, this loop is disordered as a consequence of the conformation of the loop from Ala39p–Leu43p at the pro/mature junction (Fig. 1b). In hGSI, this loop is ordered, and residues Ser129–Val130 have moved into the approximate position that is left vacant by Leu43p–Ser1 of hPGC. The conformation of this loop in hGSI is very similar to the corresponding region in human pepsinogen[9] and mature human pepsin.[10] The segment between residues Tyr125–Thr134 of hGSI and human pepsin superimpose with a root-mean-square (rms) deviation of 0.6Å for the 40 backbone atoms. In addition, the first 8 residues (Ser1–Ala8) of the mature segment have moved significantly from their position in hPGC. This segment has an extended conformation, directed away from the body of the enzyme, and is involved in interactions with a symmetry-related molecule (not shown). In the absence of symmetry interactions, we predict that this region of hGSI is flexible. In comparison with human pepsin, the N-terminal region of mature gastricsin would likely form the first β-strand as part of the six-stranded β-sheet at the back of the enzyme, replacing Ala1p–Lys9p in hGSI. In the present model of hGSI, Ser1 is located 30.5Å from its position in hPGC (C^α–C^α distance). The distance was calculated by first superimposing the backbone atoms of hPGC onto the model of hGSI. The loop Pro124–Thr135, Ser1–Ala8, Thr70–Leu80 and residues Gln159–Ser162 were excluded from this calculation. A total of 1272 backbone atoms were used to calculate the least-squares superposition, resulting in an rms deviation of 0.7Å.

There is also a major conformational change in the active site flap of hGSI, relative to hPGC (residues Thr70–Leu80; Fig. 1b). The structure observed for hGSI is unusual for aspartic proteinases, in that this loop is bent towards the active site Asp residues. However, there is no electron density visible for Tyr75 in the structure, as well as the side chain of the preceding residue (Gln74), so that this region is less certain than the remainder of the structure. The tip of the flap from residues Tyr75–Ser79 is also disordered in hPGC,[2]

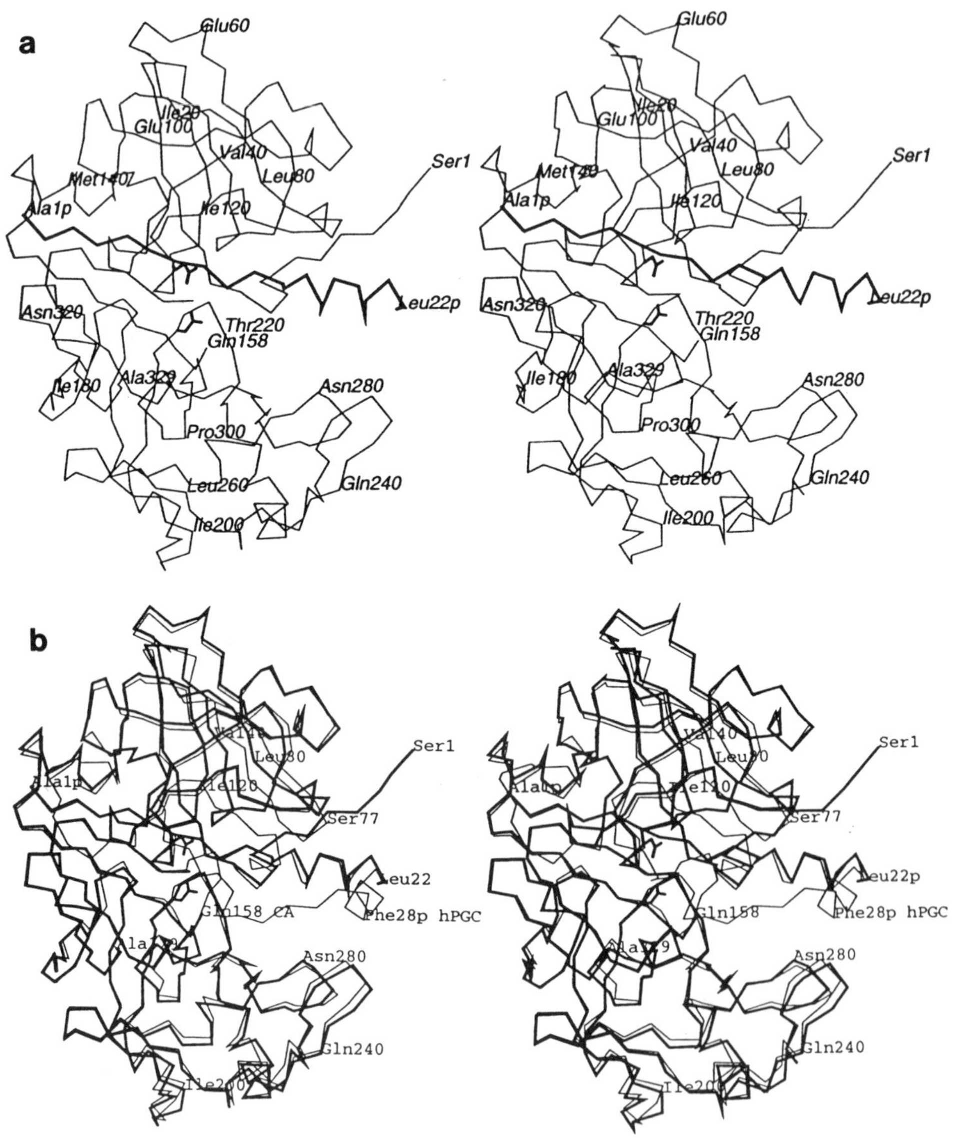

Figure 1. Stereoviews of the C^{α} tracing of hGSI. (a) Residues shown are Ala1p–Leu22p in the prosegment, Ser1–Gln158, and Gly163–Ala329. Amino acids are identified every 20 residues, and at key junctions. The prosegment is emphasized using thick lines. (b) C^{α} vectors of hPGC are shown in thin lines, superimposed onto the C^{α} atoms of hGSI (thick lines). The superposition was performed as described in the text. Labels that correspond to residues in hPGC are identified. hGSI is labelled at every 40 residues, as well as at key junctions, as before. In both (a) and (b), the side chains of Asp32 (above) and Asp217 (below) of hGSI are shown.

having poor side-chain density and high *B*-factors. It is important to note that the loop conformation of hPGC is not permissible in hGSI, since there would be steric clashes between residues at the tip of the flap (Gly76–Ser77) and the N-terminal region of the mature segment (Glu5–Pro6; Fig. 1b).

We have also modeled the binding of substrate in the active site of hGSI. Modeling was performed by superimposing the structure of the pepstatin-pepsin inhibitor complex[10] onto the model of hGSI. In the complex of pepstatin and pepsin, most of the interactions of the inhibitor with pepsin are made with residues in the P1, P2 and P3 position.[9] When pepsin is substituted by hGSI, these residues sterically clash with the segment Tyr9–Ala12 of the mature part of gastricsin. The side-chain of Tyr9 makes a hydrogen bond with Asp32 at the active site, as it does in the intact zymogen, and interferes with the S1 binding pocket (Fig. 2). However, we observe that the S1′–S3′ binding sites of hGSI are accessible to substrates. Tyr38p, which forms a hydrogen bond in the S1′ binding site that is pseudo 2-fold symmetrical to the Tyr9–Asp32 hydrogen bond, has been proteolytically removed as part of the prosegment Leu28p–Leu43p. In addition, the inhibitory salt bridge between Lys37p and the active site Asp residues has been eliminated. A water molecule now resides in the active site cleft between the Asp residues (Fig. 2), indicative of a mature active site that is poised for catalytic activity.

Implications for the Activation Process. Exposure of hPGC to conditions of low pH results in protonation of Asp residues at the active site. The immediate consequence is that the inhibitory salt bridge between Lys37p and the active site Asp residues (Asp32, Asp217) is disrupted, leading to the uncovering of the active site. Lys37p is found on a short but well-ordered 3_{10}-helix of the prosegment.[2] However, the helix is flanked by highly mobile segments, suggesting that disruption of the salt bridge may destabilize the entire region.[2] Conformational changes result in susceptibility of the Phe26p–Leu27p peptide bond to proteolysis by the newly uncovered active site. Following this intramolecular cleavage, proteolysis occurs at the pro-mature junction, likely by an intermolecular mechanism,[1,3] leading to the formation of hGSI. Taking into consideration the present

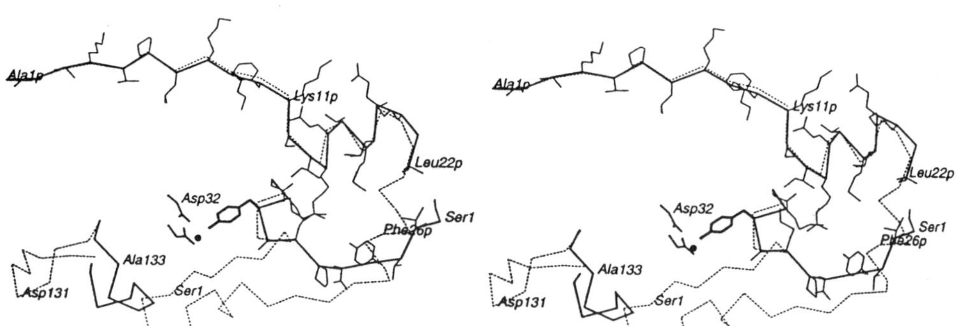

Figure 2. Stereoview of the active site area of hGSI. The C^{α} vectors of GSI are represented as solid lines, while those corresponding to hPGC are shown as dotted lines. Residues shown are Ala1p–Leu22p (hGSI), Ala1p–Leu43p (hPGC), Ser1–Asp11 (hGSI and hPGC), and Tyr125–Thr135 (hGSI and hPGC). The side-chains of Asp32, Tyr9 and Asp217 (hGSI) are shown with thick lines, while the remaining side chains of hGSI are thin lines. The position of the water molecule between the catalytic Asp residues is represented as a black sphere.

structure, we suggest that the strand Ala1p–Lys11p does not dissociate from the β-sheet during the events preceding the formation of hGSI. It is unlikely that the strand could re-insert back into the β-sheet, following the initial cleavages, to form hGSI. This model agrees with the prediction by Bent Foltmann[11] that the intermediate complex has the prosegment bound in its original position in the six-stranded β-sheet. Once the N-terminal peptide of the prosegment has been dissociated from gastricsin, the intermediate complex is unlikely to be reconstituted.[12]

The structure of hGSI reveals why the intermediate is inactive at neutral pH. The position and conformation of the mature segment from Tyr9–Ala12 sterically restricts entry of substrates to the P1–P3 sites of the catalytic cleft in the intermediate II. However, hGSI must necessarily have catalytic activity at low pH for completing the activation process. In order for hGSI to adopt an active conformation, Tyr9 is required to move out of its position at the P1 site. At low pH, this movement may be mediated by the disruption of the Asp11–Arg14p salt bridge, as well as the salt bridge between Lys11p and Asp19p. Weakening of these electrostatic interactions are likely to destabilize the helix from Ile13p–Lys20p, and facilitate movement of the segment Tyr9–Ala12 out of the active site. Dissociation of the N-terminal peptide of the prosegment is probably aided by further intermolecular cleavages by the newly forming active site. In the final step of activation, the N-terminal segment of hGSI (Ser1–Tyr9) replaces the β-strand in the prosegment, as it dissociates from gastricsin.

Finally, we have obtained a data set from a second crystal of hGSI at low temperature. The data are greater than 95% complete to 2.36Å. Initial work using this improved data set suggests that the structure of hGSI is identical to the model described here.

ACKNOWLEDGMENTS

This work was supported by the MRC of Canada through a grant to the Group in Protein Structure and Function. A.R. Khan was supported by a Medical Research Council of Canada Studentship. Research was also sponsored in part by the National Cancer Institute, DHHS, under contract with ABL.

REFERENCES

1. Foltmann, B. & Jensen, A.L. (1982) *Eur. J. Biochem.* **128**: 63–70.
2. Moore, S.A., Sielecki, A.R., Chernaia, M.M., Tarasova, N.I. & James, M.N. (1995) *J. Mol. Biol.* **247**: 466–485.
3. Kageyama, T., Ichinose, Miki, K., Athauda, S.B., Tanji, M. & Takahashi, K. (1989) *J. Biochem.* **105**: 15–22.
4. Navaza, J. (1994) *Acta Cryst* **A50**: 157–163.
5. Pannu, N.S. & Read, R.J. (1996) *Acta Cryst.* **A52**: 659–668.
6. Brunger, A.T. (1993) "XPLOR: a system for X-ray crystallography and NMR", Yale University press, New Haven, CT.
7. CCP4 (1994) The CCP4 Suite: Programs for Protein Crystallography. *Acta Cryst.* **D50**: 760–763.
8. Jones, T.A. & Kjeldgaard, M. (1995) "O - the Manual", Version 5.11, Uppsala, Sweden.
9. Bateman, K.S., Chernaia, M.M., Tarasova, N.I. & James, M.N.G. (1997) in *Structure and Function of Aspartic Proteinases: Retrovral and Cellular Enzymes.* (M.N.G. James, ed.) Plenum Press Corp., New York.
10. Fujinaga, M., Chernaia, M.M., Tarasova, N.I., Mosimann, S.C. & James, M.N.G. (1995) *Protein Sci.* **4**: 960–972.
11. Foltmann, B., Harlow, K., Houen, G., Nielsen, P.K. & Sangild, P. (1995) in *Aspartic Proteinases: Structure, Function, Biology and Biomedical Implications.* (K. Takahashi, ed.) Plenum Press Corp., New York, pp. 41–51.
12. Foltmann, B. (1988) *Biol. Chem. Hoppe-Seyler* **369**: 311–314.

ABSTRACTS

FUNCTIONS OF THE PROPEPTIDE OF YEAST PROTEINASE A

J. R. Winther, H. B. van den Hazel,* V. Westphal, and M. C. Kielland-Brandt

Carlsberg Laboratory, Department of Yeast Genetics, Gamle Carlsberg Vej 10, DK-2500 Copenhagen Valby, Denmark

The yeast vacuole is equivalent to the mammalian lysosome and as such hosts a number of hydrolytic activities. Proteinase A (PrA) plays a central role in the yeast vacuole as initiator of zymogen activation. PrA is itself synthesized as a zymogen (proPrA), which activates by the autoproteolytic removal of a 54 residue N-terminal propeptide. The mature part of the protein is very similar to that of other aspartic proteases (e.g. 46% identity to human cathepsin D). The propeptide, however, is not homologous to those of even closely related aspartic proteases.

The most obvious function of the PrA propeptide is to maintain the zymogen in its inactive state, thus preventing proteolytic activity from being unleashed prior to arrival in the vacuole. Studies *in vivo* and *in vitro* have shown that autoactivation is initiated by acidic pH and is stimulated by high ionic strength. In contrast to pepsinogen activation at low pH, it occurs primarily via a bimolecular, product-catalysed mechanism. These cellular and mechanistic factors seem to be well adapted for *in vivo* function.

Deletion of the PrA propeptide sequence results in accumulation of the glycosylated polypeptide (PrAÆpro) in the ER and its slow degradation. It is known from other cells that the "quality control" system of the ER ensures that only correctly folded proteins are allowed to exit the ER. Furthermore, while PrA is normally a very stable enzyme, PrAÆpro is degraded and no activity can be detected in cells producing PrAÆpro. PrA and PrAÆpro have the same primary sequence and we thus conclude that *in vivo* folding is compromised by deletion of the propeptide. Interestingly, the propeptide does not require covalent linkage to the enzyme region for function. The requirement for sequence conservation of the propeptide has been investigated by substitution of large parts of the propeptide by random sequence. In a genetic approach, either of the two halves of the propeptide were substituted

* Present address: Unité de Biochemie Physiologique, Université Catholique de Louvain, Place Croix du Sud, 2/20, B-1348 Louvain-la-Neuve, Belgium

with randomly generated sequence. The resulting mutants where then screened for PrA activity. A surprisingly high number of active mutant proteins (Å1%) were identified.

In addition to these functions, the C-proximal half of the propeptide contains information able to confer vacuolar sorting when fused to the normally secreted enzyme invertase the vacuole. However, that this signal is not the only vacuolar sorting signal in native proPrA.

ASPARTIC PROTEINASE ZYMOGENS FROM A STRUCTURAL PERSPECTIVE

Michael James,[1] Katherine S. Bateman,[1] Amir R. Khan,[1] Stanley Moore,[1,2] Maia M. Chernaia,[1] Masao Fujinaga[1] and Nadya Tarasova[3]

[1]MRC Group in Protein Structure and Function, Department of Biochemistry, University of Alberta, Edmonton, Alberta T6G 2H7, Canada
[2]Department of Biochemistry, Massey University, Palmerston North, New Zealand
[3]NCI-Frederick Cancer Research and Development Center, P.O. Box B, Frederick, Maryland 21702-1201

Unwanted proteolysis by cellular aspartic proteinases is achieved by the biosynthesis of these enzymes as inactive precursor molecules or *zymogens*. In the mammalian aspartic proteinases this consists of N-terminal extensions of 40–45 amino acids that are removed auto-catalytically, most often in *cis*, by a unimolecular conversion to the mature proteinases (1–4). Our group has undertaken a project to define the molecular events of this activation process from a structural viewpoint. Structures of porcine pepsin, porcine pepsinogen (4), human pepsin A, human pepsinogen A, human progastricsin (5) and an activation intermediate of human gastricsin have been done at relatively high resolution in order to define these events. Human pepsinogen and porcine pepsinogen have essentially the same tertiary structures; they differ in conformational details with respect to the tertiary structure of progastricsin. Differences are confined to residues at the C-terminus of the prosegment, the first six residues of the mature enzyme, residues at the tip of the flap and a region around residues 125 to 134 of the mature enzyme. For these gastric proteinases the conformational change is triggered by a drop in pH; this results in protonation of carboxyl groups and concomitant loss of ion pairs stabilizing the conformation of the zymogen prosegment. The zymogens all have highly mobile (disordered in the electron density maps) regions of the prosegment that contributes to ease of unfolding in the conversion process. A susceptible peptide (Leu16p–Ile17p in the case of porcine pepsinogen, Phe26p–Leu27p in the case of progastricsin) finds its way to the active site residues Asp32 and Asp215 and cleavage of the bond takes place unimolecularly. In the case of gastricsin an activation intermediate can be isolated and the structure of this cleaved intermediate form has been determined. These several structures have now provided an experimental basis that supports a proposal for the activation pathway and shows an inhibited form of the active enzyme that has a sterically blocked active site. Details of the activation pathway of the mammalian aspartic proteinase zymogens will be presented and discussed.

Research supported by the Medical Research Council of Canada, the Alberta Heritage Foundation for Medical Research, and the NCI, Frederick.

REFERENCES

1. Al-Janabi, J. *et al.* (1972). *J. Biol. Chem.* **247**:4628–4632.
2. Foltmann, B. & Jensen, A.L. (1982). *Eur. J. Biochem.* **128**:63–70.
3. Kageyama, I. *et al.* (1989). *J. Biochem.* **105**:15–22.
4. James, M.N.G. & Sielecki, A. (1986). *Nature* **319**:33–38.
5. Moore, S.A. *et al.* (1995). *J. Mol. Biol.* **245**:466–485.

Aspergillus niger ACID PROTEINASE A

Structure and Function

Kenji Takahashi,[1] N. Kagami,[2] X.-P. Huang,[1,3] M. Kojima,[1] and H. Inoue[1]

[1]School of Life Science
Tokyo University of Pharmacy and Life Science
Hachioji, Tokyo 192-03, Japan
[2]Graduate School of Science
[3]Graduate School of Agriculture
University of Tokyo
Bunkyo-ku, Tokyo 113, Japan

1. INTRODUCTION

Aspergillus niger var. *macrosporus* produces two kinds of extracellular acid proteinases, i.e., proteinase B (aspergillopepsin I or proctase B) and proteinase A (aspergillopepsin II or proctase A).[1] Proteinase B is a typical pepsin-type aspartic proteinase, whereas proteinase A is a non-pepsin type acid proteinase rather insensitive to specific inhibitors of aspartic proteinases, including pepstatin, DAN and EPNP.[2] Proteinase A has a molecular mass of 22,265 dalton, and is composed of two peptide chains, namely Lchain (39 residues) and Hchain (173 residues), which are noncovalently bound to each other.[3,4] It has no similarity in amino acid sequence with ordinary pepsin-type aspartic proteinases and appears to have fairly different substrate specificity. The active site residues have not been identified in the previous studies, although the results of chemical modification as well as its pH-activity profile indicated that certain carboxyl groups are involved in the activity.

In order to define the specificity more clearly and to identify the active site residues, we have carried out the following experiments. Firstly, we have investigated the action of the enzyme toward various peptides including oxidized insulin B chain and several physiologically active peptides, and then toward oxidized ribonuclease A. Secondly, we have constructed an expression system of proroteinase A using *E. coli*. Using this system, we have investigated the effect of mutation of individual aspartic and glutamic acid residues on the activity. The results obtained in these experiments will be described in this paper.

Aspartic Proteinases, edited by James.
Plenum Press, New York, 1998.

2. EXPERIMENTAL PROCEDURES

2.1. Hydrolysis of Peptides and Protein and Analysis of the Resulting Peptides

Oxidized bovine pancreatic ribonuclease A (500 µg) was digested with proteinase A (about 5 µg) at 37°C in 100 µl of 0.1 M sodium phosphate buffer, pH 1.8, for 30 min or 2 h, or 0.1 M sodium acetate buffer, pH 5.5, for 2 h or 20 h. Part of each digest corresponding to approximately 200 µg of protein was separated by HPLC on a TSKgel ODS-120T column (0.46 × 25 cm) in a Hitachi model 655A-11/L-5000 HPLC apparatus. Peptides were eluted with a gradient of acetonitrile (0 to 50% in 60 min) in 0.1% trifluoroacetic acid at a flow rate of 0.8 ml/min and monitored by the absorbance at 215 nm. Part of each isolated peptide was analyzed for amino acid composition in an Applied Biosystems derivatizer/analyzer (420A/130A–920A) with an on-line automatic acid hydrolysis apparatus. Amino acid sequence analysis of peptides was done by an Applied Biosystems pulse-liquid protein sequencer (477A) connected online to a phenylthiohydantoin amino acid analyzer (120A).

2.2. Plasmid Construction and Site-Directed Mutagenesis

The expression system consisted of a pAR2113 expression vector containing a T7 promotor and an expression host *E. coli* strain, BL21(DE).[5] The cDNA for the enzyme (ANA) was modified for expression of the putative proform by PCR and inserted into pAR2113 to form pAR–ANA as described previously.[6] The mutants of proteinase A were prepared in which a single aspartic acid residue was replaced with an asparagine residue, or a single glutamic acid residue with a glutamine residue by the method of Kunkel.[7] In addition, the glutamine residue at position 24 in the H chain was replaced with a glutamic acid rsidue. Although proteinase A mainly formed inclusion bodies when the cells were cultured at 37°C, soluble and active proteinase A could be obtained from the cytoplasm by lowering the culture temperature during expression to 30°C. In the present study, the activity of this soluble fraction was primarily examined for the wild and mutant proenzymes without further purification since the proenzyme showed a tendency to be activated partly to the active form during purification. The assay was done by activity staining with acid-denatured hemoglobin as a substrate on the gel after polyacrylamide gel electrophoresis of the soluble fraction) essentially as described by Furihata et al.[8] The details of these procedures will be reported elsewhere (manuscript in preparation). In addition, the inclusion bodies were isolated in some cases, and the proenzyme was partially purified, refolded at various pHs[6] and examined for the potential activity after acidification with acid-denatured hemoglobin as a substrate essentially by the method of Anson.[9]

3. RESULTS AND DISCUSSION

3.1. Specificity of Peptide Bond Cleavage of Proteinase A

3.1.1. Oxidized Insulin B Chain as a Substrate. The action of proteinase A was first investigated by Iio and Yamasaki.[10] They found that proteinase A cleaves at pH 1.5 the three peptide bonds, Asn–Gln (residues 3–4), Glu–Ala (residues 13–14) and Tyr–Thr (residues 26–27), fairly specifically, and Gly–Glu bond (residues 20–21) much more slowly. In the present study, we reinvestigated the specificity by analyzing the hydrolysis products obtained at pH 1.5 and 5.5 by HPLC and confirmed their results, except for the cleavage at the Gly–Glu bond. The cleavage of this bond could scarcely be detected in our experiments. The specificity of proteinase A is unique as compared with those of other related proteinases.[11] The cleavages at the Asn–Gln and Tyr–Thr bonds are especially notable because those peptide bonds have not previously been found to be cleaved by pepsin-type aspartic proteinases except for the cleavage of the Asn–Gln bond by endothiapepsin. Furthermore, it is noteworthy that the specificity is different even from that of the homologous enzyme scytalidopepsin in that proteinase A cleaves the Asn–Gln bond (residues 3–4) which is not cleaved by scytalidopepsin and that the specificity appears to be narrower than the latter.

3.1.2. Other Smaller Peptides as Substrates. The data were obtained partly by Ido et al.[12] and partly by us using sustance P and related physiologically active peptides as detailed elsewhere.[11] In these cases, major cleavages were observed with certain Tyr–X and Phe–X bonds, but cleavages were also observed with Asn–X and Glu–X bonds. Most peptides were cleaved at one site, and the rate of hydrolysis differed markedly among them. The results indicated that a stretch of at least four residues (position P1 to P4) at the NH_2-terminal side of the cleavage site is recognized by the enzyme and that the residue at P1 position is most important. On the other hand, several peptidyl 4-methylcoumaryl-7-amide (MCA) derivatives have been tested, but in none has the X–MCA bond been cleaved so far, indicating that the residues at the COOH-terminal side of the cleavage site (P1′, P2′, etc.) are also important for hydrolysis.

3.1.3. Oxidized Ribonuclease A as a Substrate. To investigate further the specificity, performic acid-oxidized ribonuclease A was used as a substrates.[13] We analyzed by HPLC the products of digestion at pH 1.8 or 5.5 for different periods. Figure 1 shows the amino acid sequence of oxidized ribonuclease A together with the sites of cleavage of peptide bonds. At pH 1.8, near the optimal pH, major cleavages occurred at six peptide bonds in 0.5 h among the total 123 peptide bonds and thirty peptide bonds were cleaved in 2 h to varying extents. On the other hand, at pH 5.5, major cleavages took place at five peptide bonds in 2 h and thirteen peptide bonds were cleaved in 20 h. In total, the enzyme cleaved the following peptide bonds: 3 of the 6 Tyr–X, 2 of the 3 Phe–X, 5 of the 10 Asn–X, 4 of the 5 Asp–X, 4 of the 7 Gln–X, 4 of the 5 Glu–X, 4 of the 4 His–X, 1 of the 4 Met(O_2)–X, 2 of the 10 Lys–X, and 1 of the 12 Ala–X bonds under the conditions used. Thus, the cleavages occurred mainly at the peptide bonds Tyr–X, Phe–X, Asn–X, Asp–X, Gln–X, Glu–X, and His–X. The resistance of some of these peptide bonds may be largely explained by the assumption that Pro or Cys(O_3H) (this residue is thought to be negatively charged under the conditions used)at the P_2 , P_1, P′ or P_2' position is inhibitory to the action of the enzyme as observed for the action of several endopeptidases and/or by rapid hydrolysis at the P4–P_3, P_3–P_2, and P_1 ′–P2′ sites, respectively, which would inhibit the subsequent cleavage at the P1–P_1' bonds. The cleavage of the Ala–Ala (residues 4–5), Met(O_2)–Ser (residues 79–80) and Lys–Tyr (residues 91–92) bonds is notable; however, most of the other Ala–X, Met(O_2)–X and Lys–X bonds were not cleaved.

The major cleavage sites at pH 1.8 and 5.5 are significantly different, and the number of minor cleavage sites was markedly reduced at pH 5.5. These results are thought to be partly due to dissociation at pH 5.5 of Asp and Glu residues at the P_1 or P_1' position which should both be protonated at pH 1.8 and have no charge, thus resembling Asn and Gln residues, respectively. On the other hand, some of the His–X bonds appeared to be more susceptible to the enzyme at pH 5.5 than at pH 1.8. This should be due to partial

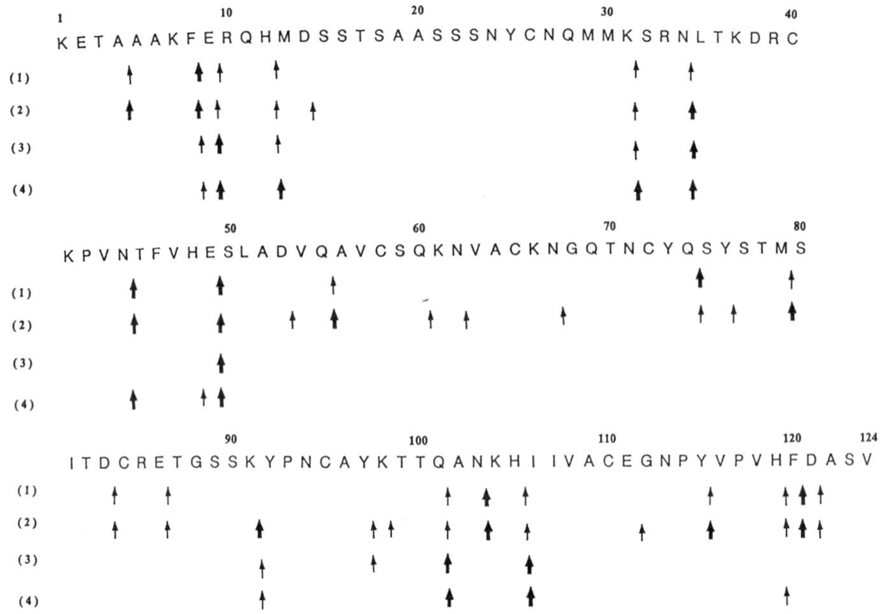

Figure 1. The sites of cleavage of oxidized ribonuclease A by proteinase A. The amino acid sequence of oxidized ribonuclease A is shown using one-letter code for amino acids. C and M stand for cysteic acid and methionine sulfone, respectively. Large and small arrows indicate major (over 40% hydrolysis) and minor (less than 40% hydrolysis) sites of cleavage, respectively under the conditions used. (1) pH 1.8, 30 min; (2) pH 1.8, 2h; (3) pH 5.5, 2h; (3) pH 5.5, 20h.

deprotonation of His residues at this pH to the uncharged form, which the enzyme appears to prefer; the enzyme may recognize His as one of the aromatic amino acid residues.

Taken together, it is concluded that proteinase A has the primary specificity at the P_1 position for Tyr, Phe, His, Asn, Asp, Gln, and Glu residues and that it has certain additional subsite specificity which was not clearly defined in the present study. The enzyme is thus thought to have an extended active site where multiple interactions occur between the enzyme and the substrate as in the case of ordinary aspartic proteinases. The cleavage specificity toward Asn–X and Gln–X bonds is fairly unusual. There is no endopeptidase known so far that has such a specificity together with specificity toward aromatic residues.

3.1.4. Comparison with the Cleavage Sites in Proproteinase A. As shown in Fig. 2, in the conversion of proproteinase A to the mature two-chain enzyme, specific cleavages should take place at at least three sites. They are the Asn–Glu bond between the prosegment and the L chain, the Tyr–Gly bond between the L chain and the intervenig sequence and the Arg–Gln bond between the intervening sequence and the H chain.[3,14] The cleavages of the Asn–Glu and Tyr–Gly bonds are consistent with the specificity of the enzyme. On the other hand, the cleavage of the Arg–Gln bond does not seem to fit the specificity of the enzyme. To shed light on this point, we have expressed the proenzyme in the *E. coli* system and purified the proenzyme.[6] When the expressed proenzyme was kept under acidic conditions, it was found to be autocatalytically converted to the active, two-chain enzyme. The NH_2-terminal sequence analysis of its L chain indicated that the cleavage between the prosegment and the L chain occurred as expected at the Asn–Glu bond, whereas the NH_2-terminal sequence analysis of the H chain revealed that the cleavage

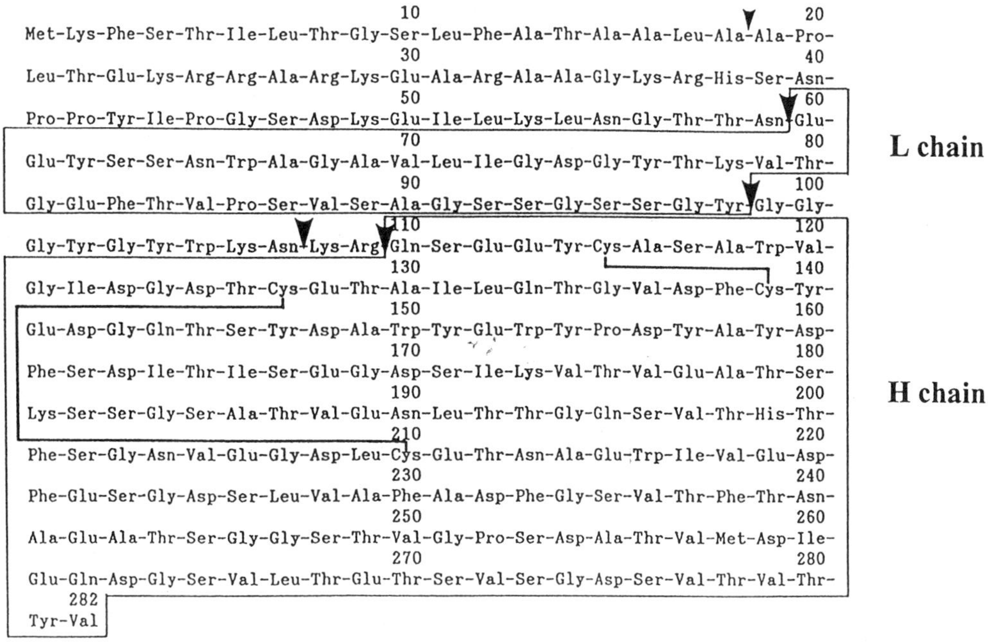

Figure 2. The cleavage sites in preproproteinase A upon conversion to the mature two-chain form. The amino acid sequence of preproproteinase A is shown, in which the regions corresponding to the L chain and the H chain in the mature enzyme are enclosed separately. Large arrowheads indicate the sites of cleavage upon conversion of proproteinase A to proteinase A. The small arrowhead indicates the assumed site of cleavage by a signal peptidase upon conversion of preproproteinase A to proproteinase A.

took place not at the Arg–Gln bond, but at the Asn–Lys bond, two residues ahead of the NH_2-terminus of the mature L chain.[6] The cleavage of this bond, however, is consisitent with the specificity of the enzyme. Presumably the Arg–Lys sequence will be removed later by the action of some other proteinase. The most probable candidate for this enzyme is proteinase B (aspergillopepsin I) which is secreted by the same mold simultaneously with proteinase A in much larger quantity. This proteinase B is known to be able to cleave Arg–X bonds.[15]

3.2. Site-Directed Mutagenesis Studies on the Active Site Residues

In order to elucidate the active site residues in proteinase A, site-directed mutagenesis studies have been performed. Proteinase A contains 18 Asp and 19 Glu residues, and 17 Asp and 16 Glu are present in the H chain (Fig. 2). Since some of the carboxyl groups of these residues, especially those in the H chain, are thought to be involved in catalysis, attempts were made to mutate aspartic acid and glutamic acid residues into asparagine and glutamine residues, respectively. For this purpose, the mutant proenzyme was expressed in the *E. coli* system, and their potential activity was measured. Since we had difficulty in purifying the expressed proenzyme due to its partial activation during the purification procedure, we employed a simplified method as described in the "Experimental Procedures" section. In brief, the expressed proenzyme was obtained from the supernatant fraction of the destrupted cells and submitted without further purification to native polyacrylamide gel electrophoresis, then to activity staining in the gel using acid-denatured hemoglobin as a substrate. In addition, we mutated the glutamine residue at position 24 in the H chain to a glutamic acid residue to know something about the role of the residue at this position since scytalidopepsin B was reported to have Glu16,17 rather than Gln18 at this position which was suggested to be part of the active site of the enzyme from chemical modification studies.[16,17] Although the h24Q→E mutant was inactive, it was found to be structurally too unstable to be activated autocatalytically. The Gln at position 24 in the H chain of proteinase A may be important for the correct folding of the peptide chain. Very recently, this residue in scytalidopepsin has been reported to be Gln from cDNA sequencing (Fig. 3).[19]

In the *E. coli* expression system, the wild-type proproteinase A was expressed mostly as inclusion bodies. This inactive proenzyme could be correctly refolded from the denatured state in 8 M urea by dilution and incubation at an appropriate pH at 4°C, and the refolded proenzyme could be activated to the mature enzyme at pH 2.0 autocatalyti-

```
Putative signal sequence
  A. niger proteinase A      MKFSTILTGSLFATAALA
  C. parasitica EapB         ..YTAALAALVTLA..AP
  C. parasitica EapC         ..YATVVAALLG.N...G
  Scytalidopepsin B
Propeptide
    APLTEKRRARKEARAAGKRHSNPPYIPGSDKEILKL------NGTTN
    TDGIIDIGDGVKLVPREP.AHTRLERLRTFRRG.MEGLESGERNSSD
    .RF.....E.N...L.RRSG.VRLPATN.EGVAI-DAAESR-.D...

Mature protein
    EEYSSNWAGAVLIGDGYTKVTGEFTVPSVSAGSSGSSGYgggygywknkr
    VS.D.......K..T.LND...TIV..TP.VP.G..--------------
    V.............S..KS...I.V..TPKSPG.----------------
    -TVE...G..I...SDFDT.SATAN...A.GQ..----------------
    1                                                50
   <QSEEYCASAWVGIDGDTCETAILQTGVDFCYEDGQTSYDAWYEWYPDYAY
    STAK.A............TS.........YAGR.GV.F........N...
    GNT..A...........AQNS........YV.GSSVA.............
    ----AAGT..........Q.......F.W-.G..--T.........EVS-
                                                    100
    DFSDITISEGDSIKVTVEATSKSSGSATVENLTTGQSVTHTFSGNVEGDL
    ...GFSV.A..T.VM.AS.S.LKA.TV.L..S...KK..QS..AESSE-.
    ...G.S..A..T.....T..TTT..T.V...V.K.TT.....T.QSAA-.
    ...G.........QMS.T...DT.....L.......K.SKS..NESS.S.

    CETNAEWIVEDFE-----SGD-SLVAFADFG-SVTFTNAEATSGGSTVGP
    ..Y..........-----..S-...N....D-T...KDCSPSVS..----
    Q.L..........-----E..-E..P..N..-T....G....TSSG..TA
    .R....F.I....ECNSNGS.CEF.P..S.SPA.E..DCSV..D.ES.SL
        150                   173
    SDATVMDIEQDGSVLTETSVSGDS-VTVTYV
    ---.IV..R.SLE....C.TT.TTT..CE..G
    A...LI....N.E...SVT...-.T...K..
    D..QITQVIINNQDV.DC....-TT.SCS..
```

Figure 3. Comparison of the amino acid sequence of proteinase A with those of *Cryphonectria parasitica* acid proteinases and scytalidopepsin, and the acidic residues in proteinase A mutated by site-directed mutagenesis. Highly conserved regions among the four enzymes are enclosed. The aspartic and glutamic acids in the mature protein are shown in boldface. The residues mutated with loss of the potential activity of proproteinase A are double-underlined, and those mutated without loss of activity are single-underlined. Dots indicate residues identical with those of proteinase A, and dashes indicate deletions.

cally.[6] The optimum pH for the refolding was 5.3 for the wild-type proenzyme. These results obtained with various mutants were consistent with those obtained by the direct in-gel activity-staining assay with the soluble proenzyme and its mutants.

In this way, we have so far examined the effect of mutation on the potential activity of the proproteinase A for 22 aspartic and glutamic acid residues in the H chain (data not shown). These include all acidic residues in the enzyme which are common to both proteinase A and scytalidopepsin B (Fig. 3). As shown in Fig. 3, only Asp14 and Glu 110 in the H chain have been found so far to be crucially important for the activity. These two residues are fairly apart from each other in the amino acid sequence, but they are thought to be able to come close to each other because there is a disulfide bond connecting both regions.

Very recently, the eap B and eap C genes of *Cryphonectria parasitica* encoding two new acid proteinases have been cloned.[20] Their sequences are also aligned in Fig. 3, together with those of proteinase A and scytalidopepsin B. The Cryphonectria enzymes are both single-chain enzymes and show significant sequence homology with proteinase A and scytalidopepsin B. In the H chain region, there are three most conservative regions. Asp14 and Glu 110 in the H chain of proteinase A are located in two of these regions and conserved in the other two enzymes. This is consistent with the supposition that these two residues are the most probable candidates for the catalytic residues in the active site of proteinase A. However, further studies are necessary to draw a more definite conclusion on this point.

At present, attempts are being made to identify the active site residues more clearly in a quantitative manner using purified samples of the proenzyme and its mutants of proteinase A using various methods.

REFERENCES

1. Koaze, Y., Goi, H., Ezawa, K., Yamada, Y. & Hara, T. (1964) Fungal proteolytic enzymes. Part I. Isolation of two kinds of acid-proteases excreted by *Aspergillus niger* var. *macrosporus*. *Agr. Biol. Chem.* 28, 216–223.
2. Chang, W.-J., Horiuchi, S., Takahashi, K., Yamasaki, M. & Yamada, Y. (1976) The structure and function of acid proteases. VI. Effects of acid protease-specific inhibitors on the acid proteases from *Aspergillus niger* var. *macrosporus*. *J. Biochem. (Tokyo)* 80, 975–981.
3. Takahashi, K., Inoue, H., Sakai, K., Kohama, T., Kitahara, S., Takishima, K., Tanji, M., Athauda, S.B.P., Takahashi, T., Akanuma, H., Mamiya, G. & Yamasaki, M. (1991) The primary structure of *Aspergillus niger* acid proteinase A. *J. Biol. Chem.* 266, 19480–19483.
4. Takahashi, K., Tanokura, M., Inoue, H., Kojima, M., Muto, Y., Yamasaki, M., Makabe, O., Kimura, T., Takizawa, T., Hamaya, T., Suzuki, E. & Miyano, H. (1991) Structure and function of a pepstatin-insensitive acid proteinase from *Aspergillus niger* var. *macrosporus*. In: *Strucure and Function of the Aspartic Proteinases* (Dunn, B. M., ed.) New York, Plenum Press, pp. 203–211.
5. Studier, F. W., Rosenberg, A. H., Dunn, J. J. & Dubendorff, J. W. (1990) Use of T7RNA polymerase to direct expression of cloned genes. *Methods in Enzymol.* 185, 60–88.
6. Kagami, N., Inoue, H., Kimura, T., Makabe, O. & Takahashi, K. (1995) Expression in *E. coli* of *Aspergillus niger* var. *macrosporus* proteinase A, a non-pepsin type acid proteinase. In: *Aspartic Proteinases. Structure, Function, Biology, and Biomedical Implications* (Takahashi, K., ed.) New York, Plenum Press, pp. 597–603.
7. Kunkel, T. A. (1985) Rapid and efficient site-specific mutagenesis without phenotypic selection. *Proc. Natl. Acad. Sci. U.S.A.* 82, 488–492.
8. Furihata, C., Kawachi, T. & Sugimura, T. (1972) Premature induction of pepsinogen in developing rat gastric mucosa by hormones. *Biochem. Biophys. Res. Commun.* 47, 705–711.
9. Anson, M. L. (1938) The estimation of pepsin, trypsin, papain, and cathepsin with hemoglobin. *J. Gen. Physiol.* 22, 79–89.
10. Iio, K. & Yamasaki, M. (1976) Specificity of acid proteinase A from *Aspergillus niger* var. *macrosporus* towards B-chain of performic acid oxidized bovine insulin. *Biochim. Biophys. Acta* 429, 912–924.

11. Takahashi, K. (1995) Proteinase A from *Aspergillus niger*. *Methods Enzymol.* 248, 146–55.
12. Ido, E., Saito, T. & Yamasaki, M. (1987) Substrate specificity of acid proteinase A from *Aspergillus niger* var. *macrosporus*. *Agr. Biol. Chem.* 51, 2855–2856.
13. Takahashi, K. (1997) The specificity of peptide bond cleavage of acid proteinase A from *Aspergillus niger* var. *macrosporus* toward oxidized ribonuclease A.*Biosci. Biotech. Biochem.* 61, 381–383.
14. Inoue, H., Kimura, T., Makabe, O. & Takahashi, K. (1991) The gene and deduced protein sequences of the zymogen of *Aspergillus niger* proteinase A. *J. Biol. Chem.* 266, 19484–19489.
15. Inoue, H., Hayashi, T., Huang, X.-P., Lu, J.-F., Athauda, S. B. P., Kong, K.-H., amagata, H., Udaka, S. & Takahashi, K. (1996) Heterologous expression and site-directed mutagenesis studies on the activation mechanism and the roles of the basic residues in the prosegment of aspergillopepsinogen I. *Eur. J. Biochem.*, 237, 719–725.
16. Tsuru, D., Shimada, S., Maruta, S., Yoshimoto, T., Oda, K., Murao, S., Miyata, T. & Iwanaga, S. (1986) Isolation and amino acid sequence of a peptide containing an epoxide-reactive residue from the thermolysin-digest of *Scytalidium lignicolm* acid protease B. *J. Biochem.(Tokyo)* 99, 1537–1539.
17. Tsuru, D., Naotsuka, A., Kobayashi, R., Yoshimoto, T. Oda, K. & Murao, S. (1989) Inactivation of *Scytalidium lignicolum* acid protease B with 1,2-epoxy-3-(4′-azido-2′- nitrophenoxy)propane. *Agr. Biol. Chem.* 53, 2751–2756.
18. Maita, T., Nagata, S., Matsuda, G., Maruta, S., Oda, K. & Tsuru, D. (1984) Complete amino acid sequence of *Scytalidium lignicolum* acid protease B. *J. Biochem. (Tokyo)* 95, 465–475.
19. Kakimori, T., Yoshimoto, T., Oyama, H., Oda, N., Gotoh, Y., Oda, K., Murao, S. & suru, D. (1996) Nucleotide sequence of the gene encoding pepstatin-insensitve acid protease B, scytalidopepsin B, of *Scytalidium lignicolum*. *Biosci. Biotech. Biochem.* 60, 1210–1211.
20. Jara, P., Gilbert, S., Delmas, P., Guillemot, J. C., Kaghad, M., Ferrara, P. & Loison, G. (1996) Cloning and characterization of the eapB and eapC genes of *Cryphonectria parasitica* encoding two new acid proteinases, and disruption of eapC. *Mol. Gen. Genet.* 250, 97–105.

CRYSTAL STRUCTURE OF THE *Rhizomucor miehei* ASPARTIC PROTEINASE

J. Wilson Quail,[1] Jian Yang,[1] Palle Schneider,[2] and Zongchao Jia[3]

[1]Department of Chemistry
University of Saskatchewan
110 Science Place
Saskatoon, Saskatchewan S7N 5C9, Canada
[2]Department of Enzyme Research
Novo Nordisk, DK-2880 Bagsvaerd, Denmark
[3]Department of Biochemistry
Queen's University
Kingston, Ontario K7L 3N6, Canada

1. INTRODUCTION

The composition of milk has been reviewed by Jenness (1). It is an extremely complex mixture of components, consisting of water, lipids, carbohydrates, proteins, salts and many miscellaneous compounds. The largest component is water. The lipids are 98% triglycerides, mostly as globules. Carbohydrate is mainly as lactose. There are several classes of proteins. The largest group (80%) are phosphoproteins called "caseins". They are associated with calcium in "casein micelles", which are 20–300 μm in diameter. The α_{s1}-, α_{s2}-, β- and κ-caseins are in the ratio 3 : 0.8 : 3 : 1. The other milk proteins are called "whey proteins" and include β-lactoglobulin, α-lactalbumin, blood serum albumin and immunoglobulins. In total there are probably about 10^5 kinds of molecules in milk.

One of the first steps in cheese making is clotting of the milk (2). The curd is formed in one of two ways: acid or enzymatic coagulation. In acid curd cheeses (cottage, cream), the curd is formed by the direct addition of acid to the milk or by lactic acid produced by the fermentation of lactose. Milk clotting enzymes obtained from animal, plant and microbial sources have been used since antiquity for the manufacture of cheese and other foods. Many proteolytic enzymes will clot milk. Milk clotting enzymes have been obtained from virtually every class of living organisms.

Selective cleavage of κ-casein along with relatively low proteolytic activity and low thermal stability are the essential characteristics of an effective cheese making enzyme. General proteolytic activity results in many components that have unpleasant taste. After

Aspartic Proteinases, edited by James.
Plenum Press, New York, 1998.

the aspartic proteinase has formed the cheese curd, the mixture is heated to destroy the enzyme and stop the proteolytic activity.

Calf-chymosin (rennin), a pepsin-like aspartic proteinase, is traditionally used in the cheese industry because it possesses high milk-clotting activity. Chymosin is the enzyme of choice and the standard against which other enzymes are measured. Chymosin splits κ-casein specifically at the Phe105–Met106 bond (2), with little other cleavage. Other milk-clotting enzymes are less specific, but have the same general effect. Splitting of κ-casein destroys the stability of the casein micelles of the milk system, and casein begins to aggregate, forming a curd. Fungal aspartic proteinases have been used as substitutes for making cheese (2). Both *Mucor pusillus* aspartic proteinase (MPP) and *Rhizomucor miehei* aspartic proteinase (RMP) (EC 3.4.23.23) possess high milk-clotting activity.

Aspartic proteinases are a class of proteolytic enzymes that cleave protein chains. Some, from mammalian sources, are digestive enzymes of broad specificity, such as pepsin. Pepsin is most active at pH 1–2 (hydrochloric acid in the stomach). Only renin works at neutral pH. All other aspartic proteinases are most effective at acidic pH (pH < 5.0). All aspartic proteinases have a binding site that can accommodate a polypeptide chain. There are two aspartate residues on the aspartic proteinase that can interact with a peptide bond of the bound polypeptide chain. The two aspartate groups plus a water molecule activate the peptide bond, resulting in hydrolysis of the bond.

Due to a worldwide shortage of calf chymosin, RMP has been used as a substitute for calf chymosin, making use of its milk-clotting activity. It exhibits high milk-clotting activity due to selective cleavage of κ-casein, along with relatively low proteolytic activity (3). However, it is the most thermally stable enzyme among the aspartic proteinases (4) and therefore is of limited use as a milk coagulant; the high thermal stability results in the survival of the enzyme activity after cooking of the curd and thus may cause ill flavor in cheese during long maturating periods (5). RMP is also the most glycosylated of the aspartic proteinases with 6% carbohydrate (6). It can be modified with hydrogen peroxide or peroxyacetic acid (7) to decrease its heat stability. Some enzymatic activity is lost, but the modified enzyme has about the same thermal stability as chymosin (8,9). The high thermal stability of some proteins has been attributed to a network of ionic pairs in the hydrophobic core of the protein (10). From denaturation studies of RMP (11) and deglycosylation studies of MPP (12), the high thermal stability of RMP has been attributed to the presence of N-linked carbohydrate chains on RMP.

Why is RMP effective as a milk-clotting enzyme? Why does RMP have the greatest thermal stability of the milk-clotting aspartic proteinases? How could RMP be modified to improve its milk-clotting ability and decrease its thermal stability? We believed that the solution of the crystal structure of the enzyme might answer the above questions and should provide a rationale for further modifications of RMP to improve its milk-clotting activity and decrease its thermal stability.

2. METHODS

Native RMP was obtained by fermentation of *R. miehei* and is identical to the enzyme that can be obtained as Rennilase (13,14). The protein was purified by the following protocol: ion-exchange chromatography on SP-Sephadex C50 in 0.05 *M* citrate buffer, pH 3.2, with 0.1 *M* sodium chloride and elution by a 0.1–0.3 *M* sodium chloride gradient; concentration of the fractions with clotting activity by ultrafiltration using DDS membrane type 800 (M_r cutoff 8–10 kDa); exchange of buffer on a Sephadex G50 medium eluted with 0.02 *M* piperazine/HCl buffer, pH 5.3; ion-exchange chromatography on DEAE-Sephadex A50 in 0.02 *M* piperazine/HCl buffer, pH 5.3, and elution by a 0–0.6 *M* sodium chloride gradient; concentration by ultrafiltration using a DDS membrane type 800; gel-filtration on Sephadex G75 superfine using 0.1 *M* acetate buffer, pH 6.0, with 0.1 *M* sodium chloride; concentration by ultrafiltration and continuous washing with distilled water; freeze drying.

Crystals of RMP were first grown by Jia *et al.* (15). Better quality crystals were obtained under improved conditions by a repeated seeding method (16). Large crystals suitable for data collection were obtained after several cycles of repeated seeding by putting the seed crystals into 5 mg/mL protein drops (5 mM citrate buffer, pH 4.0, 10% PEG 4000) to equilibrate over 13% PEG 4000 reservoir solutions (10 mM citrate buffer, pH 4.0). RMP crystallizes in the orthorhombic space group $P2_12_12_1$ [a = 41.86 Å, b = 51.21 Å, c = 174.24 Å], with Z = 4 (one enzyme molecule per asymmetric unit). The molecular mass of the protein is 38.7 kDa and the estimated solvent content is 49% (15).

The RMP X-ray data were collected to a maximum resolution of 2.15 Å at beamline X11 (EMBL beam line), DORIS ring, DESY synchrotron in Hamburg. The wavelength used was 0.927 Å. The data were processed using DENZO and SCALEPACK (17) and the CCP4 suite (18). The data set is 93.8% complete from 15 to 2.15 Å with a merging R factor of 0.058 for symmetry-equivalent reflections.

Solution of the structure by Molecular Replacement (MR) was more difficult than we had anticipated. None of the aspartic proteinases in the Protein Data Bank (PDB) gave a solution. Sequence alignment studies show that RMP has 83% identity with MPP and about 22–24% identity with other aspartic proteinases (19). We became aware that MPP had just been solved by Newman *et al.* (20). A molecular replacement solution for RMP was obtained by using the program AMoRe (21) with the 2.0 Å MPP structure as the search model. The model was refined by rigid body refinement using 10 to 2.8 Å data, followed by forty cycles of refinement using the slow-cooling protocol in X-PLOR (22) with 6.0 to 2.15 Å data. In the last seven cycles of refinement, water molecules were added to the model in regions with density > 3.5 σ in the SIGMAA weighted (23) ΔF maps. The "free-R-factor" (24) was also calculated for every cycle except the last cycle of refinement using a randomly selected data set comprising 10% of the observed data. The final coordinates of RMP have been deposited in the Brookhaven Protein Data Bank with ID code 2ASI.

3. RESULTS AND DISCUSSION

The final results of the refinement are listed in Table 1. In the final model, 2644 protein non-hydrogen atoms, 140 water molecules and 39 carbohydrate atoms are included. Missing from the final model are the four N-terminal residues (Ala1 to Asp4), Gly85 and the sidechains of 18 surface residues, all of which were omitted from the refinement due to insufficient indication of their positions in the final electron density map. The main chain torsion angles (ϕ, ψ) of the model were examined by the program PROCHECK (25). The Ramachandran plot (26) is shown in Figure 1. Except for the glycine and proline residues, all the residues are in the allowed regions. Contrary to Goldman's suggestion (10) to explain the high thermal stability of proteins, no network of salt bridges was found in the structure.

RMP contains two domains of approximately the same number of residues (N-terminal domain 1–186; C-terminal domain 187–361). A ribbon representation of the enzyme secondary structure and overall folding is given in Figure 2. The folding of RMP is very similar to that of MPP. The attachments of the carbohydrates are shown in Figure 2. A

Table 1. X-PLOR refinement results for RMP

Number of protein non-hydrogen atoms	2644
Number of water oxygen atoms	140
Number of carbohydrate non-hydrogen atoms	39
R.m.s. coordinate error (Å) from a Luzzati plot	0.2
Resolution range (Å)	2.15–6.0
Number of unique refs. used in refinement	18842
Conventional R-factor (%)	21.5
R-free (%)	28.1
R.m.s.d. for bond distances (Å)	0.01
R.m.s.d. for bond angles (°)	1.7
R.m.s.d. for dihedral angles (°)	27
R.m.s.d. for impropers (°)	1.5
Average B-factor for protein ($Å^2$)	39.1
Average B-factor for N-terminal domain ($Å^2$)	29.1
Average B-factor for C-terminal domain ($Å^2$)	48.0
Average B-factor for water molecules ($Å^2$)	45.9
Average B-factor for carbohydrates ($Å^2$)	63.7
Average B-factor for residues 82–88 ($Å^2$)	66.6

large substrate binding cleft is clearly visible between the two domains, with the two catalytic residues Asp38 and Asp237 in the middle of the cleft. The active site of RMP is shown in more detail in Figure 3. The two aspartates are bridged by a water molecule, Wat421. This active site configuration is highly conserved in aspartic proteinases. Quite a few polar residues are within 10 Å of the active site and they are involved in a hydrogen bonding network that makes the active site rigid. In Figure 4, we show the plots of average B factors of main chain and side chain atoms versus the residue numbers. The C-terminal domain has a much higher average B-factor than the N-terminal domain. This indicates that the C-terminal domain is more flexible or disordered. After examining the intermolecular interactions, we came to the conclusion that the difference in thermal fac-

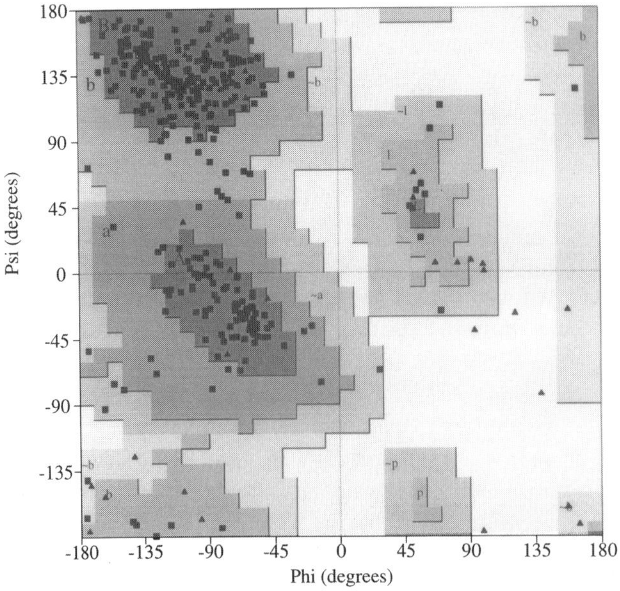

Figure 1. Ramachandran plot of RMP, showing the main chain torsion angles. Glycines are shown as triangles.

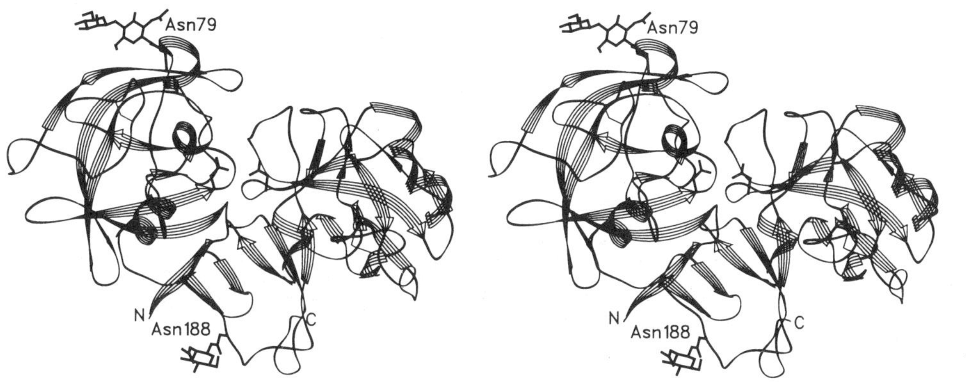

Figure 2. The ribbon representation of the secondary structure and overall folding of *Rhizomucor miehei* aspartic proteinase. The enzyme is glycosylated at Asn79 and Asn188.

tors between the two domains is due to crystal packing. The intermolecular contacts in the crystal structure of RMP show many more contacts for the N-terminal domain with adjacent molecules than found for the C-terminal domain. Residues 82 to 88 show much higher average B-factors than any other region in the N-terminal domain. These residues form a highly mobile flap over the substrate binding cleft and become more rigid upon the binding of inhibitors to the enzyme, due to the formation of hydrogen bonds between the enzyme and the inhibitors (27,28).

To get a detailed knowledge of the substrate binding site, we used molecular modeling to dock the inhibitor pepstatin A into the active site of RMP. The complex model shows that pepstatin A fits into the substrate binding cleft in an extended conformation. The possible residues involved in substrate binding are summarized in Table 2.

We submitted the α-carbon coordinates of RMP for three-dimensional structural alignments with the DALI program (29) at EMBL, Heidelberg, Germany. The result shows that RMP has an r.m.s. deviation of 0.6 Å with MPP and r.m.s. deviations of more than 1.6 Å with the other aspartic proteinases for 215–226 equivalent residues. These observations indicate that RMP and MPP are structurally similar and diverged from other

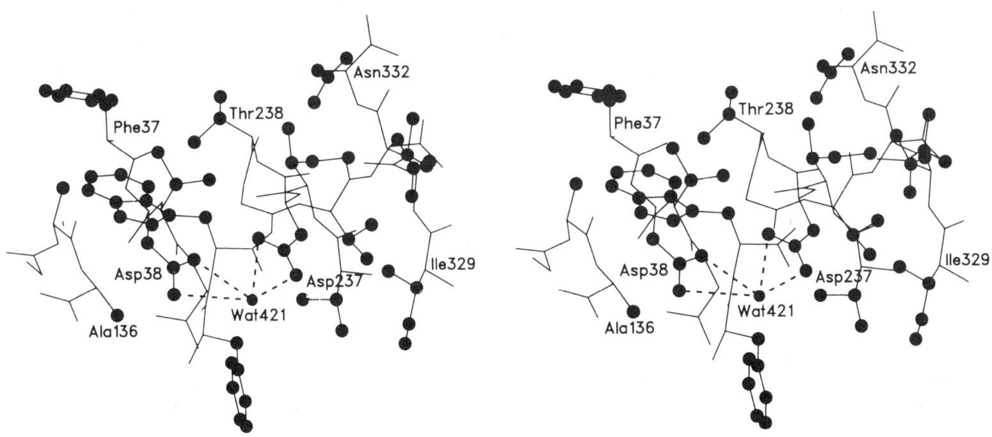

Figure 3. The active site of RMP.

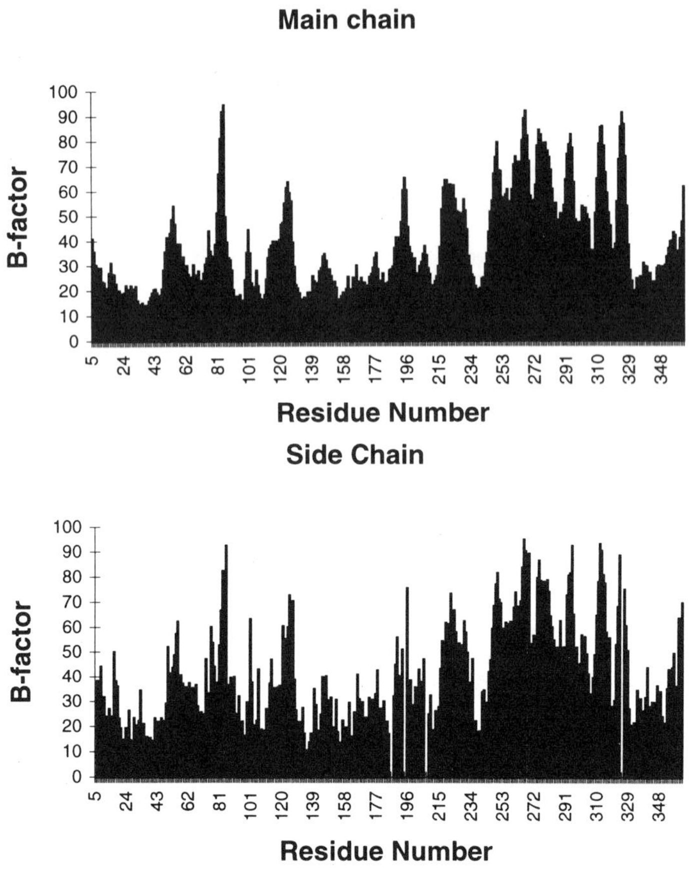

Figure 4. B-factors versus residue numbers for both main chain and side chain atoms.

aspartic proteinases at an early stage of evolution. Further studies on the comparison of the aspartic enzymes may provide an insight of the molecular evolution of these enzymes.

All the aspartic proteinases except renin show maximum activity at acidic pH (i.e., pH < 5.0). To carry out the functions at low pH, the enzymes must be stable in an acidic environment. In 1991, Andreeva and James (30) reported that pepsin has a high ratio of potentially negatively charged groups to positively charged groups and this high ratio

Table 2. Substrate binding sites of RMP

Substrate	Component	Subsite residues
P4	IVA	Asn241, Phe242, Ile244, Leu321
P3	Val	Thr240, Asn241, Glu19, Leu17
P2	Val	Thr240, Ile244, Gly325, Ile329, Gly83, Thr84
P1–P1′	STA	Asp38, Asp237, Tyr82, Gly83, Gly239, Thr240, Gly40, Leu36, Leu132, Pro117
P2′	Ala	Tyr82, Ser41, Gly40, Phe210
P3′–P4′	STA	Tyr82, Thr81, Ile80, Phe210, Asn140

The pepstatin A sequence from P4 to P4′ corresponds to residues 102 to 108 from k-casein. IVA represents Isovaleric acid and STA represents statine.

gives the enzyme an unusually stable native structure at low pH values. Pepsin has several acidic residues with depressed pKa values and these residues cause pepsin to remain slightly negatively charged even at very low pH values.

Most aspartic proteinases have an aspartate at residue 303 (porcine pepsin numbering). Exceptions are renin (alanine) and MPP and RMP (asparagine). The activity at acidic pH has attracted the attention of scientists for many years, but the reasons for it are still not clear. Changes in residue 303 have been proposed to contribute to differences in the pH optimum of catalysis (31,32). The changes occur through a mechanism that involves the degree of protonation of the active site aspartates, which are affected by polarization of peptide bond 216–217 by Asp303 through a hydrogen bond (OD_{303}–H....O_{216}). The polarization effect of this peptide bond is transferred to the carboxyl group of Asp215 *via* a second hydrogen bond between the main chain N atom of residue Gly217 and one carboxyl oxygen atom of Asp215, which in turn affects the degree of protonation of Asp215. Newman *et al.* (20) proposed that for MPP the presence of an asparagine at residue 303 in place of the usual aspartate results in a more basic pH optimum. This likely results from the reduced ability of asparagine to polarize peptide bond 216–217 through a hydrogen bond (ND_{303}–H....O_{216}).

To find out why most aspartic proteinases are most active at acidic pH values, we examined the structures of three enzymes; chymosin b (opt. pH = 3.7), RMP (opt. pH = 4.7) and renin (opt. pH = 7.0), which possess significantly different pH optima for activity. The ratios of the potentially negatively charged groups to positively charged groups are 1.6, 1.9 and 1.2 for chymosin b, RMP and renin, respectively. Since there are several acidic residues that are buried inside the proteins in the three enzymes, we think the ratio of potentially negatively charged groups to positively charged groups on the surface of the proteins might be more important in stabilizing the proteins at low pH values. The ratios of the surface charged groups are 1.5, 1.6 and 1.0, respectively, for chymosin b, RMP and renin. Therefore, chymosin b and RMP are stable at low pH values while renin is only stable at a pH near 7.0. We also examined the hypothesis of Newman *et al.* (20) described above, and we conclude that this hypothesis is unlikely to be valid.

Since the bond energy difference between the hydrogen bonds of O–H....O (chymosin b) and N–H....O (RMP) is likely very small (less than 0.5 kcal/mol), the difference in the polarization of the peptide bond 216–217 caused by switching one of the oxygen atoms in the hydrogen bond to a nitrogen atom will have little significance. We believe that this small polarization difference would not cause significant difference of the ionization state of Asp215 between chymosin b and RMP.

Three acidic residues, six polar residues and no basic residues are within 10 Å of the active site in chymosin b; three acidic residues, nine polar residues and one basic residue are within 10 Å of the active site in RMP; there are twelve polar residues and no acidic or basic residues within 10 Å of the active site in renin. The active site of each aspartic proteinase has a hydrogen bonding network and all these acidic, basic and polar residues, together with the two catalytic aspartates are involved in the network. Therefore, the influence of these acidic or basic residues on the two catalytic aspartates is significantly affected by other residues in the hydrogen bonding network. To correlate the effect of all the residues close to the active site on the catalytic aspartate residues, we calculated the surface electrostatic potentials of chymosin b, RMP and renin by the program DelPhi (33). The results show that chymosin b and renin have the most and least negative electrostatic potential at the active sites, respectively. Because the negative electrostatic potentials will decrease the pKa's of the two aspartates and the optimum pH of each aspartic proteinase is approximately the average of the pKa's of the two aspartate residues, we propose that the optimum pH of each aspartic proteinase is determined by the electrostatic potential at the active site. This, in turn, is determined by the positions and orientations of all the residues near the active site.

RMP possesses about 6% carbohydrate and is the most glycosylated among the aspartic proteinases (6). However, from the electron density map we found only a disaccharide NAG-MAN (NAG: N-acetyl-D-glucosamine; MAN: D-Mannose) and a monosaccharide NAG at Asn79 and Asn188, respectively. This suggests that the carbohydrates are very flexible. In contrast to the suggestion by Brown and Yada (11) that these two asparagines are on the surface β-turns, we found that Asn79 is in the middle of a β-strand and is close to the flap region (residues 82–88), and Asn188 is on the surface loop which connects the two domains. Denaturation studies on glycosylated and deglycosylated RMP by Brown and Yada (11) and the deglycosylation studies on MPP by Aikawa *et al.* (12) suggest that the N-linked carbohydrates confer high thermal stability on the enzymes. Our explanation for this phenomenon is that the highly mobile carbohydrates act as heat reservoirs to stabilize the enzymes against high temperature. We predict that deglycosylation of RMP by the enzymes N-glycopeptidase F or endo-β-N-acetylglucosaminidase H will result in decreases thermal stability for the enzyme. There are several asparagine residues on the surface of RMP. However, the enzyme is only glycosylated at Asn79 and Asn188. This suggests that the flap region and the connecting loop are important regions of the molecule in determining the stability of the enzyme. The carbohydrates may mask these regions from being denatured by other proteolytic proteinases or by heating. Denaturation of the connecting loop may cause the unfolding of the tertiary structure of the enzyme, resulting in a loss of activity. The flap may also play an important role as a guide for substrates. It might guide substrates into the binding cleft, direct the orientation of substrates at the binding site and stabilize the enzyme-substrate complex. The denaturation of the flap region may reduce the affinity of the active site for the substrate and consequently result in loss of enzyme activity. Further mutagenesis studies on the residues within or close to this flap region may provide a better understanding of substrate binding and the function of the carbohydrate on Asn79.

ACKNOWLEDGMENTS

We thank the Natural Sciences and Engineering Research Council of Canada for providing an operating grant (to JWQ) to support this work and the University of Saskatchewan for the award of a graduate scholarship (to JY). The authors are also indebted to Margaret Vandonselaar for her help in finding the conditions for crystallization of RMP and to Alexei Teplyakov (EMBL, Hamburg Outstation) for his help in collecting the data.

REFERENCES

1. R. Jenness, Composition of Milk, *In:* Fundamentals of Dairy Chemistry, N.P. Wong, ed., Van Nostrand Reinhold Co., New York, 1–38 (1988).
2. R.J. Brown & C.A. Ernstrom, Milk-Clotting Enzymes and Cheese Chemistry, Part I--Milk-Clotting Enzymes, *In:* Fundamentals of Dairy Chemistry, N.P. Wong, ed., Van Nostrand Reinhold Co., New York, 609–633 (1988).
3. K. Arima, S. Iwasaki & G. Tamura, Milk clotting enzyme from microorganisms Part I. Screening test and the identification of the potent fungus, *Agr. Biol. Chem.*, **31**:540–545 (1967).
4. D.B. Hyslop, A.M. Swanson & D.B. Lund, Heat inactivation of milk-clotting enzymes at different pH hydrogen-ion concentration, *Mucor miehei*, *Mucor pusillus* proteases, and rennet and rennet-pepsin, *J. Dairy Sci.*, **62**:1227–1232 (1979).

5. T. Yamashita, S. Higashi, T. Higashi, H. Machida, S. Iwasaki, M. Nishiyama, & T. Beppu, Mutation of a fungal aspartic proteinase, *Mucor pusillus* renin, to decrease thermal stability for use as a milk coagulant, *J. Biotech.*, **32**:17–28 (1994).
6. W.S. Rickert & P.A. McBride-Warren, Structural and functional determinants of *Mucor miehei* protease. III, isolation and composition of the carbohydrate moiety, *Biochim. Biophys. Acta,* **336**:437–444 (1974).
7. S. Branner-Jorgensen, P. Eigtved & P. Schneider, Reduced thermostability of modified Mucor rennet, *Neth. Milk and Dairy J.,* **35**: 361–364 (1981).
8. S. Branner-Jorgensen, P. Schneider & P. Eigtved, A method of modifying the thermal destabilization of microbial rennet and a method of cheese making using rennet so modified, U.K. Pat. Appl. 2,045,772A (1980).
9. D.A. Cornelius, Process for decreasing the thermal stability of microbial rennet, U.S. Pat 4,348,482 (1982).
10. A. Goldman, How to make my blood boil, *Structure*, **3**:1277–1279.
11. E.D. Brown & R.Y. Yada, A kinetic and equilibrium study of the denaturation of aspartic proteinases from fungi, *Endothia parasitica* and *Mucor miehei*, *Biochemica et Biophysica Acta,* **1076**:406–415 (1991)
12. J.L. Aikawa, T. Yamashita, Nishiyama, S. Horinouchi & T. Beppu, Effects of glycosylation on the secretion and enzyme activity of Mucor renin, an aspartic proteinase of *Mucor pusillus*, produced by recombinant yeast, *J. Biol. Chem.*, **265**:13955–13959 (1990).
13. A.-M. Bech & B. Foltmann, Partial primary structure of *Mucor miehei* protease milk-clotting enzyme, *Neth. Milk Dairy J.*, **35**:275–280 (1981).
14. E. Boel, A.-M. Bech, K. Randrup, B. Draeger, N.P. Fiil & B. Foltmann, Primary structure of a precursor to the aspartic proteinase from *Rhizomucor miehei* shows that the enzyme is synthesized as a zymogen, *Proteins Struct. Funct. Biol.*, **1**:363–369 (1986).
15. Z. Jia, M. Vandonselaar, P. Schneider & J.W. Quail, Crystallization and preliminary X-ray structure solution of *Rhizomucor miehei* aspartic proteinase, *Acta Crystallogr. Sect. D*, **51**:243–244 (1995).
16. C. Thaller, L.H. Weaver, G. Eichele, E. Wilson, R. Karlsson & J.N. Jansonius, Repeated seeding technique for growing large single crystals of proteins, *J. Mol. Biol.*, **147**, 465–469 (1981).
17. Z. Otwinowski, Oscillation data reduction program. *In:* Data collection and processing: Proceedings of the CCP4 study weekend, 29–30 January, 1993. L. Sawyer, N. Issacs & S. Bailey, eds., Science and Engineering Research Council, Daresbury Laboratory, Warrington, UK (1993).
18. Collaborative Computational Project #4. The CCP4 suite: Programs for protein crystallography. *Acta Crystallogr. Sect. D,* **50**:760–763 (1994).
19. M. Baudys, S. Foundling, M. Pavlik, T. Blundell & V. Kostka, Protein chemical characterization of *Mucor pusillus* aspartic proteinase. Amino acid sequence homology with the other aspartic proteinases, disulfide bond arrangement and site of carbohydrate attachment, *FEBS Lett.*, **235**:271–274 (1988).
20. M. Newman, F. Watson, P. Roychowdhury, H. Jones, M. Badasso, A. Cleasby, S.P. Wood, I.J. Tickle & T.L. Blundell, X-ray analysis of aspartic proteinases V. Structure and refinement at 2.0 Å resolution of the aspartic proteinase from *Mucor pusillus*, *J. Mol. Biol.,* **230**:260–283 (1993).
21. J. Navaza, AMoRe: an automated package for molecular replacement, *Acta Cryst. Sect. A,* **50**:157–163 (1994).
22. A.T. Brünger, *XPLOR Manual Version 3.1*. Yale University, New Haven, CT 06511, USA. (1992).
23. R.J. Read, Improved Fourier coefficients for maps using phases from partial structures with errors, *Acta Crystallogr. Sect. A*, **42**:140–149 (1986).
24. A.T. Brünger, Free R value: A novel statistical quantity for assessing the accuracy of crystal structures, *Nature,* **55**:472–475 (1992).
25. R.A. Laskowski, M.W. MacArthur, S.G. Hutchinson & J.M. Thornton, PROCHECK: a new program to check the stereochemical quality of protein structures, *J. Appl. Cryst.,* **26**:283–291 (1993).
26. G.N. Ramachandran, & V. Sasisekharan, Conformation of polypeptides and proteins, *Advan. Protein Chem.,* **23**:283–437 (1968).
27. R. Bott, E. Subramanian & D.R. Davies, Three-dimensional structure of the complex of the *Rhizopus chinensis* carboxyl proteinase and pepstatin at 2.5 Å resolution, *Biochemistry*, **21**:6956–6962 (1982).
28. M.N.G. James, A. Sielecki, F. Salituro, D.H. Rich & T. Hofmann, Conformational flexibility in the active sites of aspartyl proteinases revealed by pepstatin fragment binding to penicillopepsin, *Proc. Natl. Acad. Sci.,* **79**:6137–6141 (1982).
29. L. Holm & C. Sander, Protein structure comparison by alignment of distance matrices, *J. Mol. Biol.,* **233**:123–138 (1993).
30. N.S. Andreeva & M.N.G. James, *Structure and Function of the Aspartic Proteinases: Genetics, Structure, and Mechanisms. Advances in Experimental Medicine and Biology*, **Vol. 306,** Plenum Press, New York, USA. (1991).
31. T. Yamauchi, M. Nagahama, H. Hori & K. Murakami, Functional characterization of Asp-317 mutant of human renin expressed in COS cells. *FEBS Lett.*, **230**:205–208 (1988).

32. D. Mantafounis & J. Pitts, Protein engineering of chymosin: modification of the pH optimum of enzyme catalysis, *Prot. Eng.*, **3**:605–609 (1990).
33. K.A. Sharp & A. Nicholls, DelPhi V3.0 Manual. Department of Biochemistry and Molecular Biophysics, Columbia University, USA. (1989).

40

Rhizomucor miehei ASPARTIC PROTEINASES HAVING IMPROVED PROPERTIES

Marianne K. Harboe

Chr. Hansen A/S, 10-12 Boege Allé, Hoersholm
Denmark

1. INTRODUCTION

Aspartic proteinases are widely spread in most organisms, where they carry out many different functions. The oldest and most well-known use of aspartic proteinases is as coagulants in cheese-making. The coagulants are probably still the only commercial products containing aspartic proteinases as the active substance. The main commercial interest in coagulants is focused on how to enhance the enzyme activity and its performance during cheese-making. In contrast to this, most commercial interests in other aspartic proteinases deal with how to inhibit their activity in order to control their function. The main commercial coagulants can be divided into three groups according to their origin, animal rennet, fermentation produced chymosin (FPC) and microbial coagulants.[1] Of animal rennets, calf rennet is the traditional and the most desired coagulant for cheese-making, but FPC products made by recombinant organisms and microbial coagulants, the natural aspartic proteinases of the fungi, make up a considerable part of the coagulant market. Fermentation produced chymosin is the product of the future, and the great interest in microbial coagulants is due to the fact that they constitute a significant part of the market because of their lower prices. *Rhizomucor miehei* aspartic proteinase (EC 3.4.23.23) constitutes the active component of the most widely used microbial coagulant. The molecule consists of 361 amino acids, has a molecular weight of 38701 and contains in addition approx. 6% carbohydrate.[2] The enzyme is commercially available in its natural form, in various heat labile forms made by post-treatment of the natural form (oxidation) and as recombinant coagulant produced in *Aspergillus oryzae*.

Glycosylation is the main post-translational modification of proteins in eucaryotic cells. The glycosylation sites and the actual glycosylation vary among the aspartic proteinases. *Rhizomucor miehei* has three potential sites Asn_{79}, Asn_{188} and Asn_{313} (no 72,171 and 279 according to Porcine pepsin numbering) out of which only the two first sites have

Aspartic Proteinases, edited by James.
Plenum Press, New York, 1998.

been found to be actually glycosylated.[2] The degree, with which each site is glycosylated, varies. The recombinant enzyme is more highly glycosylated than the natural one.

The aim of this paper is to present new results on how to improve the activity of *Rhizomucor miehei* coagulants by deglycosylation as well as to discuss protein glycosylation in general and of aspartic proteinases in particular.

2. EXPERIMENTS AND RESULTS

Different commercial coagulants, like Hannilase™ (containing the native *R. miehei* aspartic proteinase), Modilase S™ (containing the oxidized native *R. miehei* aspartic proteinase) and Novoren™ (*R. miehei* aspartic proteinase heterologously produced in *Asp. oryzae*) were subjected to deglycosylation by endo-β-N-acetylglucosamidase H, also called endo H (EC 3.2.1.96). Various preparations of Endo H were used like e.g. Sigma no E7642 1 U/ml. No difference was observed between the preparations except for the differences due to the declared strength. The milk-clotting activity measured according to IDF Standard 157 [3] relative to a calf rennet standard was followed during the experiments.

It was found that the milk-clotting activity is increased significantly by deglycosylation of the coagulants.[4] The following percentages of increase in activity was found after complete deglycosylation: 30% increase for the native *R. miehei* coagulants Hannilase™ and Modilase S™ and 50% for heterologous *R. miehei* coagulant Novoren™.

The reaction of deglycosylation was characterized by varying the pH, temperature, time, solvent and the ratio between Endo H and aspartic proteinase. The deglycosylation was found to depend relatively little on pH within the range 4–8, on temperature within the range 5–40°C and on the solvent in the range 0–20% sodium chloride, which means these parameters are not critical for the reaction. Time and ratio Endo H:*R. miehei* proteinase had a significant influence but the deglycosylation was found to proceed extremely easily. Complete deglycosylation (measured when obtaining maximum stable activity) varied from seconds to weeks — as an example, the deglycosylation is complete within 1–2 days using a ratio (mg:mg) between Endo H (1 U equals approx 0.025 mg) and *R. miehei* proteinase (calculated as pure enzyme disregarding other glycoproteins present) 1:250,000.

The deglycosylated coagulants were characterized with respect to their properties of importance to cheese-making. All important characteristics were found to be improved, more or less, upon deglycosylation of the *R. miehei* proteinase. The pH dependency of the milk-clotting activity was reduced to values closer or identical to that of calf chymosin which has a desirable low value. The general proteolytic activity of deglycosylated enzyme does not increase as much as the milk-clotting activity. This means higher ratio between milk-clotting and general proteolytic activity, or in other words more specific enzyme, the most highly estimated property of milk coagulants. Furthermore, the deglycosylated coagulants were found to become more heat labile, another advantageous characteristic, however not enough to omit the destabilization (oxidation) step used for making Modilase S™ from Hannilase™. Deglycosylation was found to have no effect on the storage stability. Finally, cheese produced by using the two types of coagulants (conventional and deglycosylated) confirmed that the deglycosylated coagulant can be used in much lower dosages, it showed that more deglycosylated coagulant was retained in the curd than non-deglycosylated and it indicated that cheese made with deglycosylated coagulant is of at least as high a quality as the conventional one and causing no defects.

3. DISCUSSION

The results that *Rhizomucor miehei* coagulants get significantly higher milk-clotting activity and improved in vitro properties when deglycosylated, are new and surprising, although some information was already known. Gray et al.[5] found that *Rhizomucor miehei* heterologous produced in *Aspergillus niger* almost doubled its activity when deglycosylated. For the closely related enzyme, Aikawa et al.[6] found that the highly glycosylated *Rhizomucor pusillus* proteinase expressed in *Saccharomyces cerevisiae* increased its milk-clotting activity when deglycosylated, whereas deglycosylation had no effect on the less glycosylated native *Rhizomucor pusillus* proteinase. Christensen[7] has for Novoren, *R. miehei* heterologously produced in *Aspergillus oryzae*, found the proteinase to be overglycosylated but that deglycosylation had no effect on its milk-clotting activity which is in contrast to our findings. Apparently, the more carbohydrate an enzyme has, the higher is the risk of it affecting the activity. Looking at the three-dimensional structure of *Rhizomucor miehei*[8] which gives the position of the two carbohydrate chains it may be suggested that the effect could be due to the carbohydrate chain on Asn_{79} which, if too bulky, partly blocks access to the active site.

Looking at the general rules for N-glycosylation, the potential glycosylation sites are Asn in the sequences Asn–X–Ser/Thr. However, not all Asn which fulfill this requirement are glycosylated and the degree of glycosylation of each site can vary. Thr seems to facilitate glycosylation much more than Ser, and Glu or Pro in the X-position seems to prevent glycosylation.[9] Furthermore, several other factors are important, like e.g. the position (susceptibility) of the Asn in the three-dimensional structure, different production organisms glycosylate differently, and even the fermentation condition has a significant influence.

When applying these rules to the known sequences of aspartic proteinases as aligned by Foltmann[10] the rules seem to be in agreement with most of the published information concerning actual glycosylation with a few exceptions. Lamb chymosin, which is 95% homologous to calf chymosin and has the same glycosylation site (Asn_{295}–His–Ser — Porcine pepsin numbering), is said to be glycosylated[11] but the same site in calf chymosin is not glycosylated, and both sequences have the less easily glycosylated Ser residue. Chicken pepsin has two potential sites (Asn_{68}–Glu–Thr and Asn_{173}–Tyr–Thr) the first is said to be glycosylated[12] but the general rule is in favor of glycosylation at the second site.

The biological functions of the carbohydrate are many and varied, all proposed theories seem to be correct.[13] Glycosylation may have an effect on as different features as activity, recognition, folding, secretion, solubility, growth, antibody recognition, resistance to proteolytic, heat or alkaline attack etc, and when looking at one glycoprotein the role of the carbohydrate is often unpredictable. For many glycoproteins full glycosylation means full biological activity and many examples are known where carbohydrate seems to have no function especially in vitro. Glycosylation is most likely not vital to survival of the organism, if it were, then more conserved glycosylation would have been seen.

The carbohydrate has for fungi in general been found to increase secretion significantly which again may be part of the tool to regulate growth. This is probably one of the main functions of the carbohydrate on *Rhizomucor miehei* proteinase. What is the function of the carbohydrate on coagulants during cheese-making? There seems to be none or no positive effect. At least our characterization of the deglycosylated enzyme relative to the native one shows that the deglycosylated enzyme has the better properties. Therefore, it is difficult to see the need for, or advantage of, having carbohydrate on the *Rhizomucor miehei* proteinase when it performs its proteolytic action.

REFERENCES

1. Harboe, M. Coagulants of different origin. Eur. Dairy Magazine 2, 6–15, 1992.
2. Boel, E., Bech, A.-M., Randrup, K., Draeger, B., Fiil, N.P. and Foltmann, B. Primary structure of a precurser to the aspartic proteinase from *Rhizomucor miehei* shows that the enzyme is synthesized as a zymogen. Proteins: Structure, Function and Genetics 1:363–369, 1986.
3. International Dairy Federation. Bovine rennets. Determination of total milk-clotting activity. IDF Provisional Standard 157:1992.
4. PCT International Application. Microbially derived rennin having enhanced milk-clotting activity and method of producing same. WO 96/19582.
5. Gray, G.L:, Hayenga, K., Cullen, D., Wilson, L.J. and Norton, S. Primary structure of *Mucor miehei* aspartyl protease: evidence for a zymogen intermediate. Gene, 48, 41–53, 1986.
6. Aikawa, J., Yamashita, T., Nishiyama, M., Horinouchi, S. And Beppu, T. Effects of glycosylation on the secretion and enzyme activity of *Mucor* rennin, an aspartic proteinase of *Mucor pusillus*, produced by recombinant yeast. J. Biol. Chem., 265, 13955–13959, 1990.
7. Christensen, T., Woeldike, H., Boel, H., Mortensen, S.B., Hjortshoej, K., Thim, L. And Hansen, M.T. High level expression of recombinant genes in *Aspergillus oryzae*.. Biotechnology, 6, 1419–1422, 1988.
8. Zongchao, J., Vandonselaar, M., Schneider, P. and Quail, J.W. Crystallization and preliminary X-ray structure solution of *Rhizomucor miehei* aspartic proteinase. Acta Cryst. D51, 243–244, 1995.
9. Meldgaard, M. Glycosylation and stability of (1–3,1–4)-β-glucanases. Ph.D. Thesis November 1995, Carlsberg Laboratory and University of Copenhagen.
10. Foltmann, B. General and molecular aspects of rennets. Cheese: Chemistry, physics and microbiology (ed. P.F.Fox). vol 1, 37–68, 1993.
11. Baudys, M., Erdene,G., Kostka, V., Pavlik, M. And Foltmann, B. Comparison between prochymosin and pepsinogen from lamb and calf.. Comp. Biochem. Physiol. Vol. 89B, No. 2. 385–391, 1988.
12. Baudys, M., Kostka, V. Covalent structure of chicken pepsinogen. Eur. J. Biochem. 136, 89–99, 1983.
13. Varki, A. Biological roles of oligosaccharides: all of the theories are correct. Glycobiology Vol. 3 no. 2, 97–130, 1993.

41

STRUCTURE OF SECRETED ASPARTIC PROTEINASES FROM *Candida*

Implications for the Design of Antifungal Agents

Cele Abad-Zapatero,[1*] Robert Goldman,[2] Steven W. Muchmore,[1] Charles Hutchins,[3] Tetsuro Oie,[3] Kent Stewart,[3] Sue M. Cutfield,[4] John F. Cutfield,[4] Stephen I. Foundling,[5] and Thomas L. Ray[6]

[1]Laboratory of Protein Crystallography, D-46Y, AP-10, LL-07
[2]Anti-infective Group, D-47M, AP-9A
[3]Molecular Modeling Group, D-46Y, AP-10
100 Abbott Park Rd.
Abbott Laboratories
Abbott Park, Illinois 60064-3500
[4]Biochemistry Department
University of Otago
P.O. Box 56, Dunedin, New Zealand
[5]Oklahoma Medical Research Foundation
Oklahoma City, Oklahoma 73104
[6]Department of Dermatology
University of Iowa College of Medicine
Iowa City, Iowa 52242

1. ABSTRACT

Pathogens of the genus *Candida* can cause life threatening infections in immunocompromised patients. The three-dimensional structures of two closely related secreted aspartic proteinases from *C. albicans* complexed with a potent (Ki=0.17 nM) inhibitor, and an analogous enzyme from *C. tropicalis* reveal variations on the classical aspartic proteinase theme that dramatically alter the specificity of this class of enzymes. The novel fungal proteases present: i) an 8 residue insertion near the first disulfide (**Cys45–Cys50**,

* To whom correspondence should be addressed: Dr. Cele Abad-Zapatero, Department D-46Y, AP-10, LL-07, 100 Abbott Park Rd., Abbott Laboratories, Abbott Park, IL, 60064. Phone: 847-937-0294; fax: 847-937-2625; e-mail: abad@abbott.com, abad@mozart.pprd.abbott.com

Aspartic Proteinases, edited by James.
Plenum Press, New York, 1998.

pepsin numbering) that results in a broad flap extending towards the active site; ii) a seven residue deletion replacing helix h_{N2} (Ser110–Tyr114), which enlarges the S_3 pocket; iii) a short polar connection between the two rigid body domains that alters their relative orientation and provides certain specificity; and iv) an ordered 12 residue addition at the carboxy terminus. The same inhibitor (A-70450) binds in an extended conformation in the two variants of *C. albicans* protease, and presents a branched structure at the P_3 position. However, the conformation of the terminal methylpiperazine ring is different in the two crystals structures. The implications of these findings for the design of potent antifungal agents are discussed.

2. INTRODUCTION

The determination of the thee-dimensional structure of fungal aspartic proteinases from *Rhizopus chinensis*,[1] *Endothia parasitica*[2] and *Penicillium janthinellum*[3] revealed the archetypical 'bilobal' fold classically associated with pepsin and the entire aspartic proteinase family. The structures of these enzymes however did not exhaust the specificity repertoire of the family of fungal aspartic proteases. We describe in this paper a subset of fungal aspartyl proteases, those from the *Candida* family, that also exhibit the archetypal 'bilobal' fold, but possess a significant structural variation within the active site when compared with the *R. chinensis*, *E. parasitica* and *P. janthinellum* structures.

During the last decade, several members of the genus *Candida* have been associated with the rising incidence of serious, life-threatening fungal infections seen in immunocompromised and debilitated patients. Prominent members of this group are *Candida albicans*, *Candida tropicalis* and *Candida parapsilosis*. Secreted aspartic (acid) proteinases (SAPs) from these and other fungal pathogens with unusually broad substrate specificity, have been implicated as virulence factors.[4–8] Although initially *Candida* strains were believed to express a single SAP, further research has documented the existence of at least seven distinct genes in *C. albicans* grouped into two subfamilies represented by SAPs 1–3, and SAPs 4–6, with SAP7 being the most divergent in sequence.[9]

Structural studies have focused on SAP2 which is the most abundantly secreted protein in vitro when bovine serum albumin (BSA) is used as the sole nitrogen source in fermentations. The gene for SAP2 codes for a 398 residue preproprotein which is cleaved to a mature single polypeptide chain of 342 residues with a deduced molecular mass of 35,880 Da.[10]

Detailed reports of the three-dimensional structures of SAP2[11] and a closely related clinical isolate[12] (referred to as SAP2X) of the SAP2, complexed with the same potent inhibitor A-70450 (Ki of 0.17 nm) have been reported. In addition, Foundling and co-workers have reported the molecular structure of the secreted aspartic protease from *C. tropicalis* in complex with an unknown tetrapeptide tentatively identified as Thr–Ile–Thr–Ser (SAPT).[13] The structures of these new fungal aspartic proteinases have revealed novel and subtle variations on the classical aspartic proteinase fold that result on significantly different specificites. In this work, we review the common features of all these new proteinases in the context of the classical aspartic protease fold, compare the two different conformations found in the binding of A-70450 to the two variants of the *Candida* SAP2, and make some suggestions for the homology modeling of other members of the secreted aspartic protease family in other pathogenic fungi. The implications of these findings for the design of novel antifungal agents are discussed.

3. VARIATIONS ON THE ASPARTIC PROTEASE THEME

The overall architecture of the SAPs from *Candida* conforms with the classical aspartic protease fold typified by pepsin[14] and first described in detail by James and co-workers in the fungal enzyme penicillopepsin,[3] Subramanian in rhizopuspepsin[1] and by Blundell and collaborators for endothiapepsin from *E.* parasitica.[2] The superposition of the three-dimensional structure of the two liganded SAP2s from *C. albicans* with complexed pepsin (PDB entry 1PSA)[15] and the liganded structures of submaxillary gland renin (1SMR),[16] *Endothia parasitica* pepsin (1EED), rhizopuspepsin (3APR) and penicillopepsin (3APP) revealed that the closest structural homologue among the fungal enzymes is rhizopuspepsin (~295 pairs with approximately 1.5 Å rms). Porcine pepsin is the closest structural relative within the mammalian proteases with 283 C_α pairs and an rms of 1.70 Å (Table 2). As is the case with many other aspartic proteinases, the amino terminal lobe superimposes better (176 C_α pairs, rms 1.34 Å) than the carboxy lobe (116 pairs, rms 1.60 Å after overall superposition). Although the rms value is similar to the one obtained with pepsin in the amino terminal domain (176 C_α pairs, rms 1.36 Å), the corresponding one for the carboxy domain (116 pairs, rms 1.88 Å) is larger. The position of the C_α carbons corresponding to the fully conserved disulfide bond among all monomeric aspartic proteinases (**Cys249–Cys282** pepsin numbering in bold) is analogous in rhizopuspepsin and SAP2X. After overall superposition, the distance between Cys256 (SAP2X) and Cys253 (3APR) was 0.7 Å, and 1.7 Å between the structurally equivalent residues Cys294 (SAP2X) and Cys285 (3APR). The corresponding values when compared with pepsin were 1.7 and 4.1 Å respectively. However, the structures of the SAPs from *Candida* present several unique features in both the amino and carboxy lobes that put them into a different subclass among the monomeric proteinases.

The connection between the first two β–strands of the first β-sheet (**I**) (Val3–Gln9 and Ala15–Val20) is shorter due to a deletion of 2 residues. A classical γ-turn involving residue Gln11 in a unfavorable conformation (ϕ=65°, ψ=–52°) connects these two strands allowing a larger entrance into the active site

None of the members of the SAP family in *Candida* have proline at the position corresponding to the cis–**Pro23** in pepsin which forms the second β-hairpin. The β-turn presents the same number of residues in the *Candida* proteinases and the turn follows very closely the classical β-hairpin (Table 1, Fig. 1a).

The first disulfide bridge (**Cys45–Cys50**) is more common among the mammalian aspartic proteinases (pepsin, renin, chymosin and cathepsin D) than among the fungal enzymes; neither endothiapepsin nor penicillopepsin have this feature. The polypeptide chain contained within this first disulfide normally consists of a short helical or turn-like segment. The amino acid sequence of the SAPs from *Candida* have a long insertion between Cys47 and Cys59 (Table 1) and the structure of this loop is completely unique for the *Candida* enzymes among the known aspartic proteinases (Figs. 1, 2a,b). The segment between Cys47 and Cys59 forms a well defined, broad flap that extends towards the active site and defines a very wide entrance to the active site from the nonprime side (Fig. 2a,b). This flap is stabilized by hydrophobic interactions with Phe58 as well as hydrogen bonding from Gln54 to the main chain atoms (Gln54OE1...NPhe-51 2.85 Å; Gln54OE1...NThr50 2.53 Å). Since Gln54 is not conserved in SAP4–6 (replaced by Asp, Table 1; *vide infra*) it is likely that the conformation of this loop is different for various members of the *Candida* secreted aspartic protease gene family.

Table 1. Structural alignment of pepsin, rhizopuspepsin and the SAPs from *Candida*

PNO	1 7	17	27	32 37	45
1PSA	IGDEPLE	NYLDTEYFGT	IGIGTPAQDF	TVIFDTGSSN	LWVPSVYCSS
SAP4	GPVAVKL	DNEIITYSAD	ITIGSNNQKL	SVIVDTGSSD	LWVPDSNAVC
SAP5	GPVAVTL	HNEAITYTAD	ITVGSDNQKL	NVIVDTGSSD	LWIPDSNVIC
SAP6	GPVAVKL	DNEIITYSAD	ITVGSNNQKL	SVIVDTGSSD	LWIPDSKAIC
SNO	1 7	17	27	32 37	47
SAPT	SDVPTTL	INEGPSYAAD	IVVGSNQQKQ	TVVIDTGSSD	LWVVDTDAEC
SAP1	QAIPVTL	NNELVSYAAD	ITIGSNKQKF	NVIVDTGSSD	LWVPDASVTC
SAP2	QAVPVTL	HNEQVTYAAD	ITVGSNNQKL	NVIVDTGSSD	LWVPDVNVDC
SAP3	QTVPVKL	INEQVSYASD	ITVGSNKQKL	TVVIDTGSSD	LWVPDSQVSC
3APR	AGVGTVPMTD	YGNDIEYYGQ	VTIGTPGKKF	NLDFDTGSSD	LWIASTLCTN
PNO	48	50 58	68	77	87
1PSA	LA........	.CSDHNQFNP	DDSSTFEATS	QELSITYG T	GSMTGILGYD
SAP4	IPKWPGDRGD	FCKNNGSYSP	AASSTSKNLN	TPFEIKYADG	SVAQGNLYQD
SAP5	IPKWRGDKGD	FCKSAGSYSP	ASSRTSQNLN	TRFDIKYGDG	SYAKGKLYKD
SAP6	IPKWRGDCGD	FCKNNGSYSP	AASSTSKNLN	TRFEIKYADG	SYAKGNLYQD
SNO	48 57	59 67	77	87	97
SAPT	QVTYSGQTNN	FCKQEGTFDP	SSSSSAQNLN	QDFSIEYGDL	TSSQGSFYKD
SAP1	DKPRPGQSAD	FCKGKGIYTP	KSSTTSQNLG	SPFYIGYGDG	SSSQGTLYKD
SAP2	QVTYSDQTAD	FCKQKGTYDP	SGSSASQDLN	TPFKIGYGDG	SSSQGTLYKD
SAP3	QAG.QGQDPN	FCKNEGTYSP	SSSSSSQNLN	SPFSIEYGDG	TTSQGTWYKD
3APR	....	CGSGQTKYDP	NQSSTYQADG	RTWSISYGDG	SSASGILAKD
PNO	88 97	107	110	127	136
1PSA	TVQVGGISDT	NQIFGLSETE	PGSFLYYAPF	DGILGLAYPS	ISA SGATPV
SAP4	TVGIGGVSVR	DQLFANVRST	SAH.......	KGILGIGFQS	NEATRTPYDNL
SAP5	TVGIGGVSVR	DQLFANVWST	SAR.......	KGILGIGFQS	GEATEFDYDNV
SAP6	TVGIGGASVK	NQLFANVWST	SAH.......	KGILGIGFQT	NEATRTPYDNV
SNO	98 107	117		121 130	140
SAPT	TVGFGGISIK	NQQFADVTTT	SVD.......	QGIMGIGFTA	DEAGYNLYDNV
SAP1	TVGFGGASIT	KQVFADITKT	SIP.......	QGILGIGYKT	NEAA GDYDNV
SAP2	TVGFGGVSIK	NQVLADVDST	SID.......	QGILGVGYKT	NEAG GSYDNV
SAP3	TIGFGGISIT	KQQFADVTST	SVD.......	QGILGIGYKT	HEAE GNYDNV
3APR	NVNLGGLLIK	GQTIELAKRE	AASFA.SGPN	DGLLGLGFDT	ITTVRGV KTP
PNO	137 143	153	163	173	183
1PSA	FDNLWDQ	GLVSQDLFSV	YLSSNDDSGS	VVLLGGIDSS	YYTGSLNWVP
SAP4	PITLKKQ	GIISKNAYSL	FLNSPEASSG	QIIFGGIDKA	KYSGSLVDLP
SAP5	PISLRNQ	GIIGKAAYSL	YLNSAEASTG	QIIFGGIDKA	KYSGSLVDLP
SAP6	PISLKKQ	GIIAKNAYSL	FLNSPEASSG	QIIFGGIDKA	KYSGSLVELP
SNO	141 147	157	167	177	187
SAPT	PVTLKKQ	GIINKNAYSL	YLNSEDASTG	KIIFGGVDNA	KYTGTLTALP
SAP1	PVTLKNQ	GVIAKNAYSL	YLNSPNAATG	QIIFGGVDKA	KYSGSLIAVP
SAP2	PVTLKKQ	GVIAKNAYSL	YLNSPDAATG	QIIFGGVDNA	KYSGSLIALP
SAP3	PVTLKNQ	GIISKNAYSL	YLNSRQATSG	QIIFGGVDNA	KYSGTLIALP
3APR	MDNLISQ	GLISRPIFGV	YLGKAKNGGG	EYIFGGYDST	KFKGSLTTVP
PNO	184 193	201	206 210	215 223	233
1PSA	VSVE.GYWQIT	LDSITMDG	ETIACSGGCQAI	VDTGTSLLTG	PTSAIANIQS
SAP4	ITSD RTLSVG	LRSVNVMG	QNVNV..NAGVL	LDSGTTISYF	TPNIARSIIY
SAP5	ITSE KKLTVG	LRSVNVRG	RNVDA..NTNVL	LDSGTTISYF	TRSIVRNILY
SAP6	ITSD RTLSVG	LRSVNVMG	RNVNV..NAGVL	LDSGTTISYF	TPSIARSIIY
SNO	188 195	198 205	210	218 226	236
SAPT	VTSS VELRVH	LGSINFDG	TSVST.N ADVV	LDSGTTITYF	SQSTADKFAR
SAP1	VTSD RELRIT	LNSLKAVG	KNIN..GNIDVL	LDSGTTITYL	QQDVAQDIID
SAP2	VTSD RELRIS	LGSVEVSG	KTINT.DNVDVL	LDSGTTITYL	QQDLADQIIK
SAP3	VTSD NELRIH	LNTVKVAG	QSINA.D.VDVL	LDSGTTITYL	QQGVADQVIS
3APR	IDNSRGWWGIT	VDRATVGT	STVA SS FDGI	LDTGTTLLIL	PNNIAASVAR

Table 1. (*Continued*)

PNO 234	241	249	260	270	279	
1PSA	DIGASE..NS	DGEMVIS C	SSIDSLPDIV	FTINGVQYPL	SPSAYILQDD	
SAP4	ALGGQVHYDS	SGNEAYVADC	KTSGTVDFQF	DRNLKISVPA	SEFLYQLYYT	
SAP5	AIGAQMKFDS	AGNKVYVADC	KTSGTIDFQF	GNNLKISVPV	SEFLFQTYYT	
SAP6	ALGGQVHFDS	AGNKAYVADC	KTSGTVDFQF	DKNLKISVPA	SEFLYQLYYT	
SNO 237	246	256	266	276	286	
SAPT	IVGATWDSR	NEIYR LPSC	DLSGDAVFNF	DQGVKITVPL	SELILK....	
SAP1	AFQAELKLDG	QGHTFYVTDC	QTSGTVDFNF	DNNAKISVPA	SEFTAPLSYA	
SAP2	AFNGKLTQDS	NGNSFYEVDC	NLSGDVVFNF	SKNAKISVPA	SEFAASLQGD	
SAP3	AFNGQETYDA	NGNLFYLVDC	NLSGSVDFAF	DKNAKISVPA	SEFTAPLYTE	
3APR	AYGASDNGDG	TYTIS C	DTSAFKPLVF	SINGASFQVS	.PDSLV.F.E	
PNO	282 285 289		299		310	320 326
1PSA	DS CTS	GFEGMDVPTSSGELW		ILGDVFIRQY	YTVFDRANNK	VGLAPVA
SAP4	NGEPYPKCEI	RVRES	EDN	ILGDNFMRSA	YIVYDLDDRK	ISMAQVK
SAP5	SGKPFPKCEV	RIRES	EDN	ILGDNFLRSA	YVVYNLDDKK	ISMAPVK
SAP6	NGKPYPKCEI	RVRES	EDN	ILGDNFMRSA	YIVYDLDDKK	ISMAQVK
SNO 287	294		304	314	324	331
SAPT	DSDSSICYF	GISRN	DAN	ILGDNFLRRA	YIVYDLDDKT	ISLAQVK
SAP1	NGQPYPKCQL	LLGIS	DAN	ILGDNFLRSA	YLVYDLDDDK	ISLAQVK
SAP2	DGQPYDKCQL	LFDVN	DAN	ILGDNFLRSA	YIVYDLDDNE	ISLAQVK
SAP3	DGQVYDQCQL	LFGTS	DYN	ILGDNFLRSA	YIVYDLDDNE	ISLAQVK
3APR	EF QGQ CIA	GFGYGNW	GFA	IIGDTFLKNN	YVVFNQGVPE	VQIAPVAE
PNO						
1PSA	......	28%				
SAP4	YTSQSNIVGIN	50%				
SAP5	YTSESDIVAIN	50%				
SAP6	YTSESNIVAIN	51%				
SNO 332	342					
SAPT	YTSSSDISALT	—				
SAP1	YTSASNIAALT	56%				
SAP2	YTSASSISALT	63%				
SAP3	YTTASNIAALT	62%				
3APR	......	26%				

PNO denotes the pepsin numbering system (residues renumbered 1–326); abbreviations as in the text.
SNO denotes the sequential numbering system of the SAPs based on SAP2 (342 residues number 1–342); for clarity SAP2X has been omitted.[12]
Percentage numbers indicate percent identity with SAPT.

Table 2. Overall structural superposition of pepsin, rhizopuspepsin, and the SAPs from *Candida*

	SAPT	1PSA	SAP2X	3APR	SAP2
SAPT	—	1.62 (271)	0.87 (327)	1.52 (300)	0.80 (324)
1PSA		—	1.68 (278)	1.55 (296)	1.67 (282)
SAP2X			—	1.49 (292)	0.52 (333)
3APR				—	1.51 (291)
SAP2					—

SAPT: Secreted Aspartic Proteinase from Candida tropicalis. PDB entry 1PCT.[13]
1PSA: Pepsin/A62095 complex.[15]
SAP2X: Secreted Aspartic Protease from clinical isolate (1ZAP).[12]
3APR: Rhizopus/inhibitor complex.[25]
SAP2: Secreted Aspartic Proteinase from *Candida albicans* (1EAG).[11]
Each entry of the table is the root mean square deviation in Å between the structures denoted by the corresponding row and column of the matrix. The integer number in parentheses specifies the number of C_α pairs resulting in superposition as described in Materials and Methods. Adapted from reference 12.

The specificity displayed at the P_3 position by several members of the aspartic proteinase family resides in the residues in the S_3 subsite extending from helix hN2 (pepsin **Ser110–Tyr114**).[16] In the structure of the SAPs from *Candida*, a seven residue deletion in this region dramatically alters the enzyme pocket S_3 in all the members of this secreted gene family[9] (Table 1, Figs. 2a,b, 3). The major effect is the expansion of the physical size

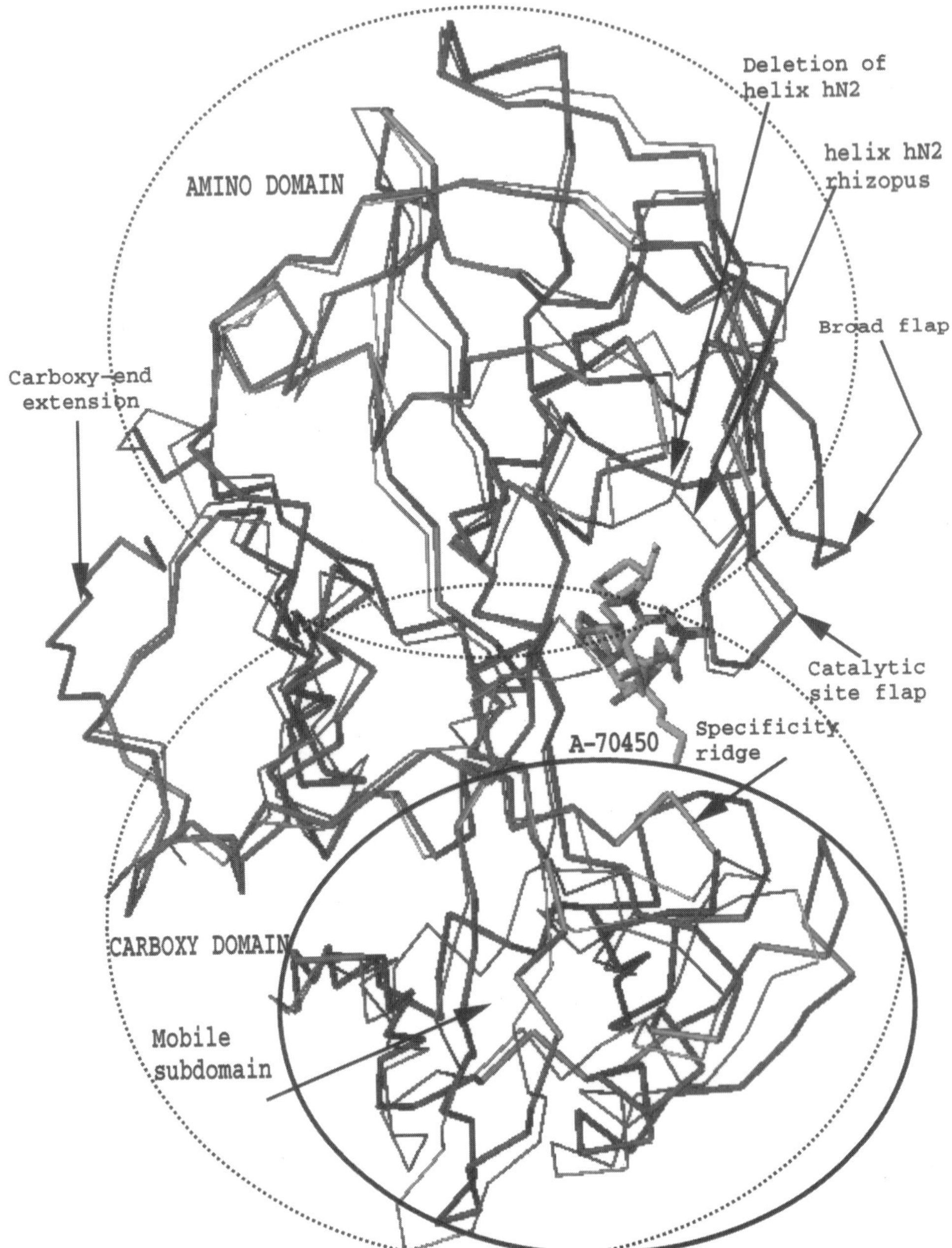

Figure 1. C_α tracing of SAP2X (thick) with the bound inhibitor A-70450 compared to the tracing of rhizopuspepsin (thin, 3APR) highlighting the features discussed in the text.

of this pocket giving rise to a funnel-like active site cavity with the wide opening in the nonprime side of the enzyme, which narrows down as it approaches the catalytic aspartates at the core of the active site.

The amino acid sequence of all the members of the SAP(1–6) gene family from *C. albicans* and SAPT from *C. tropicalis* have 342 amino acid residues. Sequence alignment would predict a carboxy terminal extension relative to the prototypical fold of unknown conformation. In the structures of the SAPs described thus far, these additional residues 'stitch' around the 176–182 region that connects the amino- and carboxy domains and is stabilized by hydrogen bonds and a β-sheet like conformation (Fig. 2a,b, 3). This extended tail has been found well ordered in the three structures documented to date[11,12]; yet, it is unknown what function this extended tail plays *in vivo*.

In relation to the other aspartic proteinases, a deletion of 4–7 residues in the loop connecting Phe298 to Ile305 contributes to a different orientation of the two different rigid body domains[17,18,14] and forms a 'specificity ridge' with the amino acid residues Asn301, Ala303, Ile305 in the three SAPs discussed here. There is considerable variation among these residues among the different SAPs from *Candida* (Table 1).

A summary of the differences between the structure of the SAPs and the cannonical pepsin fold have been presented in a pictorial fashion in Fig. 1 and can be seen in detail in Fig. 2a,b.

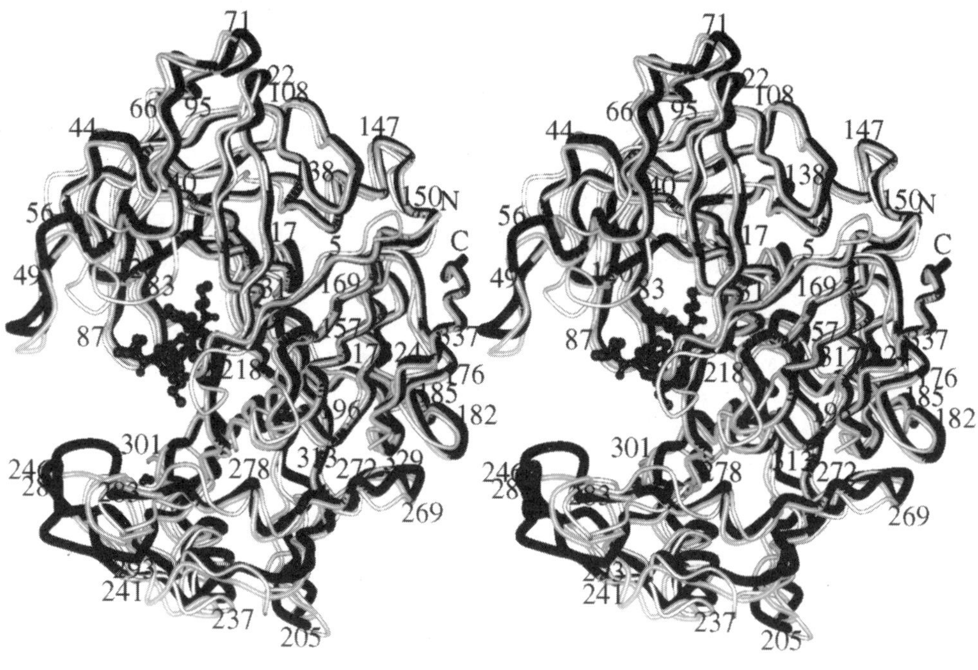

Figure 2. Stereo diagram of the polypeptide backbone fold for the SAPs from *Candida* superimposed to the rms given in Table 2 with porcine pepsin (1PSA) and its closest homolog rhizopuspepsin (3APR). a) View of SAP2X (black), SAPT (dark grey) and A-70450 bound in the funnel-like active site, and compared to rhizopuspepsin (light grey); b) SAP2 (black), SAPT (light grey) superimposed with porcine pepsin (dark grey) with the ligand A-70450 removed in a view emphasizing the differences in the cannonical aspartic proteinase fold that enlarge the active site cavity in the nonprimed side. The wavy lines in the active site identify the location of the inhibitor in SAP2 and the tetrapeptide Thr–Ile–Thr–Ser in SAPT. The N- and C-termini for SAP2s have been identified and residues numbers have been intercalated along the polypeptide chain.

4. STRUCTURAL DIFFERENCES AMONG SAP2, SAP2X, AND SAPT

As expected, the refined structures of the complexes SAP2 and SAP2X with A-70450 are very similar. After least squares superposition the rms deviation between 333 pairs C_α carbons is 0.52 Å. The amino acid sequences are highly homologous (~96% identity) and the only major variations on the polypeptide structure are located in the regions of weak electron density where the chain is highly mobile.[10] In addition, eight single amino acid replacements are well defined in the electron density maps of the refined structure and can be reassigned with confidence: Val45→Ile45, near the first disulfide; Lys81→Ser on the enzyme surface; Ala164→Ser near the inter-domain hinge; Glu278→ Asp and Leu283→Thr on the well-defined portions of the variable loops; Asp302→Lys, Leu327→Ile and Ile338→ Thr near the carboxy terminus. Furthermore, residues Ala177, Lys206, Asn257 and Ala280 of the SAP2 sequence have been refined as Gly due to the lack of side chain density beyond the C_α carbon. In relation to the SAP2 gene product, the most striking difference is the deletion of residue Phe251. Although this residue is strictly conserved in the SAP1–3 genes it is substituted by Ala or Val in SAP4–6 and absent in SAP7.[9] The definitive amino acid sequence in SAP2X cannot be assigned with confidence in the regions of the carboxy lobe where the electron density is very weak and the polypeptide chain appears to be disordered, namely Ser245–Asn250 and Gln284–Tyr291; in these areas the amino acid sequence of SAP2 from strain ATCC 10261[10] has been assumed. The structural superposition of the two enzymes is presented in Figs. 2a,b, 3 and the corresponding sequence alignment in Table 1.

SAPT is indeed more distant than the two SAP2 variants from each other. The rms deviation between 327 C_α carbon pairs is 0.87 Å. In the amino terminal domain, the differences seem to be more dramatic because of two factors. First, although the conformation is essentially the same, the insertion between the first disulfide pair (Cys47–Cys59) is slightly displaced in SAPT with respect to the position it occupies in the structures complexed with A-70450. This portion of the polypeptide chain is less rigid in SAPT probably due to the lack of an effective ligand on the nonprime side of the active site or possibly due to the absence of any crystal contacts. A second significant conformational difference in the amino-terminal domain is after Gly134 where a tyrosine residue has been inserted (Tyr135). This alteration is concomitant with the replacement of Leu94 in the SAP2s by a bulkier Phe94 in the β-strand immediately after the β-hairpin of the flap. The end result is the looping out of the polypeptide chain four residues prior to the conserved helix comprising 140–147.

SAPT also differs quite markedly from SAP2 or SAP2X in the nonprime, wide, peripheral part of the funnel shape binding cleft. In relation to SAPT, there are insertions in the amino acid sequence of the SAP2 enzymes at the 'mobile loops' (244–248 and 284–290) resulting in different surface characteristics in the walls lining the binding site. The electron density for the regions corresponding to these shorter loops in SAPT is very well defined even in the absence of crystal contacts.[13] Whether the size, their extension towards the active site, and mobility of these loops are determinants of specificity in SAPT is still an open question (Fig. 3).

At the carboxy terminus, the flexible domain (residues **190–301**) is rotated approximately 4.4° and displaced by 0.7 Å in relation to the position it occupies in rhizopuspepsin (3APR). This is comparable to the value obtained for the two variants of SAP2 and is consistent with the enviroment of the interdomain interface found in all of these enzymes and discussed elsewhere[12] (Table 1, Figs. 2a,b, 3).

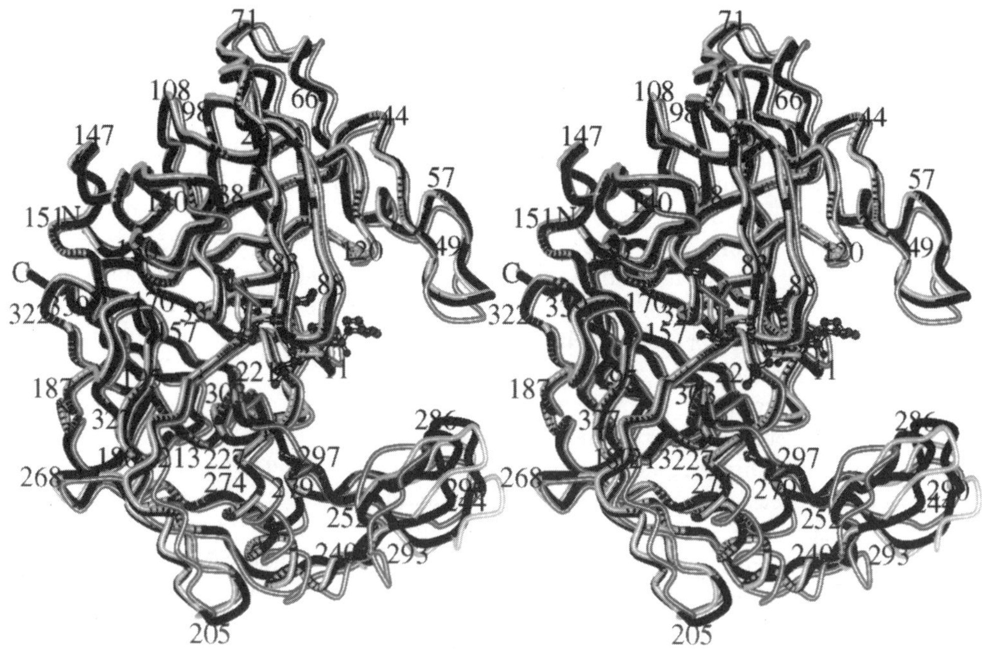

Figure 3. Stereo diagram of the polypeptide backbone fold for the SAPs from *Candida* superimposed to the rms given in Table 2 to emphasize the common features summarized in Fig. 1. SAP2X (black), SAP2 (light grey) and SAPT (dark grey). The conformation of the ligand A-70450 corresponds to the one found in the crystals of SAP2X. The N- and C- termini for SAP2X have been identified and residues numbers have been marked along the polypeptide chain.

5. IMPLICATIONS FOR THE STRUCTURE OF OTHER SAPS FROM DIFFERENT FUNGI

The grouping of different SAPs from different *Candida* species in two separate families based on the amino acid sequence has been presented by Monod and coworkers.[9] The sequences showed a wide range of sequence identity ranging from 53% between the different families (SAP2/SAP5) to 70% between members of the same family (SAP2 /SAP3). In terms of their amino acid sequence, the different SAPs from *C. albicans* could be clustered into two separate families with three members each (SAP1–3 and SAP4–6) with the longest and most divergent SAP7 being closer to the other *Candida* species. The structural alignment between the clinical isolate SAP2X and the more homologous SAP1–6 proteinases has been presented previously.[12] Table 1 shows the structural aligment between SAP1–6 and the secreted proteinase from *C. tropicalis* (SAPT) in relation to pepsin.

Although similar, the sequences in the vicinity of the unique structural features of the SAPs described above are distinctly different between SAP1–3 and SAP4–6. For instance, the sequence between the first disulfide bond (Cys47–Cys59) retains Phe58 as fully conserved but the intervening amino acids clearly segregate into two groups. The SAP1–3 enzymes have Thr50 and Gln54 providing hydrogen bonding interactions to support the rigid conformation of the loop. However, in the other family (SAP4–6) the replacements Thr50→Lys and Gln54→Asp could suggest the role of an electrostatic interaction to retain a rigid loop conformation (Table 1). The deletion of helix hN2 will probably be observed in all the SAPs although the flanking sequences clearly fall into two separate groups (Table 1).

From the viewpoint of specificity in the domain interface a clear dichotomy also exists between the SAP1–3 and SAP4–6. Although Asn304 and Ile305 are strictly conserved among the six SAPs, the sequence of residues 301 and 303 segregates the family basically into two groups. Residue 303 is Ala in SAP1–2, Tyr in SAP3, but is replaced uniformly by Asp in SAP4–6; Asn301 is also replaced by Ser in all SAP4–6 and for SAP1 and SAP3. More uniform and intriguing are the alterations at the 185–195 region where SAP1–3 preserve the characteristic sequence Glu193–Leu194–Arg195, probably critical to force the conformation of Asn304, while the SAP4–6 present either Thr–Leu–Ser or Lys–Leu–Thr. Whether or not this latter sequence will also alter the conformation of Asn304 is still unknown.

The carboxy terminal extension (residues 327–342), thus far unique to this family of enzymes, will probably have the same three dimensional conformation in the SAP1–6 enzymes due to the high sequence homology of the region of the polypeptide chain comprising Gln329–Ser336 (QVKYTS) (Table 1). This hypothesis is also supported by the existence of an identical extension in the structure SAPT from *C. tropicalis*. Since the sequence of *C. parapsilosis* is identical in this region, it is most likely that the carboxy terminal extension will have a very similar three-dimensional conformation.[9]

6. INHIBITOR CONFORMATION

In both the SAP2 and SAP2X structures, the electron density corresponding to the inhibitor A-70450 was readily interpretable as an extended conformation[11] illustrated in Fig. 4a. This extended conformation has a unique branched structure at the P3 position. Our discussion below will first identify the major points of similarity in the orientation of A-70450 as found in the SAP2 and SAP2X structures, and secondly identify a point of difference, found at the P3 branch position.

Compound A-70450 was originally prepared as part of the Abbott renin inhibition program (US patent 5,164,388) and was later discovered to inhibit SAP2Xs from *C. albicans* (US patent 5,120,718). It possesses a hydroxyethylene peptide bond isostere as a transition state mimic with the hydroxy group located between Asp32 and Asp218 (pepsin **Asp32** and **Asp215**). The inhibitor conformation in the P_2, P_1, P_1' and P_2' positions has been discussed previously[11] and is similar to that found in other studies of peptide-based inhibitors.[19] The SAP2 and SAP2X structures are in full agreement for the observed orientation of the P_2–P_2' portion of the inhibitor.

In the P_3 unit, the backbone structure of A-70450 differs significantly from previous peptide-based aspartic proteinase inhibitors. The P_3 amide is cyclized into a ketopiperazine ring. The benzylic substituent projecting from the ketopiperazine ring, formally a phenylalanine unit, is in the unnatural R-configuration. Even with the opposite configuration, this benzyl group occupies a traditional S_3 pocket, due to the backbone cyclization of this P_3 unit. This ketopiperazine ring is oriented similarly within the SAP2 and SAP2X crystal structures. The non-peptide character of the inhibitor A-70450 becomes fully evident in understanding the orientation of the terminal N-methyl piperazinyl urea unit. *A priori*, it is not obvious if this terminal unit might mimic a P_4 substituent or not. In fact, no portion of A-70450 is observed to occupy the S_4 subsite in either of the SAP2 or SAP2X structures. Instead, the N-methyl piperazine unit is directed toward the S_3 subsite in close contact with the Cys47–Cys59 loop. This S_3 subsite of SAP2 and SAP2X (and presumably of the entire SAP family as discussed above) is unusual within the aspartic proteinase family due to its large size. While both the SAP2 and SAP2X structures place the terminal unit of A-70450 within this enlarged S_3 pocket, the two structures differ in the direction and conformation of the piperazine ring. In the SAP2 structure the piperazine ring is observed to be in a chair conformation and directed so that the N-methyl group is in contact with the side chain of Ser88 (in the flap) and also the carbonyl group of Ser118 (in the pocket). Asp86 in the flap makes van de Waals contacts with one side of the ring itself, but overall there are few contacts stabilising the methylpiperazine group. In the SAP2X structure, the piperazine ring is observed to be in a twist-boat conformation and directed so that the N-methyl group is only near a chain of five water molecules (*vide infra*). The nearest approach to the protein atoms is to the hydroxyl of Tyr51 (4.2 Å to one of the carbons α to the nitrogen and 4.9 Å to the N-methyl). Close analysis of the electron density of each respective structure confirms this placement of the inhibitor structure consistent with the lack of detection of the alternate conformation. Thus, currently available data suggest that two conformations of this terminal unit of A-70450 exist with the observed orientation being influenced by either subtle differences in the protein amino acid sequence between SAP2 and SAP2X, or differences in the crystallization conditions. None of the positions of sequence difference between SAP2 and SAP2X discussed earlier are close to the S_3 pocket of the active site. Therefore, we favor the idea that differences in the crystallization conditions induce different inhibitor conformations. The SAP2 protease was crystallized at pH 4.5 in the tetragonal space group $P4_32_12$ while the SAP2X variant was crystallized at pH 6.5 in the monoclinic space group $P2_12_12_1$. It should be pointed out that the inhibitor A-70450 bears a basic N-methyl nitrogen within the piperazine ring. We do not have experimental data reporting the pKa of this nitrogen, but estimate its pKa at approximately 7 (±1), or close to neutrality. Under the SAP2X crystallization conditions, it is reasonable to assume that both protonated and nonprotonated forms of A-70450 exist in solution during the crystallization. Under the SAP2 conditions, the protonated form of A-70450 will likely predominate in solution.

Relevant to the inhibitor conformation in the P3 position are the water molecules present in the proximity of the S_3 pocket. In the SAP2X structure, a well-defined chain of five water molecules bridges Thr50 to Gln121 *via* hydrogen bonds (OG...2.72 Å-W702, W735, W632, W691, W614–2.7 Å....OE1). The first water (W702) is held the most tightly with additional hydrogen bonds to Gln54 (NE2...2.7 Å) and to the carbonyl oxygen of Ser118 (2.8 Å). In the middle of this chain, W632 and W735 are hydrogen-bonded to Asp120OD2 (Fig. 4b). This chain occupies part of the volume gained by the deletion of the hN2 helix in the cannonical aspartic proteinase fold.

In SAP2 and SAPT a water molecule analogous to W702 is found (W31, SAP2; W378, SAPT) approximately equidistant from Gln54 and Thr50. The remaining water chain differs in SAP2 because the twisted-boat conformation of the piperazine ring sterically hinders the presence of additional atoms (Fig. 4b). The SAP2 water chain consists of waters W31, W121, W122 and W123. These waters exhibit a different hydrogen bonding pattern from SAP2X and probably have low occupancies ($B > 50$ Å^2). In particular, W122 is hydrogen-bonded to the basic piperazine nitrogen *via* its one pair of electrons. In contrast, the protonated n-acylpiperazine nitrogen is interacting with the carbonyl group of the urea in SAP2X. A water chain also exists in the SAPT structure between residues Thr50 and Gln121 with waters W378, W396, W397 and W466. In summary, it is interesting that a water chain within the S_3 pocket exists in each of the SAP2, SAP2X, and SAPT structures, although the exact water positions and relative occupancies (as measured by

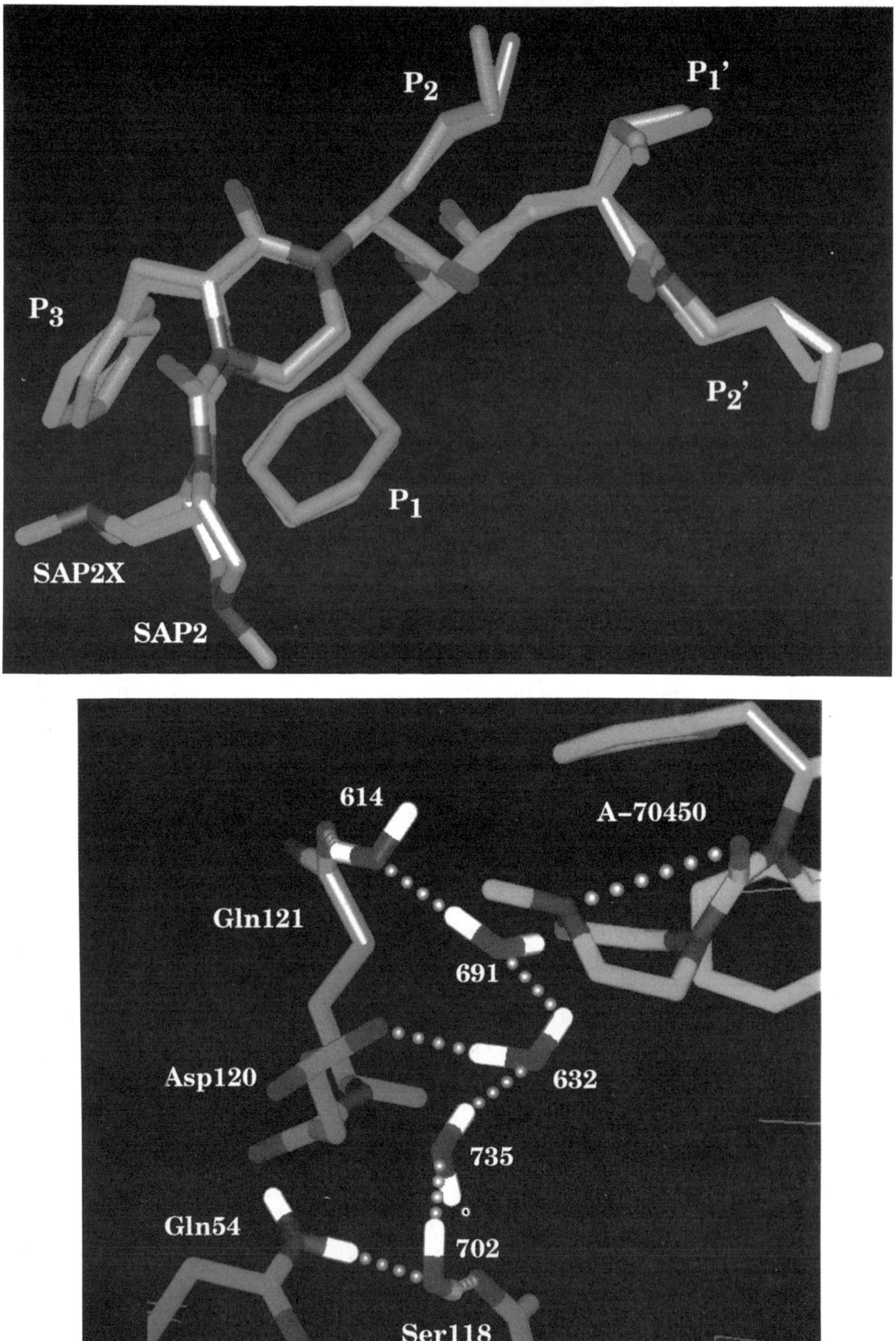

Figure 4. View of the two different conformations of the inhibitor in the complexes of A-70450 and the secreted aspartic proteinases from *C. albicans*. a) Overall view of the inhibitor. Notice the branching of the inhibitor structure into two subgroups P_{3a} and P_{3b} and the different conformations of the methylpiperazine moiety as well as the surrounding residues. b) View emphasizing the chain of five waters found in the vicinity of the methylpiperazine ring of SAP2X bridging residues Thr50 and Gln121 and occupying part of the volume corresponding to the deletion of helix hN2 in pepsin (Fig. 1).

the crystallographic B-values) differ in each structure due to the presence (absence) of an inhibitor in two different conformations.

Although we do not fully understand the interplay of ring conformation and water structure in determining the observed orientation of A-70450, we suggest that differences in the crystallization conditions have had an impact on the A-70450 conformation. It is conceivable that the protonation state of Asp120 in the two different pHs affects the microenvironment near the piperazine ring, favoring a certain hydrogen bonding network of water molecules at near neutral pH (SAP2X). The arrangement of hydrogen bonds may then affect the selection of the piperazine conformation of the inhibitor.

It is unsual to observe a twist-boat conformation for a 6-membered ring system, so we carried out *ab initio* quantum mechanics calculations to see if such a conformation could be accessed. We investigated the chair/twist-boat conformational preference for both the protonated and unprotonated forms of N-methyl-N′-(N,N-dimethylcarbamyl)piperazine. The results are schematized in Fig. 5. As expected, the unprotonated piperazine ring prefers the chair form with a energy difference of 5 kcal/mol. Interestingly, the protonated piperazine prefers the twist-boat form with an energy difference of 7 kcal/mol in part due to an intramolecular hydrogen bond between the protonated ring nitrogen and the carbonyl group of urea. These calculations correspond to a gas-phase situation, and solvent and/or enzyme active site residues would be expected to influence the degree and direction of conformational preference. With these caveats, we conclude that our calculations support the existence of both the twist-boat and chair forms of N-alkyl-N′-acylpiperazines under some set of conditions. We propose that these theoretical results are in accord with the two observed conformation alternatives as described above in the SAP2 and SAP2X crystal structures.

It is striking that the location where multiple conformations of the inhibitor exists (the P3 site) corresponds to an active site location of uniqueness for the *Candida* proteases (the S_3 site). The fact that these two different conformations fit into the active site pocket is a reflection of the large open volume on the nonprimed site of the active site for this family of enzymes.

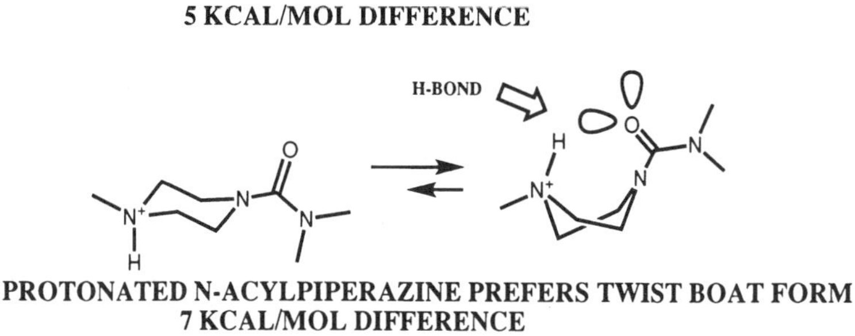

Figure 5. Conformational preferences of N-Acyl piperazines. Summary of the Quantum Mechanical studies discussed in section 6. (Inhibitor conformation). See Methods for details.

7. SUBSTRATE SPECIFICITY

Currently our knowledge of substrate sequence preference for *Candida* aspartic proteinases derives from only two published reports. One of our groups[20] published an extensive study of 30 chromogenic peptides containing a Phe–Phe (p–NO_2) scisile bond with both SAP-Strain C-74 and SAPT. This work demonstrated that the substrate specficity for cleavage is extremely broad for residues in the P_3, P_2, P_2' and P_3' positions. In another report, the SAP2X enzyme was shown to catalyze a cleavage at a His–Thr bond in a fluorogenic substrate.[21] For both the SAP and SAPT enzymes, the S_3– S_3' pockets are comprised of a mixture of hydrophilic and hydrophobic residues, with potentially subtle interactions determining residue preference. One interesting observation relevant to the discussion above is the accomodation of a large variety of residues in the P_3 position, reflecting the large S_3 binding site. Importantly however, both the SAPT and SAP enzymes strongly reject Asp residues at P_3. For SAPT (S.I. Foundling, private communication) have attributed this rejection to an electrostatic repulsion from Asp120 in SAPT. A corresponding Asp exists in SAP2 at position 120, and therefore SAP2 may be expected to exhibit a similar rejection of Asp residues in P3. A more detailed discussion of the SAPT substrate preference is available elsewhere in these proceedings.[22] Unfortunately, the lack of suitable amounts of SAP1–7 enzymes has precluded comprehensive studies of substrate specifity to date. With cloned enzymes now available in potentially larger quantities, we look forward to comparative and quantitative studies of substrate hydrolysis with all of the *Candida* enzymes. It is conceivable that the broad specificity at the P_3 position is a unique property of the secreted aspartic proteinases of this genus for the purpose cleaving a large variety of proteinaceous substrates.

8. IMPLICATIONS FOR THE DESIGN OF NOVEL ANTIFUNGAL AGENTS

Antifungal activity has been tested (systemic mouse model) for both pepstatin and A-70450 and the results were discouraging: either weak or no anti-*C. albicans* activity was detected *in vivo*. This lack of activity is possibly due to lack of potency, low specificity between fungal vs. host aspartic proteinases, and/or the inability to inhibit all members of the SAP family, or at least those SAPs most important for virulence.

The three-dimensional structure of three fungal aspartic proteinases from pathogens of the genus *Candida* can provide a completely new framework for the design of novel antifungal agents. There are specific residue replacements throughout the active site which differentiate the *Candida* fungal SAPs from mammalian aspartyl proteases such as renin, pepsin, and gastricsin. A focused chemical design effort centralizing on these specific changes might yield a clinically useful antifungal agent possessing the specificity required to be selective at impairing fungal virulence while not affecting mammalian function. More important, in our opinion, is the observation of a potentially much greater source of specificity for designing antifungal chemotherapy-the prominently large S_3 subsite which appears to be conserved in a large portion of the SAPs. Here, one can imagine entirely new structural units for the P_3 portion of aspartic proteinase inhibitors which would fill the S_{3a} and S_{3b} subsites with novel connectivities. Moreover, the S_4 site in the SAPs is currently unexplored by A-70450 suggesting that additional functional groups might bind in S_4 and improve potency.

In addition, as a last and more speculative point, it is intriguing to consider the possibility of exploiting the dual conformation of A-70450 found in the two different crystal forms as basis for the design of very large substituents at the P_3 position. It is conceivable in principle, that an inhibitor which combines both conformations of A-70450 and fully occupies the enlarged S_3 site will prove to be an extremely potent and selective inhibitor of the *Candida* SAPs.

These chemical ideas, in combination with homology modeling of the different SAPs, could provide the basis for the design of novel anticandidal agents based on two different strategies: i) broad specificity, focusing on residues in the neighborhood of the active site that are common to several SAPs; ii) narrow specificity, focusing on the homology modeling of a designated SAP from a specific pathogen. The prophylactic and therapeutic use of such agents would be a valuable addition to the limited repertoire of antifungal agents currently available. We look forward to further developments along this line.

9. METHODS

9.1. Crystallization

Full details of the protein isolation, purification, characterization and crystallization protocols of the three secreted aspartic proteases discussed in this work have been published elsewhere.[12] For the purposes of the discussion the crystallization conditions for the two complexes of SAP2 with A-70450 are summarized here.

Form A crystals of SAP2Xwere obtained in the presence of 40 mM Cacodylate Buffer pH=6.5, 5–10 mM Zn^{++} and approximately 22% PEG 8000. The crystals grew as thin plates in high concentrations of Zn^{++} (~10 mM) but developed a more uniform prismatic habit at low concentrations (~5mM) of the cation. Crystals appeared after a few days and reached full size in a few weeks. These crystals were characterized as triclinic containing eight molecules of the complex in the unit cell.[12]

Form C crystals of SAP2X were only obtained in conditions similar to form B but at low Zn^{++} concentration (~1 mM). These crystals were prismatic in habit and took more than 2 months to grow. Unfortunately, these crystals were difficult to reproduce. These crystals were orthorhombic ($P2_12_12_1$) and contained only one molecule in the asymmetric unit (Table 3).

The presence of A-70450 in the crystallization media was critical for obtaining well diffracting crystals of SAP2; pepstatin did not bind tightly enough to prevent autocatalytic cleavage at pH<6. Crystals of SAP2 from *C. albicans* strain (ATCC10261) were grown by the hanging drop method using 4 μl drops and containing 10 mg/ml of protein, in 0.1 M acetate buffer (pH 4.6) at an ammonium sulphate concentration of 2.35 M and with 0.5 M sodium chloride. Crystals grew over a period of days to weeks (up to 0.6 mm in length) as tetragonal bipyramids ($P4_32_12$, 1 mol a.u) and diffracted beyond 2.2 Å using CuKα radiation from a rotating anode X-ray source[23] (Table 3).

Crystals of SAPT were grown by the hanging drop method using a reservoir containing 0.1 M sodium acetate buffer, 20% ethanol at an approximate pH of 5.0 and 20°C. The crystals were orthorhombic ($P2_12_12_1$, 1 mol a.u) and although significant diffraction was observed to 1.4 Å on a conventional rotating anode, the working data set extended only to 1.8 Å (Table 3).[13]

9.2. Structure Determination and Refinement

Full details of the structure determination and refinement can be found in the corresponding references for SAP2,[11] SAP2X[12] and SAPT.[13] A summary of the crystallographic data for the three SAPs is given on Table 3.

Table 3. Summary of crystallographic data[1]

	SAP2	SAP2X	SAPT
Space group	$P4_32_12$	$P2_12_12_1$	$P2_12_12_1$
a (Å)	76.2	99.6	50.0
b (Å)	76.2	50.0	51.4
c (Å)	126.1	64.9	128.9
r.m.s.d bonds (Å)	0.012	0.009	0.011
r.m.s.d. angles (°)	1.97	1.66	1.76
Number of solvent molecules	123	189	231
Observations	182,260	30,743	215,435
Unique reflections	24,302	10,193	32,636
Completeness (%)	98.0	89.4	93
Rsym (%)[2]	7.2	9.9	6.8
R-factor	.186	.145	.183
Resolution (Å)	20–2.05	8–2.5	40–1.8

[1]Data summarized from references 11–13.
[2]Rsym= $\Sigma \mid I - \langle I \rangle \mid / \Sigma I$, where I is the observed intensity and $\langle I \rangle$ is the average intensity obtained from multiple of observations possibly from different crystals.

9.3. Structural Superpositions

Structural superpositions were performed using the least square superpositions routines of the package O[24] using a cutoff limit for improvement (Lsq_improve) of 3.8 Å and a consecutive stretch of three amino acid residues for the smallest fragment that can be aligned. All the superpositions were allowed to converge for these superposition parameters.

9.4. Quantum Mechanical Calculation

Calculations were carried out at the MP2/6–31G* level with total geometry optimization using the Gaussian 94 program. The estimated accuracy for the relative conformational energy diffferences is approximately ± 2 kcal/mol.

ACKNOWLEDGMENTS

We appreciate the assistance of the personnel of the research computing department at Abbott for their support during the computations required for this work. The expert assistance of Mr. John Capobianco and Ms. Jean Severin is appreciated during the process of purification, characterization and crystallization of the enzyme. We acknowledge the dedication of the members of the renin project: Drs. W. Baker, H. Stein and Mr. H.-S. Jae. Portions of this work were supported by NIH grant AI24344 (T. L. Ray). Research on SAP2 was supported by the Health Research Council of New Zealand, the N.Z. Lottery Grants Board, and the Otago Medical School Bequest Funds (S.M. Cutfield J. F. Cutfield). The SAPT work was supported by grants from the NIH and the Oklahoma Center for Applied Science and Technology to S. I. F. C. A.-Z. expresses his gratitude to S. M. Cutfield, J. F. Cutfield and S. I. Foundling for allowing him to present part of their work at the Banff VII International Aspartic Proteinases Conference.

REFERENCES

1. Subramanian, E.I.D.A.S., M. Lie, D.R. Davies, J.A. Jenkins, I.J. Tickle & T. L. Blundell. *Proc. Natl. Acad. Sci. USA* **74**, 556–559 (1977).
2. Tang, J., James, M.N.G., Hsu, I.N., Jenkins, J.A. & Blundell, T.L. *Nature* **271**, 618–621 (1978).
3. Hsu, I., Delbaere, L.T.J., James, M.N.G. & Hoffmann, T. *Nature* **266**, 140–145 (1977).
4. Douglas, L.J. *Crit. rev. Biotechnol.* **8**, 121–129 (1988).
5. Ray, T.L., Payne, C.D. & Morrow, B.J. in *Aspartic Proteinase Conference on Structure and Function of the Aspartic Proteinases: Genetics, Structures, and Mechanisms* (ed. Dunn, B.M.) 173–183 (Plenum Press, Sonoma County, California, U.S.A., 1990).
6. Cutler, E. *Ann. Rev. Microb.* **45**, 187–218 (1991).
7. Ruchel, R., de Bernardis, F., Ray, T.L., Sullivan, P.A. & Cole, G.T. *J. Med. vet. Mycol* **30**, 123–132 (1992).
8. White, T.C., Kohler, G.A., Miyadaki, S.H. & Agabian, N. *Can. J. Botany* **73** (suplement 1), *1058–1064* (1995).
9. Monod, M., Togni, G., Hube, B. & Sanglard, D. *Mol. Microbiol.* **13**, 357–368 (1994).
10. Wright, R.J., *et al. J. Bacteriol.* **174**, 7848–7853 (1992).
11. Cutfield, S.M., E.J. Dobson, B.F. Anderson, P.C.E. Moody, C.J. Marshall, P.A. Sullivan, & J. F. Cutfield. *Structure* **3**, 1261–1271 (1995).
12. Abad-Zapatero, C., R. Goldman, S. W. Muchmore, C. Hutchins, K. Stewart, J. Navaza, C. D. Payne, & T.L. Ray. *Protein Science* **5**, 640–652 (1996).
13. Symersky, J., M. Monod, & S. I. Foundling. *Biochemistry*. submitted (1997).
14. Sielecki, A.R., Fedorov, A.A., Boodhoo, A., Andreeva, N.S. & James, M.N.G. *J. Mol. Biol.* **214**, 143–170 (1990).
15. Chen, L., *et al. Acta crystallogr.* **B48**, 476–488 (1992).
16. Dealwis, C.G., *et al. J. mol. Biol.* **236**, 342–360 (1994).
17. Sali, A., *et al. Proteins: Structure, Function, and Genetics* **12**, 158–170 (1992).
18. Abad-Zapatero, C., Rydel, T.J. & Erickson, J.W. *Proteins: Structure, Function and Genetics* **8**, 62–81 (1990).
19. Bailey, D. & Cooper, J.B. *Protein Science* **3**, 2129–2143 (1994).
20. Fusek, M., Smith, E.A., Monod, M., Dunn, B.M. & Foundling, S.I. *Biochemistry* **33**, 9791–9799 (1994).
21. Capobianco, J.O., Lerner, C.G. & Goldman, R.C. *Anal. Biochem.* **204**, 96–102 (1992).
22. Koelsch, G.J.T., M. Monod, S. Foundling, X. Ling. in *VIIth International Aspartic Proteinase Conference* (ed. James, M.N.G.) (Plenum Press, Banff, Alberta, Canada, 1996).
23. Cutfield, S., Marshall, C. Moody, P., Sullivan, P. & Cutfield, J. *J. Mol. Biol.* **234**, 1266–1269 (1993).
24. Jones, T.A., Zou, J.Y., Cowan, S.W. & Kjeldgaard, M. *Acta crystallogr.* **A47**, 110–119 (1991).
25. Suguna, K., Padlan, E.A., Smith, C.W., Carlson, W.D. & Davies, D.R. *Proc. Nat. Acad. Sci. U.S.A.* **84**, 7009–7013 (1987).

42

YAPSIN1

Structure, Biosynthesis, and Specificity

Vicki Olsen, Niamh X. Cawley, and Y. Peng Loh*

Section on Cellular Neurobiology
Laboratory of Developmental Neurobiology
National Institute of Child Health and Human Development
National Institutes of Health
Bethesda, Maryland 20892

1. INTRODUCTION

The Yapsin group of enzymes is the most recently discovered subfamily of aspartic proteases. This subfamily includes: Yapsin1 (also known as Yeast Aspartic Protease 3 (Yap3p))[1] and Yapsin2 (also known as Mkc7)[2] both from *S. cerevisiae* and the mammalian Pro-opiomelanocortin Converting Enzyme (PCE), or Yapsin3 (EC 3.4.23.17), found in pituitary intermediate and neural lobe secretory granules.[3,4] Members of the Yapsin family share structural homology with Yapsin1 (Yap3p) and have a common specificity for paired or single basic residues of pro-proteins.[1–4] This is in contrast to other aspartic proteases which cleave at hydrophobic residues.[5] In the following sections the structure, biosynthesis, processing and specificity of Yapsin1, the first cloned[1] and best characterized member of the Yapsin family will be discussed.

2. STRUCTURE OF YAPSIN1

The *YAP3* gene, encoding Yapsin1, was cloned based on its ability to suppress the *kex2* null mutant phenotype, by cleaving the yeast pheromone pro-α-factor when either the *YAP3* gene product or the pheromone were overexpressed.[1] Based on the nucleotide

* Correspondence to: Y. Peng Loh, PhD., National Institutes of Health, Bldg 49 Rm 5A38, Bethesda, MD 20892, Phone: (301) 496 3239, Fax: (301) 496 9938, E-mail: ypl@codon.nih.gov

Aspartic Proteinases, edited by James.
Plenum Press, New York, 1998.

sequence, Yapsin1 is synthesized as a 569 amino acid protein with an N-terminal signal peptide, predicted to be cleaved at Gly_{21} followed by a propeptide. The C-terminal of Yapsin1 is composed of a stretch of hydrophobic residues that in the ER functions as a signal for the addition of a glycosyl-phosphatidylinositol (GPI) anchor. The anchor facilitates localization of the protein in the plasma membrane facing into the periplasmic space.[6,7] Removal of the C-terminal, including this hydrophobic region results in secretion of the truncated, yet active enzyme, to the medium.[7] Truncated Yapsin1 migrates on SDS-PAGE as three forms; an intracellular form of ~70 kDa and two extracellular forms: one with an apparent molecular mass of ~90 kDa and a heterogenous hyper-glycosylated form of ~120–150 kDa, consistent with the presence of 10 potential N-linked glycosylation sites. Treatment of Yapsin1 with Endoglycosidase H, which specifically removes N-linked sugars, reduced the size of all forms to ~65 kDa,[7] which is still bigger than that predicted for the protein backbone alone. This suggests that in addition to N-linked glycosylation, the Ser/Thr rich domain located in the C-terminal part of Yapsin1 probably contains O-linked sugar moeties, as well, similar to the aspartic protease Bar1p from *S. cerevisia*.[8]

3. BIOSYNTHESIS AND PROCESSING OF YAPSIN1

A soluble truncated form of Yapsin1 lacking the last 37 amino acid residues has been placed under the control of a galactose-inducible promotor and overexpressed in yeast.[9] Purification of the overexpressed truncated Yapsin1 has been performed both from the cell extact[9] and from the media (Cawley *et al.*, submitted). N-terminal amino acid sequence analysis, of the purified recombinant Yapsin1 from the media, revealed two different N-terminals, both from within the *YAP3* sequence. One form of Yapsin1 had Ala_{68} at the N-terminus, verifying that the pro-peptide is removed after Lys_{66}–Arg_{67} as predicted (Cawley *et al.*, submitted). The finding of a second form with the N-terminus, beginning at Asp_{145}, indicated processing of the protein into two subunits, termed α and β, each containing one of the two active site aspartic acid residues. N-terminal sequence analysis of the purified enzyme resulted in an almost equimolar ratio of the α- to β-subunit, suggesting that most of the protein was processed into the two subunits. Separation of the purified protein by SDS-PAGE under reducing conditions followed by amino acids sequencing showed that the subunits were associated by a disulfide bridge between the cysteine residue in the α subunit and one of the three cysteine residues in the β subunit. Analogous to the pepsin structure, the disulfide bridge is most likely formed between Cys_{117} and Cys_{186}.[10] The cleavage between Asn_{144} and Asp_{145} occurs within a unique loop structure not usually found among the aspartic proteases and may represent a distinctive structural property of the yeast Yapsins. The processing of Yapsin1 is summarized in Fig. 1.

Recently mature Yapsin1 and Pro-Yapsin1 were expressed in insect cells by the use of a baculovirus system. Both forms were found to be inactive. However, it was demonstrated that incubation at acidic pH resulted in activation of Pro-Yapsin1, but not of mature Yapsin1 (Cawley *et al.*, submitted). The activation of Pro-Yapsin1 occured optimally at pH 4.0 and appeared to occur by an autocatalytic process, as it does for most known aspartic proteases. In support of this is the identification of the cleavage site for removal of the propeptide as an efficient cleavage motif of Yapsin1. Thus the propeptide of Yapsin1 has at least two different functions. Firstly, it is necessary for proper folding of the nascent protein since the absence of the propeptide resulted in an incorrectly folded

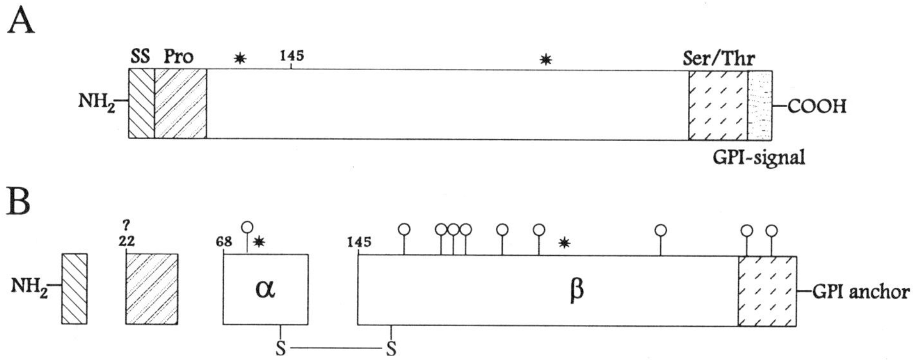

Figure 1. Diagrammatic representation of the processing of Yapsin1 showing structural domains (A) and various processing sites (B). SS: Signal peptide; Pro: Pro-peptide; Ser/Thr: Serine/Threonine rich domain; *: active site aspartic residues; O: Potential N-linked glycosylation sites; S: Cysteine residues. The numbers above the diagram represent: Lys_{22}, Ala_{68} and Asp_{145}.

enzyme without activity. Secondly, it may function by keeping the enzyme inactive until the destined compartment has been reached.

4. SPECIFICITY OF YAPSIN1

The specificity of Yapsin1 and the cleavage efficiency for a number of prohormone substrates containing mono- paired- and tetra-basic cleavage sites has been determined *in vitro*.[11,12] The pH optimum of purified Yapsin1 for these substrates were between 4.0 and 4.5. Yapsin1 was found to cleave either between, or on the carboxyl side of paired basic residue cleavage sites, and at the carboxyl side of single basic residue cleavage sites. It did not have any preference for paired, as opposed to mono basic cleavage sites. However, there was a preference for basic residue cleavage sites with additional basic residues located either up- or downstream. Evidence for this came from a thorough analysis of the efficiency of cleavage by Yapsin1 of synthetic CCK_{13-33} (Olsen *et al.*, in preparation) and other substrates[11] having basic residues flanking the cleavage site.

CCK_{13-33} was cleaved after a single lysine residue and has an arginine residue located in the P6 position. Substitution of this residue at P6 with an alanine resulted in a 50% reduction of the cleavage efficiency (k_{cat}/K_m of $CCK_{13-33} = 2.6 \times 10^4\ M^{-1}\ s^{-1}$), clearly demonstrating that this residue at P6 was important for cleavage. When additional basic residues were introduced downstream of the cleavage site, either in the P2′ or P6′ position, an enhanced cleavage rate was observed, with the P2′ being the most favorable, resulting in 19-fold increase of k_{cat}/K_m relative to CCK_{13-33}. Also, the cleavage efficiency of synthetic proalbumin peptides by Yapsin1 was enhanced when the alanine residue in the P2′ position was substituted for an arginine residue.[13] A negatively charged residue (Asp) introduced in the P2′ position of CCK_{13-33} reduced the cleavage efficiency by 60%. Recent molecular modeling studies of Yapsin1 show that the residue in the P2′ position interacts with a negatively charged residue in the S2′ pocket of the enzyme, providing an explanation for the kinetic data (K. Guruprasad and T. Blundell, unpublished data). In addition, the Yapsin1 model predicted the active site cleft to be highly electronegatively charged with negative residues close to the S6, S3, S2, S1, S1′, S2′ and S3′ pockets accounting for the high efficiency of cleavage of substrates with multiple basic residues flanking the cleavage site, such as $ACTH_{1-39}$ ($k_{cat}/K_m = 3.1 \times 10^6\ M^{-1}\ s^{-1}$).[11]

5. POSSIBLE ROLE OF YEAST YAPSINS

The exact biological function of Yapsin1 has not been determined. Nearly equal amounts of the nonessential *YAP3* gene product (Yapsin1), were immunodetected in a-haploid, α-haploid and a/α-diploid yeast, showing no mating type-specific expression.[6] However, a role under stressful conditions was suggested, in light of the presence of putative heat shock elements found in the promotor region of the *YAP3* gene.[14] Indeed, combined disruptions of the *YAP3*, *MKC7* and *KEX2* genes resulted in mutants showing slow or no growth at low and high temperatures. Specifically, a *yap3/mkc7* double mutant exhibited impaired growth at elevated temperatures, arguing that Yapsin1 and Yapsin2 are required for growth at high temperatures. At low temperatures both Yapsin1 and Yapsin2 were capable of multicopy suppression of the *kex2* cold sensitive phenotype, suggesting a function under suboptimal conditions in general.[2]

6. CONCLUSIONS

Yapsin1 is synthesized as an inactive precursor, pro-Yapsin1, which is processed to two subunits, α and β linked by a disulfide bridge. Yapsin1 has specificity for paired and mono basic residues and the efficiency of cleavage is enhanced in substrates having additional basic residues flanking the P1 site. Modeling studies on Yapsin1 predicts electronegative subsite pockets that can account for this unique specificity.

ACKNOWLEDGMENTS

V. Olsen is supported by the Danish Natural Science Research Council, Grant 9400095.

REFERENCES

1. M. Egel-Mitani, H.P. Flygenring and M. Trier Hansen, A novel aspartyl protease allowing KEX2-independent MFα propheromone processing in yeast. *Yeast* 6:127 (1990)
2. H. Komano and R.S. Fuller, Shared functions *in vivo* of a glycosyl- phosphatidylinositol-linked aspartyl protease, Mkc7, and the proprotein processing Kex2 in yeast. *Proc. Natl. Acad. Sci. USA* 92:10752 (1995)
3. Y.P. Loh, D.C. Parish and R. Tuteja, Purification and characterization of a paired basic residue-specific pro-opiomelanocortin converting enzyme from bovine pituitary intermediate lobe secretory vesicles. *J.Biol.Chem.* 260:7194 (1985)
4. N.X. Cawley, L.-P. Pu and Y.P. Loh, Immunological identification and localization of yeast aspartic protease 3-like prohormone-processing enzymes in mammalian brain and pituitary. *Endocrinology* 137:5135 (1996)
5. B.M. Dunn, M. Jimenez, B.F. Parten, M.J. Valler, C.E. Rolph and J. Kay, A systematic series of synthetic chromophoric substrates for aspartic proteases. *Biochem. J.* 237:899 (1986)
6. J. Ash, M. Dominguez, J.J.M. Bergeron, D.Y. Thomas and Y. Bourbonnais, The yeast proprotein convertase encoded by *YAP3* is a glycophosphatidylinositol- anchored protein that localizes to the plasma membrane. *J. Biol. Chem.* 270:20847 (1995)

7. N.X. Cawley, M. Wong, L.-P. Pu, W. Tam and Y.P. Loh, Secretion of yeast aspartic protease 3 is regulated by its carboxy-terminal tail: characterization of secreted YAP3p. *Biochem.* 34:7430 (1995)
8. V.L. MacKay, J. Armstrong, C. Yip, S. Welch, K. Walker, S. Osborn, P. Sheppard and J. Forstrom, Characterization of the BAR proteinase, an extracellular enzyme from the yeast *Saccharomyces cerevisiae*, in: Structure and Function of the Aspartic Proteinases, B. M. Dunn, ed., pp. 161, Plenum Press:New York, NY (1991)
9. A. Azaryan, M. Wong, T.C. Friedman, N.X. Cawley, F.E. Estivarez, H.-C. Chen and Y.P. Loh, Purification and characterization of a paired basic residue-specific yeast aspartic protease encoded by the YAP3 gene. Similarity to the mammalian pro-opiomelanocortin-converting enzyme. *J.Biol.Chem.* 268:11968 (1993)
10. J.B. Cooper, G. Khan, G. Taylor, I.J. Tickle and T.L. Blundell, X-ray analyses of Aspartic proteinases II. Three-dimensional structure of the hexagonal crystal form of porcine pepsin at 2.3 Å resolution. *J. Mol. Biol.* 214:199 (1990)
11. N.X. Cawley, H.-C. Chen, M.C. Beinfeld and Y.P. Loh, Specificity and kinetic studies on the cleavage of various prohormone mono- and paired basic residue sites by yeast aspartic protease 3. *J. Biol. Chem.* 271:4168 (1996)
12. N.X. Cawley, B.D. Noe and Y.P. Loh, Purified yeast aspartic protease 3 cleaves anglerfish pro-somatostatin I and II at di- and monobasic sites to generate somatostatin-14 and -28. *FEBS Lett.* 332:273 (1993)
13. E.C. Ledgerwood, S.O. Brennan, N.X. Cawley, Y.P. Loh and P.M. George, Yeast aspartic protease 3 (Yap3) prefers substrates with basic residues in the P2, P1 and P2′positions. *FEBS Lett.* 383:67 (1996)
14. V. Olsen, A putative prohormone convertase from *Saccharomyces cerevisiae*. Characterization of biosynthesis and cellular localization of the *YAP3* gene product. M.S. Thesis, University of Copenhagen, Denmark (1994)

43

BACTERIAL ASPARTIC PROTEINASES AS NOVEL ANTIBIOTIC TARGETS

L. H. Phylip and J. Hill

School of Molecular and Medical Biosciences
University of Wales

Bacterial infections still remain a prevalent problem world-wide. Indeed, the incidence of tuberculosis (once thought to be eradicated) is currently rising, especially amongst AIDS sufferers, with up to 20% of this population in the U.K. thought to be infected with this bacterium. The over-prescribing of antibiotics during the last few decades has resulted in the development of antibiotic resistance.[1] This is a problem of grave concern, affecting virtually all bacteria which commonly cause human illness. This concept is not restricted to a single class of antibiotics; some bacteria, e.g., multiresistant *Mycobacterium tuberculosis* and *Enterococcus faecium* have developed resistance to almost all available antibiotics. The extent of the problem is so large that a 'post-antibiotic era' may be looming where many infectious diseases will once more be almost impossible to treat.[2]

The biochemistry of prokaryotes is well-defined and the mode of action of many currently-used antibiotics is understood. These agents are targeted against specific features of the bacterium, e.g., the penicillins and cephalosporins which inhibit cell wall synthesis and the aminoglycosides which inhibit protein synthesis. In view of the ever-increasing problem of antibiotic resistance, new targets need to be identified so that novel antibacterial agents may be developed. The advent of genome sequencing projects has generated extensive databases which provide access to the sequences of the genomes from various bacteria.[3] In some cases, the bacterial sequences determined encode proteins for which no function has yet been ascribed, while in others, putative functions have been postulated on the basis of sequence similarity to proteins from other species with well-characterised activities.

Thus, the question was posed, are there genes present within bacteria that could encode aspartic proteinases? Such enzymes have not been isolated from bacteria previously and were thought not to be present in prokaryotes.[4] Database searches of a variety of bacterial genomes were performed for the occurrence together of the two hallmark sequences of aspartic proteinases, i.e., the ~Hydrophobic–Hydrophobic–Asp–Thr/Ser–Gly~ motif and the ~Hydrophobic–Hydrophobic–Gly~ sequence, which together form the secondary structural feature known as the psi loop.[5] Such sequences were found to be encoded within the

Aspartic Proteinases, edited by James.
Plenum Press, New York, 1998.

genomes of *Bacillus subtilis* (Genebank: L1S202), *Escherichia coli* (ECOUW67), *Haemophilus influenzae* (L20805) and *Neisseria gonorrhoea* (575490).

With the exception of *B. subtilis,* these bacteria are pathogenic to humans. Several outbreaks of *E.coli* poisoning in the U.K. in late '96–early '97 resulted in numerous deaths. *H. influenzae* is responsible for a wide spectrum of conditions, including meningitis, acute eppiglottitis, osteomyelitis and arthritis, with septicaemia frequently accompanying these infections. *N. gonorrhoea* is the cause of the sexually transmitted disease gonorrhoea and although genomic sequence information from the closely-related *N. meningitidis* is not yet available, it is likely that this species (which causes meningococcal meningitis) may also harbour genes with similar sequences.

Two of these bacterial species (*E. coli* and *H. influenzae*) were selected to determine whether these putative aspartic proteinase domains were capable of functioning as proteolytic entities. Genomic DNA was thus prepared from both *E. coli* and *H. influenzae* and oligonucleotide primers were designed to amplify each region of DNA encoding these domains (Fig. 1).

The minimal length for each putative aspartic proteinase was estimated by comparison with the known cleavage junctions at the N- and C-termini of retroviral aspartic proteinases which contain 100–130 amino acids.[6] Both the *E. coli* and *H. influenzae* aspartic proteinase domains were expressed in *E. coli* under the control of a T7 RNA polymerase promoter which resulted in the accumulation of both proteins as insoluble ag-

H.inf.	~F T I L L F K Q L S K P T W Q V D T L
E.coli	~G L N R G R S M D Q I N - - - - - - D T L
H.inf.	D V G Q - - - G L A T L I V K N G - - - K G I
E.coli	V S S I K I I A M M L L I I G G G G A F K Q V
H.inf.	L Y D T G S S W R G G S M A - E L E I L P Y L
E.coli	L V D S G V D K Y I A S M M H E T N I S P L L
H.inf.	Q R E G I - - V L E K L I L - - S H D D N D -
E.coli	M A W S I A A V L - R I I A L G S A T V A A I
H.inf.	H A G G A S T I L K A Y P N V E L I T P S R K
E.coli	T A G G I A A - - - - - P - - - L I A T T - G
H.inf.	N Y G E N Y R T F C **T A G** R D W H W Q G L H F
E.coli	V S P E L M V I - - **A V G** S G S V I - F S H V
H.inf.	Q - - - - I L S P H - - N V A T R A D N - - S
E.coli	N D P G F W L F K E Y F N L - T I G E T I K S
H.inf.	H S ~
E.coli	W S M ~

Figure 1. Amino acid sequences of the putative aspartic proteinase domains from *H. influenzae* and *E. coli.* The ~Hydrophobic–Hydrophobic–Asp–Thr/Ser–Gly~ motifs are underlined and the putative ~Hydrophobic–Hydrophobic–Gly~ motifs of the psi loop are shown in bold. Gaps (indicated by a hyphen) are inserted to maximise the apparent similarity.

gregates which required solubilisation in 6M urea. Both aspartic proteinase domains were refolded by a rapid 2000-fold dilution followed by reconcentration. In order to detect intrinsic activity in the recombinant aspartic proteinase domain from *E. coli*, the oligopeptide Ac–Tyr–Arg–Ala–Arg–Val–Phe*Nph–Val–Arg–Ala–Ala–Lys was used as a substrate. This peptide has been found previously to be amongst the best substrates for HIV-1 proteinase.[7] It was cleaved readily by the *E. coli* aspartic proteinase domain, under conditions comparable to those employed routinely for the assay of retroviral enzymes i.e. 0.1 M Na Acetate, pH 4.7, $\propto$=0.1 at 37°C.[8] Furthermore, the activity was completely inhibited in the presence of the diagnostic aspartic proteinase inhibitor, pepstatin. The activity associated with the recombinant aspartic proteinase domain from *H. influenzae* was investigated initially with two commonly-used protein substrates, azocasein and haemoglobin. Both of these proteins were digested by the *H. influenzae* aspartic proteinase domain, under the acidic conditions described above. The ability of the *H. influenzae* aspartic proteinase domain to cleave the synthetic peptide, Lys–Pro–Ile–Glu–Phe*Nph–Arg–Leu was examined; this peptide has been extensively used as a substrate for vertebrate and fungal aspartic proteinases e.g. cathepsin D and cathepsin E[9,10] and it was readily cleaved too by the *H. influenzae* aspartic proteinase domain. The K_m value (100 μM) obtained was not dissimilar in magnitude from the values determined for vertebrate/ fungal aspartic proteinases with this substrate.[11] Hydrolysis of this substrate by the recombinant aspartic proteinase from *H. influenzae* was completely inhibited by inclusion of pepstatin.

This is the first demonstration of a ‘typical’ aspartic proteinase activity found to be present in any bacterial species. Previous reports have described thermopsin, a thermostable acid proteinase isolated from the archaebacterium, *Sulfolobus acidocaldarius.*[12] Although the activity of this enzyme is inhibited by pepstatin, its amino acid sequence does not show any apparent similarity to those of typical aspartic proteinases, i.e., it does not contain the ~Hydrophobic–Hydrophobic–Asp–Thr/Ser–Gly~ and ~Hydrophobic–Hydrophobic–Gly~ hallmark sequences. Other enzymes from thermophilic *Bacillus* strains have been described, but the sequences of these are as yet undetermined.[13] They are not inhibited by pepstatin but are inactivated by diazoacetyl-DL-norleucine methyl ester and 1,2-epoxy-3-(p-nitrophenoxy)propane. In contrast, the *E. coli* and *H.* influenzae aspartic proteinase domains are not only susceptible to pepstatin and are capable of cleaving typical aspartic proteinase substrates but also contain the conserved structural features of the active site residues and the psi loop.

The question remains to be asked, do these proteins functions *in vivo* as aspartic proteinases? Both the *E. coli* and *H. influenzae* aspartic proteinase domains described above are found to be present within much larger proteins. The *E. coli* aspartic proteinase domain is located towards the C-terminus of a larger protein that bears homology to a gluconate transporter protein whereas the *H. influenzae* aspartic proteinase domain exists within the rec-2 protein, which has a molecular weight of ~90 kDa and is thought to be involved in the transport of foreign DNA into the *H. influenzae* cell.[14] This protein consists of nine postulated integral membrane spanning segments. The aspartic proteinase domain is located towards the C-terminus of this large protein in a hydrophilic region which is predicted to be located on the cytoplasmic side of the membrane. The exact function of the rec-2 protein remains unclear. It may interact directly with the DNA entering the cell or could aid in this process by assisting in the assembly of the pore through which the DNA gains entry into the cell. The anchoring of the *H. influenzae* aspartic proteinase domain in the cell membrane by its ‘propart’ in the much larger protein is reminiscent of the situation with the *gag–pol* polyprotein of HIV, where the proteinase is indirectly concentrated by attachment of the N-myristoylated *gag* to the host cell membrane. As in the case of the retroviral aspartic proteinases, the ~Hydrophobic–Hydrophobic–Asp–Thr/Ser–Gly~ sequence occurs only once within the bacterial aspartic proteinase domains so that dimerisation of two monomers would be required in order to form a potentially active dimeric enzyme. The anchoring of the bacterial aspartic proteinase domain in the membrane may aid in bringing together the necessary monomers into close proximity. It is unknown whether the bacterial aspartic proteinase domains are active as part of the larger protein or whether they require to be liberated from the constraints of the larger protein molecules. Indeed, both of the active bacterial aspartic proteinase domains described in this chapter were expressed as slightly larger fusion proteins with an additional 13 amino acids at the N-terminus of each.[15]

The importance of these aspartic proteinase domains in bacteria remains to be determined. This may be investigated by creating knock-out mutants deficient in both the aspartic proteinase region and in the larger protein and to determine whether this has a detrimental effect on the phenotype/pathogenicity of these organisms. If the aspartic proteinase domains are found to play a vital role, then they may present themselves as targets for the rational design of specific inhibitors. Such an approach has already been successfully adopted in the development of specific HIV proteinase inhibitors. Specific inhibitors of bacterial aspartic proteinases may lead to the generation of a novel class of much-needed antibacterial agents.

REFERENCES

1. World Health Organisation (1996) Report of the Director General HMSO, London.
2. Cohen, M. (1992) Science 257, 1050–1055.
3. Fox, J.L. (1997) Nature Biotechnology 15, 211–212.
4. Kay, J. (1985) in: Aspartic Proteinases and their Inhibitors (Kostka, V.,ed), pp. 1–17, Walter de Gruyter, New York.
5. Pearl, L.H. and Taylor, W.R. (1987) Nature 329, 351–354.
6. Gustchina, A., Kervinen, J., Powell, D.J., Zdanov, A., Kay, J. and Wlodawer, A. (1996) Protein Sci. 5, 1453–1465.
7. Griffiths, J.T., Tomchak, L.A., Mills, J.A., Graves, M.C., Cook, N.D., Dunn, B.M. and Kay, J. (1994) J. Biol. Chem. 269, 4787–4793.
8. Wilson, S.I., Phylip, L.H., Mills, J.S., Gulnik, S.V., Erickson, J.W., Dunn, B.M. and Kay, J. (1997) Biochim. Biophys. Acta, in press.
9. Beyer, B.M. and Dunn, B.M. (1996) J. Biol. Chem. 271, 15590–15596.
10. Rao-Naik, C., Guruprasad, K., Batley, B., Rapundalo, S., Hill, J., Blundell, T.L., Kay, J. and Dunn, B.M. (1995) Proteins: Structure, Function and Genetics 22, 168–181.
11. Dunn, B.M., Jimenez, M., Parten, B.F., Valler, M.J., Rolph, C.E. and Kay, J. (1986) Biochem. J. 237, 899–906.
12. Lin, X. and Tang, J. (1990) J. Biol. Chem. 265, 1490–1495.
13. Toogood, H.S., Prescott, M. and Daniel, M. (1995) Biochem. J. 307, 783–789.
14. Clifton, S.W., McCarthy, D. and Roe, B.A. (1994) Gene 146, 95–100.
15. Hill, J. and Phylip, L.H. (1997) manuscript in submission.

44

STRUCTURAL THERMODYNAMIC STUDY OF THE BINDING OF RENIN INHIBITORS TO ENDOTHIAPEPSIN

Javier Gómez, Nora Semo, and Ernesto Freire

Department of Biology and the Biocalorimetry Center
The Johns Hopkins University
Baltimore, Maryland 21218

1. INTRODUCTION

A key element in any structure-based molecular design strategy is the capability of predicting the energetics of a putative interaction from structural considerations. We have recently introduced[1] an empirical parametrization relating the structural changes induced upon formation of the complex and the energetics characterizing the binding process (ΔG, ΔH, ΔS and ΔCp), and applied it to the thermodynamic description of the binding of pepstatin A to the aspartic proteinase endothiapepsin. This study has permitted the dissection of the various enthalpic and entropic contributions to the overall free energy change upon binding and the identification of the forces driving the binding reaction. Moreover, the structure-based thermodynamic calculations allowed the estimation of the contribution of each residue to the total free energy of binding, identifying those regions of the protein and the inhibitor that contribute the most to the Gibbs free energy.

Here we present a structural thermodynamic study of the interaction between endothiapepsin and several potent renin inhibitors (H-142, H-77, H-189 and H-261). These compounds are based on the sequence of the natural renin substrate angiotensinogen (see Table 1). The scissile peptide bond between residues Leu-10 and Val-11 in angiotensinogen has been modified in order to preclude its hydrolysis by the enzyme. In H-142 and H-77 the amide bond has been transformed into its reduced isostere ($-CH_2-NH-$) and into its hydroxyethylene isostere ($-CHOH-CH_2-$) in H-261. Finally, in H-189 the two central residues were substituted by the non natural amino acid statine. These four renin inhibitors are analogues of the putative tetrahedral noncovalent transition state formed during the catalytic hydrolysis of the substrate peptide bond ($-C(OH)_2-NH_2^+-$).

Aspartic Proteinases, edited by James.
Plenum Press, New York, 1998.

Table 1. The inhibitors used in this study were based on the sequence of angiotensinogen

Angiotensinogen	~	His	Pro	Phe	His	Leu–Val	Ile	His	~	
H-189	Pro	His	Pro	Phe	His	**Sta**	Val	Ile	His	Lys
H-261	Boc	His	Pro	Phe	His	**Leu$\overset{OH}{—}$Leu**	Ile	His		
H-77		*D*-His	Pro	Phe	His	**Leu$\overset{R}{—}$Leu**	Val	Tyr		
H-142	Pro	His	Pro	Phe	His	**Leu$\overset{R}{—}$Val**	Ile	His	Lys	

$\overset{R}{—}$ the peptide bond was reduced to $-CH_2-NH-$.
$\overset{OH}{—}$ the peptide bond was reduced to $-CHOH-CH_2-$.

2. THERMODYNAMIC CHARACTERIZATION OF THE BINDING PROCESS

The energetics characterizing the formation of the inhibitor-enzyme complex has been studied by isothermal titration calorimetry as a function of temperature at pH 3.1. Due to limitations in the instrumentation currently available, the calorimetric titrations could not be performed at concentrations low enough for the accurate determination of dissociation constants smaller than 50 nM. For such tight-binding reactions, however, isothermal titration calorimetry provides a precise determination of the stoichiometry of the reaction as well as the enthalpy change upon binding, ΔH. The temperature dependence of the measured ΔH yields the heat capacity change upon binding, ΔCp. For the four inhibitors studied, the binding reaction is characterized by favorable enthalpic contributions to the Gibbs free energy (see Table 2), ranging from -5.06 kcal·mol^{-1} for H-77 to -10.46 kcal·mol^{-1} for H-189 at room temperature. The formation of the complex between the enzyme and any of the inhibitors studied is characterized by large and negative heat capacity changes (evaluated from ΔH measurements obtained at different temperatures), varying from -263 cal·K^{-1}·mol^{-1} for H-77 to -416 cal·K^{-1}·mol^{-1} for H-261.

The mechanism of the inhibition of endothiapepsin by the transition-state analogue inhibitors was investigated by analyzing the time-course of enzymatic hydrolysis of the chromogenic substrate Pro–Thr–Glu–Phe–Phe(4-NO_4)–Arg–Leu[2] in the presence of increasing concentrations of inhibitor. It was observed that inhibition of endothiapepsin by all the transition-state analogues considered in this study was time dependent (steady-state between protein and ligand was only achieved after several minutes). Time dependent inhibition has been demonstrated for several aspartic proteinases and has been explained in terms of a rapid formation of an initial transient-state and its subsequent slow isomerization to form the final tightened complex.[3] The overall inhibition constants were evaluated from the global fitting of each set of time course of enzyme reaction curves obtained at different inhibitor concentrations. The results are presented in Table 2. The four inhibitors considered in this study are shown to bind very tightly to endothiapepsin with inhibition constant in the sub-nanomolar range (see Table 2).

Taking into account that the inhibitors contain several basic groups (His, Lys and the amine in the reduced isostere in H-142 and H-77) that become buried upon complex formation, we investigated whether the binding of the inhibitor to the enzyme induces the exchange of protons with the reaction media, by determining the enthalpy of binding in different buffers at 25°C and pH 3.1. In general, the apparent enthalpy of binding, ΔH_{bind}, is the sum of the intrinsic enthalpy of reaction, ΔH_{react}, independent of the buffer used in the experiment and a term proportional to the enthalpy of ionization of the buffer, ΔH_{ion}:

$$\Delta H_{binding} = \Delta H_{react} + N_{H+}\, \Delta H_{ion} \tag{1}$$

where N_{H+} is the number of protons released ($N_{H+} < 0$) or absorbed ($N_{H+} > 0$) upon complex formation.

Not surprisingly, our results indicate that, at pH 3.1, the formation of the enzyme/inhibitor complex induces the deprotonation of some of the basic groups that become buried in the binding interface (see Table 2). It seems clear that the deprotonation of this groups would contribute unfavorably to the free energy change upon binding. This contribution would be maximal for the two inhibitors containing the reduced isostere (H-77 and H-142) since their binding to the enzyme requires the release of 1.5 protons. At this pH, the binding of pepstatin A to the protein does not involve the exchange of protons with the reaction media (at higher pH, the binding process induces the protonation of one of the aspartates in the binding site),[1] strongly suggesting that those groups involved in the ionization process belong to the inhibitor molecule. Taking into account that the Lys side-chains in H-189 and H-142 are completely exposed in both complexes, it seems reasonable to identify the His side-chain as the group responsible for the deprotonation event coupled to the binding (the pKa of the amine in the reduced isostere is presumably too high to be effectively affected at this very low pH).

3. STRUCTURE-BASED THERMODYNAMIC ANALYSIS

The formation of an enzyme/inhibitor complex involves the dehydration of the interacting surfaces and the formation of new internal interactions (hydrogen bonding, van der Waals and electrostatic, etc.). On the other hand, such a process would generate a decrease in the conformational freedom of the side chains from both protein and peptide located at the interface as well as the peptide backbone which becomes restricted into a unique conformation in the bound state. For the purpose of this discussion, the apparent free energy change upon binding can be considered as constituted by the intrinsic free energy of binding, ΔG_{int}, and the contribution due to any protonation/ionization process coupled to the binding reaction, ΔG_{ion}.

$$\Delta G = \Delta G_{int} + \Delta G_{ion} \tag{2}$$

The intrinsic free energy of binding contains contributions typically associated with changes in hydration, van der Waals interactions, hydrogen bonding, and conformational entropy as well as the change in traslational degrees of freedom upon binding.

We have recently presented a structural parametrization of the binding energetics[1] capable of predicting the intrinsic free energy change upon binding from the known three dimensional structure of the complex. This approach calculates separately the enthalpy and entropy contributions to ΔG together with their temperature dependence (ΔCp). It has been shown[1] that the enthalpy[4] and heat capacity[5] changes upon binding, ΔH and ΔCp, can be estimated from the changes in solvent accessible surface area (ΔASA) and in the interatomic distances between the interacting groups upon binding. The entropy change is calculated considering that it is composed by two major contributions: the entropy associated with the changes in hydration and the changes in conformational degrees of freedom (from both side chains and backbone). While the solvation entropy change is temperature dependent and proportional to the changes in accessible surface area, the calculation of the conformational entropy change requires the computational analysis of the probability of different conformers as a function of the dihedral and torsional angles.[6]

The availability of the crystallographic structure of endothiapepsin free in solution, and its complexes with H-189, H-261, H-77 and H-142, allowed the structure-based thermo-

Table 2. Thermodynamical characterization of the binding of several renin inhibitors to endothiapepsin at 25°C and pH 3.1. The intrinsic free energy of binding was calculated applying the structural parametrization briefly explained in the text. The ionization contribution to the free energy was calculated according to eq. 2

	Experimental[a]						Intrinsic[a]		Ionization[a]	
Inhibitor	ΔH	ΔCp	−T ΔS	ΔG	K_i	N_{H+}[b]	ΔG	K_i	ΔG	DG/N_{H+}[c]
H-189	−10.46	−327	−2.94	−13.40	1.5×10^{-10}	0.96	−15.81	2.6×10^{-12}	2.41	2.51
H-261	−9.51	−416	−4.31	−13.82	7.4×10^{-11}	0.75	−15.93	2.1×10^{-12}	2.11	2.81
H-77	−5.06	−263	−6.83	−11.89	1.9×10^{-9}	1.47	−16.26	1.2×10^{-12}	4.37	2.91
H-142	−5.68	−380	−6.98	−12.66	5.2×10^{-10}	1.48	−4.62[g]	4.1×10^{-4} [d]	—[d]	—[d]

[a] $\Delta G_{exp} = \Delta G_{int} + \Delta G_{ion}$. [b] n_{H+} : number of proton released upon binding. [c] $\Delta G/n_{H+}$: free energy of ionization per proton released. [d] the large disagreement between ΔG_{int} and ΔG_{exp} precludes an estimation of ΔG_{ion}. ΔH, −T ΔS and ΔG ω in kcal/mol. ΔCp values in cal/(K mol). K_i values in M.

dynamic modeling of the binding process. The intrinsic free energy change upon complex formation was calculated applying the structural parametrization of the binding energetics briefly explained above. The results are presented in Table 2. For all the inhibitors except H-142, the intrinsic dissociation constant is smaller than the one experimentally observed. As discussed before, the difference between experimental and intrinsic free energy is equal to the free energy of ionization. This difference amounts to 2.7 ± 0.2 kcal·mol^{-1} per proton released. Despite this protonation effect observed at pH 3.1, the binding constant is predicted to be largely independent of pH due to the required protonation of one of the members of the aspartic diad at higher pH,[1] which will contribute unfavorably to the free energy. There is a large disagreement between intrinsic and experimental free energy for H-142 that can not be attributted to the ionization event. The unusually low temperature factors found for many enzyme atoms in the crystal strcture of its complex with H-142 (some of them negative) could be an indication of possible errors in the refinement and could explain the large difference between the expected free energy change upon binding and the intrinsic one (calculated from the structure).

The results presented here and those obtained previously,[1] indicate that the structure-based thermodynamic analysis presented here accounts well for the binding energetics of peptides to proteins and that this approach can be successfully applied in molecular design algorithms as discussed recently.[7]

ACKNOWLEDGMENTS

We are indebted to Dr. D. Michael Jones (Ferring Research Institute, Southampton, U.K.) for generously providing us with the samples of the inhibitors used in this study. This work was supported by grants from the National Institutes of Health (RR04328 and GM1362).

REFERENCES

1. J. Gómez and E. Freire. (1995) *J. Mol. Biol.* **252**, 337–350.
2. B.M. Dunn, Jimenez M., Parten B.F., Valler M.J., Rolph C.E., Kay J. (1986). *Biochem. J.* **237**, 899–906.
3. D.H. Rich and Sun E.T. (1980) *Biochem. Pharmacol.* **29**, 2205–2212.
4. V.J. Hilser, Gómez and E. Freire. (1996) *Proteins.* **26**, 123–133.
5. J. Gómez, V.J. Hilser, D. Xie and E. Freire. (1995). *Proteins.* **22**, 404–412.
6. J. A. D'Aquino, J. Gómez, V.J. Hilser, K.H. Lee, L.M Amzel and E. Freire. *Proteins* **25**, 143–156.
7. J. Gómez and E. Freire. (1996). In Thermodynamic and Structure-Based Drug Design. Connely, P. and Ladbury J. Eds. R.G. Landes Co. Publishers. In press.

45

DEVELOPMENT AND TESTING OF INHIBITORS OF *Candida* ASPARTIC PROTEINASES

Iva Pichová,[1] Kateřina Brožková,[1] Libuše Pavlíčková,[1] Martin Fusek,[3] Ivana Křížová,[1] Tomáš Ruml,[2] and Milan Souček[1]

[1]Institute of Organic Chemistry and Biochemistry
Academy of Sciences of the Czech Republic
Prague, Czech Republic
[2]Department of Biochemistry and Microbiology
Institute of Chemical Technology
Prague, Czech Republic
[3]Sigma-Aldrich Ltd.
Prague, Czech Republic

1. INTRODUCTION

Candida is a typical opportunistic pathogen which is often present in healthy individuals but under certain conditions it causes fungal infections.[1] In addition to superficial infections, *Candida* is a major cause of deep systemic infections in immunocompromised patients such as HIV patients, transplant recipients, or cancer patients undergoing chemotherapy. An aspartic proteinase secreted by many members of the genus *Candida* has been often suggested to take part in the invasive character of the microorganism.[2] The aspartic proteinases secreted by *Candida* represent a potential target for the drug intervention of the disease and the studies devoted to the understanding of these enzymes span from genetic studies[3] to substrate specificity studies[4] or crystallographic studies.[5,6] It has been shown that there are differences in the specificity of secreted aspartic proteinases (SAP) from different *Candida* strains and that these proteinases are in general similar to eukaryotic aspartic proteinases with a deep active site cleft which accommodates at least eight residues of a substrate or inhibitor.

Our goal is to find effective inhibitors of *Candida* SAP. In this study we have synthesized and tested two sets of transition state inhibitors as tools of specificity determination as well as a background for future inhibitor based therapeutic design. We isolated SAP from *C. albicans* (CAAP) and *C.tropicalis* (CTAP) according to methods described by Fusek et al.[4] and analysed their inhibition by two sets of inhibitors. The pepstatin A (Iva–Val–Val–Sta–Ala–Sta–OH) was chosen as a lead structure for the first set of inhibitors. The second set contained inhibitors which were previously synthesized and tested as

Aspartic Proteinases, edited by James.
Plenum Press, New York, 1998.

inhibitors of retroviral proteinases and included the reduced amido or hydroxyethylamine or hydroxyethylene isosteres.

2. METHODS

2.1. Cultivation

C. albicans and *C. tropicalis* were grown in 1.2% YCB medium containing 0.4% bovine serum albumin. pH of the medium was adjusted to 4.0. After three days of cultivation at 30°C with shaking at 200 rpm cells were harvested and supernatant containing SAP was used for isolation of CAAP and CTAP.

2.2. Isolation of Proteinases

Proteinases were isolated from supernatant using an anion exchange chromatography on DEAE-Sephadex A25 as described by Fusek et al.[4]

2.3. Synthesis of Inhibitors

All inhibitors were synthesized by solid phase methodology with tert-butoxycarbonyl/benzyl protection. Reduced amide bonds were introduced by reductive amination of a terminal amino group of a tripeptide by appropriate N-protected α-aminoaldehyde in the presence of sodium cyanoborohydride.[7,8,9] Phenylstatine (PheSta) was prepared using a modified method of Jouin and Castro.[10] Inhibitors containing hydroxyethylene isostere (2S,4S,5S)-5-amino-2-benzyl-4-hydroxy-6-phenylhexanoic acid were synthesized with the modified method of Evans.[11] The purity of inhibitors was determined by amino acid analysis and FAB mass spectroscopy.

2.4. Inhibition Studies

IC_{50} values of inhibitors were determined by a spectrophotometric assay with chromogenic substrate Lys–Pro–Ala–Glu–Phe–Nph–Ala–Leu. In a standard experiment, 1.5 nM SAP was incubated in 0.1 M sodium acetate buffer pH 3.3 with 40 μM substrate and various concentrations of an inhibitor. Inhibitors were dissolved in DMSO, and the final concentration of DMSO in the assay was lower than 2.5%.

2.5. Inhibition of the Growth of *C. albicans* and *C. tropicalis*

C. albicans and *C. tropicalis* strains were grown in conditions described above and in the presence of various concentrations of pepstatin A. The samples were taken at the time intervals between 0 and 90 h and the OD at 600 nm was measured in a diluted samples.

3. RESULTS AND DISCUSSION

3.1. Pepstatin A Derived Inhibitors

Pepstatin has been reported to be a potent inhibitor of both, CAAP and CTAP.[12,13] We have modified the pepstatin structure at the statine (Sta) residues. Phenylstatine (Pst), i.e. (3S,4S)-4-amino-3-hydroxy-5-phenylpentanoic acid and two diastereomeres of norphenylstatine (nor Pst), i.e. (2S,3S)-3-amino-2-hydroxy-4-phenylbutanoic acid, were introduced instead of Sta. These compounds were further modified at the N-terminal P4 group. Isovaleryl residue (Iva) was replaced with both, tert-butoxycarbonyl (Boc) and t-butylacetyl (Tba) groups. We have tested also modified inhibitors containing only one Sta residue. The results are summarized in Table 1. Inhibitors which contain phenylstatine in the statine positions display about five times and three times lower activity compared to pepstatin A for CAAP and CTAP, respectively. These results are in a good agreement with the preference of both proteinases for smaller residues in the P1 positions as described by Fusek et al.[4]

Relatively small changes in the structure of N-terminal P4 group cause significant changes in the IC_{50} value for each proteinase. Apparently, S4 subsites of proteinases of *C. albicans* and *C. tropicalisis* are different. IC_{50} value of inhibitors with Tba and acetyl in the P4 positions for CAAP and CTAP differ by one and two orders of magnitude. This might implicate a sensitivity of the S4 subsite for the fine tuning of P4 residue of the inhibitor. IC_{50} values of truncated inhibitors which contain only one Pst residue increase by two orders of magnitude in comparison to those with two Sta derivatives. Results of our study are in general agreement with previously published results on the substrate specificity of CAAP and CTAP.[4]

3.2. Inhibitors with Modified Peptide Bond in the P1–P1′ Position

To map the subsite specificities of CAAP and CTAP, we have used the inhibitors with the structure Boc–F–[X]–F–Y–NH_2 where X represents an isostere moiety. Further we have modified the length of inhibitors containing peptide reduced bond. The results are summarised in Table 2.

The inhibitors containing either the –CH_2NH–, the –$CH(OH)CH_2NH$– or the –$CH_2(OH)CH_2$– groups have an inferior inhibition activity then those derived from pepstatin A. IC_{50} values reach micromolar concentration. Introduction of Val in the P2′ posi-

Table 1. Pepstatin derived inhibitors

Inhibitor	CAAP IC_{50} [μM]	CTAP IC_{50} [μM]
Iva–Val–Val–Sta–Ala–Sta–OH (pepstatin A)	0.009	0.016
Iva–Val–Val–Pst–Ala–Pst–OMe	0.049	0.044
Tba–Val–Val–Pst–Ala–Pst–OMe	0.404	3.39
Boc–Val–Val–Pst–Ala–Pst–OMe	0.03	0.035
Ac–Val–Val–Pst–Ala–Pst–OMe	0.09	0.359
Iva–Val–Val–(2S,3S)norPst–Ala–NH_2	96.5	53.0
Iva–Val–Val–(2R,3S)norPst–Ala–NH_2	72.7	80.2
Iva–Val–Val–Pst–Ala–NH_2	24.5	52.8
Iva–Val–Val–Pst–Ala–OMe	1.5	1.5
Boc–Val–Val–Pst–Ala–OMe	1.2	1.2
Boc–Pst–Glu–Phe–NH(3-quinolyl)	64.8	85.5
Boc–Pst–Glu–Phe–ol	160.0	230.0
Boc–Pst–Ile–Ile–Phe–OMe	97.7	110.0
Boc–Pst–Glu–Phe–NH$(CH_2)_2$NH–Boc	35.2	70.1
Pro–Pro–Cys–Val–Pst–Ala–Met–Thr–Met–NH_2	7.6	2.3

Abbreviations: Sta, statine; Pst, phenylstatine, i.e. (3S,4S)-4-amino-3-hydroxy-5-phenylpentanoic acid; Boc, t-butoxycarbonyl; norPst, norphenylstatine, i.e. (2S,3S)-3-amino-2-hydroxy-4-phenylbutanoic acid; Iva, isovaleryl; Tba, t-butylacetyl; Ac, acetyl; Phe-ol, phenylalaninol.

Table 2. Inhibitors with different isosteres

Inhibitor	CAAP IC_{50} [μM]	CTAP IC_{50} [μM]
Boc–Phe–[CH_2NH]–Phe–Cys–Phe–NH_2	23.5	23.5
Piv–Phe–[CH_2NH]–Phe–Asp–Phe–NH_2	112.0	128.0
Boc–Phe–Phe–[CH_2NH]–Phe–Glu–Phe–OMe	27.8	31.9
Boc–Phe–[CH_2NH]–Phe–Val–Phe–OMe	131.0	—
His–Pro–Glu–Ala–Phe–[CH_2NH]–Phe–Val–Arg–Lys–Ala–NH_2	1.7	25.2
Cys–Thr–Leu–Asn–Phe–[CH_2NH]–Pro–Ile–Ser–Pro–Ile–NH_2	17.0	17.0
Boc–Phe–[CH(OH)CH_2NH]–Phe–Ile–Phe–NH_2	38.9	43.0
Boc–Phe–[CH_2(OH)CH_2]–Phe–Gln–Phe–NH_2	27.8	38.8

[CH_2NH] depicts reduced amide bond; [CH(OH)CH_2NH] hydroxyethylamine isostere; [CH_2(OH)CH_2] hydroxyethylene isostere.

tion suppress the inhibition of CTAP, however in the case of CAAP the IC_{50} value of this inhibitor increased by one order of magnitude in comparison with Glu, Gln or Ile in the same position. We have demonstrated the preference of active sites of both proteinases for peptide molecules of more than four amino acid residues. The IC_{50} value of inhibitor HPEAF[CH_2NH]FVRKANH_2 decreases by one to two orders of magnitude compared to short inhibitors specifically for CAAP. This type of inhibitor is therefore one of the possible alternatives in the design of inhibitors of *Candida* SAP.

3.3. Testing of Inhibitors during the Growth of the *Candida* Strains

To introduce a rapid method for testing of inhibitors of *Candida* SAP without the isolation of proteinases, we have tested various inhibitors for their ability to inhibit the growth of both *Candida* strains directly in the growth medium. The growth of *C. albicans* and *C. tropicalis* was slowed down by addition of pepstatin A. While 1 μM pepstatin A completely inhibited the growth of *C. albicans*, for *C. tropicalis* this concentration of inhibitor prolonged the lag phase of the growth of the yeast by 20 h. After 90 h, the growth of *C. tropicalis* was inhibited by 26% compared to the control cultivation. This method seems to be an effective tool for rapid screening of large amounts of inhibitors of SAP.

4. CONCLUSIONS

We have tested *in vitro* a series of inhibitors derived from the structure of pepstatin A with proteinases of *C. albicans* and *C. tropicalis.* The replacement of Sta residue in the structure of pepstatin by both Pst and norPst does not improve the activity of inhibitors. We have observed a sensitivity of S4 subsite of both CAAP and CTAP to interaction with P4 residue of inhibitors. Inhibitor containing Boc at the P4 position has lower IC_{50} values for CAAP and CTAP than that with Iva at the same position. On the other hand, introduction of Tba which is by one methylene group longer than Iva, increases the IC_{50} by one and two orders of magnitude for CAAP and CTAP, respectively. The length of inhibitors plays an important role for their effectiveness. Extension of a pentapeptide into a decapeptide improves the inhibition by two orders of magnitude for CAAP. Even more dramatic effect was found for CTAP. The same modification converted an ineffective compound into a micromolar inhibitor.

We have developed the method for testing inhibitors based on the evaluation of growth rate of *Candida* strains.

ACKNOWLEDGMENT

This work was supported by grant 203/95/1028 from the grant Agency of the Czech Republic.

REFERENCES

1. Verfaillie, C., Weisdorf, D. and Haake, R. (1991) Bone. Marrow. Transplant. 8, 177.
2. Cutler, J.E. (1991) Annu. Rev. Microbiol. 45, 187–218.
3. Monod, M., Togni, G., Hube, B., Sanglard, D. (1994) Mol. Microbiol.13, 357–368.
4. Fusek, M. Smith, E.A., Monod, M., Dunn, B.M., Foundling, I.S. (1994) Biochemistry 33, 9791–9799.
5. Cutfield S.M., Dodson, E.J., Anderson, B.F., Moody, P.C.E., Marshall, C.J., Sullivan, P.A., Cutfield, J.F. (1995) Structure 3, 1261–1271.
6. Abad-Zapatero, C., Goldman, R., Muchmore, S.W., Hutchins, C., Stewart, K., Navaza, J., Payne, C.D., Ray T.L. (1996) Protein Science 5, 640–652.
7. Urban, J., Konvalinka, J., Stehlíková, J., Gregorová, E., Majer, P., Souček, M., Andreanský, M., Fábry, M., and Štrop, P. (1992) FEBS Lett. 298, 9–13.
8. Richards, A. D., Phylip, L. H., Farmerie, W. G., Scarborough, P. E., Alvarez, A., Dunn, B.M., Hirel, P. H., Konvalinka, J., Štrop, P., Pavlíčková, L., Kostka, V., and Kay, J. (1990) J. Biol. Chem. 265, 7733–7736.
9. Souček, M., and Urban, J. (1995) Collect. Czech. Chem. Commun. 60, 693–696.
10. Pavlíčková, L., Štys, D., Souček, M., Urban, J., Hrušková, O., Sedláček, J., and Štrop, P. (1992) Arch. Biochem. Biophys. 298, 753–756.
11. Evans, B. E., Rittle, K. E., Homnick, C. F., Springer, J. P., Hirshfield, J., and Veber D.F. (1985) J. Org. Chem. 50, 4615–4625.
12. Capobianco, J.O., Lerner, C.G. and Goldman, R.C. (1992) Anal. Biochem. 204, 96–102.
13. Olert, M.V., Sohnchen, R., Korting, H.C., Ollert, U., Brautigam, S. and Brautigam,W. (1993) Infect. Immun. 61, 4560.

PRIMARY SUBSTRATE SPECIFICITIES OF SECRETED ASPARTIC PROTEASES OF *Candida albicans*

Gerald Koelsch, Jordan Tang, Michel Monod, Stephen I. Foundling, and Xinli Lin

Protein Studies Program
Oklahoma Medical Research Foundation
Oklahoma City, Oklahoma, 73104
Laboratoire de Mycologie
Service de Dermatologie
Centre Hospitalier Universitaire Vaudois
1011 Lausanne, Switzerland

Candidiasis is an opportunistic infection of a pathogenic yeast of the genus *Candida*. These otherwise benign commensals become opportunistically invasive in response to impaired host defense mechanisms. Prominent among putative virulence factors are the secreted aspartic proteases, or Saps (Goldman *et al.*, 1995; Ruchel *et al.*, 1992; Cutler, 1991) which are implicated by biochemical, genetic, and immunochemical evidence. Multiple genes encoding Saps have been identified in the genome of clinical isolates of *Candida* species (Monod *et al.*, 1994). In the most virulent species, *C. albicans*, seven genes have been cloned (*SAP1–SAP7*; Monod *et al.*, 1994). The induction of *SAP* gene expression is strictly regulated in accord with changes in cell phenotype and morphology (Hube *et al.*, 1994), implying that *Candida albicans* may require stage-specific proteases for its life cycle (Odds, 1994). Of these enzymes, *SAP6* was found to be expressed exclusively in the virulent form of these cells, implying a direct invasive function (Hube *et al.*, 1994; White and Agabian, 1995).

Of the Saps of *C. albicans*, only Sap1, Sap2, and Sap3 have been detected in liquid culture supernates of clinically isolated strains, and only Sap2 has been isolated and characterized. As the specificity of the Saps is valuable for the development of potent inhibitors for therapeutic purposes and investigation of the role of each Sap in virulence (Abad-Zapatero *et al.*, 1996; Cutfield *et al.*, 1995; Goldman *et al.*, 1995), the major purpose of this work is to make a complete characterization of the primary specificity of the Saps of *C. albicans*.

Aspartic Proteinases, edited by James.
Plenum Press, New York, 1998.

Aspartic proteases Sap1, Sap2, Sap3, and Sap6 of *C. albicans* have been expressed in *E. coli* from genes encoding their putative zymogen forms. The primary specificities of each Sap toward peptide substrates has been determined by a competitive assay of defined substrate mixtures (McGeehan *et al.*, 1995). Briefly, the relative specificity of a protease toward competing substrates is the ratio of their catalytic efficiencies, k_{cat}/K_m (Fersht, 1985). Where [S] << K_m, the ratio of initial velocities of relative cleavage of each substrate equals their relative k_{cat}/K_m (Fersht, 1985). Sets of defined substrate mixtures were designed based upon a consensus cleavage site in the insulin B-chain (LVEA-LYLV). These mixtures were used to measure the relative k_{cat}/K_m values in competitive assays with each Sap at pH 4, with quantitiation by HPLC-mass spectrometry. A total of 38 unique substrates which vary by amino acid side chains present in the P_1 and P_1' positions were characterized.

Results of P_1 specificity determinations are ranked in Table 1. The preferred amino acid side chain in the P_1 position was Phe for Sap1, Sap2, and Sap6 and was Leu for Sap3. However, the preference for these residues was not exclusive and was closely followed by Leu for Sap1, Sap2, and Sap6 or by Arg for Sap3. As a whole, Saps preferred hydrophobic amino acids in the P_1 site. With relative k_{cat}/K_m values over a 50-fold range, Sap1 and Sap2 accommodated 3 to 4 residues in the P_1 site, whereas Sap3 and Sap6 accepted nearly 10 different residues, including those with polar side chains (data not shown). In accord with other fungal aspartic proteases (Lowther *et al.*, 1995; Oka & Morihara, 1974), the Saps were cleaved peptides with positively-charged amino acids in P_1.

The results of P_1' specificity determinations are ranked in Table 1. The highest preferences are Tyr for Sap1, Sap2, and Sap3, and Ala for Sap6. However, the major preference of each Sap for the P_1' site is not as significant as for the P_1 site, given the breadth of specificity. All Saps were permissive to both the largest (Trp) and the smallest (Ala) amino acid side chain at this site, with some cleavage of the P_1' Gly-containing substrate (data not shown). The trend toward larger hydrophobic amino acids in P_1' was prevalent for all Saps. Intermediate-sized, β-branched aliphatic amino acids Ile and Val were accommodated at the P_1' position by all Saps. Additionally, hydrophilic side chains were more amenable to cleavage in the P_1' position than they were in the P_1 position.

The broad specificity evinced by Saps was analyzed by examination of the crystal structure of Sap2 in complex with a transition-state inhibitor (Abad-Zapatero *et al.*, 1996). Table 2 lists those residues which compose the S_1 and S_1' subsites of the Saps, pepsin, and rhizopuspepsin. Very little difference is seen in the S_1 composition among Saps, although they exhibit varied specificities toward the P_1 site. Sap2 and Sap6 which were generally more tolerant in S_1 possess differences at positions 30 and 119 (Sap2 numbering). Yet Sap6, which is tolerant of smaller amino acids, does not possess a "larger" amino acid in this subsite. The presence of Asp^{86} in S_1, which is Ile in pepsin, has been correlated with the specificity of fungal proteases toward positively-charged amino acid side chains (Lowther *et al.*, 1995). Residues not in the subsite definition yet which are within 8.0 Å of the P_1' side chain may have an indirect effect on specificity (Ridky *et al.*, 1996). Residues within that sphere are highly conserved; however, adjacent to the conserved Tyr^{84} (Tyr^{75} in

Table 1. P_1 and P_1' ranked specificities

Enzyme	P_1 specificities	P_1' specificities
Sap1	Phe > Leu, Tyr	Tyr, Ala > Nle, Phe, Trp, Glu
Sap2	Phe, Leu, Lys, Tyr	Tyr, Phe, Trp, Glu, Leu, Ile, Nle, Asp, Ala
Sap3	Leu, Arg > Phe, Asp, Glu, Lys	Tyr, Phe, Trp, Nle, Ala, Ile
Sap6	Phe, Leu > Tyr	Ala, Nle, Glu, Tyr

Table 2. Amino acid side chains which form subsites S_1 and S_1'

Enzyme	Subsite S_1 definition[a]	Subsite S_1' definition[a]
Sap1	Ile30[b], Asp32, Gly34, Tyr84, Asp86, Ile119, Ile123, Asp218, Gly220, Thr221	Gly34, Tyr84, Gly85, Glu193, Leu216, Asp218, Ile305
Sap2	Ile30, Asp32, Gly34, Tyr84, Asp86, Ile119, Ile123, Asp218, Gly220, Thr221	Gly34, Tyr84, Gly85, Glu193, Leu216, Asp218, Ile305
Sap3	Val30, Asp32, Gly34, Tyr84, Asp86, Val119, Ile123, Asp218, Gly220, Thr221	Gly34, Tyr84, Gly85, Glu193, Leu216, Asp218, Ile305
Sap6	Ile30, Asp32, Gly34, Tyr84, Asp86, Ala119, Ile123, Asp218, Gly220, Thr221	Gly34, Tyr84, Ala85, Thr193, Leu216, Asp218, Ile305
Rpep	Asp30, Asp32, Gly34, Tyr84, Asp86, Xxx119[c], Leu123, Asp218, Gly220, Thr221	Gly34, Tyr84, Gly85, Trp193, Ile216, Asp218, Ile305
Ppep	Ile30, Asp32, Gly34, Tyr84, Thr86, Xxx119[c], Ile123, Asp218, Gly220, Thr221	Gly34, Tyr84, Gly85, Tyr193, Ile216, Asp218, Ile305

[a]Amino acid side chains of Sap2 which interact with those side chains in positions P_1 and P_1' of a substrate-analog inhibitor observed in the crystal structure of its complex with Sap2 (Cutfield et al., 1995). Residues in subsites of other Saps were identified from alignment of their implied amino acid sequences to that of Sap2 and numbered according to the Sap2 sequence. Rpep, rhizopuspepsin; Ppep, porcine pepsin.

[b]Amino acids (three-letter code) appearing in bold-faced print are those which vary between one or more Sap amino acid sequences.

[c]P_1 - Xxx119 - position of 7 amino acid deletion in all Saps

pepsin numbering), Gly^{83} in Saps 1 and 2 is Glu in Sap3 and is Lys in Sap6. Still, a simple "building-block" analysis with respect to size and charge differences of amino acids of the S_1 subsite cannot indicate the extent of their contribution to specificity.

The composition of the S_1' subsite reveals only two differences between Sap6 and the other Saps. Arg^{195} in Saps 1, 2, and 6 is Asn in Sap3, and may account for the slightly reduced specificity of Sap3, relative to the others, for carboxyl moieties of Asp and Glu in the P_1' position (Table 1). Using the crystal structure of Sap2 in complex with a transition-state inhibitor (Abad-Zapatero *et al.*, 1996), molecular models were built of Sap2 complexed with substrates which differ at the P_1' position. Starting coordinates were refined using 100 picoseconds of molecular dynamics and energy minimization of time-averaged structures using AMBER 4.0 (Pearlman *et al.*, 1991). The ability of Sap2 to accommodate a variety of sizes of hydrophobic amino acids was evident in the change of solvent accessible surface area of residues which form the S_1' subsite. With Ala, Ile, or Tyr in the P_1' position, the S_1' pocket expanded by approximately 20 Å^2 of solvent-accessible surface area for 39 Sap2 atoms to accommodate an additional 60 Å^2 increase in each substrate size. The *cis* interactions of the P_1' side chain with P_2 and P_3' side chains of the substrate account for the balance of the surface area increase in the substrate. Atoms of Leu^{216}, Glu^{193}, and Ile^{304} exhibited the largest contributions to increased exposure, by approximately 5, 3, and 1 Å^2, respectively. These residues also exhibited the greatest RMS differences between the P_1' Ala, Ile, or Tyr models. Notably, these positions vary among the Saps and other aspartic proteases (Table 2). The participation of these particular residues in formation of the enzyme-substrate complex may contribute to the versatility of the Saps to accommodate a wide range of amino acid side chains in S_1'.

ACKNOWLEDGMENTS

We thank Dr. Cele Abad-Zapatero and Abbott Laboratories for kindly providing the coordinates of Sap2 complexed with A-70450.

REFERENCES

Abad-Zapatero, C., Goldman, R., Muchmore, S.W., Hutchins, C., Stewart, K., Navaz, J., Payne, C.D., & Ray, T.L. (1996) *Protein Science 5*, 640–652.

Cutfield, S.M., Dodson, E.J., Anderson, B.F., Moody, P.C.E., Marshall, C.J., Sullivan, P.A., & Cutfield, J.F. (1995) *Structure 3,*1261–1271.

Cutler, J.E. (1991) *Annu. Rev. Microbiol. 45*, 187–218.

Fersht, A. (1985) *Enzyme Structure and Mechanism*, Second Edition, W.H. Freeman and Company, New York.

Goldman, R.C., Frost, D.J., Capobianco, J.O, Kadam, S., Rasmussen, R.R., & Abad-Zapatero, C. (1995) *Inf. Agents Dis. 4,* 228–247.

Hube, B., Monod, M., Schofield, D.A., Brown, A.J., & Gow, N.A. (1994) *Mol. Microbiol. 14,* 87–99.

Lowther, W.T., Majer, P., & Dunn, B.M. (1995) *Protein Science 4*, 689–702.

McGeehan, G.M., Bickett, D. M., Wiseman, J.S., Green, M., & Berman, J. (1995) *Meth. Enzymol. 248,* 35–46.

Monod, M., Togni, G., Hube, B., & Sanglard, D. (1994) *Mol. Microbiol. 13,* 357–368.

Odds, F.C. (1994) *J. Am. Acad. Dermatol. 31,* S2–S5.

Oka, T., & Morihara, K. (1974) *Arch. Biochem. Biophys. 165*, 65–71.

Pearlman, D.A., Case, D.A., Caldwell, J.C., Seibel, G.L., Singh, U.C., Weiner, P., & Kollman, P.A. (1991) AMBER 4.0, University of California, San Francisco.

Ridky, T.W., Bizub-Bender, D., Cameron, C.E., Weber, I.T., Wlodawer, A., Copeland, T., Skalka, A.M., & Leis, J. (1996) *J. Biol. Chem. 271,* 10538–10544.

Ruchel, R., de Bernardis, F., Ray, T.L., Sullivan, P.A., & Cole, G.T. (1992) *J. Med. Vet. Mycol. 30 Suppl. 1,* 123–132.

White, C.T., & Agabian, N. (1995) *J. Bacteriol. 177,* 5215–5221.

47

FUNCTIONAL ASPECTS OF SECRETED *Candida* PROTEINASES

Bernhard Hube,[1] Reinhard Rüchel,[2] Michel Monod,[1,3] Dominique Sanglard,[3] and Frank C. Odds[4]

[1]Institut für Allgemeine Botanik
Angewandte Molekularbiologie III
Universität Hamburg
Ohnhorststr. 18, D-22609 Hamburg, Germany
[2]Hygiene Institut der Universität Göttingen
Medizinische Mikrobiologie
Kreuzbergring 57, D-37075 Göttingen, Germany
[3]Institut de Microbiologie
Centre Hospitalier Universitaire Vaudois
CH 1011 Lausanne, Switzerland
[4]Janssen Research Foundation
Turnhoutseweg 30, B-2340 Beerse, Belgium

INTRODUCTION

The yeast *Candida albicans* is adapted to the mucosal milieu of man and is a major cause of superficial and invasive infections in immunocompromised patients.[1] Among the virulence attributes of *C. albicans*, adhesins, yeast-to-hyphal transition, reversible changes in phenotypic characteristics of the cells ("phenotypic switching") and proteinase secretion are most widely studied.[2] Extracellular proteolytic activity of *C. albicans* was discovered by Staib[3] already thirty years ago and the activity was later identified as due to typical aspartic proteinases.[4,5] Subsequently, production of *Candida* proteinase during candidosis was demonstrated by eliciting specific antibodies in patients,[6] and the detection of proteinase antigen in the blood stream and infected tissue of patients.[7,8] Circumstantial evidence supported the view that *Candida* proteinases may play a role in the pathogenesis of candidosis: i) highly proteolytic strains were reported to be more pathogenic than less proteolytic strains;[9–11] ii) chemically induced non-proteolytic mutants were found to be less pathogenic than the wild type strains;[13–15] iii) treatment of infected mice with pepstatin, an inhibitor of aspartic proteinases, caused a discrete protective effect.[16] In an attempt to establish the role played by *Candida* proteinase, a molecular genetic approach

Aspartic Proteinases, edited by James.
Plenum Press, New York, 1998.

was chosen. Surprisingly, a whole family of differentially regulated genes (*SAP1–8*) was detected coding for related secretory proteinases.[17] The *SAP1–6* genes were disrupted by molecular genetic means, and the mutants indeed proved less virulent during experimental infections than the wild type strain.[18,19] Since we now have evidence for the critical role of proteinases in the development of candidosis, this review will emphasize the expression and the specific functions of these enzymes.

INDUCTION OF SAP ACTIVITY

Proteolytic activity of *C. albicans* was induced *in vitro* with a number of proteins or peptides as the sole source of nitrogen at acidic pH values.[17] To clarify which genes are responsible for the proteolytic activity of *C. albicans* the expression of *SAP* genes was studied.[17,20–22] Seven of eight *SAP* genes were found to be regulated differentially *in vitro* (Figure 1): *SAP1* and *SAP3* were regulated during "phenotypic switching." *SAP2* mRNA was the main transcript in log-phase yeast cells grown in protein-containing medium. The expression pattern of *SAP2* and comparison of amino-terminal amino acid sequences of purified Saps suggest that *SAP2* is principally responsible for the proteolytic activity observed in most strains growing in the yeast form *in vitro*. *SAP4–SAP6* were expressed during yeast-to-hyphal transition at neutral pH, while *SAP7* was transcriptionally silent under all conditions tested *in vitro*. Expression of *SAP8* was increased at 25°C compared to expression levels at physiological temperatures. This differential expression of *SAP* genes suggests that Sap isoenzymes may play different roles in the invasion of the host.

PROCESSING OF SAP ISOENZYMES

In addition to a regulation on the mRNA-level, Sap isoenzymes of *C. albicans* are synthesized as zymogens. All *SAP* genes encode a signal sequence, a propeptide with

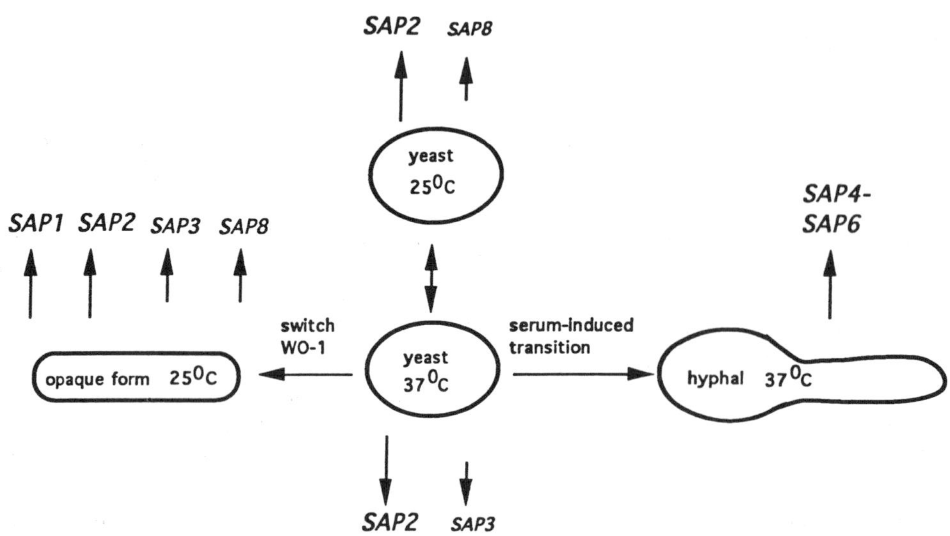

Figure 1. Differential expression of *SAP* genes by different morphological forms of *C. albicans*. "opaque" is a switching type of strain WO-1. *SAP7* was silent or weakly expressed only. Adapted from Hube, 1996.

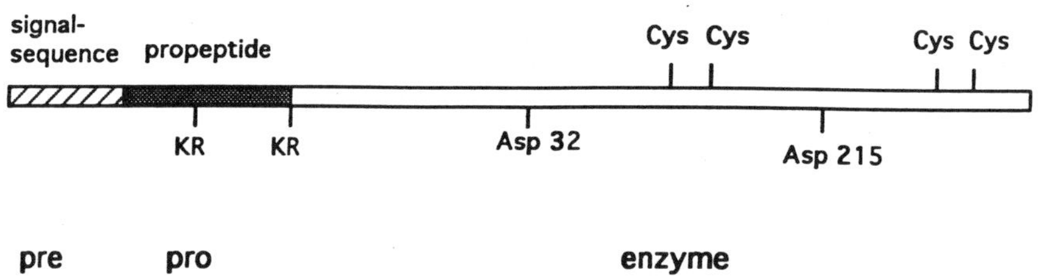

Figure 2. Characteristics of the primary structure of Saps. All *SAP* genes encode a signal sequence, a propeptide with putative Lys–Arg (KR) processing sites and a mature protein. Conserved aspartate residues which constitute the active site of aspartic proteinases are homologous to Asp-32 and Asp-215 of mammalian pepsin. Conserved cysteine residues implicate the maintenance of the three-dimensional structure.

putative Lys–Arg processing sites for a Kex2-like regulatory proteinase[23] and a mature protein with sequences highly conserved among all aspartic proteinases (Figure 2). The removal of both the signal sequence and the proregion was shown to be a prerequisite for the secretion of Saptl, the major proteinase isoenzyme of the closely related yeast *C. tropicalis*.[24]

ARE SAP ISOENZYMES ACTIVE IN THE INFECTED HOST?

The optimum of activity of Sap2, which is the Sap most thoroughly investigated, is in the range of pH 3–4. A good, though unusual target bond is His–Thr.[25] Near the pH-optimum, great hydrolytic activity versus several proteins was observed, but restricted activity was detected up to approx. pH 6. By means of secretion of organic acids from incomplete metabolization of carbohydrates, it is conceivable, that *C. albicans* may acidify its vicinity and hence favour acid proteolytic activity. Among the recently discovered Sap isoenzymes, Sap4–6 are probably induced even in the neutral pH-range.[21,22] Therefore, the previous notion, that activity of Saps and induction of more invasive germ tubes (which occurs at neutrality) are mutually exclusive, cannot be maintained anymore. However, Saps, as aspartic proteinases, may undergo irreversible "alkaline denaturation" in the neutral pH-range. When tested *in vitro,* various Sap2 proteinases from different *C. albicans* strains proved to be comparatively stable, maintaining half of their activity after 30 min. exposure to neutral pH values at 37°C.[26]

WHICH ARE POSSIBLE TARGETS OF SAPS?

The substrate specificity of Sap2 is broad. Immunoglobulins which had been extracted from *Candida* abscesses in mice were largely degraded.[27] The Fc-portion was particularly prone to hydrolysis. Likewise, complement activity was impeded by destruction of the C3 protein.[28] Other components of human defense, like the α 1-proteinase inhibitor, epidermal cystain-A and α 2-macroglobulin were also degraded *in vitro.* The latter was found (as an universal proteinase inhibitor) to bind to Sap, only to be degraded within the complex.[7] Furthermore, proteins acting as a substrate *in vitro* were the proteinaceous portions of mucins[29] β-lactoglobulin, lactoperoxidase, dentine collagen, keratin, haemoglobine, serum albumine, fibronectin, the proteinase inhibitor cystatin A and big endothelin.[17]

WHICH ROLE COULD SAPS PLAY IN PATHOGENESIS?

There is no evidence, that Saps could cause major destruction of infected tissue. However, Saps have been identified as allergens in asthma,[30] which suggests some interference with antigen processing. In addition, Sap isoenzymes may help *C. albicans* to survive the attack of cellular defence of the host. If ingested by phagocytic leukocytes, *C. albicans* cells may be killed or in turn kill the phagocyte. Tolerance of intracellular yeasts has never been observed. *In vitro* experimentation has suggested, that the ingested yeasts are indeed proteolytic and may effectively interfere with the attack of lysosomal enzymes[31] or with the enzymes responsible for the generation of oxygen radicals during phagocytosis.[32] Saps may also be involved in the attachment of the fungus to mucosal and epidermal surfaces, as has been repeatedly demonstrated.[33–36] Pepstatin-sensitive cavitation of the mucosal surface occurs during adhesion of the yeast and therefore suggests the unmasking of binding sites by Saps. In the following step of fungal pathogenesis the proteinases may facilitate invasion into deeper tissue by degrading the extracellular matrix (e.g., fibronectin and laminin). Finally, Saps may activate human proenzymes, for example by a "trypsinogen kinase" effect: clotting factor X is converted as is prekallikrein.[37,38]

WHAT ARE POSSIBLE ROLES OF SAPS OF OTHER *Candida* SPECIES?

C. tropicalis and *C. parapsilosis* are opportunistic pathogens of lesser medical importance than *C. albicans*. The two fungi do not usually colonize mucosa, but, in contrast to other *Candida* species, both can be induced to secrete Saps. In general, a participation of these proteinases in pathogenesis of mycoses seems to be unlikely,[39,40] particularly, since the profiles of alkaline denaturation of their major isoenzymes suggest an instability in the host.

CONCLUSIONS

In the light of the recently established role of the Saps as factors of candidal virulence, and with the new deletion mutants available[18,19] it should be possible to find more precise answers to the questions raised above on the roles played by the Saps during pathogenesis of candidosis. New targets of antifungal therapy may be unveiled in the course of such investigations, and specific inhibitors may be developed.

ACKNOWLEDGMENTS

Our own work was supported by the Deutsche Forschungsgemeinschaft, the EEC, the Royal Society, Janssen Research Foundation, Smith-Kline Beecham, and the Office Fédéral de la Santé Publique.

REFERENCES

1. J. E. Ewards, Invasive *Candida* Infections, *New Engl. J. Med.* **324**:1060–1062 (1991).
2. J. E. Cutler, Putative virulence factors of *Candida albicans*, *Annu. Rev. Microbiol.* **45**:187–218 (1991).
3. F. Staib, Serum-proteins as nitrogen source for yeastlike fungi, *Sabouraudia* **4**:187–193 (1965).

4. F. MacDonald and F. C. Odds, Inducible proteinase of *Candida albicans* in diagnostic serology and in the pathogenesis of systemic candidosis, *J. Med. Microbiol.* **13**:423–435 (1980a).
5. R. Rüchel, Properties of a purified proteinase from the yeast *Candida albicans*, *Biochim. Biophys. Acta* **659**:99–113 (1981).
6. F. Macdonald and F. C. Odds, Inducible proteinase of *Candida albicans* in diagnostic serology and in the pathogenesis of systemic candidosis, *J. Am. Med. Assoc.* 243:2409–2411 (1980b).
7. R. Rüchel and B. Böning, Detection of *Candida* proteinase by enzyme immunoassy and interaction of the enzyme with alpha-2-macroglobulin, *J. Immun. Methods* **61**:107- 116 (1983).
8. R. Rüchel, F. Zimmermann, B. Böning-Stutzer and U. Helmchen, Candidosis visualized by proteinase-directed immunoflourescence, *Virchow's Archiv A Pathological Anatomy* **419**:199–202 (1991).
9. M. Ghannoum and K. Abu Elteen, Correlative relationship between proteinase production, adherence and pathogenicity of various strains of *Candida albicans*, *J. Med. Vet. Mycol.* **24**:407–413 (1986).
10. A. Cassone, F. DeBernardis, F. Mondello, T. Ceddia and L. Agatensi, Evidence for a correlation between proteinase secretion and vulvovaginal candidosis, *J. Infect. Dis.* **156**:777–783 (1987).
11. A. Louie, D. M. Dixon, E. A. El-Maghrabi, J. W. Burnett, A. L. Baltch and R. P. Smith, Relationship between *Candida albicans* epidermolytic proteinase activity and virulence in mice, *J. Med. Vet. Mycol.* **32**:59–64 (1994).
12. M. W. Ollert, C. Wende, M. Görlich, C. G. McMullan-Vogel, M. Borg-von Zepelin, C.-W. Vogel and H. C. Korting, Increased expression of *Candida albicans* secretory proteinase, a putative virulence factor, in isolates from human immunodeficient virus- positive patients, *J. Clin. Microbiol.* **33**:2543–2549 (1995).
13. F. MacDonald F and F. C. Odds, Virulence for mice of a proteinase-secreting strain of *Candida albicans* and a proteinase-deficient mutant, *J. Gen. Microbiol.* **129**:431–438 (1983).
14. K. J. Kwon-Chung, D. Lehman, C. Good and P. T. Magee, Genetic evidence for role of extracellular proteinase in virulence of *Candida albicans*, *Infec. Immun.* **49**:571–575 (1985).
15. I. K. Ross, F. DeBernardis, G. W. Emerson, A. Cassone, and P. A. Sullivan, The secreted aspartate proteinase of *Candida albicans*: physiology of secretion and virulence of a proteinase-deficient mutant, *J. Gen. Microbiol.* **136**:687–694. (1990).
16. R. Rüchel, B. Ritter and M. Schaffrinski, Modulation of experimental systemic murine candidosis by intravenous pepstatin, *Zbl. Bakteriol.* **273**:391–403 (1990).
17. B. Hube, *Candida albicans* secreted aspartyl proteinases, *in*: Current Topics in Medical Mycology **7**:55–69, F. C. Odds, R. J. Hay, and M.G. Rinaldi, eds., J.R. Prous Science Publisher, Barcelona, (1996).
18. B. Hube, D. Sanglard, F. C. Odds, D. Hess, M. Monod, W. Schäfer, A. J. P. Brown and N. A. R. Gow, Gene disruption of each of the secreted aspartyl proteinase genes *SAP1, SAP2* and *SAP3* in *Candida albicans* attenuates virulence, (submitted).
19. D. Sanglard, B. Hube, M. Monod, F. C. Odds and N. A. R. Gow, A triple deletion in the *SAP4, SAP5,* and *SAP6* secretory aspartyl proteinase genes of *Candida albicans* causes attenuated virulence, (submitted).
20. B. Morrow, T. Srikantha and D. R. Soll, Transcription of the gene for a pepsinogen, *PEP1*, is regulated by white-opaque switching in *Candida albicans*, *Mol. Cell Biol.* **12**:2997–3005 (1992).
21. B. Hube, M. Monod, D. A. Schofield, A. J. P. Brown and N. A. R. Gow, Expression of seven members of the gene family encoding secretory aspartyl proteinases in *Candida albicans*, *Mol. Microbiol.* **14**:87–99 (1994).
22. T. C. White and N. Agabian, *Candida albicans* secreted aspartyl proteinases: isoenzyme pattern is determined by cell type, levels are determined by enviromental factors, *J. Bacteriol.* **177**:5215–5221 (1995).
23. D. Julius, A. Blake, L. Blair, R. Kunisawa and J. Thorner, Isolation of the putative structural gene for the lysine-arginine-cleaving endopeptidase required for processing of yeast prepro-alpha-factor, *Cell* **37**:129–132 (1984).
24. G. Togni, D. Sanglard, M. Quadroni, S. I. Founding and M. Monod, Acid proteinase secreted by *Candida tropicalis*: functional analysis of preproregion cleavages in *C. tropicalis* and *Saccharomyces cerevisiae*, *Microbiology* **142**:493–503 (1996).
25. J. O. Capobianco, C. G. Lerner and R. C. Goldman, Application of a fluorogenic substrate in the assay of proteolytic activity and in the discovery of a potent inhibitor of *Candida albicans* aspartic proteinase, *Anal. Biochem.* **204**:96–102 (1992).
26. T. Wagner, M. Borg-v. Zepelin and R. Rüchel, pH-dependent denaturation of extracellular aspartic proteinases from *Candida* species, *J. Med. Vet. Mycol.* **33**:275–278 (1995).
27. R. Rüchel, A variety of *Candida proteinases* and their possible targets of protepolytic attack in the host, *Zbl. Bakt. Hyg. A* **257**:266–274 (1984).
28. H. Kaminishi, H. Miyaguchi, T. Tamaki, N. Suenaga, M. Hisamatsu, I. Mihashi, H. Matsumoto, H. Maeda and Y. Hagihara, Degradation of humoral host defence by *Candida albicans* proteinase, *Infect. Immun.* **63**:984–988 (1995).

29. A.-R. Colina, F. Aumont, N. Deslauriers, P. Belhumeur and L. De Repentigny, Evidence for degradation of gastrointestinal mucin by Candida albicans secretory aspartyl proteinases, *Infec. Immun.* **64**:4514–4519 (1996).
30. K. Akiyama, T. Shida, H. Yasueda, H. Mita, T. Yamamoto and H. Yamaguchi, Atopic asthma caused by *Candida albicans* acid protease: case reports, *Allergy* **49**:778–781 (1994).
31. M. Borg and R. Rüchel, Demonstration of fungal proteinase during phagocytosis of *Candida albicans* and *Candida tropicalis*, *J. Med. Vet. Mycol.* **28**:3–14 (1990).
32. M. Borg-v.Zepelin and P. Schuff-Werner, Chemiluminescence of polymorphonuclear granulocytes in the presence of selected *Candida* species, *Mycoses* **35**:121–129 (1992).
33. M. Borg and R. Rüchel, Expression of extracellular acid proteinase by proteolytic *Candida* spp. during experimental infection of oral mucosa, *Infec. Immun.* **56**:626–631 (1988).
34. T. L. Ray and C. D. Payne, Scanning electron microscopy of epidermal adherence and cavitation in murine candidiasis: a role for *Candida acid* proteinase, *Infec. Immun.* **56**:1942–1949 (1988).
35. E. A. El-Maghrabi, D. M. Dixon and J. W. Burnett, Characterization of *Candida albicans* epidermolytic proteases and their role in yeast-cell adherence to keratinocytes, *Clin. Experiment. Dermatol.*, **15**:183–191 (1990).
36. M-W. Ollert, R. Söhnchen, H. C. Korting, U. Ollert, S. Bräutigam and W. Bräutigam, Mechanisms of adherence of *Candida albicans* to cultured human epidermal keratinocytes, *Infec. Immun.* **61**: 4560–4568 (1993).
37. H. Kaminishi, M. Tanaka, T. Cho, H. Maeda and Y. Hagihara, Activation of the plasma kallikrein-kinin system by *Candida albicans* proteinase, *Infec. Immun.* **58**:2139–2143 (1990).
38. H. Kaminishi, H. Hamatake, T. Cho, T. Tamaki, N. Suenaga, T. Fujii, Y. Hagihara and H. Maeda, Activation of blood clotting factors by microbial proteinases, *FEMS Microbiol. Lett.* **121**:327–332 (1994).
39. P. A. De Viragh, D. Sanglard, G. Togni, R. Falchetto and M. Monod, Cloning and sequencing of two *Candida* parapsilosis genes encoging acid proteases, *J. Gen. Microbiol.* **139**:335–342 (1993).
40. G. Togni, D. Sanglard and M. Monod, Acid proteinase secretion by *Candida tropicalis*: virulence of a proteinase negative mutant, *J. Med. Vet. Mycology* **32**:257–265 (1994).

48

SUBSTRATE SPECIFICITY OF NON-PEPSIN-TYPE ACID PROTEINASE, *Aspergillus niger* PROTEINASE A

Shinji Komatsu,[1] Wataru Nishii,[2] Hiroshi Sasaki,[1] Tomonari Muramatsu,[2] and Masaru Tanokura[1]

[1]Biotechnology Research Center
University of Tokyo, Bunkyo-ku, Tokyo 113
[2]National Cancer Center Research Institute
Chuo-ku, Tokyo 104, Japan

INTRODUCTION

Proteinase A secreted by the fungus *Aspergillus niger* var. *macrosporus* (Aspergillopepsin II, EC 3.4.23.19) has been considered to be a non-pepsin-type acid proteinase, because the enzyme has various unique features as compared with typical pepsin-type aspartic proteinases.[1,2] It is insensitive to inhibitors specific and prevalent to pepsin-type aspartic proteinases, such as pepstatin, diazoacetyl-DL-norleucine methyl ester (DAN), and 1,2-epoxy-3-(*p*-nitrophenoxy)propane (EPNP).[3] Its molecular weight is about half of those of pepsin-type aspartic proteinases, consisting of two polypeptide chains, a light chain of 37 amino residues and a heavy chain of 173 residues, bound non-covalently to each other.[4,5] No protein in the GenBank database is homologous in the primary structure with the enzyme except for *Scytalidium lignicoloum* proteinase B.[4] Proteinase A has no consensus sequence around catalytic Asp residues in aspartic proteinases, Asp–Thr/ Ser–Gly, and it shows distinct substrate specificity to oxidized insulin B chain and oxidized RNase A.[2,6] In some studies it was reported that the enzymatic cleavage occurs after an aromatic residue, and it prefers a longer peptide chain,[6,7] but its substrate specificity has still been vague. In this study, we applied the peptide library method[8–10] to the enzyme to investigate its P1 and P2 site specificities, with two synthetic peptide libraries, Arg–Gly–Phe–Phe–Xaa–Thr–Pro–Arg–Ala and Arg–Gly–Phe–Xaa–Tyr–Thr–Pro– Arg–Ala.

MATERIALS AND METHODS

Proteinase A was purified from the lyophilized culture medium of *Aspergillus niger* var. *macrosporus* by two step chromatography on DEAE-Toyopearl columns.[11] After

Aspartic Proteinases, edited by James.
Plenum Press, New York, 1998.

purification, the enzyme was dissolved in 50 mM sodium acetate buffer at pH 3.0, and stored at −80°C.

Peptide libraries were synthesized by the ordinary solid-phase peptide synthesis method at 0.1 mmol scale using Fmoc-chemistries and HOBT activation. The starting peptide sequence, Arg–Gly–Phe–Phe–Tyr*Thr–Pro–Arg–Gly (* is the cleaved bond), is a mimic nonapeptide decided from one of the cleavage site of oxidized insulin B chain.[6] To synthesize Xaa substituted site, we used the equimolar mixture of 20 protected Fmoc-amino acids. Cleavage of the peptides from the resin and deprotection of the side-chain protecting groups was performed using a mixture of phenol (0.05 g), ethanedithiol (0.05 ml), thioanisole (0.10 ml), water (0.10 ml), and trifluoroacetic acid (TFA) (1.0 ml) with stirring for 100 minutes. Peptides were separated from the resin with a glass filter, precipitated in 15 ml of ice-cold diethyl ether, and collected by centrifugation at 2500 g. Peptide libraries were purified with reverse-phase HPLC using a Shimadzu LC-10 system with a YMC PROTEIN-AP column (250 × 20 mm) with mobile phases of water and acetonitrile, each containing 0.1% (v/v) trifluoroacetic acid (TFA). Peptides in the libraries were separated into individual peptides and characterized by FAB-MS to confirm that peptide libraries contain 20 objective peptide.

Peptide digestion was performed in 50 mM glycine-HCl buffer at pH 1.8 and 37°C. The reaction was started by mixing a peptide library solution and an enzyme solution to make the mole ratio approximately 5000:1. To stop the reaction, the reaction solution was added to twice the volume of 0.5M phosphate buffer, pH 8.0. Then the solution was directly injected into a reverse-HPLC system to separate peptides and to analyze them. To separate the peptides of libraries, we used Shimadzu LC-10 system with two YMC ODS-M80 columns (150 × 4.6 mm) connected in series. Chromatograms were developed over 120 minutes using linear gradients. Flow rates for all analyses were 1 ml/min. Peak heights were used to calculate relative k_{cat}/K_M values after chromatograms.

Under the condition that a number of substrates compete against one another, $v_0/[S]_0$ values are proportional to k_{cat}/K_M values.[8,12] The peak heights are also proportionate to concentrations of substrates ($[S]_0$), the initial velocities of the fraction of peptide hydrolyzed are directly proportional to k_{cat}/K_M values of each peptide. Relative k_{cat}/K_M values were calculated from the decrease in the HPLC peak heights of individual peptides, in this manner.

RESULTS AND DISCUSSION

The peptide library RGFFXTPRA was subjected to proteinase A digestion, and analyzed with a reverse-phase HPLC system (Figure 1). After HPLC analysis, some peaks contains two or more peptides. In such a case, we treated a complexed peak as one "co-eluted" peptide mixture. Each peak height was measured to determine the fraction of peptide hydrolyzed, and relative k_{cat}/K_M values were calculated from the initial velocities of the fraction of peptide hydrolyzed. To investigate the specificity at P2 site, the peptide library RGFXYTPRA was used and k_{cat}/K_M values were calculated in the same way.

The results on P1 and P2 site investigation were summarized in Tables 1 and 2. At first glance, it is clear that the specificity at P1 site is higher than that at P2 site. It shows that the enzyme prefers methionine, phenylalanine, and histidine at P1 site, which cannot be categorized as "aromatic residues." It is remarkable that the peptide, which contains tryptophan, a large aromatic residue, at P1 site, was not hydrolyzed, nor the peptides contains branched chain amino acid residues, such as leucine or isoleucine, neither. These evidences mean that the specificity at P1 site of the enzyme is considerably complicated.

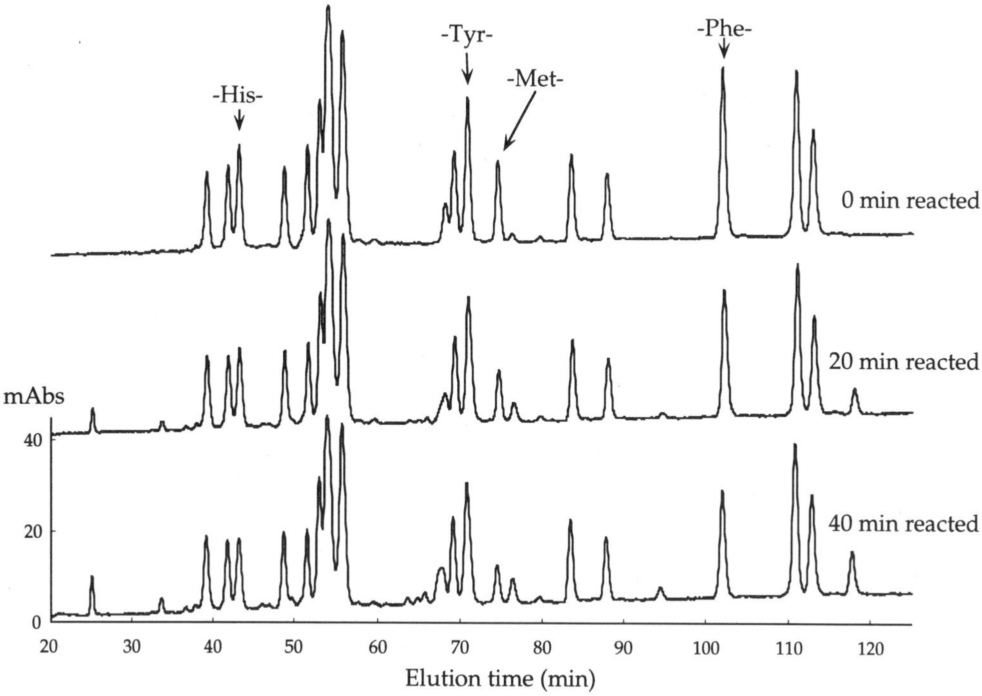

Figure 1. HPLC chromatograms of P1 substituted peptide library, RGFFXTPRA, analyses. Chromatography was performed using a linear gradient, from 6 to 16% acetonitrile over 120 minutes, and tracing at 215 nm.

Table 1. Relative k_{cat}/K_M values of the P1 substituted peptide library, RGFFXTPRA, for proteinase A

Amino acid substitution at P1	Relative k_{cat}/K_M [†]
Lys	0.0
Arg	1.0
His	3.4
Gly	0.0
Asn	2.2
Ser,Ala (co-eluted)	0.7
Asp,Gln (co-eluted)	1.7
Glu,Thr (co-eluted)	0.8
Pro,Val (co-eluted)	0.0
Tyr	1.0
Met	6.4
Ile or Leu	0.0
Ile or Leu	0.0
Phe	3.8
Trp	0.0
Cys(tBut)	0.0

†Values are relative to that of RGFFYTPRA.

Table 2. Relative k_{cat}/K_M values of the P2 substituted peptide library, RGFXYTPRA, for proteinase A

Amino acid substitution at P2	Relative k_{cat}/K_M *
Lys	0.7
His	0.3
Arg	1.0
Gly	0.0
Ser,Asn,Gln (co-eluted)	0.2
Thr	0.2
Ala,Asp,Glu (co-eluted)	0.3
Cys	0.4
Val	0.4
Pro,Tyr	1.5
Met	1.1
Ile or Leu	0.3
Ile or Leu	0.7
Phe	1.0
Trp	1.1

*Values are relative to that of RGFFYTPRA.

Table 3. The kinetic parameters and k_{cat}/K_M values determined for individual peptides, and compared with the relative values objected by the peptide library method

Sequence	k_{cat} (sec^{-1})	K_M (mM)	Relative k_{cat}/K_M Individual	Relative k_{cat}/K_M Library
RGFFMTPRA	0.26	0.016	9.9	3.2
RGFFFTPRA	0.48	0.070	4.2	2.3
RGFFHTPRA	0.19	0.016	7.3	2.1
RGFFYTPRA	0.54	0.33	1.0	1.0

To confirm that the peptide library method evaluates k_{cat}/K_M values correctly, some peptides, which RGFFXTPRA contains, were synthesized separately, and the k_{cat} and K_M values were measured individually. The result certifies the correctness of this method (Table 3).

The result on P2 site specificity showed that the enzyme seems to tend to cleave only the peptides which contain large residue at P2 site, hardly influenced by the charges of the residues.

REFERENCES

1. K. Takahashi, M. Tanokura, H. Inoue, M. Kojima, Y. Muto, M. Yamasaki, O. Makabe, T. Kimura, T. Takizawa, T. Hamaya, E. Suzuki, and H. Miyano, Structure and function of pepstatin-insensitive acid proteinase from *Aspergillus niger* var. *macrosporus*, *in*: Structure and Function of the Aspartic Proteinases: Genetics, Structures, and Mechanisms, B. Dunn, ed., 203–211, Plenum Press (New York) (1991)
2. K. Takahashi, Proteinase A from *Aspergillus niger* , *Methods in enzymology* **248**,146–154 (1995)
3. W.-J. Chang, S. Horiuchi, K. Takahashi, M. Yamasaki, and Y. Yamada, The structure and function of acid proteases. VI. Effects of acid protease-specific inhibitors on the acid proteases from *Aspergillus niger* var. *macrosporus*, *J. Biochem. (Tokyo)* **80**, 975–981 (1976)
4. K. Takahashi, H. Inoue, K. Sakai, T. Kohama, S. Kitahara, K. Takashima, M. Tanji, S. B. P. Athauda, T. Takahashi, H. Akanuma, G. Mamiya, and M. Yamasaki, The primary structure of *Aspergillus niger* acid proteinase A, *J. Biol. Chem.* **266**,19480–19483 (1991)
5. H. Inoue, T. Kimura, O. Makabe, and K. Takahashi, The gene and deduced protein sequence of the zymogen of *Aspergillus niger* acid proteinase A, *J. Biol. Chem.* **266**,19484–19489 (1991)
6. K. Iio, and M. Yamasaki, Specificity of acid proteinase A from *Aspergillus niger* var. *macrosporus* towards B-chain of performic acid oxidized bovine insulin, *Biochem. Biophys. Acta* **429**,912 (1976).
7. E. Ido, T. Saito, and M. Yamasaki, Substrate specificity of acid proteinase A from *Aspergillus niger* var. *macrosporus*, *Agric. Biol. Chem.* **51**,2855–2856 (1987)
8. J. R. Petithory, F. R. Masiarz, J.F. Kirsch, D. V. Santi, and B. A. Malcolm, A rapid method for determination of endoproteinase substrate specificity: Specificity of the 3C proteinase from hepatitis A virus, *Proc. Nati. Acad. Sci. USA* **88**, 11510–11514 (1991)
9. J. Bermann, M. Green, E. Sugg, R. Anderegg, D. S. Millington, D. L. Norwood, J, McGeehan, and J. Wiseman, Rapid Optimization of Enzyme Substrate Using Defined Substrate Mixtures, *J. Biol. Chem.* **267**, 1434–1437 (1992)
10. A. J. Birkett, D. F. Soler, R. L. Wolz, J. S. Bond, J. Wiseman, J. Berman, R. Harris, Determination of Enzyme Specificity in a Complex Mixture of Peptide Substrates by N-Terminal Sequence Analysis, *Anal. Biochem.* **196**, 137–143 (1991)
11. M. Tanokura, H. Matsuzaki, S. Iwata, A. Nakagawa, T. Hamaya, T. Takizawa, and K. Takahashi, Crystallization and Preliminary X-ray Investigation of Proteinase A, a Non-Pepsin-type Acid Proteinase from *Aspergillus niger* var. *macrosporus*, *J. Mol. Biol.* **223**, 373–375 (1992)
12. A. Firsht, in *Enzyme Structure and Mechanism*, Freeman (New York), 111–112 (1985)

PEPSTATIN-INSENSITIVE CARBOXYL PROTEINASES FROM PROKARYOTES

Catalytic Residues and Substrate Specificities

Kohei Oda,[1] Saori Takahashi,[1] Masaaki Ito,[1] and Ben M. Dunn[2]

[1]Department of Applied Biology
Kyoto Institute of Technology
Matsugasaki, Sakyoku, Kyoto 606, Japan
[2]Department of Biochemistry and Molecular Biology
University of Florida
College of Medicine
Gainesville, Florida 32610-0245

Pseudomonas sp. 101 carboxyl proteinase (PCP)[1] and *Xanthomonas* sp. T-22 carboxyl proteinase(XCP)[2] are the first and second pepstatin-insensitive carboxyl proteinases(CPs) from prokaryotes. These enzymes are not inactivated by pepstatin, DAN, and EPNP. Recently, we have cloned, sequenced, and expressed both PCP[3] and XCP[4] genes. The primary structures of PCP and XCP show no similarity to pepstatin-sensitive CPs (aspartic proteinases), whereas about 52% identity exists between PCP and XCP. Moreover, the conserved catalytic residues (–Asp*–Thr–Gly–) for pepstatin-sensitive CPs do not exist in both PCP and XCP.

As a part of understanding the structure-function relationships of PCP and XCP, site-directed mutagenesis was carried out for the identification of catalytic residues, and the substrate specificities using synthetic chromogenic substrates were also investigated.

MATERIALS AND METHODS

Materials

Taq DNA polymerase was from Perkin Elmer and restriction endonucleases from Nippon Gene. DNA ligation kit was from Takara. Authentic PCP and XCP were purified by the method of Oda *et al.*[1,2] The chromogenic peptide substrates used in this experiments were synthesized and characterized as described previously.[5,6]

Aspartic Proteinases, edited by James.
Plenum Press, New York, 1998.

Site-Directed Mutagenesis and Western Blotting

Site-directed mutagenesis was performed using polymerase chain reaction (PCR) technique.[7] The newly developed expression vector, pUK223-3[4] was used for the construction of PCP and XCP mutants and expressed in *E. coli* JM109 cells. For the construction of XCP mutants, pUKXCPΔ192, a C–Pro region deletion plasmid[4] was used to adjust the activation conditions of XCP to PCP. Western blot analyses of the expressed PCP and XCP mutant proteins were performed by the reported methods.[3,4]

Purification of Recombinant PCP and XCPs

Cell cultures containing recombinant plasmids were grown at 30°C with aeration in 2 x YT medium (1.6% bactotryptone, 1.0% yeast extract, and 0.5% NaCl, pH 7.0) supplemented with 1.0% glucose and 0.1 mg/ml ampicillin. When the A_{660} reached 2.0, IPTG was added at a final concentration of 1mM. Growth was continued at 30°C for 3 h. Cells from 1,000 ml cultures were collected by centrifugation and sonicated in 100 ml of 20 mM sodium phosphate buffer, pH 6.5. The sonicate was centrifuged at 25,000 x g for 30 min and the supernatant was fractionated with ammonium sulfate (80% saturation). The precipitate was dissolved in 50 mM sodium acetate buffer, pH 4.8 and dialyzed to activate enzymes against the same buffer at 25°C overnight. The dialysate was applied to a column of DEAE-Sepharose CL-6B and the adsorbed proteins were eluted with NaCl gradient. The active fractions were pooled and further purified by Sephadex G-75 column chromatography to homogeneity.

Determination of Kinetic Parameters for Chromogenic Substrates

The general structure of the substrate used in this study is P5–P4–P3–P2–Phe–Nph–P2′–P3′ (Nph is *p*-nitro-L-phenylalanine). The hydrolysis of a substrate between Phe and Nph was monitored at 37°C in 0.1 M sodium formate buffer, pH 3.5 by a decrease in absorbance at 300 nm. The K_m and V_{max} values were estimated by Lineweaver-Burk plotting, with at least six initial substrate concentrations. The k_{cat} values were obtained with the equation: $V_{max} = k_{cat}$ [E], where [E] is the enzyme concentration.

RESULTS AND DISCUSSION

We presumed catalytic residues based on the sequence similarity between PCP and XCP (Fig. 1) and constructed several Ala mutants of Asp or Glu residues. PCP mutants of D84A, D124A, D170A, E217A, E222A, D225A, D265A, and D328A, and XCP mutants of XD79A, XD169A, XE230A, XE235A, and XD348A were constructed. When wild-type PCP and XCP were expressed in *E. coli*, inactive precursors were accumulated in the cells. These precursors were autocatalytically converted to active mature enzymes under the acidic conditions[3,4] (Fig. 2). The D170A, E222A, and D328A mutants for PCP and XD79A, XD169A, and XD348A mutants for XCP were not converted to mature form after activation, and catalytic activity could not be detected in these mutants. The active mutants of PCP were purified and the kinetic parameters were determined using synthetic substrates. The specificity constants of the purified PCP mutants are nearly the same value to that of wild-type PCP except for D84A mutant (Table 1). The specificity constant of D84A mutant was four orders of magunitude lower than that of wild-type PCP. These

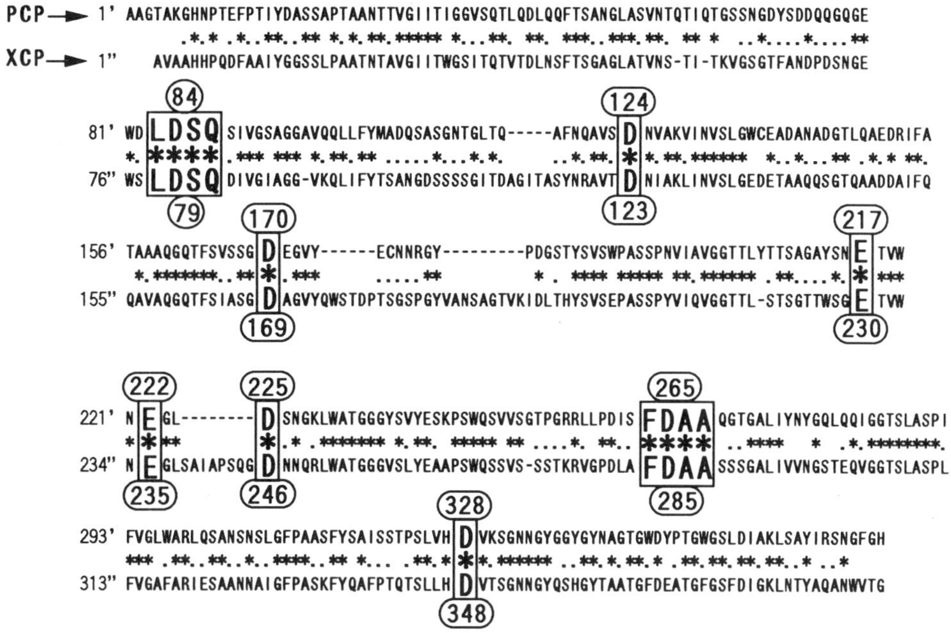

Figure 1. Candidates for catalytic residues of PCP and XCP.

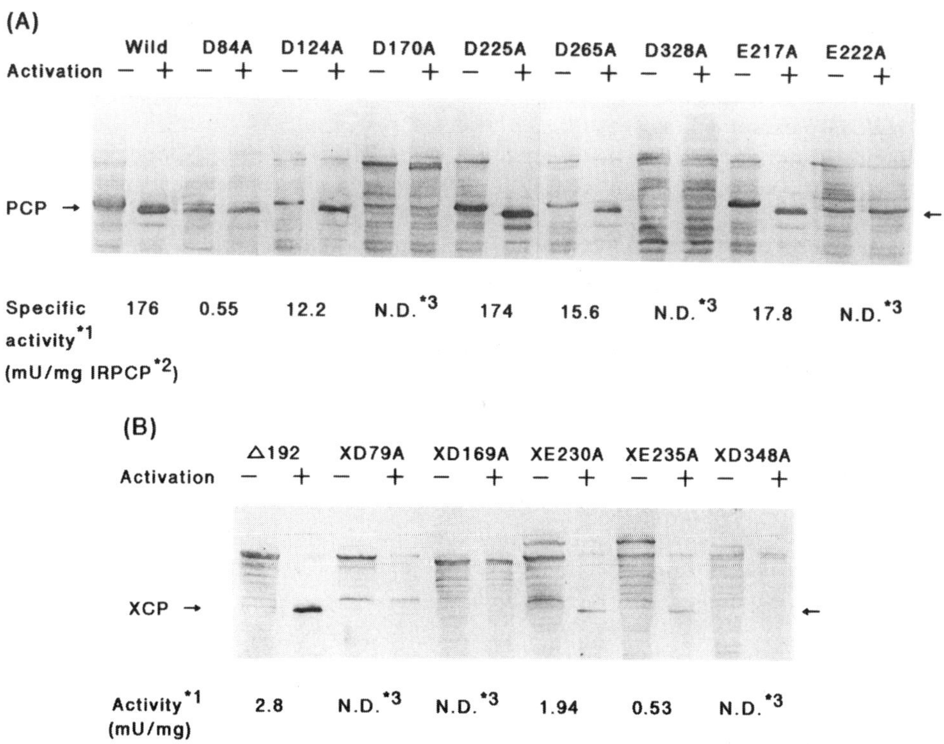

Figure 2. Western blot analyses of PCP (A) and XCP (B) mutants expressed in *E. coli* JM109 cells. *E. coli* extract before (–) and after (+) activation were resolved in SDS-PAGE (12.5% acrylamide) and analyzed by Western blotting. The arrow heads indicate the authentic PCP (A) and XCP (B), respectively. *[1] Proteinase activity was measured using casein as a substrate. *[2] The concentrations of immuno-reactive PCP (IRPCP) in *E. coli* extract were determined by enzyme immunoassay. *[3] N.D., Proteinase activity was not detected.

Table 1. Hydrolysis of chromogenic substrate by purified PCP mutants

	Wild-type	D84A	D124A	D217A	D225A	D265A
K_m (μM)	7.4	149	6.7	10.1	29.6	9.1
k_{cat} (s^{-1})	59.3	0.055	30.9	56.0	65.2	59.5
k_{cat}/K_m ($\mu M^{-1}s^{-1}$)	8.0	0.0009	5.8	5.5	2.2	6.5

Substrate; Lys–Pro–Ala–Leu–Phe–Nph–Arg–Leu (Nph, p-nitro-L-phenylalanine)

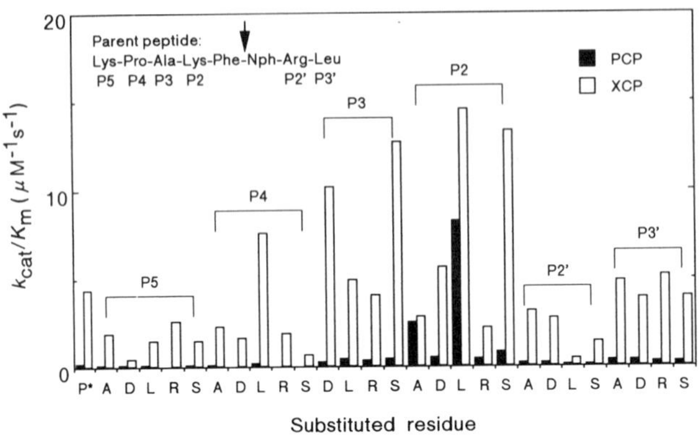

Figure 3. Effect of substitutions at P5 to P3′ of chromogenic peptide substrates on specific constants for PCP and XCP. P*, parent peptide.

results strongly suggest that Asp170 and Asp328 of PCP and Asp169 and Asp348 of XCP are catalytic residues. Asp 84 and Glu222 of PCP and Asp79 of XCP were also suggested to be involved in their catalytic functions.

We also examined the substrate specificities of PCP and XCP using 28 chromogenic synthetic substrates with a general structure of P5–P4–P3–P2–Phe–Nph–P2′–P3′.[5,6] The rates of hydrolysis by PCP and XCP were influenced by the replacements at P5, P4, P3, P2, and P2′ positions in a substrate. These results suggest that PCP and XCP have at least 7 subsites (S5–S2′). PCP and XCP were shown to hydrolyze a synthetic substrate, Lys–Pro–Ala–Leu–Phe–Nph–Arg–Leu, most effectively among the substrates tested. The kinetic parameters of this peptide for PCP were K_m=6.3 μM, k_{cat}=51.4 s^{-1}, and k_{cat}/K_m=8.16 $\mu M^{-1} s^{-1}$. The kinetic parameters for XCP were K_m=3.6 μM, k_{cat}=52.2 s^{-1}, and k_{cat}/K_m= 14.5 $\mu M^{-1} s^{-1}$. As shown in Fig. 3, PCP showed a stricter substrate specificity than XCP. That is, the specificity constant (k_{cat}/K_m) of each substrate for PCP was in general <0.5 $\mu M^{-1} s^{-1}$, but was drastically improved by the replacement of Lys by Leu at the P2 position. Similar characteristics were observed for human cathepsin E.[8] On the other hand, XCP showed a less strict substrate specificity, with most of the peptide exhibiting reasonable k_{cat}/K_m values (>1.0 $\mu M^{-1} s^{-1}$). These characteristics are similar to those of rhizopuspepsin.[9] Thus it was found that the substrate specificities of PCP and XCP differ considerably, in spite of the high similarity in their primary structures.

REFERENCES

1. K. Oda, M. Sugitani, K. Fukuhara, and S. Murao, *Biochim. Biophys. Acta* 923, 463–469 (1987)
2. K. Oda, T. Nakazima, T. Terashita, K. Suzuki, and S. Murao, *Agric. Biol. Chem.* 51, 3073–3080 (1987)
3. K. Oda, T. Takahashi, T. Tokuda, Y. Shibano, and S. Takahashi, *J. Biol. Chem.* 269, 26518–26524 (1994)
4. K. Oda, M. Ito, K. Uchida, Y. Shibano, and S. Takahashi, *J. Biochem.* 120, 564–572 (1996)
5. B. M. Dunn, M. Jimenez, B. F. Parten, M. J. Valler, C. E. Rolph, and J. Kay, *Biochem. J.* 237, 899–906 (1986)

6. B. M. Dunn, P. E. Scarborough, R. Davenport, and W. Swietniki, (1994) Methods in Molecular Biology, B. M. Dunn and M. W. Pennington eds. 36, 225–243, Humana Press Inc., Totawa, NJ
7. R. Higuchi, Recombinant PCR (1990) M. A. Innis, D. H. Gelfand, J. J. Sninsky, and J. J. White eds. PCR Protocols, A Guide to Methods and Applications. Academic Press.
8. C. Rao-Naik, K. Guruprasad, B. Batley, S. Rapundalo, J. Hill, T. Blundell, J. Kay, and B. M. Dunn, *Proteins* 22, 168–181 (1995).
9. W. T. Lowther, Z. Chen, X. L. Lin, J. Tang, and B. M. Dunn, (1991) Structure and function of the aspartic proteinases, B. M. Dunn ed. pp.275–279, Plenum Press, New York.

OVERCOMING THE UNFAVOURABLE ENTROPIC CONTRIBUTION OF LIGAND BINDING WITH A MACROCYCLIC INHIBITOR BOUND TO PENICILLOPEPSIN

Marie E. Fraser,[1] J. Hoyt Meyer,[2] Paul A. Bartlett,[2] and Michael N. G. James[1]

[1]MRC Group in Protein Structure and Function
Department of Biochemistry
University of Alberta
Edmonton, Alberta T6G 2H7, Canada
[2]Department of Chemistry
University of California
Berkeley, California 94720-1460

1. INTRODUCTION

One way to improve the binding of a ligand to its protein receptor is by forcing the unbound ligand to adopt the same conformation as it has when bound to the receptor. The conformational constraint then reduces the entropic disadvantage of ordered binding to the protein. The constraint should change the structure of the ligand as little as possible, so as to retain the favourable enthalpic interactions of the ligand with the receptor. Including a ring structure in the design of a ligand can reduce its conformational flexibility and it may be possible to introduce the ring without changing how the ligand binds to the receptor.

1.1. Design of the Inhibitor

Meyer and Bartlett have designed a macrocyclic inhibitor of penicillopepsin based on the structure of the phosphonate inhibitor Iva–Val–Val–Leu^{P}(O)PheOMe [Iva=isovaleryl, Leu^{P}=the phosphinic acid analogue of L-leucine, (O)Phe=L-3-phenyllactic acid, OMe= methyl ester] (PP2) in a complex with penicillopepsin [1]. In the complex, the backbone of PP2 is in an extended conformation, but the side chains of the P1′ and the P2 residues (Phe and Val) and of the P1 and P3 residues (Leu^{P} and Val) share extended binding pockets on opposite sides of the peptide (Figure 1). It should be possible to constrain the conformation

Aspartic Proteinases, edited by James.
Plenum Press, New York, 1998.

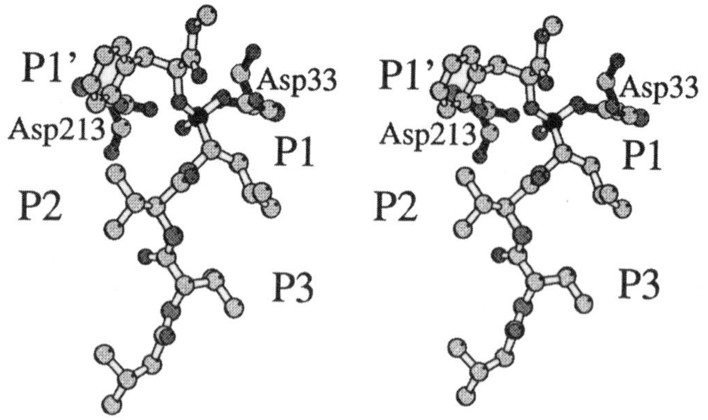

Figure 1. PP2 and two aspartic acid residues in the active site. In this ball-and-stick drawing, the atoms are shaded according to their types, with the intensity of the shading increasing from carbon to nitrogen to oxygen to phosphorus. The inhibitor is drawn with unshaded sticks representing the bonds whereas the aspartic acid residues have shaded sticks. This stereo figure was drawn using the program MOLSCRIPT [2].

of the inhibitor by cyclizing it, and still have a good inhibitor. The cyclization in our work involved linking the atoms of the residues in the P1 and P3 positions.

CAVEAT [3] was used to search a 3D-database for a structural motif that could be used as the bridging unit. The database was searched for a motif having bonds that would overlap with the Cα–Cβ bonds of the P1 and P3 residues of PP2 and would link the two side chains. Any of the resulting motifs that also overlapped with atoms in the active site of the enzyme were rejected. The search suggested using a *trans*-decalin or a naphthalene motif as the bridging unit. Since naphthalene is the easier synthetic target, it was chosen as the bridging unit. The macrocyclic inhibitor, **1**, as well as the two acyclic analogues, **2** and **3**, (Figure 2) were synthesized. Availability of the macrocyclic inhibitor and the acyclic analogues makes it possible to evaluate the entropic effect of the cyclization—provided that they bind to penicillopepsin in the same manner.

1.2. Inhibition of Penicillopepsin

Inhibition constants were determined at 25ºC, pH 3.5, in 100 mM sodium acetate using the substrate Ac–Ala–Ala–Lys–(p–NO_2)Phe–Ala–Ala–NH_2 [4]. They are presented in Table 1.

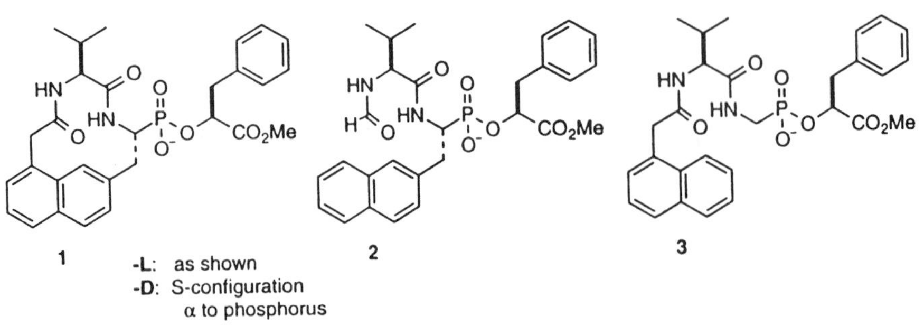

Figure 2. Schematic diagram of the macrocyclic inhibitor, **1**, and its two acyclic analogues, **2** and **3**.

Table 1. Inhibition constants for the compounds

Compound	K_i (μM)
1-L	0.80
1-D	>1000
2-L	7.6
2-D	>1000
3	110

Compounds **1** and **2**, the compounds with naphthalene in the P1 position, have an extra chiral centre. It is not surprising that the analogues with the natural chirality α to the phosphorus (**1-L**, **2-L**) bind better than those with opposite chirality (**1-D**, **2-D**).

The macrocyclic inhibitor **1** is the analogue with the least conformational freedom and it is the best inhibitor. Analogue **3** would be expected to have more conformational freedom than **2** and it does bind more weakly. But these interpretations of variations of the inhibition constants in terms of conformational freedom assume that the three inhibitors bind in the same way to penicillopepsin. To confirm or deny this, the structure of each inhibitor in complex with the enzyme should be solved and refined.

2. CRYSTALLOGRAPHIC DATA

The macrocyclic inhibitor **1-L** has been cocrystallized with penicillopepsin. Diffraction quality crystals were grown by seeding small crystals into hanging drops. The hanging drops were prepared by mixing 6 ml of a solution 24 mg ml^{-1} protein containing a five-fold molar excess of inhibitor with 8 ml of the reservoir solution (0.1 M sodium acetate, pH 4.6, 40% saturated ammonium sulfate); and allowing them to equilibrate overnight. The crystals diffract to ~1.6 Å resolution. Information about the unit cell, the data set and the refinement is listed in Table 2. Two drawings showing the electron density for the inhibitor are presented in Figure 3.

Table 2. Crystallographic data

Space group	C2
Cell dimensions	a = 97.88 Å, b = 46.64 Å, c = 66.59 Å, β = 116.14°
Resolution range	60 Å – 1.8 Å
Number of observations	73775
Number of unique data	23690
Completeness of data set	94%
Number of atoms in:	
Current model	2616
Penicillopepsin	2364
Inhibitor	39
Crystallographic residual	14.5%
Root mean square deviation (rmsd) from ideal:	
Bond lengths	0.011 Å
Bond angles	1.317º
Planes (triagonal, other)	0.010 Å, 0.020 Å
rmsd for B-factors of bonded atoms	2.257 $Å^2$

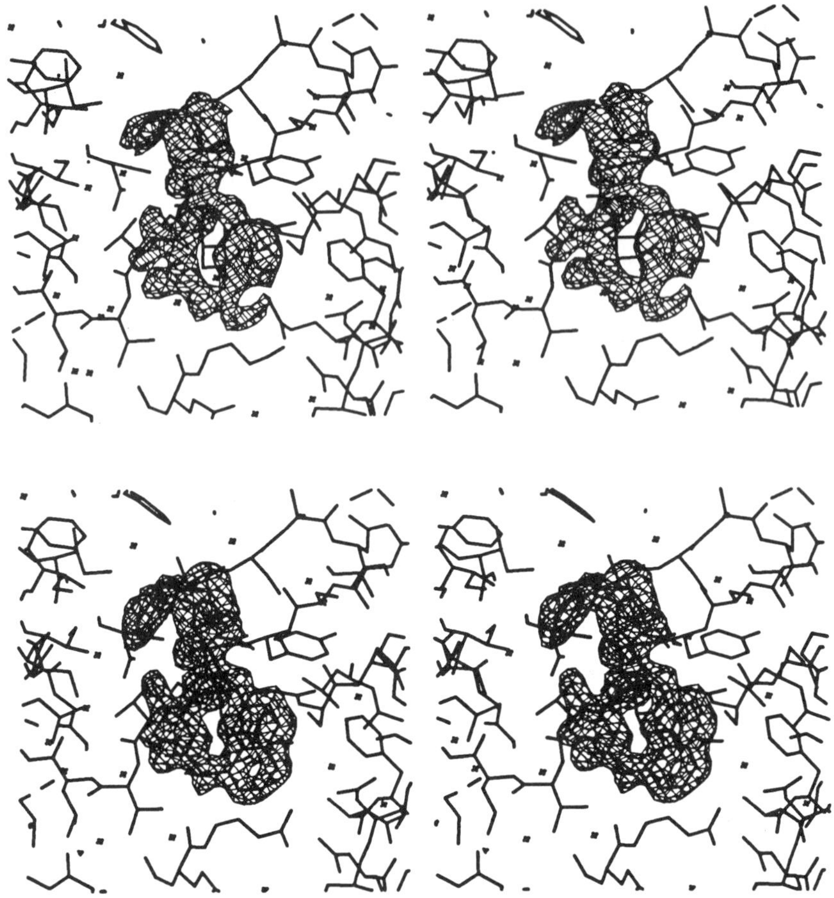

Figure 3. Electron density for the inhibitor. (a) Initial Difference Fourier Map (F_o–F_c, α_c). The map was calculated using the observed structure factor amplitudes (F_o) and structure factor amplitudes and phases (F_c, α_c) calculated from the model of the PP2 complex without the PP2 inhibitor. It is contoured at the 3σ level. (b) Current Difference Fourier Map ($2F_o$–F_c, α_c). The map was calculated using the observed structure factor amplitudes (F_o) and structure factor amplitudes and phases (F_c, α_c) calculated from the current model. It is contoured at the 1σ level in the region of the inhibitor. The thicker lines in each figure represent the model, including the water molecules that are shown as crosses. Figures 3 and 4 were drawn using the program TOM and ancillary plotting programs [5].

3. RESULTS

The peptide backbone of the macrocyclic inhibitor superposes well with that of PP2 (Figure 4). In the P1 position, there is good superposition of the atoms of the naphthalene bridge with the Cα, Cβ and Cγ atoms of PP2's leucine residue. This is one of the two Cα–Cβ bonds that were used in the search for a bridging fragment. The superposition of the naphthalene motif with the Cα–Cβ bond in the P3 position of the PP2 complex is less good. While the torsion angle N–C–Cα–Cβ of the two valine residues of PP2 is −83°, the equivalent angle in the macrocyclic inhibitor is −7°. The plane of the naphthalene rings is at about a 130° angle to the peptide backbone, sticking out of the binding pocket so that naphthalene displaces residues of the flap of penicillopepsin. In the PP2 complex, the

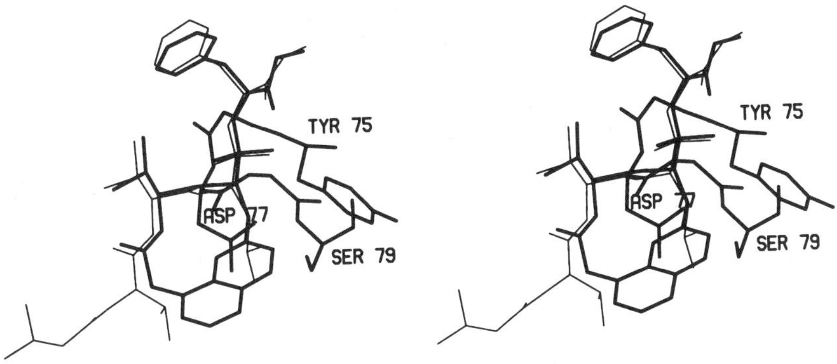

Figure 4. Superposition of the macrocyclic inhibitor and PP2. The macrocyclic inhibitor is drawn in bold lines and PP2 is drawn in fine lines. Also shown are five residues of the flap of penicillopepsin: Tyr75–Gly76–Asp77–Gly78–Ser79. They are shown in the conformations they adopt when the macrocyclic inhibitor is bound to penicillopepsin.

amide nitrogen of the residue in P2 (Val) forms a hydrogen bond with Oδ2 of Asp77, one of the flap residues. With the macrocyclic inhibitor bound, the side chain of Asp77 forms hydrogen bonds only with atoms in residues of the flap: the amide nitrogen and Oγ of Ser79. There appear to be no hydrogen-bonding interactions made by the amide nitrogen of the macrocyclic inhibitor which is equivalent to the amide nitrogen of the residue in P2 (Val) of the PP2 inhibitor.

The macrocyclic inhibitor does not bind to penicillopepsin exactly as expected, but it still may be a suitable example for evaluation of the entropic effect. For it to be a good example, first, the ring structure would have to hold the inhibitor in the same conformation in solution as it has when bound to the enzyme. The solution structure of an analogue of the macrocyclic inhibitor (with (O)Me replacing (O)PheOMe) has been determined by NMR. The data indicate that the conformation of the macrocyclic inhibitor in solution is similar to that in this crystal structure. Second, the acyclic analogues should bind to penicillopepsin in the same way. It will be interesting to see if they do.

ACKNOWLEDGMENTS

We are indebted to Theo Hofmann for the supply of penicillopepsin. The work in Edmonton has been supported by a grant to the MRC Group in Protein Structure and Function and by grant MRC MT12831, both from the Medical Research Council of Canada. The work in Berkeley is supported by grant GM-30759 from the National Institutes of Health.

REFERENCES

1. Fraser, M.E., Strynadka, N.C.J., Bartlett, P.A., Hanson, J.E. and James, M.N.G., *Biochemistry* **31**, 5201–5214 (1992).
2. Kraulis, P.J., *J. Appl. Cryst.* **24**, 946–950 (1991).
3. Lauri, G. and Bartlett, P.A., *J. of Computer-Aided Mol. Design* **8**, 51–66 (1994).
4. Hofmann, T. and Hodges, R.S., *Biochem. J.* **203**, 603–610 (1982).
5. Jones, T.A. Interactive Computer Graphics: FRODO, in *Methods in Enzymology* **115**, 157–171 (1985). Modifications for TOM/FRODO Albert/Caltech by C. Cambillau, M. Israel, G. Griffin and A. Chirino.

ABSTRACTS

X-RAY STRUCTURES OF FIVE INHIBITOR COMPLEXES WITH YEAST PROTEINASE A

N. Cronin,[1] M. Badasso,[1] J. Cooper,[1] S Wood,[1] A. Doherty,[2] C. Humblet,[2] E. Lunney,[2] T. Dreyer,[4] D. J. Hoover,[3] and T. L Blundell[1]

[1]Department of Crystallography, Birkbeck College, University of London, WC1E 7HX
[2]Department of Chemistry and Pharmacology, Parke-Davis Pharmaceutical Research Division, Warner-Lambert Company, Ann Arbor, Michigan 48106-1047
[3]Department of Medicinal Chemistry, Pfizer Inc., Central Research Division, Groton, Connecticut 06340
[4]Department of Chemistry, Carlsberg Laboratory, Gamle Carlsberg Vej 10, DK-2500, Copenhagen, Denmark

The activation process of some vacuolar (lysosomal) proteinases in the yeast *Saccharomyces cerevisiae* is initiated by the PRA1 (PEP4) gene product, yeast proteinase A. Pro-proteinase yscB and procarboxypeptidase yscY are converted to the mature enzyme forms in the vacuole by hydrolysis of the pro-part by yeast proteinase A. The intracellular control of yscA is fascinating in that its protease activity is inhibited by a unique peptide inhibitor IA_3 (Dreyer *et al.*, 1985). YscA, although exhibiting N-linked glycosylation, is not targeted to the vacuole (lysosome) via the mannose 6-phosphate mechanism characterized for its mammalian counterpart cathepsin D (45% sequence homology).

The structure of native yscA has been previously solved to 3.5 Å and an inhibitor complex with the renin difluorostatone Pfizer inhibitor **CP-81,282** (Aguilar *et al.*, 1996). A previous study of 21 inhibitor complexes with endothiapepsin helped to develop modelling methodologies to dock ligands in models of renin (Bailey *et al.*, 1994). The search for orally active inhibitors of human renin remains a challenge to medicinal chemists.

Peptide inhibitors, peptido-mimetic ligands, pro-peptides and macromolecular natural inhibitors such as IA_3 furnish important information on the nature of protein-ligand interactions in the aspartic proteinase family. The structures of a number of inhibitors bound to proteinase A have been solved by x-ray analysis. Of outstanding interest is **PD-129,541,** a peptide inhibitor where cyclization was introduced into the design in order to lock the peptide in the highest affinity conformation (to give a net gain in affinity for entropic reasons) and to potentially enhance bioavailability. **PD-133,450** comes from a series based on the general structure 4-(morpholinylsulfonyl)–L–Phe–P_2–(cyclohexyl)Ala[isostere]–P_1'–P_2', the transition-state mimetic being a glycol analogue. This inhibitor bound in an ideal extended conformation in the active site as already noted for its complex with endothiapepsin. **CP-108,420**, a hydroxylethylene transition-state analogue, was of interest due to a non-peptide residue at P_3 which occupies the P_3 subsite very well and forms a hydrogen bond to Thr 77-OG, which normally forms a standard hydrogen bond to P_2–N in the inhibitor's main chain. A Pfizer inhibitor, **CP-72,647**, a homostatine with a basic histidine residue at P2, was also designed to optimize binding to renin, as was **CP-81,198**, a difluorostatine, analogous to the difluorostatone above. Investigation of the affinity of these inhibitors revealed the difluorostatone to have the highest affinity (Ki = 90 nM) and the cyclic inhibitor to have the least (Ki = 5 M). However, none of these inhibitors has a binding potency comparable to IA_3, the natural inhibitor of proteinase A (Ki = 7 nM).

REFERENCES

Aguilar CF, Badasso M, Cronin NB, Newman MP, Cooper JB, Wood SP, Dreyer T, Johnson MS and Blundell TL (1996) *Journal Molecular Biology* (in press).

Bailey D and Cooper JB (1994) *Protein Science* **3**, 2129-2143.

Dreyer T, Valler MJ, Kay J, Charlton P and Dunn BM (1985) *Biochem. J.* **231**, 777-779.

THE HIGH RESOLUTION STRUCTURE OF THE EXTRACELLULAR ASPARTIC PROTEINASE FROM *Candida tropicalis* YEAST

Jindrich Symersky[2] and Stephen I. Foundling[1]

[1]Laboratory of Protein Crystallography, Protein Studies Program, Oklahoma Medical Research Foundation, 825 N.E. 13th Street, Oklahoma City, Oklahoma 73104, USA

[2]Center for Macromolecular Crystallography, University of Alabama, 1918 University Boulevard, Brimingham, Alabama 35294, USA

The crystal structure of the secreted aspartic proteinase from *Candida tropicalis* yeast (SAPT) has been determined to 1.8 Å resolution. This yeast is an opportunistic pathogen in immuno-compromised patients. The classic aspartic proteinase bilobal structure and domain topology is conserved in SAPT and the substrate binding cleft is situated between the two domains. Structural comparisons with pepsin and fungal aspartic proteinases show that insertions and deletions in the primary sequence, modify the SAPT binding cleft to make it more spacious, particularly in the non-primed side. An unexpected tetrapeptide has been found to occupy binding subsites S_1'–S_3' and this suggests the order of release of peptide products in the catalytic mechanism of these enzymes.

STRUCTURAL ANALYSIS OF PLASMEPSIN II

A Comparison with Human Aspartic Proteases

Abelardo M. Silva,[1*] Angela Y. Lee,[1] John W. Erickson,[1] and Daniel E. Goldberg[2]

[1]Structural Biochemistry Program
National Cancer Institute/SAIC
P.O. Box B, Frederick, Maryland 21702
[2]Howard Hughes Medical Institute
Departments of Molecular Microbiology and Medicine
Washington University School of Medicine
Box 8230, 660 South Euclid Avenue
St. Louis, Missouri 63110

INTRODUCTION

Malaria is one of the world's most devastating diseases, afflicting several hundred million people a year and killing an estimated two million of them, mostly children.[1] The etiologic agent is a parasitic protozoan of the genus *Plasmodium*. The organism causes disease in its intraerythrocytic phase. A prominent feature of its development inside red blood cells is the degradation of hemoglobin. During its short cycle, the parasite consumes nearly all of the host hemoglobin, generating amino acids for its growth and maturation.[2] This is a vast catabolic process since hemoglobin constitutes 95% of the soluble red blood cell protein. Degradation occurs in an acidic digestive vacuole and involves a series of proteases. Two aspartic proteases, termed plasmepsins,[3,6] and one cysteine protease, called falcipain,[4] have been purified from the digestive vacuole and characterized.[5] The plasmepsins appear to make an initial attack on the native hemoglobin molecule, which is followed by proteolysis of the large fragments into small peptides by falcipain.[5,7] Intact hemoglobin cannot be cleaved by falcipain unless the substrate is first denatured.[8] Cysteine protease inhibitors cause osmotic swelling of the digestive vacuole and ultimately death of the parasite,[4] perhaps through accumulation of plasmepsin diges-

* Send correspondence to: Abelardo M. Silva, National Cancer Institute, Building 322, Frederick, Maryland 21702. Phone : (301) 846 1977; Fax : (301) 846 6066; E-mail: abelardo@ncifcrf.gov

Aspartic Proteinases, edited by James.
Plenum Press, New York, 1998.

tion products that are too large to exit the vacuole.[8] The products of digestive vacuole proteolysis are small peptides, which appear to be exported for terminal degradation by cytoplasmic exopeptidases.

Both plasmepsins are capable of making an initial cleavage in the hemoglobin alpha chain between Phe 33 and Leu 34.[5,7] This bond is also the first to be cleaved when hemoglobin is incubated with an extract of purified digestive vacuoles.[7] It is believed that this strategic initial cleavage leads to unraveling of the hemoglobin molecule, allowing further proteolysis to proceed efficiently. The plasmepsins are capable of several other cleavages after the initial event. Plasmepsin I has a preference for Phe in the P1 position, while plasmepsin II likes hydrophobic residues on both sides of the scissile bond.[5] The genes for both plasmepsins have been cloned.[3,6] They are 73% identical and the mature coding regions share about 30–35% identity with the mammalian aspartic proteases. The plasmepsin propieces are substantially longer than mammalian counterparts (124–125 amino acids instead of 40–50) and have a membrane-spanning domain[3] (Table 1). This allows targeting of newly synthesized plasmepsins as integral membrane proteins, probably via the cell surface, before a cleavage and activation event that appears to be mediated by a novel acidic processing protease (Francis, Banerjee and Goldberg, submitted). Inhibitors of the plasmepsins have been identified as extensions of mammalian aspartic protease inhibition programs. One such compound, SC-50083, blocks plasmepsin I with a ki of 140 pM and plasmepsin II at 110 nM. In culture it stops hemoglobin degradation and kills parasites with an IC50 of 2–5 μM.[3] It is not clear whether blockade of plasmepsin I, plasmepsin II or both is necessary for drug action. When the organisms are grown in a medium depleted of most exogenous amino acids, forcing them to rely more heavily on hemoglobin degradation, the parasites are more sensitive to inhibitors. These data gave rise to the hypothesis that plasmepsins play a key role in the essential process of hemoglobin degradation. The search for more potent and more selective inhibitors was begun.

The plasmepsins have varied degree of amino acid sequence homology with the aspartic proteases of humans. After amino acid sequence alignment of plasmepsin II with all five human aspartic proteases the proportion of identical amino acids are: 32.8% for cathepsin D, 31.7% for cathepsin E, 31.4% for renin, 30.9% for pepsin, and 27.3% for gastricsin, Table 1. We present here a structural comparison of plasmepsin II with two of the human enzymes, cathepsin D and renin. Such analysis is relevant for the development of new antimalarial drugs based on the inhibition of the plasmepsins.

RESULTS AND DISCUSSION

Recombinant plasmepsin II expresses well in *E.* coli,[9] and this has allowed large-scale purification and crystallization of the enzyme. Thus recombinant plasmepsin II, in complex with pepstatin A, was used for structure determination as described.[10] The crystals were very small, hexagonal rods having less than 0.05 mm in diameter, and were very sensitive to radiation damage. Because of that, diffraction data, which extended to 2.7 Å resolution, was collected in house at liquid nitrogen temperature. The crystals belong to space group $P3_121$, with a = b = 142.1 Å, c = 97.6 Å, and contain two molecules in the asymmetric unit. The structure was determined by molecular replacement,[11] using the structure of human cathepsin D[12] as a search model, and was refined. The atomic coordinates have been deposited with the Brookhaven Protein Data Bank, accession code 1SME.

The plasmepsin II molecule is formed by a single chain of 329 amino acids and is folded into two topologically similar N- and C-terminal domains, Figure 1. The domains

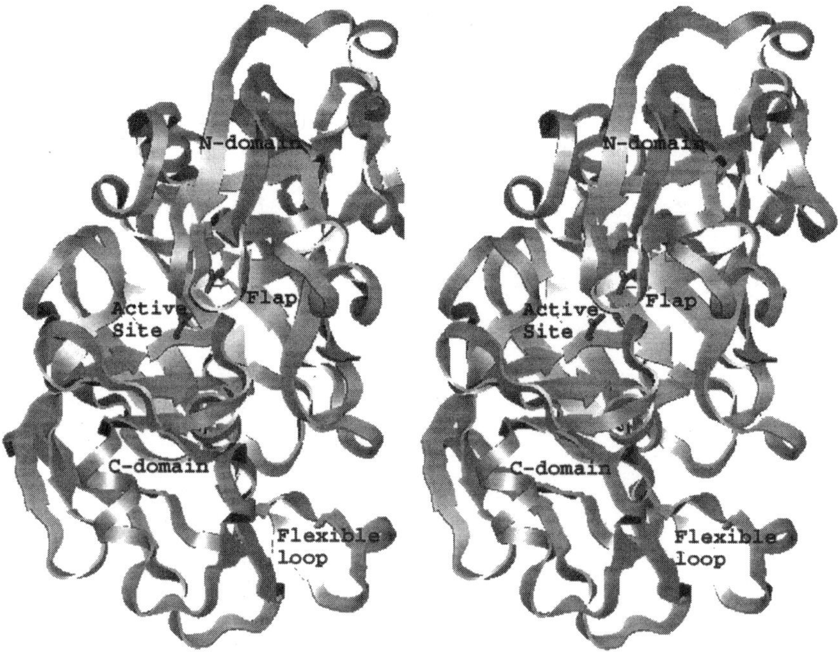

Figure 1. Stereo ribbon model of plasmepsin II. Side chains of Asp 34 and Asp 214, which form the active site, are indicated. The "flexible loop" displays the largest displacement relative to the N-domain when the two independent molecules found in the crystals are compared.

contact each other along the bottom of the binding cleft that contains the catalytic dyad, Asp 34 and Asp 214. A β-hairpin structure, known as the "flap," lies perpendicular over the binding cleft and interacts with pepstatin A. The amino and carboxyl ends of the polypeptide chain of plasmepsin II are assembled into the characteristic six-stranded inter-domain β-sheet, which serves to anchor the domains together. An unusual feature of plasmepsin II is the substitution of serine in place of threonine at residue 215 in the Asp–Thr–Gly signature sequence for the C-domain. Pepstatin A is bound in an extended conformation, with its central statine hydroxyl interacting with both catalytic aspartic acid residues. The main chain of the inhibitor makes hydrogen bonds with the main chain of the flap, as well as with amino acid residues that surround the catalytic center, Figure 2.

A comparison of the two crystallographically-independent plasmepsin II molecules reveals a large inter-domain displacement, which does not affect the interactions of pepstatin A with the enzyme. The two conformations can be related by a relative rotation of 5.2° between the N and C domains. After superposition of both molecules on their N-terminal domains, equivalent C_α atoms of a "flexible loop" on the C-domain are displaced up to 3.8 Å, Figure 1. Such flexibility may be important for the ability of plasmepsin II to recognize and bind specific sequences on the native hemoglobin molecule. In fact, interdomain motion of the magnitude observed here has also been detected in a number of proteins. In particular, inhibited and uninhibited forms of human renin also exist in crystals in two conformations[13,14]. The structural variability observed in renin and plasmepsin II involve large movements at the topologically equivalent flexible-loop region of the C-domain. However, the conformation of those loops in themselves are quite different in renin and plasmepsin II, Figure 3.

Table 1. Amino acid sequence alignment of the pro-form of all human aspartic proteases and plasmepsin II[a]

```
               1                                                                              80
Cathepsin E    .......... .......... .......... ....MKTLLL LLLVL....L ELGEAQGSLH RVPLRRHPSL KKKLRARSQL
Pepsin         .......... .......... .......... ....MKWLLL LGLVA....L ....SECIMY KVPLIRKKSL RRTLSERGLL
Gastricsin     .......... .......... .......... .......MKW MVVVL....V CLQLLEAAVV KVPLKKFKSI RETMKEKGLL
Cathepsin D    .......... .......... .......... ....MQPSSL LPLAL....C LLAAPASALV RIPLHKFTSI RRTMSEVGGS
Renin          .......... .......... ........MD GWRRMPRWGL LLLLWGSCTF GLPTDTTTFK RIFLKRMPSI RESLKERGVD
Plasmepsin II  MDITVREHDF KHGFIKSNST FDGLNIDNSK NKKKIQKGFQ ILYVLLFCSV MCGLFYYVYE NVWLQRDNEM NEILKNSEHL
               ---------- ---------- ---------- ---------- ---------- ---------- ---*------ ----------

               81                                                                            160
Cathepsin E    S......... .EFWKSHNLD MIQ..FTESC S......... ..MDQSAK.E PLINYLDMEY FGTISIGSPP QNFTVIFDTG
Pepsin         K......... .DFLKKHNLN PARKYFPQWE A......... ..PTLVDE.Q PLENYLDMEY FGTIGIGTPA QDFTVVFDTG
Gastricsin     G......... .EFLRTHKYD PAWKY...RF G......... ..DLSVTY.E PM.AYMDAAY FGEISIGTPP QNFLVLFDTG
Cathepsin D    V......... .EDLIAK..G PVSKYSQAVP A......... ..VTEGPIPE VLKNYMDAQY YGEIGIGTPP QCFTVVFDTG
Renin          M......... .ARLGPEWSQ PMKRL..... T......... ..LGNTTSSV ILTNYMDTQY YGEIGIGTPP QTFKVVFDTG
Plasmepsin II  TIGFKVENAH DRILKTIKTH KLKNYIKESV NFLNSGLTKT NYLGSSNDNI ELVDFQNIMF YGDAEVGDNQ QPFTFILDTG
               ---------- ---------- ---------- ---------- ---------- ---------- -*----*--- *-*----***

               161                                                                           240
Cathepsin E    SSNLWVPSVY CT..SPACKT HSRFQPSQSS TYSQPGQSFS IQYGTGSLSG IIGADQVSV. .......... EGLTVVGQQF
Pepsin         SSNLWVPSVY CS..SLACTN HNRFNPEDSS TYQSTSETVS ITYGTGSMTG ILGYDTVQV. .......... GGISDTNQIF
Gastricsin     SSNLWVPSVY CQ..SQACTS HSRFNPSESS TYSTNGQTFS LQYGSGSLTG FFGYDTLTV. .......... QSIQVPNQEF
Cathepsin D    SSNLWVPSIH CKLLDIACWI HHKYNSDKSS TYVKNGTSFD IHYGSGSLSG YLSQDTVSVP CQSASSASAL GGVKVERQVF
Renin          SSNVWVPSSK CSRLYTACVY HKLFDASDSS SYKHNGTELT LRYSTGTVSG FLSQDIITV. .......... GGITVT.QMF
Plasmepsin II  SANLWVPSVK CT..TAGCLT KHLYDSSKSR TYEKDGTKVE MNYVSGTVSG FFSKDLVTV. .......... GNLSLPYKFI
               *-*-****-- *------*-- --------*- -*-------- --*--*---* ----*---*- ---------- ----------

               241                                                                           320
Cathepsin E    GESVTEP.GQ TFVDAEFDGI LGLGYPSLAV GGVTPVFDNM MAQNLVDLPM FSVYMSSNPE GGA..GSELI FGGYDHSHFS
Pepsin         GLSETEP.GS FLYYAPFDGI LGLAYPSISS SGATPVFDNI WNQGLVSQDL FSVYLS..AD DQS..GSVVI FGGIDSSYYT
Gastricsin     GLSENEP.GT NFVYAQFDGI MGLAYPALSV DEATTAMQGM VQEGALTSPV FSVYLS.NQQ GSS..GGAVV FGGVDSSLYT
Cathepsin D    GEATKQP.GI TFIAAKFDGI LGMAYPRISV NNVLPVFDNL MQQKLVDQNI FSFYLSRDPD AQP..GGELM LGGTDSKYYK
Renin          GEVTEMP.AL PFMLAEFDGV VGMGFIEQAI GRVTPIFDNI ISQGVLKEDV FSFYYNRDSE NSQSLGGQIV LGGSDPQHYE
Plasmepsin II  EVIDTNGFEP TYTASTFDGI LGLGWKDLSI GSVDPIVVEL KNQNKIENAL FTFYLPVHDK H....TGFLT IGGIEERFYE
               ---------- ------***- -*-------- ---------- ---------- *--*------ ---------- -**-------
```

Table 1. (*Continued*)

	321							400
Cathepsin E	GSLNWVPVTK	QAYWQIALDN	IQVGGTVM.F	CSEGCQAIVD	TGTSLITGPS	DKIKQLQNAI	GAAP.VDGEY	AVECANLNVM
Pepsin	GSLNWVPVTV	EGYWQITVDS	ITMNGEAI.A	CAEGCQAIVD	TGTSLLTGPT	SPIANIQSDI	GASENSDGDM	VVSCSAISSL
Gastricsin	GQIYWAPVTQ	ELYWQIGIEE	FLIGGQASGW	CSEGCQAIVD	TGTSLLTVPQ	QYMSALLQAT	GAQEDEYGQF	LVNCNSIQNL
Cathepsin D	GSLSYLNVTR	KAYWQVHLDQ	VEVAS.GLTL	CKEGCEAIVD	TGTSLMVGPV	DEVRELQKAI	GAVPLIQGEY	MIPCEKVSTL
Renin	GNFHYINLIK	TGVWQIQMKG	VSVGSSTL.L	CEDGCLALVD	TGASYISGST	SSIEKLMEAL	GAKKRLF.DY	VVKCNEGPTL
Plasmepsin II	GPLTYEKLNH	DLYWQITLDA	.HVGNIML..	..EKANCIVD	SGTSAITVPT	DFLNKMLQNL	DVIKVPFLPF	YVTLCNNSKL
	*---------	---**-----	----------	--------**	-*-*------	----------	----------	----------
	401							479
Cathepsin E	PDVTFTINGV	PYTLSPTAYT	LLDFVDGMQF	CSSGFQGLDI	HPPAG.PLWI	LGDVFIRQFY	SVFDRGNNRV	GLAPAVP..
Pepsin	PDIVFTINGV	QYPVPPSAYI	L....QSEGS	CISGFQGMNL	PTESG.ELWI	LGDVFIRQYF	TVFDRANNQV	GLAPVA...
Gastricsin	PSLTFIINGV	EFPLPPSSYI	L....SNNGY	CTVGVEPTYL	SSQNGQPLWI	LGDVFLRSYY	SVYDLGNNRV	GFATAA...
Cathepsin D	PAITLKLGGK	GYKLSPEDYT	LKVSQAGKTL	CLSGFMGMDI	PPPSG.PLWI	LGDVFIGRYY	TVFDRDNNRV	GFAEAARL.
Renin	PDISFHLGGK	EYTLTSADYV	FQESYSSKKL	CTLAIHAMDI	PPPTG.PTWA	LGATFIRKFY	TEFDRRNNRI	GFALAR...
Plasmepsin II	PTFEFTSENG	KYTLEPEYYL	QHIEDVGPGL	CMLNIIGLDF	PVPT....FI	LGDPFMRKYF	TVFDYDNHSV	GIALAKKNL
	*---------	--------*-	----------	*---------	----------	**--*-----	---*--*---	*-*------

[a]The multiple alignment was performed using the program PileUp of the Wisconsin Package (Version 8, September 1994, Genetics Computer Group, Inc), which employs the method of Needleman and Wunsch (J. Mol. Biol. 48: 443-453 (1970)) for each pairwise alignment, and a clustering strategy (Sneath, P.H.A. and Sokal, R.R. (1973) in Numerical Taxonomy (pp. 230-234), W.H. Freeman and Company, San Francisco, California, USA). Results of the alignment of plasmepsin II vs the human enzymes can be summarized in terms of percentages of identity and similarity; they are: 32.8% identity and 53.4% similarity for cathepsin D; 31.7 % identity and 51.7% similarity for cathepsin E; 31.4% identity and 53.7% similarity for renin; 30.9% identity and 52.6% similarity for pepsin; and 27.3% identity and 51.8% similarity for gastricsin.

*Indicates amino acids common to all sequences.

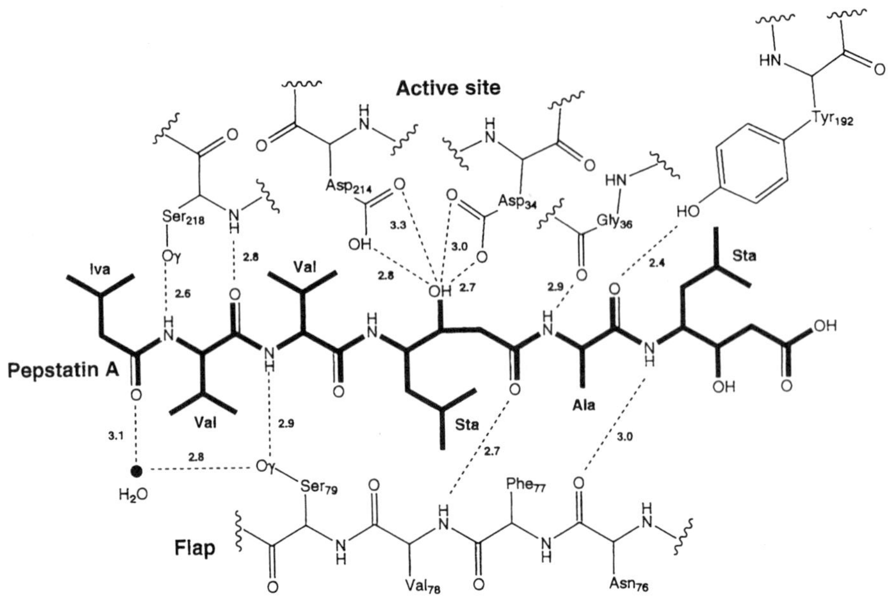

Figure 2. Hydrogen bonding interactions between pepstatin A and plasmepsin II.

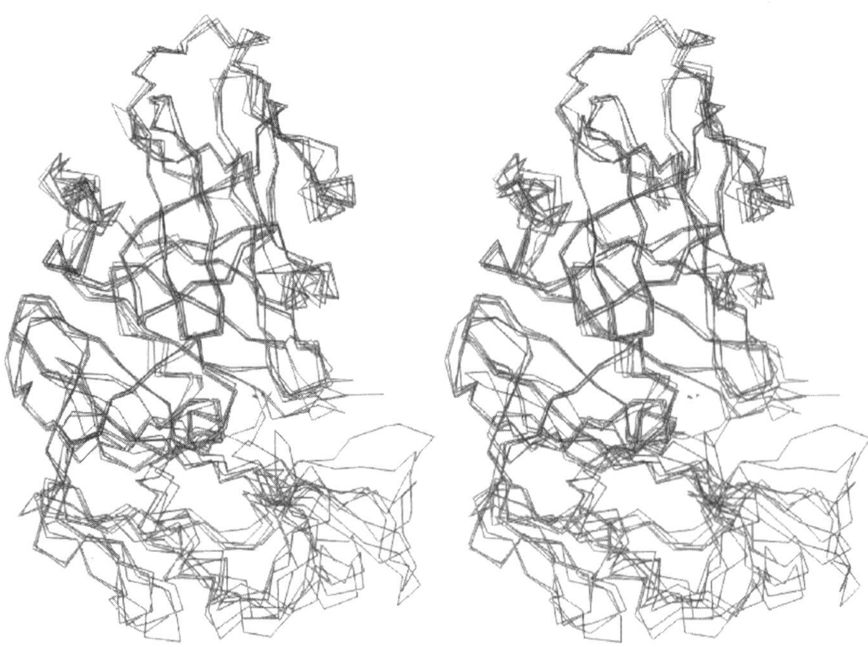

Figure 3. Stereo C_α trace superposition of cathepsin D, two independent plasmepsin II molecules found in the same crystal, and two independent renin molecules found in the same crystal. All molecules were superimposed pairwise, as described in Table 2, onto one of the plasmepsin II molecules. Note the good agreement at the N-terminal domain and active site, and also the structural variability at the "flexible-loop" region.

Plasmepsin II displays the typical topology of eukaryotic aspartic proteases.[15] To quantify their similarities plasmepsin II was structurally compared with cathepsin D[12] and renin.[14] Structural superpositions were performed in two steps using the program Quanta (Molecular Simulations, Inc.). First, the amino acid sequences were aligned according to the conserved amino acids and the C_α atoms of those pairs were superimposed. Then, in an iterative procedure, all C_α atoms pairs closer than 1.5 Å were used to compute the superposition transformation, the molecules were superimposed and pairs closer than 1.5 Å computed again. The process was finished when the number of distance-equivalenced pairs remained constant. Results of the superpositions of all pairs of molecules including cathepsin D, two conformations of plasmepsin II and two conformations of renin are in Table 2. The structural alignment process used automatically eliminates parts of the molecules that are more different to each other, either because in differences in folding or differences in relative conformations. Therefore, since the molecules being compared have a similar number of amino acids, the number of C_α atoms pairs closer than 1.5 Å would be a good indicator of relative overall structural similarities. Table 2 was sorted according to that number. First and second in Table 2 are respectively {renin (molecule 1) *vs* renin (molecule 2)} and {plasmepsin II (molecule 1) *vs* plasmepsin II (molecule 2)} which suggests that relative conformational changes are of a larger magnitude in plasmepsin II than in renin. Next in the list is cathepsin D *vs* renin, with similar scores for both conformations of renin. This is followed by cathepsin D *vs* plasmepsin II, with one conformation of plasmepsin II being more similar to the single one of cathepsin D than the other one. Finally, plasmepsin II *vs* renin shows even scores in all cases, thus detecting no cross similarity of conformations. For the best agreement of plasmepsin II *vs* cathepsin D there were

Table 2. Superposition analysis of plasmepsin II, human cathepsin D and human renin[a]

	All pairs		Identical pairs		Pairs closer than 1.5Å	
	No.	rmsd [Å]	No.	rmsd [Å]	No.	rmsd [Å]
RE^1-RE^2	333	1.02	333	1.02	298	0.54
PL^1-PL^2	329	0.97	329	0.97	281	0.53
$CD-RE^2$	332	2.16	153	1.79	248	0.72
$CD-RE^1$	331	2.31	153	1.78	247	0.70
PL^2-CD	318	1.87	108	1.34	205	0.69
PL^1-CD	318	2.07	109	1.60	204	0.68
PL^2-RE^1	321	2.74	112	2.30	180	0.72
PL^1-RE^1	321	2.64	112	2.29	179	0.72
PL^1-RE^2	321	2.65	112	2.42	179	0.76
PL^2-RE^2	321	2.68	112	2.36	177	0.72

[a]The structures were compared by pairs using an iterative procedure. The amino acid sequences were first aligned according to their identical amino acids and the C_α atoms of those pairs were superimposed. Then all C_α atoms pairs closer than 1.5 Å were used to compute a new superposition transformation, the molecules were superimposed and pairs closer than 1.5 Å computed again. This process was finished when the number of distance-equivalenced pairs remained constant. The molecules compared are: plasmepsin II in two conformations (PL^1 and PL^2), renin in two conformations (RE^1 and RE^2) and cathepsin D (CD). The first two columns indicate the total number of equivalenced C_α atoms pairs and the rmsd for those pairs; the third and fourth columns indicate the number of equivalenced C_α atoms pairs which correspond to identical amino acids and the rmsd for those pairs; and the fifth and sixth columns show the number of equivalenced C_α atoms pairs which are less than 1.5 Å apart and the rmsd for those pairs. The number of C_α atoms pairs closer than 1.5 Å is taken as the best indicator of the relative overall structural similarity between a pair of molecules, thus the table was sorted according to that number.

205 pairs of equivalent C_α atoms at a distance less than 1.5 Å (rmsd = 0.69 Å), the rmsd for all 318 pairs in the alignment was 1.87 Å and 108 pairs corresponded to identical amino acids. For the best agreement of plasmepsin II *vs* renin there were 180 pairs of equivalent C_α atoms at a distance less than 1.5 Å (rmsd = 0.72 Å), the rmsd for all 321 pairs in the alignment was 2.74 Å and 112 pairs corresponded to identical amino acids. These results indicate that, overall, plasmepsin II is structurally closer to cathepsin D than to renin. In all cases the structures have a better structural agreement on the N-terminal domain than on the C-terminal domain, Figures 3, 4, and 5. Human and Plasmodium enzymes are quite similar in several regions including the flap, active site loops, and the six-stranded interdomain beta-sheet for instance. There are important differences not only in the "flexible loop" region, but also at the "pro-loop" which is a characteristic structural feature of eukaryotic aspartic proteases, Figures 4 and 5. The pro-loop includes three consecutive proline residues in cathepsin D and renin. The pro-loop is much shorter in plasmepsin II and also its conformation is such that produces a more open binding cleft, particularly when compared with renin.

The conformation of pepstatin A is similar when bound to cathepsin D and to plasmepsin II, Figure 6, but there are some differences at both ends of the molecule. Met 307, which forms part of the S4 subsite, in cathepsin D is substituted by Ile 290 in plasmepsin II, which results in different conformations of the isovaleryl residue at the N-terminus of pepstatin A. Conformational differences at the C-terminus of pepstatin A are due to the presence of Val78 at the tip of the flap and to the substitution of Ile 311 in cathepsin D by Phe 294 in plasmepsin II. These two changes create a smaller S3′ pocket in plasmepsin II which cannot accommodate the terminal statine side chain as readily as in cathepsin D. Overall the changes in the bound conformation are such that pepstatin A displays a larger solvent accessible area in plasmepsin II than in cathepsin D, 289 Å^2 and 182 Å^2 respectively. There is no available structure for the renin/pepstatin A complex; thus the structure

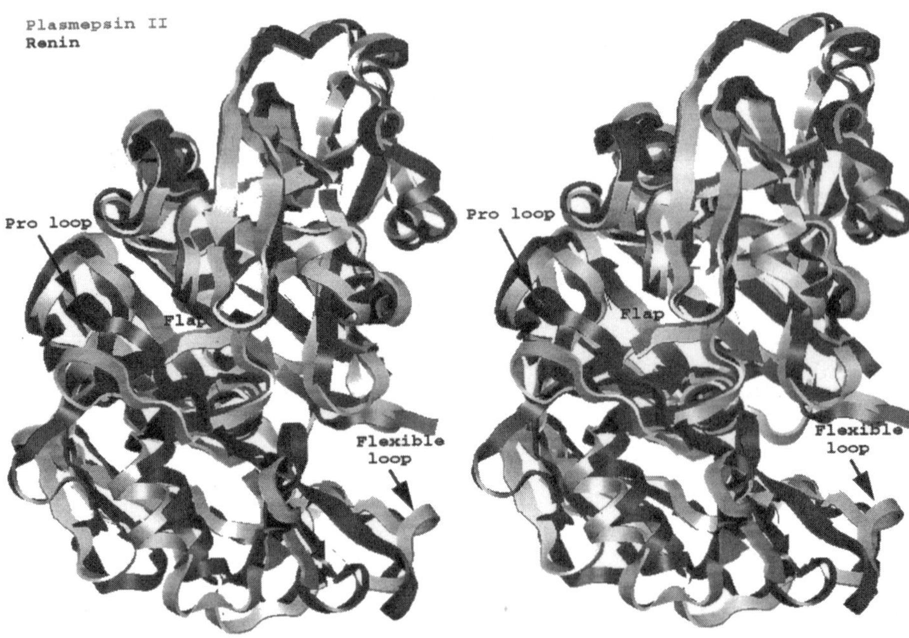

Figure 4. Stereo ribbon model superposition of plasmepsin II (grey) with renin (black).

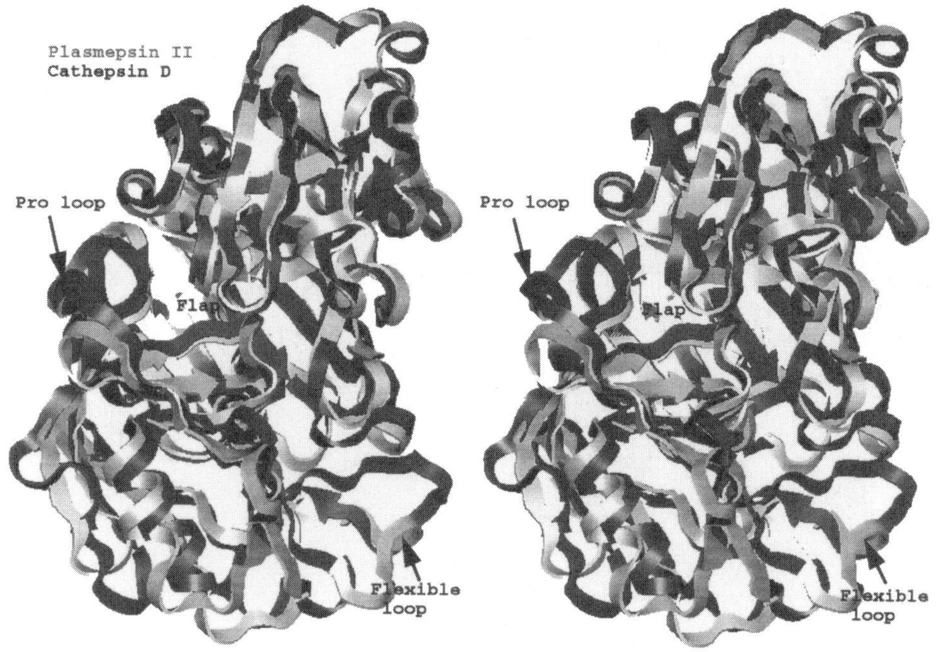

Figure 5. Stereo ribbon model superposition of plasmepsin II (grey) with cathepsin D (black).

of renin in complex with the peptidic inhibitor Bila 980 was used for structural comparison.[14] The superposition of Bila 980 with pepstatin A molecules, Figure 6, after protein structure alignment, also shows a similar form of binding and orientation of the peptide bonds. The location of the hydroxyl at the active site is also similar to the statine hydroxyl positioning of pepstatin A.

When the binding clefts of renin and cathepsin D are compared with that of plasmepsin II, Figures 7 and 8, their good agreement at the active site and flap becomes evident. At

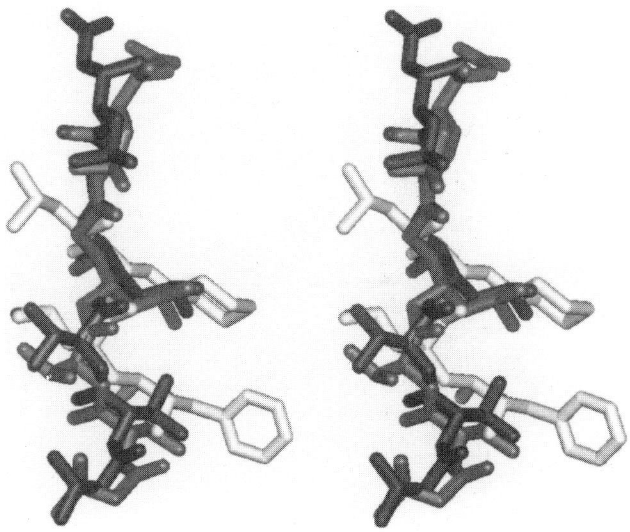

Figure 6. Stereoscopic representation of the superposition of pepstatin A from the crystal structures of plasmepsin II (black), human cathepsin D (grey), and the inhibitor Bila 980 bound to renin (white).

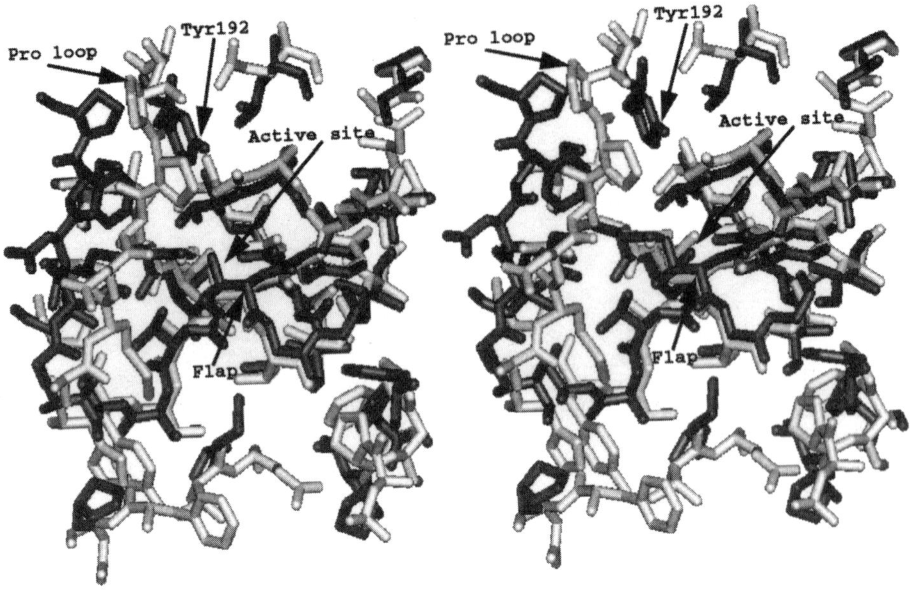

Figure 7. Stereoscopic representation of the superposition of the binding clefts of plasmepsin II (black) and renin (white).

the tip of the flap plasmepsin II displays Val 78, cathepsin D has Gly 79 and renin shows Ser 76. Opposite to it, there is Leu 292 in plasmepsin II, Met 309 in cathepsin D, and Met 289 in renin. These pairs of amino acid residues are capable of coming in contact, thus effectively wrapping the flap around inhibitors or substrates. Clearly the Val–Leu pair is bulkier and more hydrophobic than the mammalian counterpart which would affect substrate binding.

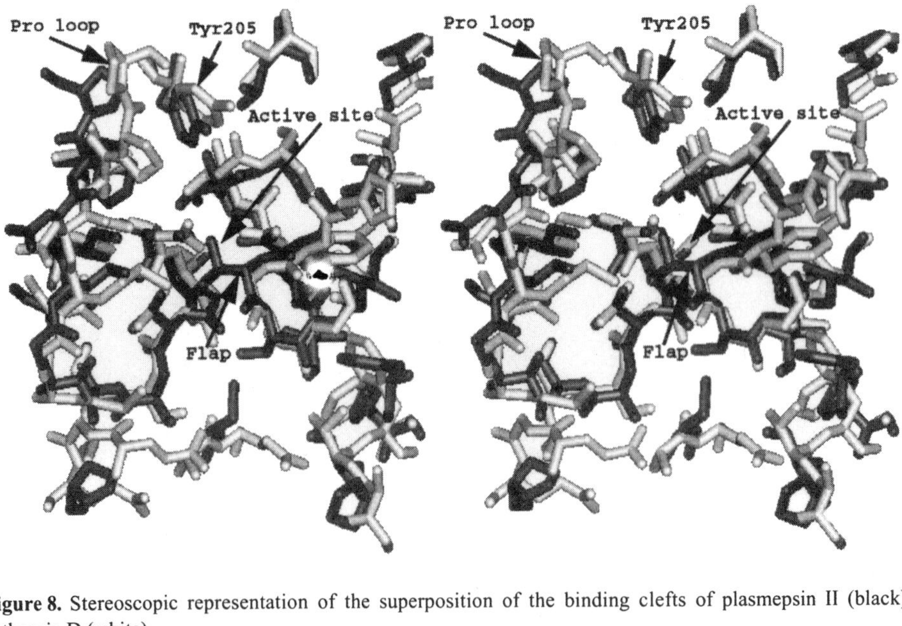

Figure 8. Stereoscopic representation of the superposition of the binding clefts of plasmepsin II (black) and cathepsin D (white).

Otherwise the binding cavities display similar distributions of mostly hydrophobic amino acid residues. Despite these overall similarities, plasmepsin II appears to be more similar to cathepsin D than to renin. The pro-loop in renin sits at a shorter distance from the tip of the flap. This creates a tighter binding cleft which allows the pro-loop to participate in subsite interactions. Thus, renin subsites are quite different form cathepsin D and also plasmepsin II subsites. The fact that the binding cleft of renin is narrower and contributed by different amino acids than in cathepsin D or plasmepsin II, could be correlated with the specificity of renin which has only one known natural substrate. Although cathepsin D has a "standard" pro-loop, it stills displays a binding site in "open" conformation similar to the plasmepsin II cavity. Another important feature of renin is the absence of Tyr 192 (Tyr 205 in cathepsin D) which directly interacts with pepstatin A, Figure 2, in both pepstatin A containing complexes. At that location there is Val 189 in renin, thus creating a much larger cavity.

Our main conclusion is that in the structure-based design of plasmepsin II inhibitors for the development of antimalarial drugs, cross inhibition with cathepsin D, rather than renin, should be the matter of main concern.

ACKNOWLEDGMENTS

This research was supported in part by the National Cancer Institute under contract NOl CO-74102; by NIH Grant AI 37977 to D. E. G. The contents of this publication do not necessarily reflect the views or policies of the DHHS, nor does mention of trade names, commercial products, or organizations imply endorsement by the U.S. Government.

REFERENCES

1. WHO (1995) In *Control of tropical diseases: malaria control.* Geneva, Switzerland, WHO Office of Information.
2. Francis, S.E., Sullivan, D.J. Jr, and Goldberg, D.E. (1997) *Ann. Rev. Mircobiol.* In press.
3. Francis, S.E., Gluzman, I.Y., Oksman, A., Knickerbocker, A., Mueller, R., Bryant, M., Sherman, D.R., Russell, D.G., and Goldberg, D.E. (1994) *EMBO J* **13**:306–317.
4. Rosenthal, P. J., Wollish, W. S., Palmer, J. T., and Rasnick, D. (1991) . *J. Clin. Invest.* **88:**1467–1472.
5. Gluzman, I.Y., Francis, S.E., Oksman, A., Smith, C.E., Duffin, K.L., and Goldberg, D.E. *J Clin. Invest* (1994) **93:**1602–1608.
6. Dame, J. B., Reddy, G. R. , G., Yowell, C. A., Dunn, B. M., Kay, J., and Berry, C. *Mol. Biochem. Parasitol.* (1994) **64:**177–190.
7. Goldberg DE; Slater AF; Beavis R; Chait B; Cerami A; Henderson GB (1991) *J Exp Med* **173:** 961–9
8. Francis, S.E., Gluzman, I.Y., Oksman, A., Banerjee, D., and Goldberg DE (1996) *Mol. Biochem. Parasitol.* **83**: 189–200.
9. Hill J; Tyas L; Phylip LH; Kay J; Dunn BM; Berry C (1994) *FEBS Lett* **352**: 155–8
10. A. M. Silva, A. Y. Lee, S. V. Gulnik, P. Majer, J. Collins, T.N. Bhat, P. J. Collins, R. E. Cachau, K. E. Luker, I. Y. Gluzman, S. E. Francis, A. Oksman, D. E. Goldberg and J.W.Erickson. (1996) *Proc.Nat.Acad.Sciences USA,* **93**: 10034–10039.
11. J. Navaza, (1993) *Acta Cryst.* **D49**: 588.
12. Baldwin, E. T., Bhat, T. N., Gulnik, S., Hosur, M. V., Sowder, R. C., Cachau, R. E., Collins, J., Silva, A. M., and Erickson, J. W. (1993) *Proc.Nat.Acad.Sciences USA*, **90**: 6796–6800.
13. J.Rahuel, J.P.Priestle and M.G.Gruetter (1991) *J.Struct.Biol.* **107**: 227.
14. Tong L; Pav S; Lamarre D; Pilote L; LaPlante S; Anderson PC and Jung G (1995) *J Mol Biol* **250**: 211–222.
15. Davies, D. (1990) *Annu. Rev. Biophys. Biophys. Chem.* **19**: 189.

52

SCREENING ASPARTYL PROTEASES WITH COMBINATORIAL LIBRARIES

Carolyn DiIanni Carroll and Marc Orlowski

Pharmacopeia, Inc.
Princeton, New Jersey 08540

1. ABSTRACT

Large numbers of pharmaceutically relevant low-molecular weight compounds can now be synthesized using combinatorial methods. Screening these large libraries of compounds requires high throughput assays. These methods are utilized to search for inhibitors of the aspartyl proteases, plasmepsin II and cathepsin D. Plasmepsin II, a protease found in the malaria parasite, hydrolyzes human hemoglobin, the nutrient source for the parasite and is a new target for anti-malaria therapy. Cathepsin D may be involved in many biological processes and inhibitors would help to clarify the role of cathepsin D in these processes. Plasmepsin II and cathepsin D are ~35% identical in amino acid sequence. Therefore, a comparison of the screening results of these two enzymes will be very useful in determining each enzyme's specificity and demonstrating the power of utilizing encoded combinatorial libraries.

2. ASPARTYL PROTEASES USED IN SCREENING ASSAYS

The malaria parasite, *Plasmodium falciparum*, avidly consumes the host hemoglobin as a source of nutrients. Plasmepsin I and II are required for the initial stages of hemoglobin digestion (1). A peptidomimetic inhibitor of plasmepsin blocked enzyme activity, prevented hemoglobin degradation and killed the parasite in culture (2). Since resistance to the known antimalarial therapies is increasing, new antimalarial therapies are desperately needed. Therefore, plasmepsin inhibition is an excellent target for antimalarial therapy.

Cathepsin D is a human protease in the endosomal-lysosomal pathway, and is involved in protein turnover. It may also be involved in antigen processing and presentation, connective tissue disease, muscular dystrophy, cancer, and possibly Alzheimer's disease. Therefore, inhibitors would help to clarify the role of cathepsin D in these processes and could be useful therapeutics in these diseases. Plasmepsin II and cathepsin D are 35%

Aspartic Proteinases, edited by James.
Plenum Press, New York, 1998.

identical in amino acid sequence, and therefore, comparison of the screening results from both proteases will identify substituents necessary for selective inhibition of these two enzymes. In addition, the X-ray crystal structures have been solved for both enzymes (3, 4) and will aid in the design of inhibitors and provide further understanding of interactions between the inhibitors and enzymes. For in depth description of these proteases please read additional articles in this volume.

3. MATERIALS AND METHODS

Dr. Daniel E. Goldberg at Washington University generously provided us with purified plasmepsin II (5). Purified cathepsin D was purchased from Athens Research. The substrates were purchased from AnaSpec. The plasmepsin substrate concentration in the assay is 3–6 times its Km value and, therefore, a change in activity is readily apparent in the presence of an inhibitor. Assay conditions were established to yield linear formation of product with time, good signal to background and very consistent results. The screening reactions are performed as end point assays. However, to determine kinetic parameters, the identical assay conditions are used and are monitored over time with various substrate and inhibitor concentrations. Ki values were determined from slope replots of Lineweaver-Burk analysis (6).

The substrates of these assays are based on principles of fluorescence energy transfer (7) and consist of an octapeptide with a fluorescent donor, EDANS (5–[(2–aminoethyl)amino]naphthalene–1–sulfonic acid), attached to the C-terminus and a quenching acceptor, DABCYL (4–(4–dimethylaminophenylazo)benzoic), attached to the N-terminus. The plasmepsin II substrate is DABCYL–γ–aminobutyric acid–Glu–Arg–Met–Phe–Leu–Ser–Phe–Pro–EDANS; the sequence of the octapeptide spans the known cleavage site in hemoglobin and hydrolysis occurs at the expected Phe/Leu site (1, 5). The cathepsin D substrate is DABCYL–γ–aminobutyric acid–Lys–Pro–Ile–Glu–Phe–Phe–Arg–Leu–EDANS, and hydrolysis occurs between the Phe/Phe residues (8, 9). The fluoresence of the unhydrolyzed substrate is very low because the EDANS fluoresence is quenched by the nearby DABCYL. However, upon hydrolysis, the fluorescence of the EDANS-linked peptide fragment is restored to the full quantum yield of EDANS. The EDANS fluorescence was measured using a fluorescence plate reader with an excitation filter of 350 nm and an emission filter of 510 nm. These substrates were chosen because the resulting assays are sensitive even with low assay volumes, have a good signal to noise ratio, and are amenable to high throughput screening. Using a robotic liquid handling system, 30 plates or 50,000 beads can be assayed and analyzed per day (see below).

4. COMBINATORIAL CHEMISTRY AND ECLIPS™ TECHNOLOGY

Combinatorial chemistry (for review see reference 10) applied to drug discovery dramatically increases the number of molecules available to test in biological assays. Thus far, over 1.7 million small molecules have been synthesized at Pharmacopeia using ECLiPS™ (Encoded Combinatorial Library on Polymeric Support) technology. Interfacing these libraries with high throughput screens has resulted in the identification of highly active compounds for enzymes (11) and receptors (12) as well as the accumulation of significant information about structure-activity relationships (11, 12). In general, using the structure-activity relationship obtained from screening several large combinatorial libraries, additional focused libraries are subsequently synthesized to enhance potency, selectivity and *in vivo* activity.

Combinatorial libraries are synthesized at Pharmacopeia using the ECLiPS™ technology using split-pool synthesis (13–15) and after each synthetic step an encoding molecule or tag is attached to the bead identifying that unique step of the reaction (16) as illustrated in Figure 1. The tags are attached to the bead through an oxidatively labile linker and the compounds are attached to the bead through a photolabile linker, allowing selective removal of either the tag or compound. The tags are polyhalogenated aromatics with a variable alkyl chain, and are analyzed by electron capture gas chromatography after cleavage from the solid support (16, 17). Thus, "reading" the tags uniquely identifies the synthetic path used to generate an individual compound. Using high throughput screening strategies, potent and selective compounds are identified as well as structure-activity relationships.

In general there are three types of libraries synthesized at Pharmacopeia. One type is the multi-targeted libraries which consists of up to 500,000 members. In addition some targeted libraries are also synthesized and have a core structure that would be predicted, based on information from the literature, to inhibit a specific class of enzyme. For example, libraries having a core structure of statine or hydroxyethylamine would be likely to contain inhibitors of aspartyl proteases (18–22). The third type of libraries are focused libraries which are synthesized after several active structures have been identified from larger libraries.

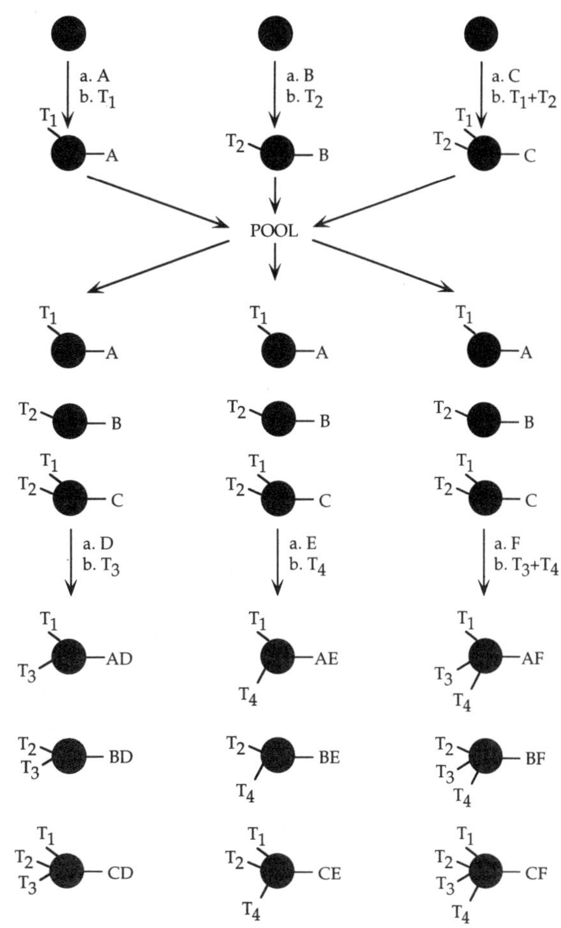

Figure 1. Schematic of split-pool synthesis followed by encoding with chemical tags. In this example, there are 3 reaction flasks (3 columns) containing beads (depicted by filled-in circles). Compounds A, B, and C are added to 3 separate reaction flasks and attached to the beads followed by an encoding step using tags 1 and 2 (depicted as T_1 and T_2). These beads are then pooled together, mixed and then redivided into a second set of flasks, each of which contains approximately equal amounts of beads carrying the compounds A, B, and C. Then the next set of reactions is performed with compounds D, E, and F and followed by encoding with tags T_3 and T_4. In this example, 9 compounds were synthesized with each bead carrying a single compound and a unique set of tags that identifies that specific compound.

5. LIBRARY ARRAYING

In order to increase assay throughput, mixtures of compounds are assayed. Approximately 10–20 beads per well are arrayed into 96 well filter bottom plates. Fifty percent of the compound is removed from the bead by UV irradiation and the compound mixture is filtered into a daughter plate. The conditions for compound removal are determined empirically by analysis of at least two quality control compounds. This analysis also indicates an approximate amount of compound released from the bead. Then a portion of the compound mixtures (typically resulting in ~1–2 μM of each compound in the assay) is transferred to 2 new granddaughter plates. One of the 2 identical granddaughter plates is assayed with plasmepsin II and the other is assayed with cathepsin D. Once an active well is identified, the 10 or 20 beads in the corresponding filter plate are rearrayed at one bead per well, and photolyzed to remove the remaining compound. The compound is then filtered into a new daughter plate, and split into two granddaughter plates in order to assay identical amounts of identical compounds with each protease. If too many actives are observed, the amount of the bead eluate is lowered as necessary on subsequent plates or in the rearray plates. In addition, the cutoff can be adjusted to obtain only the most potent compounds. After screening, comparison of identical wells assayed with plasmepsin II and cathepsin D determines if the compound inhibits plasmepsin II, inhibits cathepsin D or inhibits both proteases. After identification of an active well, the bead is obtained from the corresponding master plate. The tags from that bead are removed and subjected to electron capture gas chromatography. The tags identify the synthetic path of that bead and, therefore, the structure of the active compound.

Libraries are typically screened in several-fold redundancy since sampling of the libraries is statistical. For example, screening a single library equivalent of a library would result in evaluating only 63% of the library (11). The remaining 37% would represent duplicates or higher order multiples. Screening 4 equivalents of a library indicates that 98% of all the compounds in that library would be screened at least once and many others multiple times. When a structure is observed more than once, the possiblity of it being a false positive is dramatically reduced. Therefore, screening at multiple redundancy provides an internal control for reproduciblity. Also, the variation in activity observed for a single structure found multiple times is an indication of the consistency with which all stages of synthesis, photolysis, elution, splitting samples and assays are performed.

6. ANALYSIS OF RESULTS

The data is analyzed in several ways: 1) A selectivity plot is a graph of the cathepsin D activity versus the plasmepsin II activity in wells containing single compounds. A plot of this type may contain thousands of data points (11). Since the selectivity plot correlates the plasmepsin II and cathepsin D activity, it identifies the selective compounds. 2) Since some combinatorial libraries may be synthesized with various substituents around a core structure, inhibitory compounds can be analyzed by determining the frequency that each substituent is observed in each of the variable positions. These plots identify some general structure-activity information (12). 3) All the inhibitory structures are listed with their corresponding plasmepsin II and cathepsin D activity. Compounds which inhibit only plasmepsin II, only cathepsin D or both proteases are identified. From this list, compounds observed more than once will be identified. Similar activity is usually observed when compounds are observed more than once which indicates that all processed are consistent (12).

7. RESYNTHESIS OF COMPOUNDS

In order to confirm the active compounds observed in these libraries, compounds must be synthesized and purified. The compounds chosen to be resynthesized are typically the most potent and selective, are observed more than once and are consistent with the general structure-activity relationship observed (11, 12). The compounds are typically resynthesized on the bead, hydrolyzed from the bead and purified by preparative HPLC. Structural analysis by NMR and mass spectrometry confirms the identity of the compound. Then the IC50 and Ki values are determined for plasmepsin II and cathepsin D.

8. CONCLUSIONS

Although we can not reveal the structure of the libraries or the resynthesized compounds at this time, we were able to identify several inhibitors for both plasmepsin II and cathepsin D from screening two target libraries containing a total of 32,000 compounds. In addition, we obtained structure-activity information as well as identified a competitive inhibitor of plasmepsin II with a Ki value of 110 nM that is >100-fold selective over cathepsin D.

ACKNOWLEDGMENTS

We would like to thank Dr. Daniel E. Goldberg (Washington University) for the plasmepsin II used in these studies and many helpful discussions regarding malaria and the plasmepsin enzymes. We would also like to thank Dr. John Kay (University of Wales College of Cardiff) for many helpful discussions about cathepsin D.

REFERENCES

1. Goldberg, D.E., Slater, A.F.G., Cerami, A. and Henderson, G.B. (1991) *J. Exp. Med.* **173,** 961–969.
2. Francis, S.E., Gluzman, I.Y. Oksman, A., Knickerbocker, A., Mueller, Bryant, M.L., Sherman, D.R., Russell, D.G., and Goldberg, D.E. (1994) *EMBO J,* **13,** 306–317.
3. Baldwin, E.T., Bhat, T.N., Gulnik, S., Hosur, M.V., Sowder, R.C., Cachau, R.E., Collins, J., Sivia, A.M., and Erickson, J.W. (1993) *Proc. Natl. Acad. Sci.,* **90,** 6796–6800.
4. Silva, A.M., Lee, A.Y., Gulnik, S.V., Majer, P., Collins, J., Bhat, T.N., Collins, P.J., Cachau, R.E., Luker, K.E., Gluzman, I.Y., Francis, S.E., Oksman, A., Goldberg, D.E., and Erickson, J.W. (1996) *Proc. Natl. Acad. Sci.* **93,** 10034–10039.
5. Luker, K.E., Francis, S.E., Gluzman, I.Y. and Goldberg, D.E. (1996) *Molecular and Biochemical Parasitology* **79,** 71–78.
6. Segel, I.H. (1975) *Enzyme Kinetics*, New York, Wiley.
7. Matayoshi, E.D., Wang, G.T., Krafft, G.A., and Erickson, J. (1990) *Science* **247,** 954–958.
8. Scarborough, P.E., Guruprasad, K., Topham, C., Richo, G.R., Conner, G.E., Blundell, T.L. and Dunn, B.M. (1993) *Protein Science* **2,** 264–276.
9. Scarborough, P.E., and Dunn, B.M. (1994) *Protein Engineering* **7,** 495–502.
10. (a) Chabala, J.C. (1995) *Current Opinion in Biotechnology* **6,** 632–639. (b) Ellman, J.A. (1996) *Chem. Rev.* **96,** 555–600.
11. Burbaum, J.J., Ohlmeyer, M.H.J., Reader, J.C., Henderson, I., Dillard, L.W., Li, G., Randle, T.L., Sigal, N.H., Chelsky, D., and Baldwin, J.J. (1995) *Proc. Natl. Acad. Sci. USA* **92,** 6027–6031.
12. Apell, K.C., Chung, T.D.Y., Ohlmeyer, M.J.H., Sigal, N.H., Baldwin, J.J., and Chelsky, D. (1996) *J. of Biomolecular Screening* **1,** 27–31.

13. Sebestyen, M., Dibo, G., Kovacs, A., and Furka, A. (1993) *Biorg. Med. Chem. Lett.* **3,** 413–416.
14. Gallop, M.A., Barrett, R.W., Dower, W.J., Fodor, S.P.A., and Gordon, E.M. (1994) *J. Med. Chem.* **37,** 1233–1251.
15. Gordon, E.M., Barrett, R.W., Dower, W.J., Fodor, S.P.A., and Gallop, M.A. (1994) *J. Med. Chem.* **37,** 1385–1401.
16. Ohlmeyer, M.H.J., Swanson, R.N., Dillard, L.W., Reader, J.C., Asouline, G., Kobayashi, R., Wigler, M., and Still, W.C. (1993) *Proc. Natl. Acad. Sci. USA* **90,** 10922–10926.
17. Nestler, H.P., Bartlett, P.A. and Still, W.C. (1994) *J. Org. Chem.* **59,** 4723–4724.
18. Pearl L., and Blundell, T., (1984) *FEBS Letters* **174,** 96–101.
19. Blundell, T.L., Cooper, J., Foundling, Jones, D.M., Atrash, B., and Szelke, M., (1987) *Biochemistry* **26,** 5585–5590.
20. Abdel-Meguid, S.S. (1993) *Medicinal Research Reviews* **13,** 731–778.
21. Shewale, J. G., Takahashi, R. and Tang, J. In *Aspartic Proteinases and Their Inhibitors*, Kostka, V., Ed. Wlater de Gruyter: Berlin, 1986; pp 101–116.
22. Rich, D. H. In *Proteinase Inhibitors*; Barrett, A. J., Salvensen, G., Eds.; Elsevier Science Publishers B. V.: Amsterdam, 1986; Chapter 5 and references therein.

THERMODYNAMICS AND PROTON UPTAKE FOR PEPSTATIN BINDING TO RETROVIRAL AND EUKARYOTIC ASPARTIC PROTEASES*

Dong Xie, Sergei Gulnik, Laura Collins, Elena Gustchina, T. Nagarajan Bhat, and John W. Erickson

Structural Biochemistry Program
SAIC Frederick
National Cancer Institute
Frederick Cancer Research and Development Center
Frederick, Maryland 21702-1201, USA

1. INTRODUCTION

Aspartic proteases have been the subject of intensive physicochemical research in the areas of catalytic mechanism, three-dimensional structure, zymogen activation and substrate specificity.[1,2] The important roles of various members of this family in human diseases, including hypertension,[3] cancer,[4] malaria[5] and AIDS[6] have made these enzymes the target of rational drug design efforts. Thermodynamic characterization of inhibitor binding is important for our understanding of the mechanism and energetics of enzyme-drug interactions. Recent developments in high sensitivity isothermal titration calorimetry (ITC) have made it possible to obtain extensive thermodynamic information for molecular interactions.[9,10] The binding of pepstatin to three aspartic proteases, endothiapepsin from *Endothia parasitica*, plasmepsin II from *Plasmodium falciparum* and cathepsin D from human liver have been characterized using this technique.[7,8]

In this work, we extend these studies to a retroviral aspartic protease and report the results of ITC measurements on the association of pepstatin with HIV-1 protease. In addition, we compare the thermodynamic quantities with those obtained for pepstatin-binding to three eukaryotic aspartic proteases.

* The content of this publication does not necessarily reflect the views or policies of the Department of Health and Human Services, nor does mention of trade names, commercial products, or organization imply endorsement by the U.S. Government.

Aspartic Proteinases, edited by James.
Plenum Press, New York, 1998.

2. MATERIALS AND METHODS

2.1. Enzyme and Inhibitor

Wild type HIV-1 protease was purified as described previously.[11] Pepstatin A was purchased from Sigma (St. Louis, MO).

2.2. Isothermal Titration Calorimetry (ITC)

Reaction heats were measured using an Omega titration microcalorimeter (Microcal, Northamton, MA) whose design and operation has been described previously.[12] Protein solutions were prepared by dialysis against 10 mM of desired buffer at 4°C. Protein concentrations were determined by measuring optical absorbance at 280 nm using an extinction coefficient of 1.15 $(mg/ml)^{-1}$. Pepstatin A solution was prepared by dissolving the peptide powder with a small amount of water at pH 11.5, then further diluted with distilled water to the desired concentration. All protein and inhibitor samples were degassed for 15 minutes prior use. Previous studies have shown a lack of concentration dependence for the heat of injecting pepstatin into buffer. The heat of dilution for pepstatin was thus determined by taking the average of the last three injections after the protein was saturated with inhibitor. The solution was stirred at a rate of 400 rpm.

3. RESULTS AND DISCUSSION

Figure 1 shows a typical curve from an ITC experiment. Each peak represents injection of 6 μl of pepstatin A into a solution of HIV-1 protease at 5 min. time intervals. The area under the peak corresponds to the heat of mixing. The last few peaks represent the heat of dilution of the injected inhibitor since the protease is saturated with inhibitor and no further binding can occur. The reaction heat was determined by integrating the area under the peaks and subtracting the heat of dilution. The positive sign of the peak indicates that the overall reaction is endothermic. At 25°C and pH 6.5 in MES buffer, the reaction was characterized by an total enthalpy change of 5.3 kcal/mol and a dissociation constant of 1.5 μM. The enthalpy determined from a single calorimetry experiment is a function of both the binding enthalpy and the ionization enthalpy of the buffer and can be represented by the equation:

$$\text{Æ}H_{ITC} = \text{Æ}H_{binding} + \text{Æ}n \bullet \text{Æ}H_{ion} \qquad (1)$$

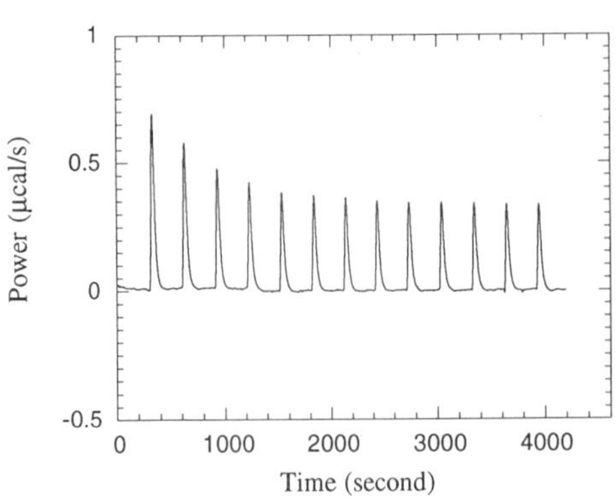

Figure 1. Isothermal calorimetric titration of HIV-1 protease with pepstatin A in 10 mM MES buffer, pH 6.5 at 25°C. The protease and inhibitor concentrations were 17 μM and 0.6 mM respectively. For each injection, 6 μL of inhibitor solution was injected in 10 seconds at 5 minute intervals.

where ÆH$_{binding}$ is the buffer-independent binding enthalpy, Æn is the net number of protons transferred between buffer and complex, and ÆH$_{ion}$ is the ionization enthalpy of the buffer. As can be seen from equation (1), the reaction enthalpy will be buffer-dependent when there is a change in the pKa of one or more ionizable groups during inhibitor binding, since this will result in the release or uptake of protons. From a plot of ÆH$_{ITC}$ vs ÆH$_{ion}$ for different buffers, values for Æn and ÆH$_{binding}$ can be calculated from the slope and intercept, respectively, of Equation (1).

Analysis of the buffer-dependent enthalpy for the association of pepstatin and HIV-1 protease resulted in a net uptake of 0.4 ± 0.1 protons/complex and a ÆH$_{binding}$ of 3.8 ± 0.3 kcal/mol (Fig. 2). These results indicate that pepstatin binding is endothermic at pH 6.5 and 25°C, and therefore must be driven completely by a favorable entropy term. This thermodynamic behavior is typical for desolvation of apolar groups and indicates the importance of hydrophobic interactions in the binding process.[13] Proton uptake by the complex indicates that the electrostatic environment of an ionizable group has to be modified upon inhibitor binding. Inspection of the three-dimensional structure of the HIV-1 protease-pepstatin complex[14] reveals that the ionizable groups in the binding cavity belong to four aspartic acid and two arginine residues: the two catalytic aspartates Asp25 and Asp125, Asp29, Asp129, Arg8 and Arg108. Among these residues, only the two catalytic aspartic acid residues make direct contacts with the inhibitor. The carboxylate groups of Asp29 and Asp129 form electrostatic interactions with the positively charged guanidinium side chains of Arg108 and Arg8, respectively, and thus should remain fully ionized at pH 6.5. We conclude that pepstatin binding results in the transfer of 0.4 proton, on average, from the buffer to one of the two catalytic aspartates. This result is consistent with a recent NMR study of the ionization state of the HIV-1 protease active site in a pepstatin complex.[19] It was shown that the pKa of one of the catalytic aspartates increases from a value below 6.0 to one above 6.5 upon pepstatin binding, and would lead to proton uptake at pH 6.5. The pKa of the second catalytic aspartate was found to be below 2.5 in the pepstatin complex, and must therefore be fully ionized at pH 6.5. Taken together, these results provide strong experimental evidence for the charge state of the active site to be −1 at pH 6.5.

Table 1 summarizes the thermodynamic quantities of pepstatin binding to four aspartic proteases under similar conditions. All four proteins exhibit a similar heat capacity change upon pepstatin A binding. This parameter corresponds to the temperature dependence of all thermodynamic parameters, ÆH, ÆS and ÆG, and its relatively large value emphasizes the importance of considering temperature-dependence when comparing experimental results. Based on the sign and magnitude of the enthalpy and entropy changes, the proteases can be classified into two groups. For the first group, plasmepsin II and

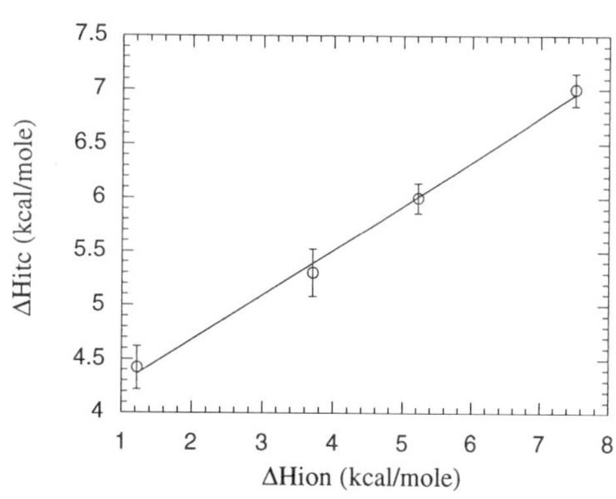

Figure 2. The calorimetry-measured enthalpy as a function of buffer ionization enthalpy at pH 6.5 and 25°C. The buffers were 10 mM sodium phosphate, ACES, MOPS and ACES. The number of transferred protons was 0.4±0.1 determined from the slope of the linear fitting. The intercept was 0.8±0.3 kcal/mol corresponding to the buffer-independent enthalpy.

Table 1. Thermodynamic properties of pepstatin binding to aspartic proteases at 36°C

	HIV-1 Protease	Cathepsin D[a]	Plasmepsin II[a]	Endothiapepsin[b]
pH	6.5	6.5	6.5	7.0
ÆC_p (cal/mol·K)	−270 ± 20	−330 ± 20	−400 ± 60	−310 ± 20
ÆH (kcal/mol)	0.8 ± 0.3	−1.0 ± 0.4	−13.6 ± 0.6	−10.3 ± 0.2
ÆS (cal/mol·K)	30.4	33.3	−10.7	−3.2
Æn[c]	0.4 ± 0.1	2.9 ± 0.1	1.7 ± 0.1	1.1 ± 0.1
Ionizable group	4ASP	2ASP, GLU, HIS	2ASP, HIS	4ASP
ÆG (kcal/mol)	−8.6	−11.3	−10.3	−9.3
K_d (nM)	900 ± 100	11 ± 90	55 ± 70	300

[a]Xie, et. al.[8]
[b]Gomex, et. al.[7]
[c]Æn is defined as the number of transferred protons between the solvent and the binding complex upon inhibitor binding. These transfer processes are due to pKa change of the ionizable groups involved in binding. See text for detail.

endothiapepsin, the binding is driven by a favorable enthalpy, indicating that polar interactions play an important role.[13] For the second group, cathepsin D and HIV-1 protease, the situation is reversed: here the entropy term is highly favorable and the enthalpy term is either small or unfavorable. This suggests that for the second group, the desolvation of apolar groups is the key contributor to the total energetics of pepstatin binding. Since the conformations of pepstatin and the polarity of the active site residues are very similar in all four complexes, we conclude that the differences in binding thermodynamics reflect structural differences between the free and complexed forms of the enzymes. For HIV-1 protease and cathepsin D, except for the flap movement commonly observed for aspartic proteases, there is little conformational change of the protein upon pepstatin binding.[15,16] For plasmepsin II and endothiapepsin, on the other hand, substantial movements between the C-terminal subdomain relative to the N-terminal domain and active site region upon pepstatin binding have been reported.[8,17,18] This movement was suggested to be responsible for the differences in pepstatin binding energetics between plasmepsin II and cathepsin D, and to result in the protonation of His164 in plasmepsin II which is located well outside of the binding cavity but between the two domains.[8] These studies suggest that the energetic difference between the two classes of aspartic protease may be due to differences in subdomain movements upon binding. The energetic consequence of this movement for plasmepsin II and endothiapepsin is a gain of enthalpy and loss of entropy, as shown in Table 1. This type of thermodynamic behavior typically results from the formation of polar interactions, such as hydrogen bonds in proteins,[13] which we assume provide the driving force for the domain movement.

From the analysis of the buffer-dependence of reaction enthalpy we can estimate the charge of the active site of aspartic proteases in their pepstatin complexes. As was shown above, the charge state of the HIV-1 protease active site is −1, where one aspartate is fully ionized and the other is protonated. In the pepstatin complex of plasmepsin II, the pKa's of the two aspartates were estimated to be 3.0 and 7.5.[8] Therefore, at pH 6.5 the protonation state of the active site must also be −1. Is this charge state common to the binding of all aspartic proteases with pepstatin? As shown in Table 1, the calorimetry experiments for cathepsin D indicate that 2.9 protons were transferred to the binding complex and thus three ionizable groups become protonated at pH 6.5. Examination of the cathepsin D structure[16] revealed that the inhibitor binding site consists of four ionizable groups: His77, Glu260 and the two catalytic aspartates. Since the binding interface is similar to that of plasmepsin II, and the ionizable groups in plasmepsin II consists of two catalytic aspartates and one histidine, these data suggest that the three protonated groups belong to His77, Glu260 and one of the catalytic aspartates, whose pKa increases when inhibitor binds. The second catalytic aspartate would remain ionized at this pH. Based on these results, we conclude that for the cathepsin D/pepstatin complex the charge state of the two catalytic aspartates is again −1. For endothiapepsin, 1.1 protons were transferred to the binding complex. Although four aspartates are involved in binding in the pepstatin complex,[20] only the two catalytic aspartates, Asp32 and Asp215, make polar contacts with the inhibitor, again suggesting that one of the catalytic aspartates becomes protonated upon inhibitor binding, consistent with a −1 charge state of the active site. The carboxylate groups of Asp30 and Asp77 make van der Waals contacts with hydrophobic regions in pepstatin. In addition, Asp30 forms hydrogen bonds to two water molecules in both the free and inhibited forms of the enzyme. Asp77, which is in the flap and changes conformation during inhibitor binding, is involved in hydrogen bonds with Ser70 OH and a water molecule in both states. Thus, we can assume that the ionization states of Asp30 and Asp77 remain unchanged upon inhibitor binding and are most likely ionized at pH 7.0 where these experiments were performed.[7] In summary, the calorimetry studies indicate that the active site carboxylates for the four examined pepstatin/aspartic protease complexes share a common charge state in the pH range from 3 to 7.

4. CONCLUSIONS

Recently, ITC has been extensively used to investigate molecular interactions (see reviews by Freire, et. al.[9] and Ladbury, et. al.[10]). The thermodynamic information provided by this technique is especially helpful for mechanistic studies. In this work, we examined the thermodynamic properties of the binding of pepstatin to four aspartic proteases. Several conclusions may be drawn from the comparison: 1) Inhibitor binding exhibits a similar and significant temperature dependence for all four enzymes; 2) Aspartic proteases from diverse structural classes, the C2 symmetric, homodimeric retroviral proteases and the asymmetric, single chain eukaryotic enzymes, can exhibit similar thermodynamic properties for inhibitor binding; 3) Although the binding free energies are of the same magnitude, related enzymes from the same structural class, for example plasmepsin II and cathepsin D which share 35% sequence identity, exhibit different binding mechanisms due to different interdomain movements; 4) The calorimetry data suggest that the energetic consequence of the interdomain movements in aspartic proteases is the formation of intramolecular polar interactions, e.g. hydrogen bonds; 5) In the pH range from 3 to 7, all of the proteases share a common charge state for the two catalytic aspartates; 6) Calorimetry and crystallography can be used together to identify pKa changes during inhibitor binding and to evaluate their structural origins as well as their energetic consequences. These conclusions are not intuitively obvious from examination of the crystallographic structures alone, and illustrate how the combination of thermodynamic analysis and structural characterization can uniquely contribute to our understanding of the mechanism and energetics of protein-inhibitor interactions.

ACKNOWLEDGMENTS

The authors thank Dr. Ernesto Freire and Dr. Javier Gomez of the Biocalorimetry Center at The Johns Hopkins University for the initiation of cathepsin D experiments.

REFERENCES

1. Fruton, J. S. *Adv. Enzymol.* 1976, *44*, 1–36.
2. Davies, D. R. *Annu. Rev. Biophys. Biophys. Chem.* 1990, *19*, 189–215.
3. Inagami, T.; Misono, K.; Chang, J.; Takii, Y.; Dykes, C. In *Aspartic proteinases and their inhibitors.*; V. Kostka, Ed.; Walter de Gruyter: Berlin/New York, 1985; pp 319–337.
4. Rochefort, H. *Semin. Cancer Biol.* 1990, *1*, 153–160.
5. Goldberg, D. E.; Slater, A. F. G.; Cerami, A.; Henderson, G. B. *Proc. Natl. Acad. Sci. USA* 1990, *87*, 2931–2935.
6. DeBouck, C. *AIDS Res. Hum. Retroviruses* 1992, *8*, 153–164.
7. Gomez, J.; Freire, E. *J. Mol. Biol.* 1995, *252*, 337–350.
8. Xie, D.; Gulnik, S.; Collins, L.; Suvorov, L.; Lee, A.; Gustchina, E.; Erickson, J. *In preparation* .
9. Freire, E.; Mayorga, O. L.; Straume, M. *Anal. Chem.* 1990, *62,* 950A-959A.
10. Ladbury, J.; Chowdhry, B. *Chemistry & Biology* 1996, *3*, 791–801.
11. Gulnik, S.; Suvorov, L.; Liu, B.; Yu, B.; Anderson, B.; Mitsuya, H.; Erickson, J. *Biochemistry* 1995, *34*, 9282–7.
12. Wiseman, T.; Williston, S.; Brandt, J.; Lin, L. *Anal. Biochem.* 1989, *179*, 131–135.
13. Murphy, K. P.; Freire, E. *Adv. Protein Chem.* 1992, *43*, 313–361.
14. Fitzgerald, P.; McKeever, B.; VanMiddlesworth, J.; Springer, J.; Heimbach, J.; Leu, C.; Herber, W.; Dixon, R.; Darke, P. *J. Biol. Chem.* 1990, *265*, 14209–19.
15. Wlodawer, A.; Erickson, J. W. *Annu. Rev. Biochem.* 1993, *62*, 543–85.
16. Baldwin, E. T.; Bhat, T. N.; Gulnik, S.; Hosur, M. V.; Sowder, R. C., II; Cachau, R. E.; Collins, J.; Silva, A. M.; Erickson, J. *Proc. Natl. Acad. Sci. USA* 1993, *90*, 6796–6800.
17. Sali, A.; Veerapandian, B.; Cooper, J.; Moss, D.; Hofmann, T.; Blundell, T. *Proteins: Struct. Func. Gen.* 1992, *12*, 158–170.
18. Silva, A.; Lee, A.; Gulnik, S.; Majer, P.; Collins, J.; Bhat, T.; Collins, P.; Cachau, R.; Luker, K.; Gluzman, I.; Francis, S.; Oksman, A.; Goldberg, D.; Erickson, J. *Proc. Natl. Acad. Sci. USA* 1996, *93*, 10034–39.
19. Smith, R.; Brereton, I.; Chai, R.; Kent, S. *Nature Struct. Biol.* 1996, *3*, 946–950.
20. Bailey, D.; Cooper, J.; Tickle, I.; Veepandian, B.; Blundell, T.; Atrash, B.; Jones, D.; Szelke, M. *Biochem, J.* 1993, *289*, 363–371.

54

EXPRESSION, PURIFICATION, AND CHARACTERIZATION OF THE RECOMBINANT PEPSIN INHIBITOR FROM *Ascaris suum*

Jeffrey Zalatoris,[1] Chetana Rao-Naik,[1] Gregory Fecho,[1] Karen Girdwood,[2] John Kay,[2] and Ben M. Dunn[1]

[1]Department of Biochemistry and Molecular Biology
University of Florida
Gainesville, Florida 32610
[2]Department of Biochemistry
University of Wales, College of Cardiff
Cardiff, Wales, United Kingdom

1. INTRODUCTION

Only a few natural inhibitors of aspartic proteinases have been discovered to date. A small number of peptides from bacterial origins have been shown to inhibit members of this protease family.[1–2] Larger proteinaceous inhibitors of aspartic proteases have been discovered in the potato[3] (cathepsin D inhibitor), in yeast,[4] and the intestinal nematode *Ascaris*.[5–7] Similar genes have been discovered in other parasitic nematodes.[8] The *Ascaris* aspartic proteinase inhibitor (PI-3) is a 16.7 kDa protein that specifically inhibits pepsin, gastricsin, and, less strongly, cathepsin E. This inhibitor possesses six cysteine residues which form three disulfide bridges, and it contains helical and strand structural motifs.

The study of the *Ascaris* inhibitor has been complicated by the lack of structural information. Although Peanasky's group made great strides in studying the native protein,[7] little structural information had been gleaned from their work. We set initial goals to develop a recombinant model of the protein that had similar properties to the native protein. Additionally, we wanted to develop an expression and purification scheme that would generate enough of the protein for crystallographic studies of the interaction between this inhibitor and pepsin. The development of a recombinant model of this inhibitor was sought because of the difficulty in obtaining the protein and the potential harmful effects of working with the roundworm. A gene had been previously cloned into a pET-3d plasmid.[9] Expression and purification from *E. coli* resulted in a recombinant protein with properties similar to those of the native PI-3 inhibitor. Such results support the use of this recombinant inhibitor as a close analog of the native form of the protein in future studies.

Aspartic Proteinases, edited by James.
Plenum Press, New York, 1998.

2. METHODS AND RESULTS

Purification of the recombinant inhibitor involved the overexpression of the gene product from *E. coli.* BL21[DE3] cells containing the pET-3d vector with the clone of the inhibitor gene were grown in M9CA media to OD_{600} of 0.6 followed by induction with 0.45 mM IPTG. Three hours following IPTG addition, the cells were harvested and lysed by French pressure cell. Inclusion bodies were recovered and denatured in a 8 M urea, 300 mM 2-mercaptoethanol solution. The suspension of denatured protein was dialyzed with five different buffer changes to allow refolding. An ammonium sulfate precipitation was performed on the sample. At 40% saturation, the centrifugation pellet was discarded. The second pellet was obtained at 70% saturation. This second pellet was resolubilized in pH 5.0 sodium phosphate buffer and loaded onto a Superdex 75 gel filtration column for chromatographic separation. The purification scheme is listed in Table 1.

SDS-PAGE analysis and mass spectrophotometric data were used to ensure purity of the isolated protein. Protein thus obtained was analyzed for inhibition behavior against various aspartic proteinases, including an activity titration against a previously quantified pepsin sample. The titration analysis of the purified protein indicated that greater than 90% of the protein obtained was active.

To study the behavior of this protein with respect to that of the native protein, K_i values were determined for the recombinant inhibitor against five of the aspartic proteinases. The spectrophotometric analyses were performed at 37°C in 50 mM buffer at the pH values listed in the table below. For comparison to the native inhibitor, the conditions used were similar to those previously utilized.[6] The results are shown in Table 2. Similarity between the inhibitory properties of the recombinant protein and the native protein was observed against most of the enzymes.[6] Only for the case of cathepsin E was there a discrepancy in the inhibitory behavior, with the recombinant inhibitor exhibiting weaker binding at two pH values.

The lack of a structure of the native protein limits the ability to make structural comparisons between the native and recombinant inhibitors. However, Martzen, *et al.* reported CD data for the native protein.[7] Those data appear quite similar to CD data for the recombinant protein obtained in this study (data not shown), providing the only evidence of secondary structures known.

Table 1. Purification scheme of the recombinant PI-3 from the *E. coli* pET-3d expression system

Stage of purification	Yield[a]
Volume of media	4 liters
Total mass of cells	5.6 grams
Inclusion bodies	430 mg
8 M urea	46 mgs
Refolded protein (dialysis)	42 mgs
70% $(NH_4)_2SO_4$ pellet	14 mgs
Gel filtration	13 mgs
Active protein	12 mgs

[a]Representative yield reflects the average values from two separate purification procedures. All mass yields were determined by protein concentration (A_{280}), with the exception of cell mass, inclusion body mass (direct weighing) and active protein (based on an activity titration of gel filtration product).

Table 2. Comparison of K_i values of the recombinant and native PI-3

pH	Enzyme	Recombinant PI-3 (nM)	Native PI-3 [5,6] (nM)
3.5	Porcine pepsin	0.2	0.5
3.1	Human pepsin	0.3	2
3.1	Human cathepsin E	>50	3
4.7	Human cathepsin E	15	2
3.1	Mouse cathepsin E	6	1
3.7	Human cathepsin D	Not inhibited	Not inhibited

The results for most of the inhibitory studies are closely related between the native and recombinant forms of the protein. Additionally, the current structural information does not suggest any gross differences between the two proteins. Therefore, the recombinant protein will be used as a model for the native protein in further studies to delineate the structural basis of inhibition. Additionally, this recombinant system will be used in the generation of site-directed mutants to study the precise interactions of amino acids between this inhibitor and pepsin.

ACKNOWLEDGMENTS

We would like to thank Tim Nilsen at Case Western Reserve University for the kind donation of the mRNA library of *Ascaris suum*, which was used to develop the cDNA of the gene of the inhibitor. We would also like to thank the University of Florida Protein Chemistry Core Facility for the synthesis of peptides used in kinetic analyses, N-terminal sequencing, and the mass spectrometer. Support for this research has been awarded through NIH grant DK18865 to Ben M. Dunn.

REFERENCES

1. H. Umezawa, T. Aoyagi, H. Morishima, M. Matsuzaki, M. Hamada and T. Takeuchi, *Journal of Antibiotics*, **23**, p. 259, (1970).
2. M. Fukumura, S. Satoi, N. Kuwana and S. Murao, *Agricultural and Biological Chemistry*, **35**, p. 1310, (1971).
3. M. Mares, B. Meloun, M. Pavlik, V. Kostka and M. Baudys, *FEBS Letters*, **251**, p. 94, (1989).
4. T. Dreyer, M. J. Valler, J. Kay, P. Charlton, and B. M. Dunn, *Biochemical Journal*, **231**, p. 777, (1985).
5. G. M. Abu-Erreish and R. J. Peanasky, *Journal of Biological Chemistry*, **249**, p. 1558, (1974a, b).
6. M. J. Valler, J. Kay, T. Aoyagi and B. M. Dunn, *J. Enzyme Inhibition*, **1**, p. 77, (1985).
7. M. Martzen, B. McMullen, N. Smith, K. Fujikawa and R. J. Peanasky, *Biochemistry*, **29**, p. 7366, (1990).
8. X. Q. Hong, J. S. Mejia, S. Kumar, F. B. Perler and C. K. S. Carlow, *Parasitology*, **112**, p. 331, (1996).
9. M. Barman, Senior thesis, University of Florida, 1995.

X-RAY CRYSTALLOGRAPHIC STUDIES OF THE COMPLEX BETWEEN PORCINE PEPSIN AND THE ASPARTIC PROTEINASE INHIBITOR PI-3 FROM THE NEMATODE *Ascaris suum*

Jens F. W. Petersen,[1] Maia M, Chernaia,[1] Chetana Rao-Naik,[2]
Jeffrey L. Zalatoris,[2] Ben M. Dunn,[2] and Michael N. G. James[1]

[1]MRC Group in Protein Structure and Function
Department of Biochemistry
University of Alberta
Edmonton, Alberta T6G 2H7, Canada
[2]Department of Biochemistry and Molecular Biology
University of Florida
Gainesville, Florida 32610

INTRODUCTION

Ascaris suum is an intestinal endoparasitic nematode found in pigs. In order to avoid proteolytic digestion by the host the nematode has developed a large battery of proteinase inhibitors. Among those is the aspartic proteinase inhibitor PI-3. PI-3 consists of a single polypeptide chain containing 149 residues with a molecular weight of 16.4 kD. It has three disulfide bridges. CD and secondary structure predictions indicate that the fold of PI-3 is mixed alpha/beta.[1] Until now only a very few larger aspartic proteinase inhibitors have been characterized. Aspartic proteinase inhibitors from the Filarie family of parasites have been reported (Dirofilaria, Brugia and Onchocerca). These inhibitors are more closely related to each other (~60% sequence identity) than PI-3 (less than 20%).[2] The inhibitor form *Ascaris suum* show a broad specificity towards members from the pepsin related family of aspartic proteinases having Ki's in the range of 0.2–15 nM (porcine and human pepsin and cathepsin E). Cathepsin D however, is not inhibited by PI-3, although it shares up to 45 identity to the above mentioned aspartic proteinases.[3] This has important aspects as it might be used to distinguish between cathepsin D and E. The following report describes the progress in the structure determination of the molecular complex between porcine pepsin and PI-3.

Aspartic Proteinases, edited by James.
Plenum Press, New York, 1998.

EXPERIMENTAL

Crystallization and Data Collection

Recombinant PI-3 is expressed as inclusion bodies, then refolded and purified as previously described.[3] Purified PI-3 was mixed in a 1.2:1 molar ratio with purified porcine pepsin in a final concentration of 10 mg/ml. Crystals were grown using PEG4000 in a concentration of 15% (w/v) buffered with 0.1 M Citrate buffer at pH 3.5. Crystals useful for X-ray diffraction analysis grew as thin plates. SDS-PAGE analysis confirms that the crystals hold a 1:1 complex of pepsin and uncleaved PI-3. X-ray diffraction data were collected at the National Synchrotron Light Source, Brookhaven National laboratory, on a Mar image plate scanner. The crystal was flash frozen at 100 K. The crystals belong to space group $C222_1$ and have unit cell parameters of: a=65.3 Å, b=97.3 Å and c=137.7 Å. There is one PI-3/pepsin complex per asymmetric unit. The data were indexed and integrated to 2.6 Å using the program DENZO.[4] Scaling, averaging and Wilson scaling were done using the programs AGROVATA, ROTAVATA and TRUNCATE from the CCP4 program package.[5] The data set has a R_{sym} 9.6% with a completeness of 93.7% for all measured data. Further statistics of the data set are shown in Table 1.

Structure Determination

A molecular replacement search was performed using the program AMORE.[6] The search model that was used was the monoclinic crystal structure of porcine pepsin.[7] Data

Table 1. Statistics of the data set

Resolution Å	Rsym %	I/σ(I)	Completeness %	Multiplicity
7.69	5.1	11.3	94.6	2.3
6.48	5.6	10.6	92.3	2.3
5.71	6.2	9.3	94.9	2.4
5.16	6.1	10.4	95.1	2.3
4.74	6.3	9.5	93.8	2.4
4.41	6.4	9.8	93.4	2.4
4.14	6.7	8.7	96.3	2.4
3.92	7.4	9.0	94.5	2.4
3.72	7.7	8.1	97.9	2.3
3.56	8.7	8.1	97.7	2.4
3.41	9.6	6.0	95.3	2.4
3.28	12.0	6.0	95.7	2.4
3.17	13.8	5.3	96.0	2.4
3.06	16.8	4.4	94.4	2.5
2.97	17.0	4.3	96.0	2.4
2.88	19.2	3.9	94.8	2.4
2.80	21.1	3.4	95.1	2.5
2.73	24.2	3.1	96.9	2.4
2.66	29.4	2.5	93.8	2.3
2.60	30.8	2.4	96.2	2.4

$$Rsym = \frac{\sum_{\bar{h}}\sum_{i}\left|I_i(\bar{h}) - \langle I(\bar{h})\rangle\right|}{\sum_{\bar{h}}\sum_{i} I_i(\bar{h})}$$

i is summed over all symmetry related in intensities.

in the range from 10–4 Å were used. Only one clear solution was found in both the rotational and transitional search. The molecular replacement solution was refined using rigid body refinement in AMORE. The correlation coefficient and Rvalue improved from 50.3% and 42.2% to 51.9% and 41.3% respectively. No severe overlap was found in the crystal packing (Fig. 1). Most of the electron density was well defined except a large region (residues 222–259 and 268–287) in the C-terminal part of the pepsin molecule which had no or poorly defined electron density. This part constitutes a separate domain of the pepsin molecule which in general has been shown to be flexible in mammalian aspartic proteinases.[8]

Map Improvement and Modelbuilding

Based on the position of porcine pepsin determined by the molecular replacement search, electron density maps were calculated with the coefficients (m|Fo| – D|Fc| or 2m|Fo| – D|Fc|), using the program SIGMAA[9] (|Fo| and |Fc| denotes observed and calculated structure factor amplitudes, m is a estimated figure of merit and D a weighting term for the calculated structure factor amplitude). The difference density clearly revealed parts of the inhibitor. Difference density could only be seen on the P′ side of the active site cleft, whereas no difference density is present at the P side of the active side cleft. The most prominent feature in the difference density map is a tripeptide close to the active site. In addition, it was possible to trace part of an alpha-helix and three strands. All attempts to optimize the initial model using simulated annealing for different types of density modification (solvent flattening, skeletonization of histogram matching) either distorted the difference electron density for the inhibitor or had no effect. The current strategy has therefore been to inspect the initial difference electron density maps and manually remove or correct the pepsin part and build a poly alanine model of the inhibitor in repeating cycles. Presently about 50 residues of PI-3 have been traced, whereas residues 222–259 and residues 268–287 have been removed from the pepsin part of the complex.

DISCUSSION

The current model is still at a very early stage in the structure determination and it is only possible at present to draw some very general and tentative observations. When the

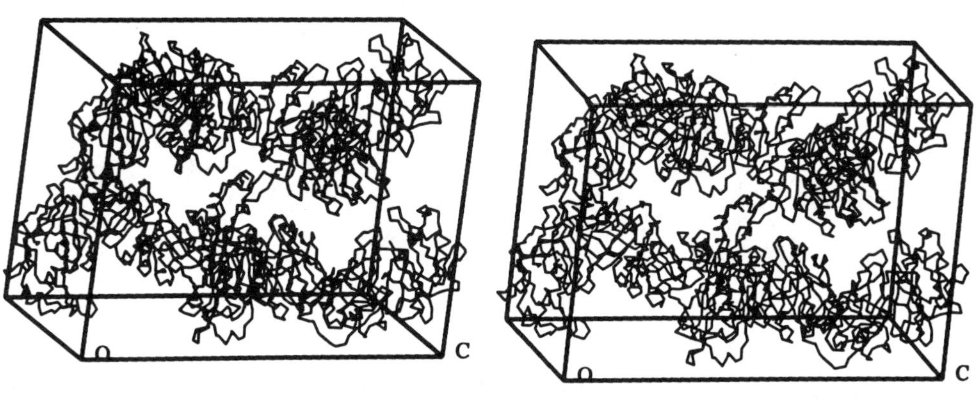

Figure 1. The packing of the porcine pepsin molecules in the unit cell. Only the C_α backbone is drawn.

packing of the porcine pepsin molecules is inspected it is clearly seen that there is sufficient space for a protein with a size of 149 residues (Fig. 1). Based on the initial difference electron density maps there is only density occupying the P′ side of the active site cleft in the pepsin molecule. No pronounced difference electron density can be seen close to the subsites S1, S2 and S3. Some of the most pronounced structural elements of PI-3, which can be modeled at this stage are shown in Figure 2. Furthermore, SDS-PAGE analysis of the complex crystals shows that PI-3 is bound as the intact inhibitor. It is therefore possible at this stage with reasonable certainty to conclude that the intact PI-3 is occupying and most likely blocking the S_n' sites of the active site cleft in pepsin. It is unlikely that PI-3 binds in the manner of a substrate spanning the subsites S1, S2 and S3. Cleavage in vitro of PI-3 by pepsin has characterized.[3] Three cleavage sites were found in that study. We propose that the cleavage of PI-3 by pepsin occurs as a competing process to the binding, and that none of the cleavage sites are associated directly with the inhibition.

At this stage two regions of pepsin molecule can be seen, in contact with the inhibitor (Fig. 2). The first region is from residue 70 to 76 and the second region from residue 290 to 294. The specific interaction between PI-3 and pepsin in these regions is difficult to characterize at present. It is interesting that the second region in pepsin is close to the so-called proline rich region in cathepsin D. Conformational differences in the S_n' region of cathepsin D might therefore be responsible for its poor affinity with PI-3.

The pronounced feature of PI-3, in the difference electron density is found close to the active site running parallel to residues Ile73, Thr74 and Tyr75 in pepsin towards Asp215 in pepsin. In Fig. 3 is shown the difference electron density (2m|Fo| – D|Fc|) in this region. The feature is presently modeled as Xxx–Phe–Arg. Two side chains, one from Asp215 and the other from Tyr189, might interact with this strand from PI-3. Asp215 is

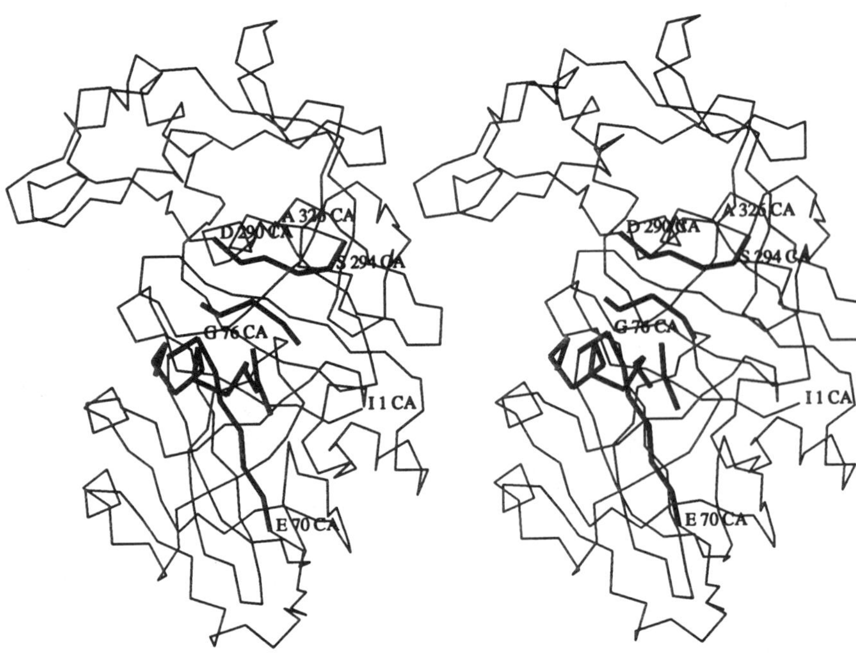

Figure 2. C-alpha backbone of porcine pepsin (thin line) and the clearest features in PI-3 (thick line). Regions in pepsin (residues 70–76 and 290–294) which are in contact with PI-3 are also drawn with thick lines.

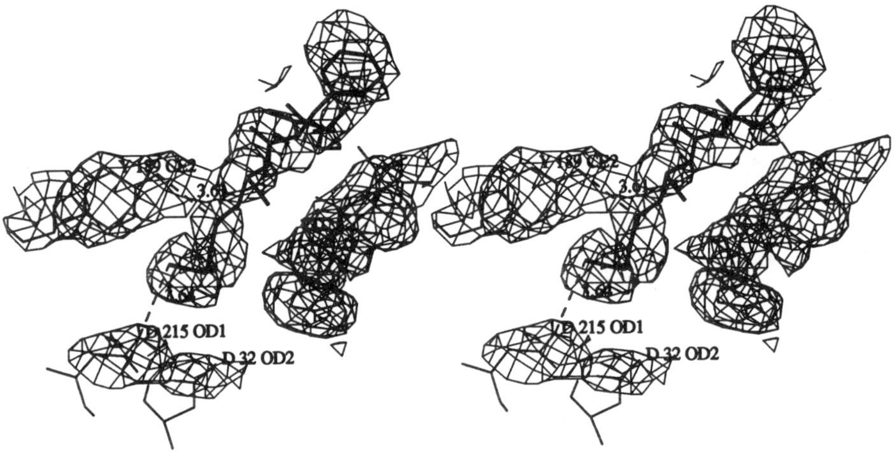

Figure 3. Electron density map (2m|Fo| – D|Fc|) close to the active site of pepsin contoured at the 1σ level. Pepsin (thin line). PI-3 model (thick line).

almost within hydrogen bond distance to the modeled arginine and Tyr189 is close to the C-gamma of the arginine. Quite interestingly Tyr189 forms a hydrogen bond to pepstatin in the human pepsin/pepstatin complex.[10] It seem plausible that Tyr189 in the porcine pepsin/PI-3 complex should interact in a similar manner.

Presently the structure determination of the complex is made difficult by the fact that the phasing originates from a model, which effectively only accounts for approximately 60% of protein content in the asymmetric unit. Work is therefore being done to produce a heavy atom derivative of the pepsin/PI-3 complex crystals, to assist in the phasing.

ACKNOWLEDGMENTS

This work was done at the MRC Group for Protein Structure and Function, which is supported by a grant from the Canadian MRC. JFWP is supported by Killam postdoctoral fellowship. Work at the University of Florida was supported by NIH grant DK18865.

REFERENCES

1. Martzen M.R., McMullen B.A., Smith N.E., Fujikawa K. and Peanasky R.J. *Biochemistry* 29:7366–7372 (1990).
2. Karen Girdwood. University of Wales. Personal communication.
3. Rao-Naik C., Zalatoris J., Barman M., Farmerie W.G., Banull M., Rogers R., Lees W.E., Girdwood K., Kay J. and Dunn B.M. (1997) In. Prep.
4. Otwinowski Z. (1986), Yale University, Connecticut, U.S.A.
5. SERC Daresbury Laboratory (1979)
6. Navaza, J. *Acta. Cryst.* A50:157–163 (1990)
7. Sielecki A.R., Fedorov A.A., Boodhoo A., Adreeva N.S. and James M.N.G. *J. Mol. Biol.* 214:143–170 (1990)
8. Sali A., Veerapandian B., Cooper J.B., Moss D.S., Hofmann T. and Blundell T.L. *Proteins* 12:158–170 (1992).
9. Read R.J., *Acta. Cryst.* A42:140–149 (1989)
10. Fujinaga M., Chernaia M.M., Tarasova N.I., Mosimann S.C, and James M.N.G. *Protein Science* 4:960–972 (1995)

56

STUDIES ON PLASMEPSINS I AND II FROM THE MALARIAL PARASITE *Plasmodium falciparum* AND THEIR EXPLOITATION AS DRUG TARGETS

Richard P. Moon,[1] Daniel Bur,[1] Hansruedi Loetscher,[1] Allan D'Arcy,[1] Lorraine Tyas,[2] Christian Oefner,[1] Fiona Grueninger-Leitch,[1] Daniel Mona,[1] Katherina Rupp,[1] Arnulf Dorn,[1] Hugues Matile,[1] Ulrich Certa,[1] Colin Berry,[2] John Kay,[2] and Robert G. Ridley[1*]

[1]F. Hoffmann-La Roche, Pharmaceuticals Division
Pharma Research Pre-Clinical
Basel, Switzerland
[2]School of Molecular and Medical Biosciences
University of Wales, Cardiff, United Kingdom

1. INTRODUCTION

1.1. The Need for New Antimalarials

Malaria is one of the major diseases of the world. Between 2 to 3 million deaths occur each year, mainly children under 5 years of age living in sub-Saharan Africa. Up to 300 million people are infected at any given time and up to 2 billion people (close to half the world population) live in malarious areas and are at risk of infection. The most virulent of the four malaria species which infect humans is *Plasmodium falciparum* and the spread of resistance by this species to the available drugs, such as chloroquine, has resulted in a critical world health situation with a desperate need to develop new drugs.

1.2. The Parasite Life Cycle and the Biology of Haemoglobin Degradation

The malarial parasite undergoes a complicated life cycle and transmission requires stages in both the mosquito and human host. Human infection is initiated by the bite of an infected mosquito and the concomitant inoculation of sporozoites into the blood stream. These sporozoites find their way to the liver, where they multiply asexually over a period

* Corresponding author: Dr. Robert G. Ridley, Dept. PRPI (70/130), F. Hoffmann-La Roche, CH-4070 Basel, Switzerland. Tel: 0044-61-688-2575; fax: 0044-61-688-2729; e-mail: robert_g.ridley@am@rbcamm

Aspartic Proteinases, edited by James.
Plenum Press, New York, 1998.

of two weeks to generate many thousands of merozoites, which are released into the blood stream. Each merozoite is capable of invading and infecting an erythrocyte and over a period of two days can generate a further 20 to 30 daughter merozoites for release into the blood stream. Some of these blood stage parasites differentiate into sexual forms which, when taken up by a mosquito, undergo sexual reproduction and develop into infective sporozoites, thus completing the life cycle.

During the asexual erythrocytic stage of development, the parasite undergoes an initial establishment phase where very little metabolic activity occurs (ring stage), followed by a growth phase (trophozoite stage), and finally a replicative phase (schizont stage) before their release from the erythrocyte for re-infection of fresh red blood cells. During the trophozoite stage up to 80% of the erythrocytes' haemoglobin is degraded by the parasite (Ball *et al.*, 1948; Groman, 1951; Roth, Jr., *et al.*, 1986). The reasons for this high level of metabolic activity are probably two-fold. First, the amino acids released are a potential source of nutrients. This is supported by evidence that parasite growth in culture requires supplementation with amino acids (methionine, cysteine, isoleucine, glutamine and glutamate) that are absent or in low abundance in haemoglobin (Divo *et al.*, 1985; Francis *et al.*, 1994). Secondly the parasite requires space to grow and develop in the erythrocyte and may need to degrade haemoglobin to achieve this. This is supported by evidence that not all products of haemoglobin degradation are required by the parasite (Vander Jagt *et al.*, 1992). Reconstitution of erythrocytes containing as little as 6% of the original haemoglobin can support parasite growth (Rangachari *et al.*, 1987) and a large proportion of the amino acids generated by haemoglobin degradation are secreted from the parasite (Zarchin *et al.*, 1986).

The degradation of haemoglobin takes place in an acidic lysosomal organelle called the digestive vacuole (Olliaro and Goldberg, 1995). Several types of proteinase are involved in this process. Using specific inhibitors to study haemoglobin degradation *in vitro*, aspartic proteinase, cysteine proteinase, serine proteinase and metalloproteinase activities have all been detected in digestive vacuole extracts (Goldberg *et al.*, 1990). However, only two aspartic proteinases (plasmepsins I and II) and one cysteine proteinase (falcipain) have been characterised.

Here we report the progress towards exploiting the plasmepsins as chemotherapeutic targets for the generation of new drugs. Recombinant gene expression, recombinant enzyme purification, crystallisation studies and inhibitory data obtained with some peptidic inhibitors are presented.

2. PLASMEPSINS AS POTENTIAL DRUG TARGETS

2.1. Aspartic Proteinase Inhibitors Inhibit Malarial Parasite Growth in Culture

The discovery of the plasmepsins immediately opened up an area of investigation for antimalarial drug discovery because so much structural information on aspartic proteinases is available. Also it is known from work on renin and the HIV aspartic proteinase that specific inhibitors of aspartic proteinases can be developed and registered for human therapy. The key issues facing us when we began to work on these enzymes were two-fold:

1. the proof of principle that inhibiting these enzymes would kill the parasite i.e. that the enzymes are essential for the parasite.
2. the generation recombinant protein for detailed enzyme kinetic studies, structural studies and compound screening.

We approached this in two ways. First, even before we had recombinant enzyme, we had access to compounds that had been designed as inhibitors of other aspartic proteinases and we also had access to a medium throughput screen against parasite growth in culture. We therefore screened several hundred compounds for inhibition of malarial growth directly. This yielded two compounds that showed a reasonable antiparasitic activity, in the sub-micromolar range. These compounds and their antiparasitic activities are given in Table 1. Significantly, the activities were similar for both a drug sensitive strain of *P. falciparum* (NF54) and a multi-drug resistant strain of *P. falciparum* (K1). For comparison, chloroquine exhibits IC_{50} values of 10 nM and 300 nM respectively against these two strains of parasite.

This initial result, coupled with data from Goldberg's laboratory demonstrating that an aspartic proteinase inhibitor of plasmepsin I was weakly active against *P. falciparum* growth (Francis *et al.*, 1994), gave us some confidence that the plasmepsins could be valid antimalarial drug targets. However, it was still necessary to demonstrate that the antimalarial activity observed resulted from plasmepsin inhibition. For this we needed to generate recombinant enzyme for further studies.

2.2. Gene Structure of Plasmepsins and Functional Expression of Plasmepsin II

The genes for both plasmepsin I and plasmepsin II have been cloned (Dame *et al.*, 1994; Francis *et al.*, 1994) and an overview of their gene structure in comparison to a typical aspartic proteinase is given in Figure 1. Like other eukaryotic aspartic proteinases it is assumed each is initially expressed as a preproenzyme that is eventually processed to a mature enzyme. The N-termini of both mature plasmepsins in the parasite food vacuole

Table 1. Structure and activity of two peptidic inhibitors of *P. falciparum* malarial parasite growth (adapted from Moon et al. 1997)

Inhibitor	Structure	Parasite strain IC_{50} (nM) K1	Parasite strain IC_{50} (nM) NF54
Ro 40-4388	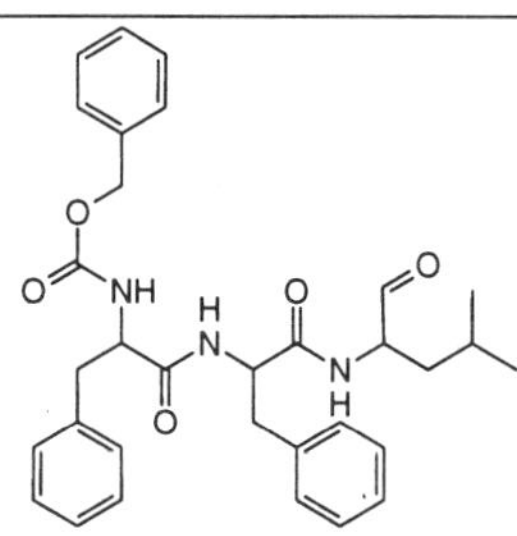	250	300
Ro 40-5576	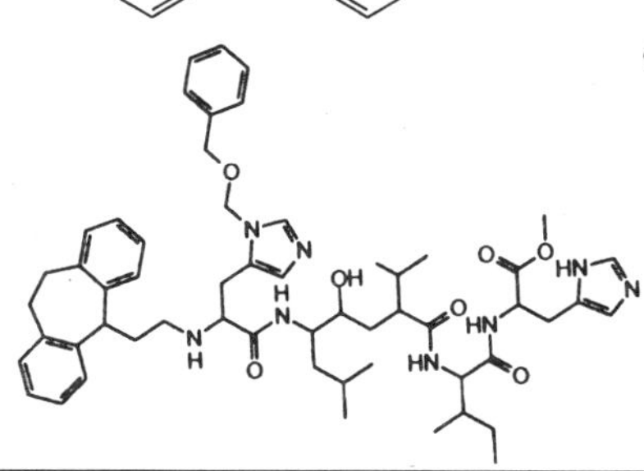	900	1700

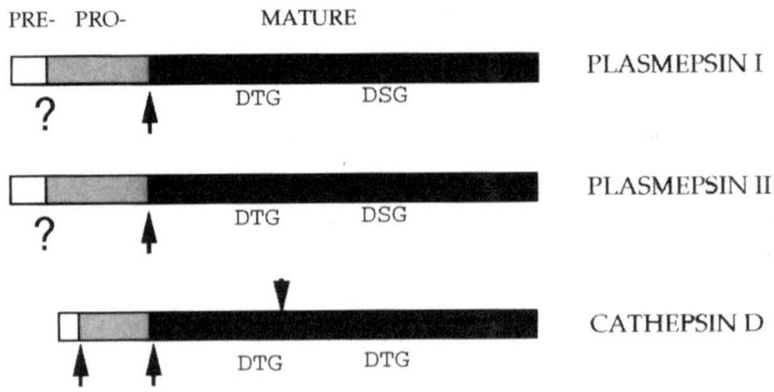

Figure 1. Schematic representation of primary protein structure of plasmepsins I and II compared to cathepsin D. Known processing sites are marked by arrows and the relative positions of the conserved catalytic motifs, DTG and DSG, are shown. Note the extended N-terminus of the plasmepsin precursors compared to cathepsin D.

have been determined (Goldberg *et al.*, 1991; Gluzman *et al.*, 1994) and they begin at positions similar to other aspartic proteinases. The pro-parts of the plasmepsins however are much longer, by about 70 amino acids, than regular aspartic proteinases and it is uncertain where, or even whether, processing of the preproenzyme to the proenzyme occurs. Toward the N-terminus of the prepro region of the enzymes there is a potential transmembrane domain and it has been postulated that the extended N-terminal domain of the plasmepsins is required to allow a membrane associated transport process of the proenzyme to its target organelle, the lysosomal food vacuole (Francis *et al.*, 1994).

The extended propart of the plasmepsins gave rise to some difficulties in functionally expressing recombinant plasmepsin. Many aspartic proteinases are expressed as zymogens that remain catalytically dormant until the environmental pH falls sufficiently to permit autocatalytic processing to the mature active enzyme (Fusek and Vetvicka, 1995). Expression of the full length preproenzymes, including the hydrophobic transmembrane segments, resulted in low yields of expression. We therefore decided to express the plasmepsins as truncated precursor proteins with proparts of approximately the same length as regular proenzymes, i.e. from Ser77P in proplasmepsin II and from Ser76P in proplasmepsin I. The truncated proenzymes were expressed from pET3a vectors with the 13 amino acid T7. Tag region as an N-terminal fusion partner. The recombinant proteins were expressed in inclusion bodies and were subjected to urea solubilisation followed by a rapid dilution refolding technique. This technique worked well for proplasmepsin II, producing zymogen capable of autoactivation and good yields of mature plasmepsin II (Hill *et al.*, 1994). However, the same procedure did not yield autoactivatable proplasmepsin I.

2.3. Functional Expression of a Truncated Form of Proplasmepsin I

Several attempts were made to functionally express proplasmepsin I in a variety of eukaryotic expression systems with no improvement. We therefore further analysed the autocatalytic processing of recombinant proplasmepsin II to see if this could provide some clues on how to proceed. The final strategy selected is outlined in Figure 2. It was known that the truncated form of recombinant proplasmepsin II did not autoactivate at the expected site on the protein, namely at the N-terminus of the mature enzyme found in the parasite. Instead, it autocatalytically activated 12 amino acids upstream from this position at the sequence –Asn111P–Phe112P*Leu113P–Asn114P–, where * designates the scissile bond (Hill *et al.*, 1994). In proplasmepsin I the analogous region is Lys110P–Phe111P*Phe112P–Lys113P.

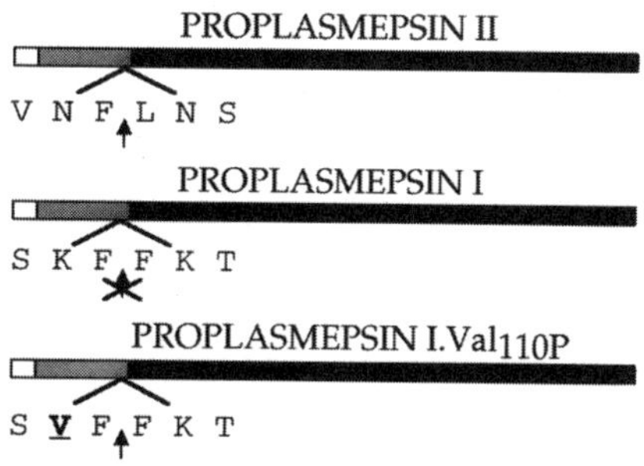

Figure 2. Strategy for generation of self-activating recombinant proplasmepsin I.

This would place Lys110P in the S2 pocket of the active site (nomenclature of Schechter and Berger (Schechter and Berger, 1967)) which is known to be unfavourable for aspartic proteinases (Scarborough and Dunn, 1994).

Modelling of this region confirmed the unfavourability of the lysine residue in the S2 pocket and suggested that mutation of Lys110P to Val110P would produce a cleavable bond (Figure 3). Note that the valine residue gives a much less constrained and better fit than the lysine residue in terms of both size and charge.

This approach yielded autoactivatable proplasmepsin I (Moon *et al.*, 1997). However, the yield was very low due to poor refolding of the truncated precursor and high levels of protein aggregation. Removal of the protein aggregates formed during the refolding step was crucial to isolating soluble processed plasmepsin I upon acidification and activation of the zymogen. Failure to do this resulted in coprecipitation of the active processed enzyme and misfolded protein aggregates at acid pH. Aggregate removal was achieved by ultrafiltration of refolded zymogen through a 100 kDa molecular mass cut-off membrane (Figure 4a). This yielded a mixture of zymogen molecules enriched for correctly folded zymogen but in which a significant proportion of the protein was not processed at the acid pH. The remaining misfolded protein was removed by anion exchange chromatography to yield preparations suitable for kinetic and inhibitor studies (Figure 4b). Typical yields of fully processable the proplasmepsin I. Val110P mutant protein were 10–15 μg/g of *E. coli* cell paste under optimal expression conditions. Approximately 50–60% was functional as measured by pepstatin titration in chromogenic peptide assays (Moon *et al.*, 1997).

Once we had established how to produce pure functional plasmepsin I we modified the procedure to obtain higher yields. This was achieved by changing the refolding conditions and replacing the two ultracentrifugation steps with hydrophobic interaction chromatography (HIC). Inclusion bodies were solubilised in Buffer A (20 mM sodium phosphate, pH 7.5; 5 mM EDTA; 0.5M NaCl) + 8M urea and then diluted 50-fold into Buffer A; 1.2 mM glutathione (oxidised:reduced = 1:5) and the mixture allowed to stand overnight. Precipitates were removed by centrifugation and the supernatant loaded on to a hydrophobic interaction column. The column was washed with a gradient of 100% Buffer A to 100% Buffer B (20 mM sodium phosphate, pH 7.5; 5 mM EDTA) in two column volumes, followed by two column volumes of Buffer B. Elution of proplasmepsin I.Val110P by Buffer C (50% Buffer B; 50% ethylene glycol (v/v), pH 8) was followed by measuring enzyme activity at pH 4.7 using the fluorogenic substrate peptide EDANS–DRLERTFLSFTTD–Dabsyl. Following further purification by anion exchange chromatography, overall yields of 90–100 ∝g of fully processable proplasmepsin I.Val110P per g of *E. coli* cell paste was

Table 2. Activity of two inhibitors of *P. falciparum* cell growth against plasmepsins I and II (adapted from Moon *et al.* 1997)

	Enzyme K_i (nM)		Parasite IC_{50} (nM)	
Inhibitor	Plasmepsin 1	Plasmepsin 2	K1	NF54
Ro 40-4388	9	700	250	300
Ro 40-5576	8	250	900	1700

achieved, i.e. 6 to 10 times the yields obtained with the ultrafiltration method. Pepstatin titration indicated that 60% of the zymogen was functional. The purity of the preparation was similar to the 100 kDa filtrate shown in Figure 4b.

2.4. Inhibitor Studies with Purified Plasmepsins I and II

The availability of functional recombinant enzyme allowed a rigorous determination of the K_i values for the two *P. falciparum* growth inhibitors Ro 40-4388 and Ro 40-5576 against the two enzymes, to assess the extent to which they were active against the enzymes. The results are shown in Table 2. They demonstrate that these compounds show a specificity of action against plasmepsin I over plasmepsin II suggesting their likely mode of action is through plasmepsin I. Significantly, these inhibitors also show a specificity against the human aspartic proteinases cathepsin D and cathepsin E (see Tyas *et al.* this volume).

These data strongly support the hypothesis that plasmepsin I is a valid antimalarial drug target on two counts. First, inhibition of the enzyme correlates with inhibition of parasite growth. Second, specificity of inhibition over human aspartic proteinases is possible.

2.5. Structural Studies with Plasmepsins I and II

The availability of large amounts of recombinant enzyme has allowed us to undertake crystallisation studies. We were able to further purify plasmepsin II to an isoelectrically pure species using a preparative isoelectric focusing apparatus (Octopus C.H.I.E.F., Dr. Weber GmbH, Munich, Germany) (Loetscher, unpublished). Crystals of plasmepsin II complexed with pepstatin A were obtained in 1.6 M ammonium sulphate at pH 7 (Figure 5). The X-ray diffraction structure of plasmepsin II obtained from these crystals was identical to that already reported by Silva *et al.* (Silva *et al.*, 1996). We have developed an improved model of plasmepsin I based on this plasmepsin II structure and compared the architecture of the active sites of both enzymes. Figure 6 illustrates that significant differences exist in the vicinity of the S1′ and S3 pockets and these might account for the differential activities of the inhibitors described. We now have structural studies in progress with plasmepsin I and it is hoped that the availability of structural information for both plasmepsins will greatly assist in the discovery of potent specific inhibitors of these enzymes.

3. BIOLOGY OF PLASMEPSINS

3.1. The Two Plasmepsins Are Not Redundant and Perform Specific Functions

If our interpretation of the inhibitor data presented in Table 2 is correct, namely that inhibition of plasmepsin I inhibits parasite growth, this clearly illustrates that the two

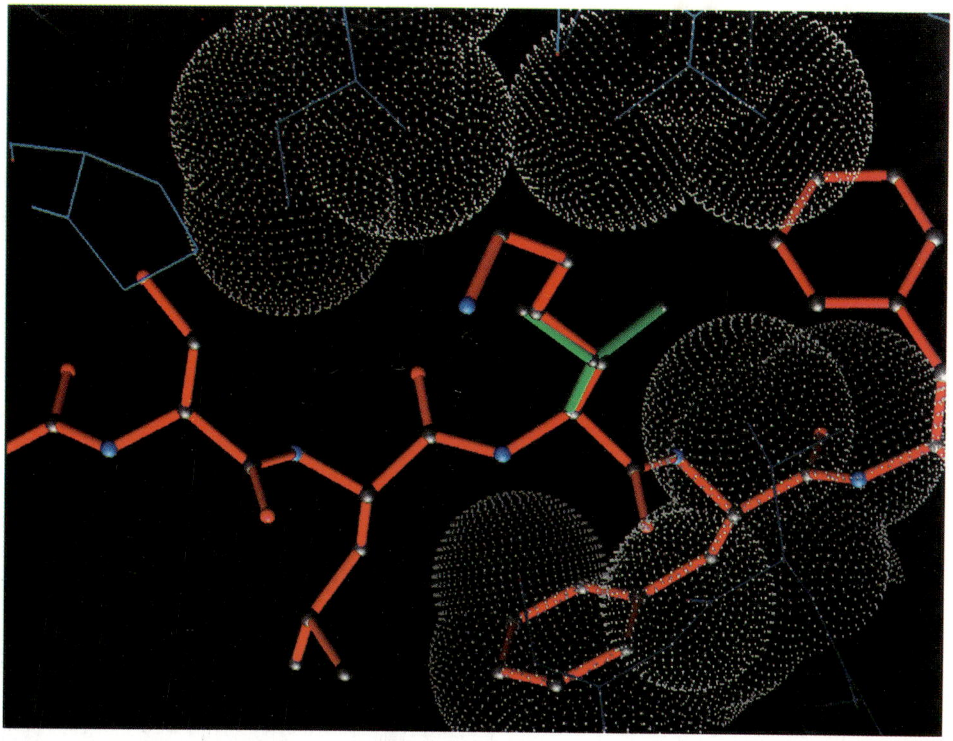

Figure 3. Computer model of mature plasmepsin I, with proregion residues Ser108P to Phe112P modelled into the active site, showing that a change of residue from Lys110P to Val110P results in an improved fit of the sidechain in the S2 pocket.

Figure 5. Crystals of activated recombinant proplasmepsin II complexed with isovaleryl pepstatin A.

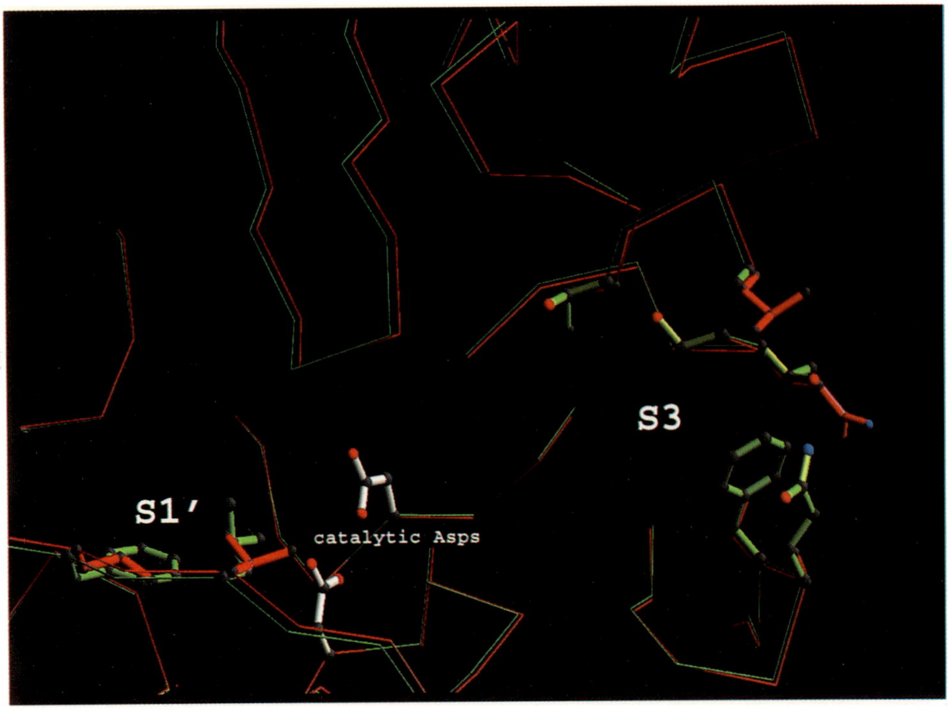

Figure 6. Superposition of plasmepsin II X-ray structure (green) and a computer model of plasmepsin I (red). Amino acid side chains differing between the two enzymes in the S1′ and S3 pockets are indicated.

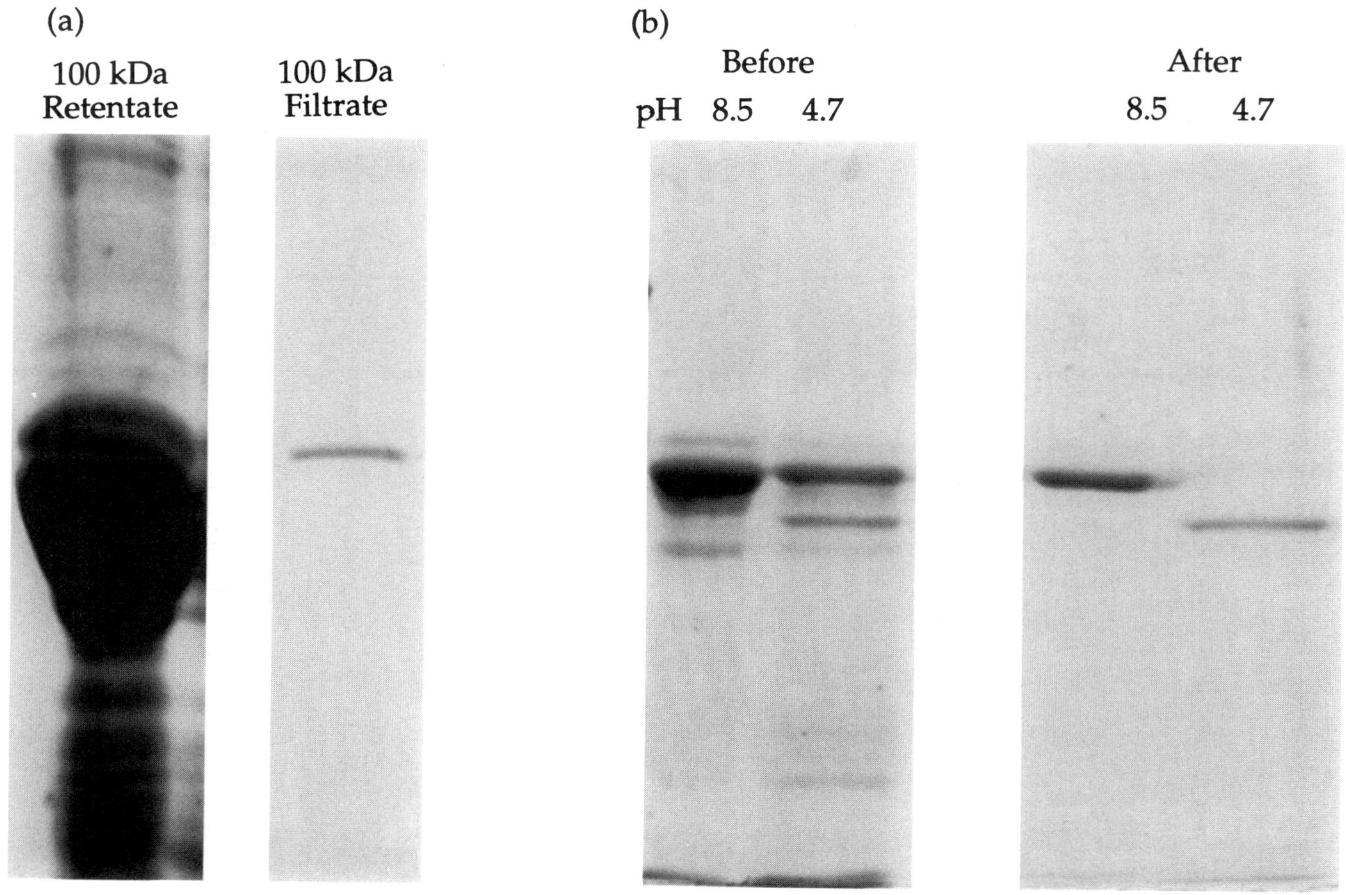

Figure 4. Purification of functional recombinant mutant proplasmepsin I.Val110P. (a) Removal of high molecular weight protein aggregates by ultrafiltration through a 100 kDa molecular weight cut-off membrane. (b) Purification of fully autoactivatable proplasmepsin I.Val110P from the filtrate by anion exchange chromatography. Adapted from Moon *et al*. 1997.

enzymes, despite their similar sequence, do not perform the same function. Certainly, plasmepsin II cannot compensate for the inhibition of plasmepsin I. We have also generated other data supporting the hypothesis that these two similar enzymes play unique roles. For instance, in kinetic studies utilising chromogenic peptide substrates based on the proposed major target α–globin cleavage site of the plasmepsins, namely Phe33*Leu34, plasmepsin I and plasmepsin II displayed different k_{cat} values suggesting a different specificity (Moon *et al.* 1997; Tyas *et al.* this volume). However, the most compelling data supporting unique biological roles for the two enzymes comes from Northern Blot data which suggest that proplasmepsin I is expressed at a much earlier stage in the erythrocytic life cycle than proplasmepsin II (Moon *et al.*, 1997).

3.2. Processing of Plasmepsins *in Vitro* and in the Parasite

The fact that a cleavage site had to be engineered into proplasmepsin I to obtain pH-dependent autocatalytic processing raises the question of how maturation of the zymogen occurs in the parasite. The possibility cannot be excluded that a third proteinase acting *in trans* might be necessary to process proplasmepsin I to mature plasmepsin I (see also Goldberg, this volume). However, no such enzyme has so far been identified. Addition of mature plasmepsin II to the authentic, non-mutant, recombinant plasmepsin I did not bring about activation of this precursor in our hands. Others (Luker *et al.*, 1996) have reported that the food vacuole malarial cysteine proteinase, falcipain, and trophozoite extract also failed to activate a similar version of recombinant plasmepsin I. A complex situation also exists for proplasmepsin II. Authentic, recombinant proplasmepsin II does autoactivate at acidic pH, but the location of the cleavage site varies depending on the conditions used. In some cases, the mature enzyme generated is extended by 12 amino acids at the N-terminus as reported previously (Hill *et al.*, 1994). Under other conditions, the recombinant enzyme obtained corresponds directly to the naturally occurring plasmepsin II as purified from the parasite (Ben Dunn, personal communication). Although it is believed that most aspartic proteinase zymogens undergo autocatalytic processing *in vivo* some, such as renin, do require another enzyme acting in *trans* (Fusek and Vetvicka, 1995).

4. CONCLUSIONS AND OUTLOOK

The data generated by ourselves (Moon *et al.*, 1997) and others (Francis *et al.*, 1994; Silva *et al.*, 1996) strongly suggest that the plasmepsins are essential for parasite growth and as such represent potential antimalarial drug targets. Other data generated by our labs confirm that inhibitors of plasmepsin I can show specificity over the human aspartic proteinases cathepsin D and cathepsin E reinforcing this conclusion (see Tyas *et al.* this volume). The availability of recombinant protein to screen for new inhibitors together with X-ray structural information should greatly assist in inhibitor design. However, it should be stressed that the inhibitors presented here are still a long way from being viable drugs. Although there are now a number reports of plasmepsin inhibitors showing activity against malarial growth in culture (Francis *et al.*, 1994; Silva *et al.*, 1996; Moon *et al.*, 1997) nobody has yet demonstrated an antimalarial activity in an animal model.

It should also be stressed that a great deal remains to be understood about the biology of the plasmepsins, in particular their biosynthesis and targeting to the food vacuole organelle, and their processing to produce mature enzyme. It may be that further investigation of these phenomena yield other potential drug targets.

ACKNOWLEDGMENTS

This investigation was supported in part by grants from the UK Biotechnology and Biological Sciences Research Council (to L. T.), from the Royal Society (to C. B.) and from the United Nations Development Program/World Bank/WHO Special Programme for Research and Training in Tropical Diseases (to J. K. and C. B.).

REFERENCES

Ball, E. G., McKee, R. W., Anfinsen, C. B., Cruz, W. O. and Geumen, Q. M. (1948) Studies on malarial parasites IX. Chemical and metabolic changes during growth and multiplication *in vivo* and *in vitro*. *J. Biol. Chem. 175*, 547–571.

Dame, J. B., Reddy, G. R., Yowell, C. A., Dunn, B. M., Kay, J. and Berry, C. (1994) Sequence, expression and modeled structure of an aspartic proteinase from the human malaria parasite *Plasmodium falciparum*. *Mol. Biochem. Parasitol. 64*, 177–90.

Divo, A., Geary, T., Davis, N. and Jensen, J. (1985) Nutritional requirements of *Plasmodium falciparum* in culture. I. Exogenously supplied dialyzable components necessary for continuous growth. *J. Protozool. 32*, 59–64.

Francis, S. E., Gluzman, I. Y., Oksman, A., Knickerbocker, A., Mueller, R., Bryant, M. L., Sherman, D. R., Russell, D. G. and Goldberg, D. E. (1994) Molecular characterization and inhibition of a *Plasmodium falciparum* aspartic hemoglobinase. *EMBO J. 13*, 306–17.

Fusek, M. and Vetvicka, V. (1995) *Aspartic Proteinases: Physiology and Pathology.* CRC Press Inc., Boca Raton, Florida.

Gluzman, I. Y., Francis, S. E., Oksman, A., Smith, C. E., Duffin, K. L. and Goldberg, D. E. (1994) Order and specificity of the *Plasmodium falciparum* hemoglobin degradation pathway. *J. Clin. Invest. 93*, 1602–8.

Goldberg, D. E., Slater, A. F., Beavis, R., Chait, B., Cerami, A. and Henderson, G. B. (1991) Hemoglobin degradation in the human malaria pathogen *Plasmodium falciparum*: a catabolic pathway initiated by a specific aspartic protease. *J. Exp. Med. 173*, 961–9.

Goldberg, D. E., Slater, A. F., Cerami, A. and Henderson, G. B. (1990) Hemoglobin degradation in the malaria parasite *Plasmodium falciparum*: an ordered process in a unique organelle. *Proc. Natl. Acad. Sci. USA 87*, 2931–5.

Groman, N. B. (1951) Dynamic aspects of the nitrogen metabolism of *Plasmodium gallinaceum in vivo* and *in vitro*. *J. Infect. Dis. 88*, 126–150.

Hill, J., Tyas, L., Phylip, L. H., Kay, J., Dunn, B. M. and Berry, C. (1994) High level expression and characterisation of Plasmepsin II, an aspartic proteinase from *Plasmodium falciparum*. *FEBS Lett. 352*, 155–8.

Luker, K., Francis, S., Gluzman, I. and Goldberg, D. (1996) Kinetic analysis of plasmepsins I and II aspartic proteases of the *Plasmodium falciparum* digestive vacuole. *Mol. Biochem. Parasitol. 79*, 71–8.

Moon, R. P., Tyas, L., Certa, U., Rupp, K., Bur, D., Jaquet, C., Matile, H., Loetscher, H., Grueninger-Leitch, F., Kay, J., Dunn, B. M., Berry, C. and Ridley, R. G. (1997) Recombinant expression and kinetic characterisation of plasmepsin I from *Plasmodium falciparum*. *Eur. J. Biochem. 244*, 552–560.

Olliaro, P. L. and Goldberg, D. E. (1995) The plasmodium digestive vacuole: Metabolic headquarters and choice drug target. *Parasitology Today 11*, 294–297.

Rangachari, K., Dluzewski, A., Wilson, R. and Gratzer, W. (1987) Cytoplasmic factor required for entry of malaria parasites into RBCs. *Blood 70*, 77–82.

Roth Jr., E., Brotman, D., Vanderberg, J. and Schulman, S. (1986) Malarial pigment-dependent error in the estimation of hemoglobin content in *Plasmodium falciparum*-infected red cells: implications for metabolic and biochemical studies of the erythrocytic phases of malaria. *Am. J. Trop. Med. Hyg. 35*, 906–11.

Scarborough, P. and Dunn, B. (1994) Redesign of the substrate specificity of human cathepsin D: the dominant role of position 287 in the S2 subsite. *Protein Eng. 7*, 495–502.

Schechter, I. and Berger, A. (1967) On the size of the active site in proteases. I. Papain. *Biochem. Biophys. Res. Commun. 27*, 157–62.

Silva, A. M., Lee, A. Y., Gulnik, S. V., Maier, P., Collins, J., Bhat, T. N., Collins, P. J., Cacheu, R. E., Kuker, K., Y., G. I., Francis, S. E., Oksman, A., Goldberg, D. E. and Erikson, J. W. (1996) Structure and inhibition of plasmepsin II, a hemoglobin-degrading enzyme from *Plasmodium falciparum*. *Proc. Natl. Acad. Sci. USA 93*, 10034–10039.

Vander Jagt, D. L., Hunsaker, L. A., Campos, N. M. and Scaletti, J. V. (1992) Localization and characterization of hemoglobin-degrading aspartic proteinases from the malarial parasite *Plasmodium falciparum*. *Biochim. Biophys. Acta 1122*, 256–264.

Zarchin, S., Krugliak, M. and Ginsburg, H. (1986) Digestion of the host erythrocyte by malaria parasites is the primary target for quinoline-containing antimalarials. *Biochem. Pharmacol. 35*, 2435–42.

PLASMEPSINS I AND II FROM THE MALARIAL PARASITE *Plasmodium falciparum*

L. Tyas,[1] R. P. Moon,[2] H. Loetscher,[2] B. M. Dunn,[3] J. Kay,[1] R. G. Ridley,[2] and C. Berry[1]

[1]School of Molecular and Medical Biosciences
University of Wales
Cardiff, CF1 3US, Wales, United Kingdom
[2]F. Hoffmann-La Roche, Pharmaceuticals Division
Pharma Research Pre-clinical
CH-4070 Basel, Switzerland
[3]Department of Biochemistry and Molecular Biology
University of Florida
Gainesville, Florida 32610-0245

During an infection of erythrocytes by the human malarial parasite *Plasmodium falciparum*, up to 80% of host cell haemoglobin is degraded to provide amino acids for parasite nutrition.[1,2] Two aspartic proteinases, plasmepsin I and plasmepsin II are believed to be responsible for initiating this catabolic process[3] and thus are suitable targets for antimalarial chemotherapy. These enzymes have 73% identity to each other and approximately 30% identity to the five known human aspartic proteinases.

Full length gene sequences encoding the precursor forms of plasmepsins I and II have been derived[3,4] and the two recombinant plasmepsins have been expressed in *E. coli* as truncated zymogens.[5,6] The recombinant precursor of plasmepsin II, termed semi-proplasmepsin II, readily converted itself into the mature plasmepsin II upon exposure to the appropriate acid pH. In contrast, mature plasmepsin I was not produced upon acidification of its recombinant precursor or when the precursor was incubated with extrinsically added enzymes including mature active plasmepsin II. Consequently, it was necessary to engineer an autoactivation sequence into the truncated propart of semi-proplasmepsin I, the details of which are described elsewhere[6] (Moon *et al.*, this volume).

A detailed assessment of the pH dependence of the two mature recombinant plasmepsins was carried out at a constant ionic strength of 100mM using the chromogenic substrate Leu–Glu–Arg–Ile–Phe*Nph–Ser–Phe, enabling hydrolysis to be followed spectrophotometrically. This peptide is related to the sequence in haemoglobin found around the Phe_{33}*Leu_{34} bond in the α-globin chain which is considered to be the position at which

Aspartic Proteinases, edited by James.
Plenum Press, New York, 1998.

haemoglobin degradation initiates.[7] Plasmepsin II is active over the pH range 3.5–6.5 whilst plasmepsin I is functional over the slightly broader range of 3.0–7.0. Therefore both enzymes show a similar pH dependence and exhibit close to optimum activity at pH 5.0, the estimated pH of the parasite food vacuole where haemoglobin digestion takes place.[8,9]

The activity of recombinant plasmepsin II was examined using synthetic peptide substrates based on the sequence that occurs at the primary Phe_{33}*Leu_{34} cleavage site in the α-globin chain of haemoglobin. The substrate Ala–Leu–Glu–Arg–Thr–Phe*Nph–Ser–Phe–Pro–Thr was synthesised to contain a threonine residue in the P_2 position in place of the Met_{32} residue present at this location in the sequence of haemoglobin from humans and several other mammals.[10,11] The kinetic parameters derived at pH 4.7 indicated that this substrate (peptide 1, Table 1) was cleaved effectively by plasmepsin II.

The Michaelis constant, K_m, was of the same order as those previously reported for hydrolysis of similar peptides by other aspartic proteinases (recombinant and naturally-occurring) from human, animal and fungal sources.[12] However, the value obtained for the catalytic constant (k_{cat}) was lower in comparison.

The ability of plasmepsin II to cleave a synthetic substrate that is not related to the sequence surrounding the Phe_{33}*Leu_{34} bond from the α-chain of haemoglobin was also investigated. This substrate Lys–Pro–Ile–Val–Phe*Nph–Arg–Leu has been shown previously to be hydrolysed readily by the human enzymes, cathepsin D[13] and cathepsin E.[14] The K_m value measured for this substrate (peptide 2, Table 1) was 10-fold lower than that determined for the haemoglobin-derived substrate (peptide 1, Table 1) whilst the k_{cat} value was comparable. By changing the Val residue in the P_2 position to a Lys (peptide 3, Table 1), hydrolysis of the resultant peptide by plasmepsin II was virtually abolished. It seems that plasmepsin II is not highly specific and is able to hydrolyse a substrate that bears no relation to the sequence of haemoglobin. Plasmepsin II has a specificity constant (k_{cat}/K_m) that is 6-fold higher for cleavage of peptide 2 compared to the value determined for the haemoglobin-based substrate (peptide 1). The major difference in the kinetic parameters derived for hydrolysis of these two peptides arises in the binding constant (K_m). Examination of the structure determined for plasmepsin II by X-ray crystallography[15] makes it possible to predict which residues contribute to the nature of individual subsites and as a result, govern interactions with each residue of a particular substrate. For instance, it has already been shown[16] that the residue at position 13 of an aspartic proteinase (numbering of porcine pepsin) has an important influence on S_3 subsite interactions. Plasmepsin II has a hydrophobic Met at this location and this would confer a preference for a small hydrophobic residue in the P_3 position of a substrate. The specificity of the S_2 subsite is affected by the sequence of residues at 287–289[13] and in plasmepsin II, the residues involved are –Ile–Gly–Leu– which are hydrophobic and similar to the corresponding –Met–Gly–Met– sequence from cathepsin D. Thus it would appear that like cathepsin D,

Table 1. Kinetic parameters for the hydrolysis of chromogenic peptide substrates by recombinant plasmepsin II

	P_6	P_5	P_4	P_3	P_2	P_1	*	P_1'	P_2'	P_3'	P_4'	P_5'	K_m (μM)	k_{cat} (s^{-1})	k_{cat}/K_m ($mM^{-1}s^{-1}$)
1	Ala	Leu	Glu	Arg	Thr	Phe	*	Nph	Ser	Phe	Pro	Thr	60	6	100
2		Lys	Pro	Ile	Val	Phe	*	Nph	Arg	Leu			6	3.6	600
3		Lys	Pro	Ile	Lys	Phe	*	Nph	Arg	Leu			n.d.	0.04	—

Assays were performed at 37°C in 0.1M sodium acetate buffer pH 4.7.
Nph = p-nitrophenylalanine. n.d. = not determined.

the S_2 and S_3 subsites of plasmepsin II are hydrophobic in nature and on this basis the haemoglobin-related substrate (peptide 1, Table 1), with Arg and Thr in the P_3 and P_2 positions respectively may not represent the optimal substrate.

Kinetic parameters were also determined for the hydrolysis by recombinant plasmepsin I and plasmepsin II of two different chromogenic peptides, in which the Met_{32} residue in the cleavage sequence around the Phe_{33}*Leu_{34} bond in haemoglobin was replaced by Val or Ile.

It can be seen that both substrates are readily cleaved (Table 2) and the specificity constants measured for plasmepsin II are directly comparable in magnitude to those obtained with peptides 1 and 2 in Table 1. For the two substrates in Table 2, similar K_m values were derived for plasmepsin I and plasmepsin II but the k_{cat} values measured for plasmepsin II were 6- to 12-fold higher than those for plasmepsin I. Consequently, the specificity constants (k_{cat}/K_m) calculated for both enzymes indicated that plasmepsin I was 5 to 9 times less effective than plasmepsin II at cleaving substrates which mimic the proposed initial cleavage site in the α-chain of haemoglobin. However, before significant emphasis is placed on the implications of these results it is vital to establish that the recombinant enzymes reflect the properties of their respective naturally-occurring plasmepsins, both in terms of activity and specificity.

Inhibition studies were also carried out with the recombinant plasmepsins using three synthetic compounds that have been shown to prevent growth of *P. falciparum* in culture.[6] The structure of these three compounds is depicted in Figure 1. All three inhibitors were found to be more potent against recombinant plasmepsin I than plasmepsin II (Table 3). Compound Ro40-4388 (Figure 1), which is the most effective in preventing parasite growth,[6] is an effective inhibitor of recombinant plasmepsin I (Table 3).

Furthermore, a separate inhibitor which has been reported by Francis *et al.*,[3] to be specific for naturally occurring plasmepsin I, is able to impair the process of haemoglobin degradation in culture. Plasmepsin II was apparently not able to compensate for the

Table 2. Hydrolysis of synthetic peptides related to the haemoglobin cleavage site: ~Leu–Glu–Arg–Met–Phe*Leu–Ser–Phe~

										Plasmepsin I			Plasmepsin II		
	P_5	P4	P3	P2	P1	*	P_1'	P_2'	P_3'	K_m (μM)	k_{cat} (s^{-1})	k_{cat}/K_m ($mM^{-1}s^{-1}$)	K_m (μM)	k_{cat} (s^{-1})	k_{cat}/K_m ($mM^{-1}s^{-1}$)
1	Leu	Glu	Arg	Ile	Phe	*	Nph	Ser	Phe	10	1.3	130	25	15	600
2	Leu	Glu	Arg	Val	Phe	*	Nph	Ser	Phe	30	1.6	50	20	9	450

Assays were performed at 37°C in 0.1M sodium formate buffer pH 4.4.

Table 3. Interaction of several compounds with plasmepsins I and II and the human aspartic proteinase, cathepsin D

Inhibitor	Plasmepsin I K_i (nM)	Plasmepsin II K_i (nM)	Cathepsin D K_i (nM)
Ro40-4388	4	500	70
Ro40-5576	6	250	2,000
Ro42-1118	300	4000	1,000

Individual assay conditions were as follows: i) recombinant plasmepsin I and plasmepsin II; 0.1M sodium formate buffer pH 4.4. ii) human cathepsin D; 0.1M sodium formate buffer pH 3.1.

loss of plasmepsin I function. Both of these studies suggest that plasmepsin I may be the key drug target.

An important consideration in the design of any inhibitor is its interaction with human enzymes. In this respect, the effect of Ro40-4388 and the two other peptidic inhibitors of *P. falciparum* growth, was investigated on cathepsin D, the human aspartic proteinase most similar to the plasmepsins (Table 3). It can be seen that as well as being a potent inhibitor of plasmepsin I, Ro40-4388 inhibits cathepsin D more effectively than plasmepsin II. From this lead compound, it may be possible to exploit structural differences between the active sites of plasmepsin I, plasmepsin II and the human aspartic proteinases to develop a more selective, potent inhibitor. The crystal structure of plasmepsin II has already been determined[15] and work is currently being undertaken to elucidate the plasmepsin I crystal structure. Together these structures will be invaluable in the design of

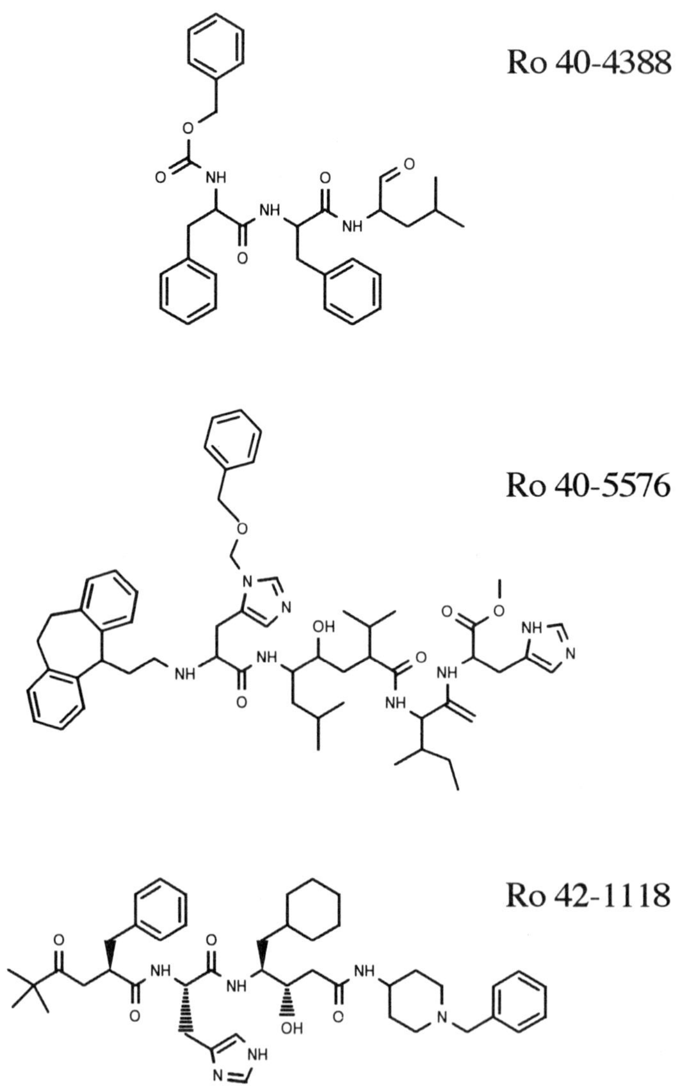

Figure 1. Inhibitors of *P. falciparum* growth in human red blood cells in culture.

improved, more specific inhibitors of the plasmepsins which have been recognised by the World Health Organisation as important targets for antimalarial therapy.

ACKNOWLEDGMENTS

This investigation received financial support from the UK Biotechnology and Biological Sciences Research Council (for L. T.), from the Royal Society (for C. B) and from the United Nations Development Programme/World Bank/WHO Special Programme for Research and Training in Tropical Diseases (TDR) (for J. K. and C. B.).

REFERENCES

1. Roth, E.F., Brotman, D.S., Vanderberg, J.P. and Shulman, S. (1986) Am. J. Trop. Med. Hyg. 35, 906–911.
2. Scheibel, L.W. and Sherman, I.W. (1988) in: Malaria: Principles and Practice of Malariology. (Wernsdorfer, W.H. and McGregor, I. Eds.) pp 219–252, Churchill Livingstone, Edinburgh.
3. Francis, S.E., Gluzman, I.Y., Oksman, A., Knickerbocker, A., Mueller, R., Bryant, M.L., Sherman, D.R., Russell, D.G. and Goldberg, D.E. (1994) EMBO J. 13, 306–317.
4. Dame, J.B., Reddy, G.R., Yowell, C.A., Dunn, B.M., Kay, J. and Berry, C. (1994) Mol. Biochem. Parasitol. 64, 177–190.
5. Hill, J., Tyas, L., Phylip, L.H., Kay, J., Dunn, B.M. and Berry, C. (1994) FEBS Lett. 352, 155–158.
6. Moon, R.P., Tyas, L., Certa, U., Rupp, K., Bur, D., Jaquet, C., Matile, H., Loetscher, H., Grueninger, F., Kay, J., Dunn, B.M., Berry, C. and Ridley, R.G. (1997) Eur. J. Biochem. 244, 552–560.
7. Goldberg, D.E., Slater, A.F.G., Beavis, R., Chait, B. and Henderson,G.B. (1991) J. Exp. Med. 173, 961–969.
8. Krogstadt, D.J., Schlesinger, P.H. and Gluzman, I.Y. (1985) J. Cell Biol. 101, 2302–2309.
9. Yayon. A., Cabantchik, Z.I. and Ginsburg, H. (1984) EMBO J. 3, 2695–2700.
10. Abbasi, A. and Braunitzer, G. (1985) J. Protein Chem. 4, 57–67.
11. Beard, J.M. and Thompson, E.O.P. (1971) Aust. J. Biol. Sci. 24, 765–786.
12. Dunn, B.M., Jimenez, M., Parten, B.F., Valler, M.J., Rolph, C.E. and Kay, J. (1986) Biochem. J. 237, 899–906.
13. Scarborough, P.E., Guruprasad, K., Topham, C., Richo, G.R., Conner, G.E., Blundell, T.L. and Dunn, B.M. (1993) Prot. Sci. 2, 264–276.
14. Rao-Naik, C., Guruprasad, K., Batley, B., Rapundalo, S., Hill, J., Blundell, T., Kay, J. and Dunn, B.M. (1995) Proteins: Structure, Function and Genetics 22, 168–181.
15. Silva, A.M., Lee, A.Y., Gulnik, S.V., Majer, P., Collins, J., Bhat, T.N., Cachau, R.E., Luker, K.E., Gluzman, I.Y., Francis, S.E., Oksman, A., Goldberg, D.E. and Erickson, J.W. (1996) Proc. Natl. Acad. Sci. USA. 93, 10034–10039.
16. Dunn, B.M., Valler, M.J., Rolph, C.E., Foundling, S.I., Jimenez, M. and Kay, J. (1987) Biochim. Biophys. Acta 913, 122–130.

ABSTRACTS

PLASMEPSIN I AND PLASMEPSIN II AS ANTIMALARIAL DRUG TARGETS

R. G. Ridley,[1] R. Moon,[1] D. Bur,[1] U. Certa,[1] H. Loetscher,[1] A. D'Arcy,[1] C. Oefner,[1] H. Matile,[1] L. Tyas,[2] J. Kay,[2] and C. Berry[2]

[1]Pharma Division, Pre-Clinical Research, F. Hoffmann-la Roche, CH-4070 Basel, Switzerland
[2]School of Molecular and Biomedical Sciences, University of Wales, Museum Avenue, Cardiff CF1 3US, Wales, United Kingdom

During its growth and development inside the erythrocyte the malarial parasite consumes up to 80% of the host cell's haemoglobin as a source of nutrients and energy. The molecular machinery responsible for this is housed in a lysosomal organelle of the protozoan parasite, commonly referred to as the digestive vacuole. Two aspartic proteases, plasmepsin I and plasmepsin II, are located in the organelle of the most virulent human malaria parasite *Plasmodium falciparum* and they play a critical role in the hemoglobin degradation process. We have monitored their gene expression in the parasite and conclude that they have complementary, non-redundant functions. The importance of hemoglobin degradation to the parasite, the significance of the plasmepsins in this process and the proven ability to obtain specific inhibitors of aspartic proteases makes these enzymes attractive drug targets. We have obtained functional recombinant gene expression for both plasmepsins and will present a preliminary comparative assessment of their kinetic properties and their susceptibility to inhibitors.

COMPARATIVE STUDIES OF THE PLASMEPSINS OF THE HUMAN MALARIA PARASITES *Plasmodium falciparum*, *Plasmodium vivax*, AND *Plasmodium ovale*

John B. Dame,[1] Charles A. Yowell, Jr.,[1] Jennifer Westling,[2] and Ben M. Dunn[2]

[1]Departments of Pathobiology and
[2]Biochemistry and Molecular Biology, University of Florida, Gainesville, Florida 32610

Plasmepsins are found in the food vacuole of malaria parasites and are recognized as novel targets for antimalarial drug development. Primary gene sequence data available from two different plasmepsins from *Plasmodium falciparum* (PfPMI and PfPMII) were used to design PCR primers to clone genes encoding plasmepsins from *Plasmodium vivax* (PvPM) and *Plasmodium ovale* (PoPM). The primary amino acid sequence of each plasmepsin was deduced from the DNA sequence, and comparative studies of the structure of these four plasmepsins were initiated. Using computer-aided modelling, guided by the recently resolved crystal structure of PfPMII (Silva *et al.*, Proc. Natl. Acad. Sci. USA, In Press), the amino acid side chains forming the substrate binding cleft of each of the plasmepsins under study were predicted. Multiple sequence alignments of these four plasmepsins identified only minor differences in residues lining the substrate binding cleft suggesting that inhibitors may have a similar effect on plasmepsins from multiple species. Active recombinant forms of PfPMII and PvPM have been produced in *E. coli*, and properties of the active site of these enzymes are being studied using several methods. Chief among these are kinetic studies using panels of synthetic, chromogenic peptide substrates systematically varied at different amino acid positions. Initial results have revealed a preference for hydrophobic amino acids in the P3 and P2 positions of substrates. These data were compared with the modelled structure of the active site to derive hypotheses about the interaction of the enzyme with the substrate. These hypotheses will be independently tested by altering the active site of the enzyme using site-directed mutagenesis followed by evaluating the effects of the mutation on substrate specificity. From the subsite specificity of the enzyme, the expected properties of various inhibitors may be predicted and more powerful inhibitors designed. Initial screening of inhibitors has identified a parallel between inhibitors of human cathepsin D and plasmepsin II. Studies are also under way to determine how many plasmepsins are encoded by the genome of each species and to assess functional similarities among the two plasmepsins from *P. falciparum* and those of *P. vivax* and *P. ovale*.

Supported by NIH Grant AI39211-01 to JBD and BMD.

TAILS AND THE UNEXPECTED: PARASITE ASPARTIC PROTEINASES

C. Berry

School of Molecular and Medical Biosciences, University of Wales College of Cardiff, Cardiff, CF1 3US, Wales, United Kingdom

The successes achieved in developing inhibitors of HIV proteinase have proved that aspartic proteinases are viable targets for therapeutic intervention. This has focussed our attention on members of this family of enzymes in parasites which infect humans and other vertebrate hosts. Genes have been sequenced encoding aspartic proteinases from human malaria parasites (*Plasmodium*), the chicken parasite *Eimeria tenella* and the human and dog hookworm *Ancylosotoma caninum.* All of the proteins encoded are clearly members of the aspartic proteinase super-family but exhibit features not seen previously in the vertebrate/fungal enzymes.

The enzymes from both *Plasmodium* (plasmepsins) and *Eimeria* have unusually long proparts consisting of 125 and 87 amino acids respectively, with no obvious signal peptide sequence in the malaria enzymes. The *Eimeria* and *A. caninum* proteins also have C-terminal extensions of 35 and 30 amino acids respectively. Secondary structure predictions indicate that the serine and alanine rich *Eimeria* extension is likely to form a potential membrane-spanning helix. These N- and C-terminal extensions may be important in localisation of the enzymes to specific intracellular compartments, membrane anchoring and/or folding.

Whilst the above features are unusual in these enzymes, the amino acid residues essential for catalysis are unchanged relative to typical aspartic proteinases. One putative enzyme (HAP) from *Plasmodium falciparum* identified from its gene sequence does however show major changes in the characteristic active site motifs; most strikingly, the first –Asp–Thr–Gly– motif has the sequence –His–Thr–Ala– in this protein and Tyr75 is replaced by a serine residue. A model was built for HAP using the structure of human cathepsin D as a target. From this it appears that the catalytic apparatus of HAP from *P. falciparum* has compensatory mutations implying that it may act by a novel catalytic mechanism. Since *P. falciparum* has little unused DNA and no known pseudogenes, the detection of specific mRNA for HAP indicates that the gene product is likely to be functional.

ASPARTIC PROTEINASES FROM THE ANIMAL PARASITES *Plasmodium berghei* AND *Eimeria tenella*

M. J. Humphreys and C. Berry

School of Molecular and Medical Biosciences, University of Wales College of Cardiff, Cardiff CF1 3US, Wales, United Kingdom

Considerable effort has been invested into the aspartic proteinases produced by *Plasmodium falciparum*, *P. vivax* and *P. ovale*, the parasites which cause malaria in humans. *P. berghei* infects rats and thus offers a potential animal model in which to evaluate the potency and safety of drugs under development for targeting against proteinases of the human malarial parasites. In the poultry and livestock industries, multi-billion dollar losses occur annually as a result of infection by *Eimeria* species, which produce a complex intestinal disease.

For these reasons, genes encoding the precursors of aspartic proteinases from *P. berghei* and *E. tenella* were cloned and their sequences were determined. The encoded proteins were predicted to consist of 451 and 467 amino acids respectively. The two proparts appear to consist of 125 and 87 amino acids respectively and, as such, are much longer than those of vertebrate or fungal aspartic proteinases, but are comparable in size to the proparts from the *Plasmodium* species that infect humans. The sequences of the two mature enzymes have (%) identities to other aspartic proteinases as follows:

	Cathepsin D	*E. tenella*	Plasmepsin II	Plasmepsin I
P. berghei	31	25	58	56
Plasmepsin I	33	27	73	
Plasmepsin II	34	26		
E. tenella	41			

The *E. tenella* protein is further distinguished by the presence of an additional 35 residues at its C terminus. Secondary structure prediction programmes suggest that this serine-alanine rich region forms a transmembrane helix.

Genes encoding truncated precursors of these two aspartic proteinases were subcloned into pET vectors. Upon induction of the recombinant plasmids in *E. coli*, high levels of recombinant proteins were detected in each case at molecular weights consistent with those predicted for the full length proteins. Both proteins were insoluble in the *E. coli* lysates and so a wide variety of solubilization and refolding conditions were evaluated in attempts to generate correctly-folded precursors that would be capable of autoactivation. However, no conversion of either zymogen to mature enzyme was observed under any of the conditions examined.

In an attempt to produce self-processing zymogens, constructs were engineered in which, for example, a potential autoactivation site was introduced immediately adjacent to the putative N-terminal residue of the mature proteinases; and the C terminal extension was deleted from the *E. tenella* recombinant protein. The expression of these constructs in *E. coli* together with characterisation of the resultant *P. berghei* and *E. tenella* proteins will be discussed.

ELUCIDATION OF S_2 AND S_3 SPECIFICITY DETERMINING RESIDUES WITHIN THE ACTIVE SITE OF PLASMEPSIN II

Jennifer Westling,[1] Howard Saft,[1] Tarak Shah,[1] John Dame,[2] John Kay,[3] and Ben M. Dunn[1]

[1]Department of Biochemistry and Molecular Biology and
[2]Pathobiology, University of Florida, Gainesville, Florida 32610-0245
[3]Department of Biochemistry, University of Wales College of Cardiff, Cardiff, Wales, United Kingdom

During the human erythrocytic infective stages, the malarial parasite relies on the degradation of the host hemoglobin to provide amino acids for further growth and development. Aspartic proteinases, such as plasmepsin II, have been shown to be involved in this essential catabolic pathway. Furthermore, addition of pepstatin blocks this degradation of hemoglobin.[1] Therefore, plasmepsin II (PMII) has become a possible target for the development of new antimalarial drugs.

A requirement of an antimalarial drug against PMII is that it be highly specific so that it does not affect any homologous proteins, such as other aspartic proteinases of the host. As a result, characterization of the subtle differences between the specificity of PMII and other human aspartic proteinases is necessary. Differences in specificity between the aspartic proteinases results from variations in the amino acids that lie within the active site pockets. Previous kinetic analysis using chromogenic octapeptides with substitutions in P_2 and P_3 positions have given insights on the binding preferences of PMII.[2] These studies, in conjunction with the three-dimensional X-ray crystal structure,[3] have revealed specific amino acids within the S_1 (Ser77), S_2 (Ile287) and S_3 (Met13) subsites that may contribute to the unique binding characteristics of this enzyme. Site-directed mutagenesis of these residues has been carried out to explore their specific roles in dictating PMII specificity and the resulting kinetic properties of the mutant enzymes ascertained.

Supported by NIH grant AI 39211 to JD and BMD.

REFERENCES

1. Gluzman, I.Y. *et al.* (1994) *J. Clin. Invest.* **93**:1602–1608.
2. Westling, J. *et al.* (1995) *Protein Science* **4**, Suppl. 2, 132.
3. Silva *et al.* (1996) *PNAS USA*, in press.

A PROTEINASE FROM *Plasmodium falciparum* WITH A NOVEL CATALYTIC APPARATUS

P. J. Matharu,[1] U. Certa,[2] R. Moon,[2] D. Bur,[2] M. J. Humphreys,[1] J. Kay,[1] R. G. Ridley,[2] and C. Berry[1]

[1]School of Molecular and Medical Biosciences, University of Wales College of Cardiff, Cardiff, CF1 3US, Wales, United Kingdom
[2]F. Hoffmann-La Roche Ltd., CH-4002, Basel, Switzerland

Two conventional aspartic proteinases, plasmepsins I and II, have been implicated in the digestion of haemoglobin within the food vacuole of the malarial parasite, *Plasmodium falciparum*. During our investigations into the genes encoding these proteins, a third related gene was identified. The protein encoded by this gene was predicted to consist of 461 amino acids; its propart, in common with those of proplasmepsins I and II, appeared to contain in excess of 120 residues. The protein sequence of the mature region (HAP) shared amino acid identities with plasmepsins I and II of 58% and 55%, respectively. The sequence around Asp215 (pepsin numbering) was –Ile–Leu–Asp–Ser–Ala–Thr–Ser–. However, Asp32 was replaced by a histidine residue in the sequence:

			32				
HAP	–Leu	Phe	His	Thr	Ala	Ser	Ser–
Plasmepsin I	–Ile	Phe	Asp	Thr	Gly	Ser	Ala–
Plasmepsin II	–Ile	Leu	Asp	Thr	Gly	Ser	Ala–
Human cathepsin D	–Val	Phe	Asp	Thr	Gly	Ser	Ser–

The sequence flanking Tyr75 in the flap was:

			75			
HAP	–Leu	Thr	Ser	Lys	Ala	Gly–
Plasmepsin I	–Met	Asn	Tyr	Val	Ser	Gly–
Plasmepsin II	–Met	Asn	Tyr	Val	Ser	Gly–
Human cathepsin D	–Ile	His	Tyr	Gly	Ser	Gly–

Since the sequence of HAP is 31% identical to that of human cathepsin D, a model was built using the structure of human cathepsin D determined by X-ray crystallography as a target. Whereas aspartic proteinase homologues reported previously by others to contain alterations in an –Asp–Thr–Gly– motif were believed to be catalytically defective, the catalytic apparatus of HAP from *Plasmodium falciparum* appears to have compensatory mutations. This implies a novel catalytic mechanism and led us to coin the term histo-aspartic proteinase (HAP) for this putative enzyme.

In order to gain an insight into its catalytic efficiency, the gene encoding HAP was subcloned into a number of vectors to enable production of recombinant protein. Northern blot analysis of *Plasmodium falciparum* maintained in culture in human red blood cells was carried out to establish that the HAP gene was transcribed together with those encoding pro-plasmepsin I and pro-plasmepsin II.

PUTATIVE ASPARTIC PROTEINASE INHIBITORS FROM PARASITIC NEMATODES

K. Girdwood,[1] F. B. Perler,[2] and C. Berry[1]

[1]School of Molecular and Medical Biosciences, University of Wales College of Cardiff, Cardiff CF1 3US, Wales, United Kingdom
[2]New England Biolabs Inc., Beverly, Massachusetts 01915

Very few protein inhibitors of aspartic proteinases are known to exist in nature. However, a protein (molecular weight »17,000) from *Ascaris lumbricoides* has been shown to be a potent pepsin inhibitor and to be effective (Ki 2–5 nM) also against cathepsin E. In contrast, it does not inhibit cathepsin D and has thus become a valuable tool to distinguish between the intracellular roles of cathepsin D and cathepsin E.

The protein sequence of one isoform of the *Ascaris* inhibitor has been reported and the gene encoding this isoform has been cloned, sequenced and expressed (see abstract from Zelatoris *et al.*). Genes encoding proteins that are similar in sequence to the *Ascaris* inhibitor have been identified in other parasitic nematodes including *Onchocerca volvulus*, *Brugia malayi* and Dirofilaria *immitis* (the causative agents of river blindness, elephantiasis and dog heartworm respectively). These genes appear to be translated in the form of pre-pro-proteins and the proteins are major antigens produced during the infection cycle of these parasites in the mammalian host. These nematode species are classified within the family Filaridae and are more closely related to one another than they are to *Ascaris*.

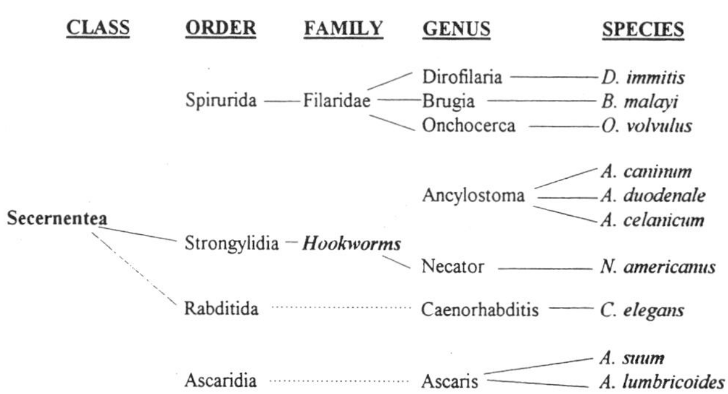

These filarial proteins show »60% similarity to one another but have considerably less similarity overall (»20%) to the known inhibitor from *Ascaris*. However, over a stretch of 15 residues within one disulphide bonded region of the *Ascaris* inhibitor, the identity with the filarial proteins rises to 73% implying that this region may form the inhibitory site that interacts with the active site of pepsin and cathepsin E. These features prompted investigations into these proteins produced by the filarial parasites in order to determine whether they function as aspartic proteinase inhibitors.

EXPRESSION, PURIFICATION AND STRUCTURAL CHARACTERIZATION OF THE RECOMBINANT PEPSIN INHIBITOR FROM *Ascaris suum*

Gregory Fecho, Chetana Rao-Naik, Jeffrey Zalatoris, and Ben M. Dunn

Department of Biochemistry and Molecular Biology, University of Florida, Gainesville, Florida 32610-0245

The *Ascaris* family of intestinal endoparasites infects both humans and pigs worldwide. These nematodes produce a variety of proteinase inhibitors which may evade the activity of host gastrointestinal enzymes, such as trypsin, chymotrypsin, elastase and pepsin, as well as the host immune system.[1,2] Each inhibitor forms a complex with its target host protein to render it inactive. In addition, presentation of the host protein in the proteinase-inhibitor complex may cause immunosuppression during parasite migration from the intestine to the liver.[2]

These inhibitors interact with pig and human pepsins in a stoichiometric molar ratio[3] and have been shown also to inhibit human cathepsin E and somewhat less potently human gastricsin.[3,4] The *Ascaris* inhibitors, in contrast, do not inhibit cathepsin D[3,4] and have thus become valuable tools that permit distinction between aspartic proteinase activities (e.g. cathepsin D and cathepsin E) in dissecting their cellular functions.[5]

The pET 3d system was utilized to produce an inactive, insoluble form of the protein as inclusion bodies in *E. coli*. The protein from inclusion bodies was solubilized, refolded, and purified to homogeneity using ammonium sulfate precipitation followed by size exclusion chromatography. The inhibitory property of the protein was determined by a functional kinetic assay. Characterization using circular dichroism, fluorescence spectroscopy and 1-D NMR studies were carried out to verify the structural authenticity of the protein. The results suggested that the recombinant protein possessed the required specificity, activity and structural characteristics of the naturally-occurring form. This work was supported by NIH grant DK 18865 to BMD.

REFERENCES

1. Martzen, M.R. *et al.* (1985) *Exp. Parasitol.* **60**, 139.
2. Peanasky, R.J. *et al.* (1987) "Molecular Paradigms for Eradicating Helminthic Parasites" (ed. Macinnis, A.J.) Alan R. Liss, New York 349.
3. Abu-Erreish, G.M. and Peanasky, R.J. (1974) *J. Biol. Chem.* **249**, 1558.
4. Valler, M.J. *et al.* (1985) *J. Enzyme Inhibition* **1**, 77.
5. Bennett, K. *et al.* (1992) *Eur. J. Immunol.* **22**, 1519.

TOWARD THE STRUCTURE OF THE PLASMEPSIN II PRECURSOR FROM *Plasmodium falciparum*

Nina Khazanovich-Bernstein,[1] Maia M. Chernaia,[1] Hansruedi Loetscher,[2] Richard Moon,[2] Robert G. Ridley,[2] and Michael N. G. James[1]

[1]Medical Research Council Group in Protein Structure and Function, Department of Biochemistry, University of Alberta, Edmonton T6G 2H7, Canada
[2]F. Hoffmann-La Roche, CH-4070, Basel, Switzerland

Malaria is a very serious infectious disease, affecting approximately 300 million people every year. Malaria is caused by the protozoan parasite *Plasmodium*, and is transmitted by the bite of the female Anopheles mosquito. Although malaria has been controlled by drugs and insecticides, development of resistance to these substances is rendering many of them useless in the fight against the disease. New strategies are needed to combat malaria, including the development of new drugs to which the parasites are not resistant.

One set of possible targets for anti-malarial drug design are the proteinases that *Plasmodium* uses to degrade haemoglobin. These enzymes include the two highly homologous aspartic proteinases, plasmepsin I and plasmepsin II. Plasmepsin I makes the initial cleavage in haemoglobin, and plasmepsin II makes subsequent cleavages of the denatured protein. Plasmepsins are produced by the parasite as inactive precursors which are activated by proteolytic cleavage at acidic pH.

We aim to elucidate the crystal structure of the plasmepsin II precursor from *P. falciparum*, the most virulent of the malaria-causing species of *Plasmodium*. The structure of the zymogen should help us understand the mechanism of activation of the plasmepsins, and may lead to the development of compounds which interfere with this process.

We have crystallized the plasmepsin II precursor in two crystal forms, in space group P1 ($a = 52.5$ Å, $b = 85.6$ Å, $c = 99.9$ Å, $\alpha = 100.6°$, $\beta = 96.8°$, $\gamma = 106.25°$), and in space group $P2_1$ ($a = 52.8$ Å, $b = 137.6$ Å, $c = 115.6$ Å, $\beta = 98.5°$). Both crystal forms contain four protein molecules in the asymmetric unit. Efforts are currently under way to solve the structure of the plasmepsin II precursor by molecular replacement.

58

STRUCTURAL AND FUNCTIONAL ASPECTS OF CARDOSINS

Carlos Faro,[1] Miguel Ramalho-Santos,[1] Paula Veríssimo[1], José Pissarra,[2] Carlos Frazão,[3] Júlia Costa,[3] Xin-Li Lin,[4] Jordan Tang,[4] and Euclides Pires[1]

[1]Departamento Bioquímica
Universidade de Coimbra
Apartado 3126, 3000 Coimbra, Portugal
[2]Centro de Citologia Experimental e Departamento Botânica
Universidade do Porto, Portugal
[3]Instituto Tecnologia Química Biológica
UNL, Oeiras, Portugal
[4]Protein Studies Program
Oklahoma Medical Research Foundation
Oklahoma City, Oklahoma

1. INTRODUCTION

Cardosins are aspartic proteinases of the flowers of *Cynara cardunculus* L. These flowers have economical relevance in Portugal since they are traditionally used in the manufacture of highly appreciated ewe cheeses such as Serra, Azeitão and Serpa. Although the milk-clotting activity of the cardoon has been exploited for centuries, the biochemistry of the process was relatively unknown until some years ago when we and other laboratories started to study both basic and applied aspects of the cardoon preparation [1–7]. In a first stage it was demonstrated that the milk-clotting activity is due to the presence of aspartic proteinases which cleave the peptide bond Phe105–Met106 of k-casein [1,3], a bond also cleaved by other milk-clotting enzymes used for cheese making [8]. Cleavage of this bond is known to induce destabilization of the casein micelle and subsequent formation of a clot, [9], and thus it is possible that the milk-clotting process induced by cardoon proteinases occurs in a similar way. However, the organoleptic properties of the products obtained with the flower of cardoon are clearly different from those of cheeses made from the same milk with chymosin or microbial rennets [10], stressing the unique characteristics of the cardoon enzymes.

Several aspartic proteinases have been isolated from species of the genus *Cynara* [1,2,6,11] Although the exact relationship between these enzymes is still unclear, we

Aspartic Proteinases, edited by James.
Plenum Press, New York, 1998.

have focused our efforts on the study of two related proteinases that have been isolated from pistils of a standard variety of the cardoon *Cynara cardunculus*L [6]. It is worthwhile to mention that in order to ensure a good correlation between molecular and biochemical data, the research work has been carried out with flowers from the same clone. The proteinases have been named cardosin A and cardosin B and although their general characteristics had already been described in previous works [6,12] a brief summary follows this section.

The aim of the present article is to present a review of our recent studies on the cardosins. The work includes cloning, determination of the 3D structure and studies on the processing and localization of the cardosins.

2. CHARACTERIZATION OF CARDOSINS

Cardosins account for the majority of the soluble protein in mature pistils and for this reason they have been isolated by a quick two step procedure that includes extraction at low pH followed by gel filtration and ion-exchange chromatography [6]. They can be distinguished by the relative mobility of their chains in SDS-polyacrylamide gels. While cardosin A migrateds as two chains with apparent molecular masses of 31 and 15 kDa cardosin B yields bands corresponding to apparent molecular masses of 34 and 14 kDa. Additional differences between the two cardosins were also observed for the amino acid sequences indicating that they are the products of different genes. In spite of such difference cardosins share a high degree of homology and for this reason it was suggested that they may have arisen during evolution by gene duplication [6,12].

Both cardosins are inhibited by pepstatin A, have proteolytic activity within the acidic range of pH and preferentially cleave peptide bonds between residues with hydrophobic side chains. [5,6,13]. However, slight differences can be found between the cardosins as concerns their specificity and kinetic properties. As judged by the hydrolytic action of each cardosin on insulin B chain [5] and on isolated bovine caseins [13,14], cardosin B has undoubtedly a broader specificity than cardosin A. Hydrolysis usually occurs in peptide bonds containing at least one amino acid with either aliphatic or aromatic side chain. In this respect cardosin A is more selective than cardosin B. However, cleavage of the peptide bonds by cardosin A occurs consistently at much slower rates. This finding is confirmed by the kinetic analysis done for the hydrolysis of two synthetic peptides, Leu–Ser–Phe(NO_2)–Nle–Ala–Leu–oMe and Lys–Pro–Ala–Glu–Phe–Phe(NO_2)–Ala–Leu. The Kcat and Kcat/Km values of cardosin B for the hydrolysis of these substrates are considerably higher than the ones of cardosin A. Both cardosins are inhibited by pepstatin A (Ki = 3 nm for cardosin A and Ki<1 nM for cardosin B), and other general aspartic proteinase inhibitors such as diazoacetyl-DL-norleucine [6] and pepstatin derivatives (unpublished results). Cardosins have also different active-site ionization constants [6]. Taken together these results suggest that although cardosins have probably similar structures, their catalytic properties are significantly different. Whether this difference is the result of a specialization in their biological functions remains to be elucidated.

3. PRIMARY STRUCTURE

About 50% of the sequences of mature cardosins have been determined at the protein level [6]. The partial amino acid sequences revealed that they have similar, but not identical sequences, and that they differ from cyprosin, another aspartic proteinase isolated from flowers of Cynara. Based on this information cloning of the two cardosins was attempted and a full-length cDNA of cardosin A is already available. Since only partial cDNAs of cardosin B have been isolated so far, data on the cloning of this enzyme are excluded from the present work.

Full-length cDNAs corresponding to cardosin A were amplified by PCR from an adaptor ligated cDNA library of immature flowers of *Cynara cardunculus* L. In a first stage an 0.8 kb internal segment of the cardosin A cDNA was amplified using degenerated primers encoding two amino acid sequences previously determined by protein sequencing. Following cloning and sequencing of the 0.8 Kb fragment, 5′ and 3′ RACE were performed with cardosin specific primers and primers for the adaptor. Most of the 5′RACE-PCR fragments contained a putative start codon whereas all 3′ RACE -PCR fragments contained three stop codons plus the poly (A) sequence. Based on these sequences, specific primers for the 5′ and 3′ ends of the open reading frame were synthesised and full lenght cDNAs of cardosin A were then PCR amplified from the cDNA library. The cDNA contains an open reading frame encoding a protein of 504 amino acid residues (Fig. 1). Sequence analysis reveals that cardosin A is synthesised as a preproenzyme apparently composed of a signal region of about 24 amino acids, a prosegment of 42 amino acids and a 438 amino acid long sequence. In common with other plant aspartic proteinases [15–17], the cardosin A cDNA contains an internal sequence of 102 amino acids with no homology to the mamalian or microbial aspartic proteinases. This sequence separates the 31 kDa chain from the 15 kDa chain and will be herein designated as the plant specific inserted sequence or PSI domain. The catalytic aspartic acid residues are both located in the 31 kDa chain at Asp–Thr–Gly and Asp–Ser–Gly. Additional residues typical of the aspartic proteinases and thought to contribute to catalysis were identified at position 80 (Tyr) and 82 (Thr). Two consensus sequences for N-glycosilation are present in the amino acid sequence at Asn–Gly–Thr and Asn–Val–Ser. According to the biochemical data these sites are indeed glycosylated and occur one in each chain [6]. Interestingly an Arg–Gly–Asp cell-attachment motif, characteristic of integrin binding proteins such as fibronectin and vitronectin [18], was identified in cardosin A sequence and is located at positions 189–191.

The alignment of the deduced amino acid sequence of cardosin A with other aspartic proteinases is shown in Fig. 1. Cardosin A has high similarity to plant aspartic proteinases (77% to cyprosin, 75% to barley proteinase and 70% to oryasin) and 40% similarity to human cathepsin D which is the aspartic proteinase of non-plant origin with higher similarity to cardosin A. According to Guruprasad et al. [19], the internal region characteristic of plant aspartic proteinases, is similar to saposins, sphingolipid-activating proteins in animals. The similarity is more evident when secondary structures are compared but still there are some common features within the amino acid sequence. They include six conserved cysteines, an invariant Tyr residue and a consensous sequence for N-glycosylation. Cardosin A shares all these features except the glycosylation site suggesting that it may not be crucial for the function of the PSI domain.

4. STRUCTURE OF THE OLIGOSACCHARIDE CHAINS

Both chains of cardosin A were found to contain covalently linked carbohydrates as they were positively stained by the Schiff reagent. In cardosin B, however, only the 34 kDa chain stained with this reagent indicating that the 14 kDa chain is non-glycosylated [12].

	1–60
Cardosin A	MGTSIKANVLALFLFYLLSPTVFSVSDDGLIRIGLKK.....RKVDRIDQLRGRRALMEG
Aspr_Horvu	MGTRGLALALLAAVLLLQTVLPAASEAEGLVRIALKK.....RPIDR.NSRVATGLSGGE
Aspr_Orysa	.MAKRHLLLVTTCLWALSCALLLHASSDGFLRVNLNK.....KRLDK.EDLTAAKLAQQG
Cyp1_Cynca	..................................LKK.....RKVNILNHPGEHAGSNDA
Catd_Human	.........MQPSSLLPLALCLLAAPASALVRIPLHKFTSIRRTMSEVGGSVEDLIAKGP

	61–120
Cardosin A	NARKDFGFRGTVRDSGSAVVALTNDRDTSYFGEIGIGTPPQKFTVIF**DTG**SSVLWVPSSM
Aspr_Horvu	EQPLLSGANPLRSEEEGDIVALKNYMNAQYFGEIGVGTPPQKFTVIF**DTG**SSNLWVPSAK
Aspr_Orysa	NRLLKTG......SSDSDPVPLVDYLNTQYYGVIGLGSPPQNFTVIF**DTG**SSNLWVPSAK
Cyp1_Cynca	NARRKYGVRGNFRDSDGELIALKNYMDAQYFGEIGIGTPPQKFTVIF**DTG**SSNLWVPSSK
Catd_Human	VSKYSQAVPAVTEGPIPEV..LKNYMDAQYYGEIGIGTPPQCFTVVF**DTG**SSNLWVPSIH

	121–180
Cardosin A	C.INSKACPAHSMYESSDSSTYKENGTSGAIIYGTGSITGFFSQDSV...........TY
Aspr_Horvu	C.YFSIACYLHSRYKAGASSTYKKNGKPAAIQYGTGSIAGYFSEDSV...........TV
Aspr_Orysa	C.YFSIACYLHSRYNSKKSSSYKADGETCKITYGSGAISGFFSKDNV...........LV
Cyp1_Cynca	C.YFSVACLFHSKYRSTDSTTYKKNGKSAAIQYGTGSISGFFSQDSV...........KL
Catd_Human	CKLLDIACWIHHKYNSDKSSTYVKNGTSFDIHYGSGSLSGYLSQDTVSVPCQSASSASAL

	181–240
Cardosin A	GDLVVKEQDFMEATDETDNVFLHRLFDGILGLSFQTIS....VPVWYNMLNQGLVKERRF
Aspr_Horvu	GDLVVKDQEFIEATKEPGITFLVAKFDGILGLGFKEISVGKAVPVWYKMIEQGLVSDPVF
Aspr_Orysa	GDQVVKNQKFIEATRETSVTFIIGKFDGILGLGYPEISVGKAPPIWQSMQEQELLADDVF
Cyp1_Cynca	GDLLVKEQDFIEATKEPGITFLAAKFDGILGLGFQEISVGDAVPVWYTMLNQGLVQEPVF
Catd_Human	GGVKVERQVFGEATKQPGITFIAAKFDGILGMAYPRISVNNVLPVFDNLMQQKLVDQNIF

	241–300
Cardosin A	SFWLNRNVDEEEGGELVFGGLDPNHFRGDHTYVPVTYQYYWQFGIGDVLIGDKSTGFCAP
Aspr_Horvu	SFWLNRHVDEGEGGEIIFGGMDPKHYVGEHTYVPVTQKGYWQFDMGDVLVGGKSTGFCAG
Aspr_Orysa	SFWLNRDPDASSGGELVFGGMDPKHYKGDHTYVPVSRKGYWQFNMGDLLIDGHSTGFCAK
Cyp1_Cynca	SFWLNRNADEQEGGELVFGGVDPNHFKGEHTYVPVTQKGYWQFEMGDVLIGDKTTGFCAS
Catd_Human	SFYLSRDPDAQPGGELMLGGTDSKYYKGSLSYLNVTRKAYWQVHLDQVEVASGLT.LCKE

	301–360
Cardosin A	GCQAFA**DSG**TSLLAGPTAIVTQINHAIGANGVMNQQCKTVVSRYGRDIIEMLRSKIQPDK
Aspr_Horvu	GCAAIA**DSG**TSLLAGPTAIITEINEKIGAAGVVSQECKTIVSQYGQQILDLLLAETQPKK
Aspr_Orysa	GCAAIV**DSG**TSLLAGPTAIVAQVNHAIGAEGIISTECKEVVSEYGEMILNLLIAQTDPQK
Cyp1_Cynca	GCAAIA**DSG**TSLLAGTTTIVTQINQAIGAAGVMSQQCKSLVDQYGKSMIEMLLSEEQPEK
Catd_Human	GCEAIV**DTG**TSLMVGPVDEVRELQKAIGAVPLI...........................

	361–420
Cardosin A	ICSHMKLCTFDGALDISSIIESVVDKNNVKSSGGIHDEMCTFCEMAVVWMQNEIKRSETE
Aspr_Horvu	ICSQVGLCTFDGTRGVSAGIRSVVDDEPVKSNGLRADPMCSACEMAVVWMQNQLAQNKTQ
Aspr_Orysa	VCSQVGLCMFDGKRSVSNGIESVVDKENLGS.....DAMCSVCEMAVVWIENQLRENKTK
Cyp1_Cynca	ICSQMKLCSFDGSHDTSMIIESVVDKSKGKSSGLPM..RCVPCARWVVWMQNQIRQNETE
Catd_Human	....

	421–480
Cardosin A	DNIINYANELCEHLSTSSEELQVDCNTLSSMPNVSFTIGGKKFGLTPEQYILKVGKGEAT
Aspr_Horvu	DLILDYVNQLCNRLPSPMGESAVDCGSLGSMPDIEFTIGGKKFALKPEEYILKVGEGAAA
Aspr_Orysa	ELILNYANQLCERLPSPNGESTVSCHQISKMPNLAFTIANKTFILTPEQYIVKLEQGGQT
Cyp1_Cynca	ENIINYVDKLCERLPSPMGESAVDCSSLSSMPNIAFTVGGKTFNLSPEQYVLKVGEGATA
Catd_Human	.................QGEYMIPCEKVSTLPAITLKLGGKGYKLSPEDYTLKVSQAGKT

	481–528
Cardosin A	QCISGFTAMD.ATLLGPLWILGDVFMRPYHTVFDYGNLLVGFAEAA..
Aspr_Horvu	QCISGFTAMDIPPPRGPLWILGDVFMGPYHTVFDYGKLRIGFAKAA..
Aspr_Orysa	VCISGFMAFDIPPPRGPLWILGDVFMGAYHTVFDFGKDRIGFAKSA..
Cyp1_Cynca	QCISGFTAMDVAPPHGPLWILGDVFMGQYHTVFDYGNLRVGFAEAA..
Catd_Human	LCLSGFMGMDIPPPSGPLWILGDVFIGRYYTVFDRDNNRVGFAEAARL

Figure 1. Alignment of cardosin A with other aspartic proteinases. The deduced amino acid sequence of cardosin A is compared with the amino acid sequences of the barley proteinase (Aspr_Horvu), oryzasin (Aspr_Orysa), cyprosin (Cyp1_Cynca) and human cathepsin D (Catd_Human). The specific inserted sequence characteristic of plant aspartic proteinases can be identified by the region to which human cathepsin D has no homologous residues. Points denote gaps in the sequence and active -site aspartates are in bold.

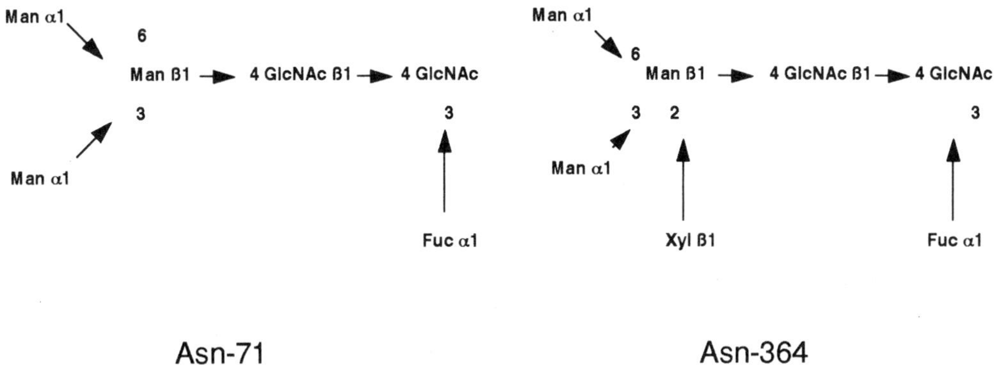

Figure 2. Proposed structures of the major oligosaccharides in each glycosylation site of cardosin A. Two N-glycosylation sites are present in cardosin A at positions Asn71 and Asn364 located in the 31 and 15 kDa chains respectively.

The structures of the glycans of cardosin A were determined by exoglycosidase sequencing and mass spectrometry [20]. Results showed that the two glycosylation sites, Asn-71 and Asn-364, contain glycans of the plant complex type with proximal fucose α1,3-linked. Various structures were identified in each site suggesting the occurence of different glycoforms of cardosin A. The predominant glycans in each site are depicted in Fig. 2. The major difference between them is that the Asn-364 oligosaccharide contains xylose β1,4-linked.

X-ray diffraction analysis of cardosin A crystals located both glycosylation sites away from the active site (fig. 4) suggesting therefore that the oligosaccharides do not play an essential role on the activity and specificity of the enzyme. However, it is possible that the oligosaccharides are important for the correct folding or to maintain the stability of the enzyme.

5. CRYSTAL STRUCTURE OF CARDOSIN A

A crystallographic structural model of Cardosin A (Fig. 3) was obtained by the molecular replacement method using cathepsinD as search model, and a diffraction data set to 2.85 A resolution, Rcryst(I)=7%. The refinement is still under way (actual R=21.5%, Rfree=29.5%) but several structural features can already be pointed out. The overall structure is clearly a typical aspartic proteinase structure with two domains separated by a large cleft [21,22]. As previously suggested [6], the characteristic plant specific insertion is missing in the crystallised protein, with the electronic density fading away by residue Asn241 and appearing again only by residue Glu351. The break in the polypeptide chain would correspond to the loop 238–244 when using the pepsin numbering. Both glycosylation sites were assigned, one in each domain, with 4 rings already modelled at the Asn-71 site and three rings at the Asn-364. As mentioned above both glycans were assigned away from the specificity defining cleft. Three disulfide bridges, one in the amino domain (Cys45–Cys50) and two in the carboxy-domain (Cys206–Cys210, Cys249–Cys282, pepsin numbering) were identified in the mature cardosin A. None of them are interchain disulfide bridges and therefore the two chains are held together by strong interactions of non-covalent nature.

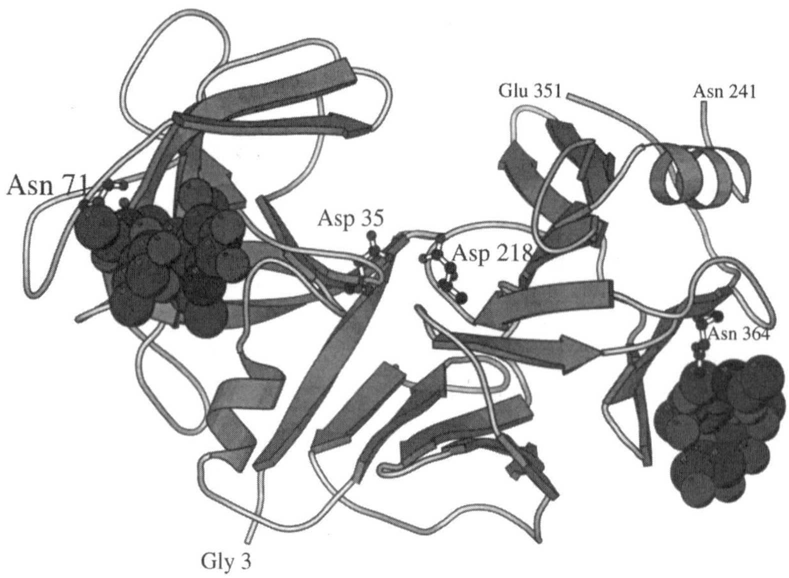

Figure 3. A ribbon diagram of the crystal structure of cardosin A showing the overall fold, elements of secondary structure and the sugar chains attached to each domain. The diagram was generated using the program MOLSCRIPT (30).

6. PROTEOLYTIC PROCESSING OF CARDOSINS

Cardosins are synthesised as single chains which undergo further processing into a two-chain mature enzymes [6]. Comparison between the deduced amino acid sequences of plant aspartic proteinases and data from protein sequencing and amino acid analysis of each chain of cardosins suggested that processing of the single chain into a two-chain enzyme is likely to occur through cleavage of two internal peptide bonds. During this process, complete removal of the PSI occurs in cardosin A whereas in cardosin B the PSI domain is believed to be partially removed. The 34 kDa chain of cardosin B is considerably longer than the corresponding chain in cardosin A and as they both have similar N-terminal sequences it is likely that differential cleavage at the C-terminal may account for the observed difference in size. In this case, a segment of the plant specific sequence remains at the C-terminal part of the 34 kDa chain of cardosin B upon conversion of the single chain precursor.

In order to study the proteolytic processing of cardosin A, a polyclonal antibody was raised against a synthetic peptide whose sequence was derived from a region in the cDNA of cardosin A corresponding to PSI. The antibody was used for immunoblotting analysis of proteins extracted from immature flowers of *C. cardunculus* L and a protein with an apparent molecular mass of 64 kDa was identified as the putative precursor of cardosin A. Upon incubation of protein extracts at pH 3.0 for one hour, the 64 kDa band dissapeared on SDS-polyacrylamide gels yielding two bands with apparent molecular masses of 31 and 15 kDa. Comparison between the deduced amino acid sequence and the results obtained on both N- and C-terminal sequencing of the cardosin A chains allowed to identify the cleavage sites that give rise to the mature form of the enzyme. A scheme of the proposed proteolytic processing of cardosin A is shown elsewhere [23] together with a detailed description of these studies on the structural and funtional relevance of PSI.

7. ORGAN-SPECIFIC EXPRESSION OF CARDOSINS

A monospecific antibody against cardosin A and a polyclonal antibody specific for cardosin B were produced in rabbits and used to study the organ specific expression of cardosins in *C. cardunculus* L. Western blotting analysis revealed an expression of both cardosins highly restricted to the pistils [24]. Protein extracts from seed, root, midrib, leave, bract, stamen and pollen did not show any reaction with these antibodies. The strongest reaction for both enzymes was found in the mature, open capitula stage, but comparable levels of expression of cardosins in the closed and senescent capitula stages were also observed [24]. Since the proteolytic activity of cardosins is quite stable, expression of cardosins is likely to occur during a short period at early stages of floral development, being then accumulated until the late stages of senescence. Expression of cardosins seems therefore to be a highly regulated process and deserves further studies.

8. ANATOMY OF THE STIGMA AND THE STYLE OF *C. cardunculus* L.

The female organ of a flower is called the pistil and can be divided into three parts: stigma, style and ovary. The stigma is the pollen receptive surface whereas the style supports pollen tube growth during reproduction to the basely located ovary where fecundation takes place. Cardosins are mainly expressed in stigmas and styles of *C. cardunculus* L. and for this reason a brief description of these two parts is given below and their histology is shown in Fig. 4.

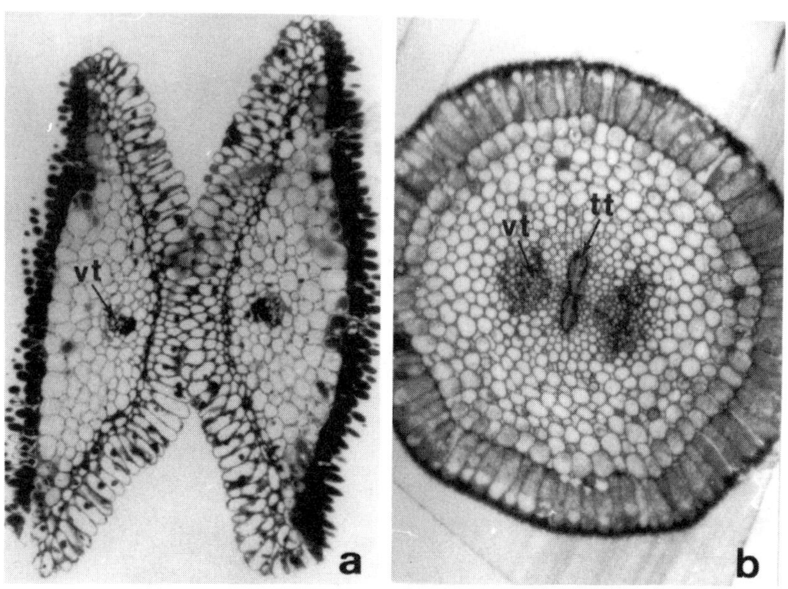

Figure 4. Anatomy of the stigma and the style of *Cynara cardunculus* L. (a) Cross section of the stigma near the apex showing two stigmatic portions not completely united that consist of a glandular papillate epidermis, thin-walled parenchyma and vascular tissues (vt). (b) Cross section of the style with a cuticle-covered epidermis surrounding parenchyma cells. The inner core is occupied by the transmitting tissue (tt), and vascular tissues (vt) with sclerenchyma.

The stigma of *C. cardunculus* flowers is very long and of the papillate type. It has two longitudinal grooves because it consists of two stigmatic portions not completely united reflecting the nature of the pistil that is composed of two united carpels. A papillate epidermis covers thin walled parenchyma cells that surround vascular tissues. The epidermic papillae cells are unicellular, smooth and the wall is covered with a cuticle (Fig. 4A).

The style, of the closed type, is composed of a single layer of epidermis with a thick cuticle surrounding several layers of vacuolated cortical parenchyma cells (Fig. 4B). The inner portion of the style is occupied by two vascular bundles side by side of a central transmitting tissue. The cortical cells appear roundish in transverse section with the thin layer of cytoplasm adpressed to the cell wall due to the presence of a large vacuole. The transmitting tissue located in the center of the style consists of roundish cells with very thick loosely fibrillar walls that provide the route for pollen tube growth.

9. TISSUE-SPECIFIC EXPRESSION AND CYTOLOGICAL LOCALIZATION OF CARDOSINS

Tissue printing analysis using the above mentioned antibodies revealed a differential expression of the two cardosins within the pistil [24]. While cardosin A is accumulated in the superficial cell layers of the stigma and style, expression of cardosin B is apparently restricted to the transmitting tissue. Further studies by immunogold electron microscopy also revealed a differential cytological localization of the two cardosins (Fig. 5). Cardosin A is accumulated in protein storage vacuoles of the stigmatic papillae and in the subepidermal cells of the upper portion of the stigma (Fig. 5A). It is also found in the vacuoles of the epidermis cells of the style although in less amount. Conversely, cardosin B is an extracellular protein that occurs in the extracellular matrix of the transmitting

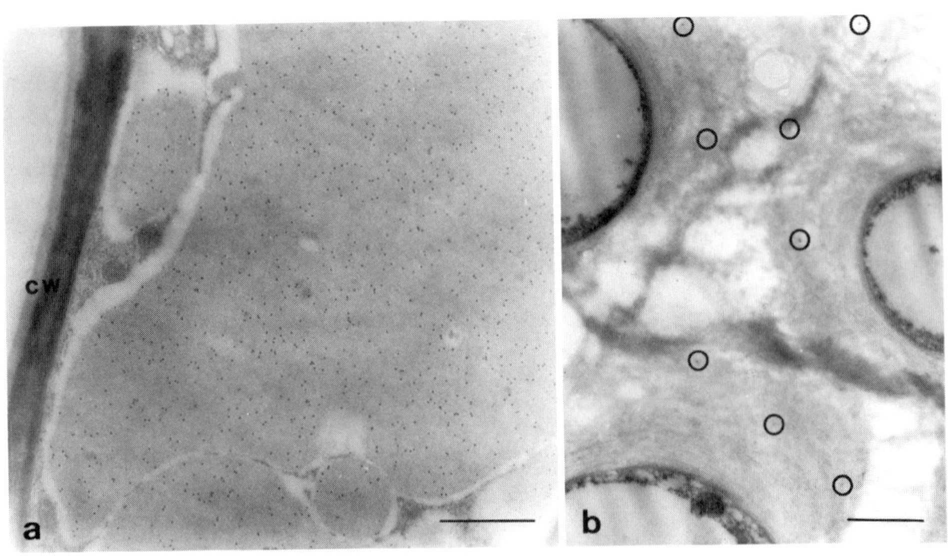

Figure 5. Immunocytochemical localization of cardosin A and cardosin B. (a). Immunocytochemical localization of cardosin A in epidermic papilla of the stigma; labelling is abundant in protein storage vacuoles. cw, cell wall. Scale bar, 1 μm. (b) Localization of cardosin B in the transmitting tissue of the style, gold labeling appeared over the intercellular matrix. Scale bar, 1 μm.

tissue (Fig. 5B). These results strongly support the idea that cardosins may play different roles in the plant.

10. POSSIBLE BIOLOGICAL FUNCTIONS FOR CARDOSINS

Cardosins are unique among plant aspartic proteinases in their abundance and organ-specific expression. Indeed, plant aspartic proteinases have been identified in several parts of plants, from seeds to leaves, but not in flowers [25]. Moreover, they usualy occur in small amounts and often are quite difficult to isolate in a pure form, which is not the case of cardosins. Plant aspartic proteinases have been suggested to be involved in protein turnover or specific proteolytic processing of other proteins [25], but with the exception of those from insectivorous plant [26] their functions are largely unknown. In what cardosins are concerned, the results above described suggest several hypothesis for their function.

Cardosin A and cardosin B are probably involved in the pollen-pistil interactions that occur during plant sexual reproduction. The stigmatic papillae is the site where pollen first contacts and interacts with the female sporophyte in its way to the ovary, being capture and adhering to the stigma surface, hydrating and germinating. All these processes certainly involve molecular events of recognition, signalling, and response [27]. The fact that cardosin A is localized in the vacuole of the papillae suggests a possible role of the enzyme in these processes. Abundant glycoproteins in the stigmatic papillae have long been thought to be involved in self incompatibility [28] but in the case of the cardoon plant, we do not have any data regarding the existence of a functional reaction.

Involvement of cardosin B in the pollen pistil interaction is likely to occur in a different context. Transmitting tissue is known to support pollen tube growth towards the ovary, a process that requires continous degradation of the extracellular matrix to ensure both feeding of pollen tube and its movement throughout the pistil. Cardosin B being an extracellular enzyme expressed at the transmitting tissue may contribute for both process and would thus fullfill an important function in sexual reproduction of cardoon.

The large accumulation of cardosin A in the peripheral cells of the stigma also suggests a possible involvement of the enzyme in defense against pathogens and/or insects. The bursting of the vacuoles in a hypersensitive reaction to invasion could lead to an arrest and eventual destruction of the pathogen through the action of the proteinase. If active in the digestive tract of herbivore insects, cardosin A could exert a pernicious effect on digestive physiology of the insect. An involvement of cardosin B in this process is less probable since expression of the enzyme is apparently restricted to the inner part of the pistil which in principle is not easly accesible to insects or other pathogens

Finally, we cannot exclude an involvement of cardosins in flower senescence, although it seems to be at least a secondary function of cardosins. The activity of the enzyme could thus contribute to the release of seeds and eventually to signal the senescence of the whole plant. Since vacuolar proteases do not participate in protein breakdown during senescence, except in a terminal stage, when the vacuolar membrane is broken [29], cardosin B as an extracellular enzyme could have a leading role in the process.

In summary, cardosin A may have a primary function in interaction with pathogens and together with cardosin B they are involved in the polen-pistil interactions that occur during sexual reproduction of the plant. Both cardosins may also have a role in flower senescence although this is for sure a secondary function of the two enzymes.

REFERENCES

1. Faro, C.J., Alface, J.S. and Pires, E.V. (1987) Purification of a protease from the flowers of *Cynara cardunculus* L. *Ciênc. Biol.*12, 5A, 201.
2. Heimgartner, U., Pietrzak, M., Geertsen, R., Brodelius, P., Figueiredo, A.C. and Pais, M.S.(1990) Purification and partial characterization of milk clotting proteinases from flowers of *Cynara cardunculus* L. *Phytochemistry* 29, 1405–1410.
3. Faro, C.J., Moir, A.G.J. and Pires, E.M.V. (1992) Specificity of a milk clotting enzyme extracted from the thistle *Cynara cardunculus* L.: action on oxidised insulin and k-casein *Biotech. Lett.* 14, 841–847.
4. Macedo, I.Q., Faro, C.J. and Pires, E.M.(1993) Specificity and kinetics of the milk-clotting enzyme from cardoon (*Cynara cardunculus* L) toward bovine k-casein". *J. Agric. Food Chem.* 41, 1537–1540.
5. Veríssimo, P.C., Esteves, C., Faro, C.J. and Pires, E. (1995) The vegetable rennet from *Cynara cardunculus* L. contains two proteinases with chymosin-like and pepsin-like specificities" *Biotech. Lett.* 17, 614–645.
6. Veríssimo, P., Faro, C., Moir, A., Lin, Y., Tang, J. and Pires, E. (1996) Purification, characterization and partial amino acid sequencing of two novel aspartic proteinases from the flowers of *Cynara cardunculus* L. *Eur. J. Biochem.* 235, 762–768.
7. Cordeiro, M., Jakob, E., Puhan, Z., Pais, M.S. and Brodelius, P.E. (1992) Milk clotting and proteolytic activities of purified cynarases from *Cynara cardunculus* L.-a comparison to chymosin. *Milchwissenschaft* 47, 683–687.
8. Drohse, H.B. and Foltman, B. (1989) Specificity of milk-clotting enzymes towards bovine k-casein Biochem. Biophys. Acta 995, 221–224.
9. Visser, S. (1981) Proteolytic enzymes and their action on milk proteins: a review. *Neth. Milk Dairy J.* 35, 65–88.
10. Sá, F.V. and Barbosa, M. (1972) Cheese-making with a vegetable rennet from cardo (Cynara cardunculus). *J. Dairy Res.* 39, 335–343.
11. Verissimo, P., Ramalho-Santos, M., Faro, C. and Pires, E. (1998) A comparative study on the aspartic proteinases from different species of Cynara. In structure and function of aspartic proteinases: retroviral and cellular enzymes (M. N. G. James, ed.) New York: Plenum Press.
12. Faro, C., Verissimo, P., Lin,Y., Tang, J. and Pires, E. (1995) Cardosins A and B, aspartic proteinases from the flowers of cardoon. In Aspartic proteinases: structure, function, biology and biomedical implications (Takahashi, K., ed.) New York: Plenum Press, pp. 374–377.
13. Ramalho-Santos, M., Veríssimo, P., Faro, C.J. and Pires, E.V.(1996) Action on bovine as1-casein of cardosins A and B, aspartic proteinases from the flowers of the cardoon *Cynara cardunculus* L. *Biochem. Biophys. Acta* 1297, 83–89.
14. Simões, I., Verissimo, P., Faro, C. and Pires, E.V. Action on bovine ß-casein of cardosins A and B, aspartic proteinases from the flowers of the cardoon *Cynara cardunculus* L. (submitted for publication).
15. Asakura T, Watanabe H, Abe K, Arai S (1995) Rice aspartic proteinase, oryzasin, expressed during seed ripening and germination, has a gene organization distinct from those of animal amd microbial aspartic proteinases. *Eur J Biochem* 232: 77–83.
16. Cordeiro MC, Xue Z-T, Pietrzak M, Pais MS, Brodelius PE (1994) Isolation and characterization of a cDNA from flowers of *Cynara cardunculus* encoding cyprosin (an aspartic proteinase) and its use to study the organ-specific expression of cyprosin. *Plant Mol Biol* 24: 733–741.
17. Runeberg-Roos P, Törmäkangas K, Östman A (1991) Primary structure of a barley-grain aspartic proteinase. *Eur J Biochem* 202: 1021–1027.
18. Leahy, D.J., Aukhil, I. and erickson H.P. (1996) 2.0 A Crystal structure of a four-domain segment of human fibronectin encompassing the RGD loop and synergy region. *Cell* 84, 155–164.
19. Guruprasad K, Törmäkangas K, Kervinen J, Blundell TL (1994) Comparative modelling of barley-grain aspartic proteinase: a structural racionale for observed hydrolytic specificity. *FEBS Lett* 352: 131–136.
20. Costa, J., Ashford, A., Nimetz, M., Bento, I., Frazão, C., Esteves, C., Faro, C., Kervinen, J, Pires, E., Veríssimo, P., Wlodawer, A., Carrondo, MA. (1997) The glycosylation of barley (*Hordeum vulgare* L.) and cardoon (*Cynara cardunculus* L) aspartic proteinases *Eur. J. Biochem.* (in press).
21. Tang, J. and Wong, R.N.S. (1987) Evolution in the structure and function of aspartic proteinases *J. Cell Biochem.* 33, 53–63.
22. Davies, D.R.(1990) The structure and function of the aspartic proteinases *Annu. Rev. Biophys. Chem.* 19, 189–215.
23. Ramalho-Santos, M., Pissarra, J., Pires, E. and Faro, C.(1997) Cardosinogen A, a precursor form of the major aspartic proteinase from cardoon. In structure and function of aspartic proteinases: retroviral and cellular enzymes (M.N.G.James.ed.) New York: Plenum Press.

24. Ramalho-Santos, M., Pissarra, J., Verissimo, P., Pereira, S., Salema, R., Pires, E. and Faro, C.(1997) "Cardosin A accumulates in protein storage vacuoles of the stigmatic pappilae of *Cynara cardunculus* L." *Planta* (accepted for publication).
25. Kervinen, J., Tormakangas, K., Runeberg-Roos, P., Guruprasad, K., Blundell, T. and Teeri, T.H. (1995) Structure and possible function of aspartic proteinases in barley and other plants In Aspartic proteinases: structure, function, biology and biomedical implications (Takahashi, K., ed.) New York: Plenum Press, pp. 241–254.
26. Tokes, Z.A., Woon, W.C. and Chambers, S.M. (1974) Digestive enzymes secreted by the carnivorous plant napenthes macferlaine L. *Planta* 119, 39–46.
27. Elleman CJ, Dickinson HG (1994) Pollen-stigma interaction during sporophytic self-incompatibility in *Brassica oleracea. In* EG Williams, AE Clarke, RB Knox, eds, Genetic Control of Self-Incompatibility and Reproductive Development in Flowering Plants. Kluwer Academic Publishers, Dordrecht, pp 67–87.
28. Nasrallah JB, Stein JC, Kandasamy MK, Nasrallah ME (1994) Signaling the arrest of pollen tube development in self-incompatible plants. *Science* 266: 1505–1508.
29. Boller T (1986) Roles of proteolytic enzymes in interactions of plants with other organisms. *In* MJ Dalling ed, Plant Proteolytic Enzymes, Vol 1. CRC Press, Boca Raton, pp. 67–96.
30. Kraulis, P. (1991) MOLSCRIPT: a program to produce both detailed and schematic plots of protein structures *J. Appl. Crystallogr.* 24, 946–950.

MOLECULAR CLONING OF ASPARTIC PROTEINASES FROM FLOWERS OF *Cynara cardunculus* SUBSP. *flavescens* CV. CARDOON AND *Centaurea calcitrapa*

P. E. Brodelius,[1] M. Cordeiro,[1] P. Mercke,[1] A. Domingos,[2] A. Clemente,[2] and M. S. Pais[3]

[1]Department of Plant Biochemistry, Lund University
P.O. Box 117, 22100 Lund, Sweden
[2]Departamento de Biotecnologia, INETI
Estrada das Palmeiras, 2745 Queluz, Portugal
[3]Centro Biotech Vegetal, FCL
Bloco C2, Campo Grande, 1700 Lisboa, Portugal

1. INTRODUCTION

Aspartic proteinases of human, animal, parasite, fungal and retroviral origin have been extensively investigated. The 3-D structures of several such enzymes have been solved and they all have a number of features in common.[1] By contrast, relatively little is known about aspartic proteinases in plants. During the last few years, however, an increasing number of studies on plant aspartic proteinases have been carried out.[2,3] We are since a few years involved in studies of milk-clotting enzymes of plant origin.

So far, the cDNA derived primary structure of ten plant aspartic proteinases has been reported. Considerable homology exists not only among the plant enzymes but also between the plant aspartic proteinases and their counterparts from other phyla. Plant aspartic proteinases may be readily distinguished by one notable feature. The cDNA sequences predict a large insert of ~100 amino acid residues that is positioned internally in the mature enzyme segment. These inserts show conservation of sequence.

2. ISOLATION AND SEQUENCING OF cDNA CLONES ENCODING PLANT ASPARTIC PROTEINASES

We have previously reported on the purification and characterization of three aspartic proteinases (cyprosins) from flowers of *Cynara cardunculus*.[4] Water extracts of these

Aspartic Proteinases, edited by James.
Plenum Press, New York, 1998.

flowers are used by farmers in Portugal and Spain to produce cheeses. We have also reported in the isolation and sequencing of one cDNA clone (*cypro1s*: EMBL no. X69193) encoding one of the cyprosins (cyprosin A).[5] This clone contained only part of a preprosequence. Using the PCR technique the missing part of this preprosequence has been obtained. Furthermore, the original cDNA library has been rescreen using *cypro1s* as probe and a second cDNA (*cypro11*: EMBL no. X81984) encoding another cyprosin (cyprosin B) was isolated and sequenced. The two cDNA clones contained open reading frames of 1515 and 1527 bp encoding 505 and 509 amino acids, respectively. The size of the putative mature proteins is 440 (47.7 kD) and 441 (47.9 kD) and the size of the insert is 103 and 104 amino acids for cyprosin A and B, respectively. The deduced amino acid sequences of these two cyprosins show an overall identity of 85 and a similarity of 92%.

Recently, we have also isolated and sequenced a clone (*cent8;* EMBL no. Y09123) encoding an aspartic proteinase from a cDNA library obtained from mRNA isolated from flowers of *Centaurea calcitrapa*. These flowers are also used by farmers to make cheeses. The open reading frame of *cent8* encoded a protein of 509 amino acids; a putative mature aspartic proteinase (441 amino acids) and a preprosequence (68 amino acids). The size of the insert is 104 amino acids.

Cynara and *Centaurea* belong to the same plant family (Compositae) and the deduced amino acid sequences of the three cDNA clones from these species show a high identity to each other as shown in Figure 1. In fact, the identity between the *Centaurea*

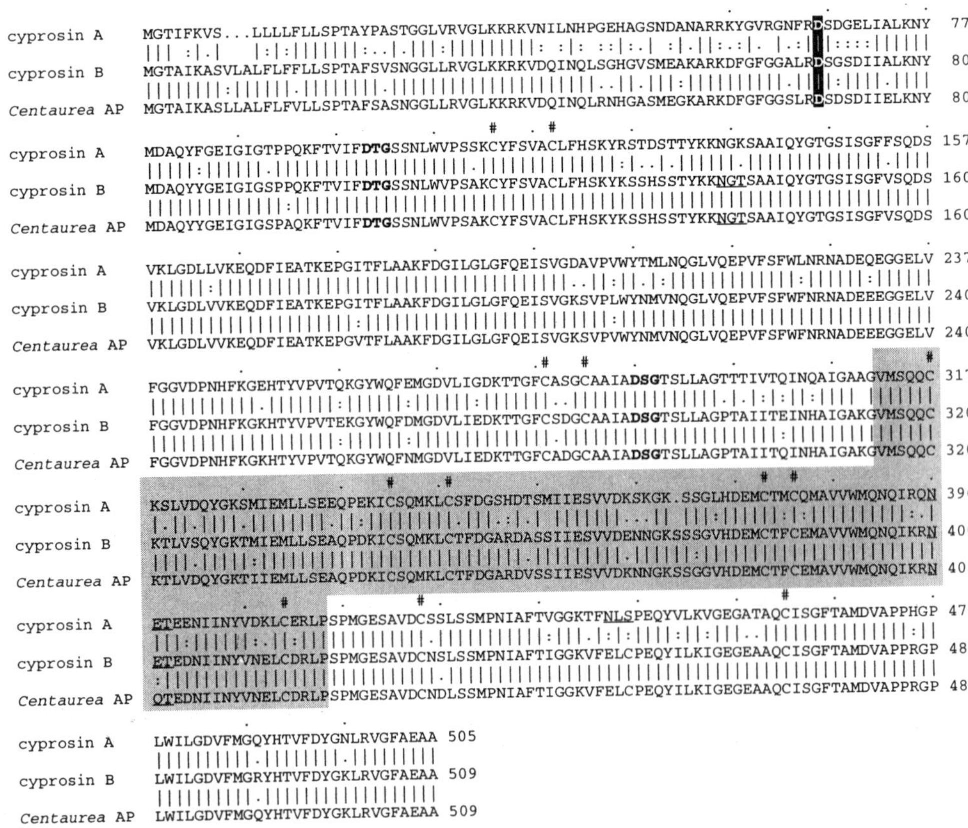

Figure 1. Alignment of the deduced amino acid sequences of cyprosin A, cyprosin B and *Centaurea* aspartic proteinase. The putative N-terminal of the mature enzymes is highlighted in black; catalytic aspartic acids are in bold; conserved cysteins are marked #; putative N-glycosylation sites are underlined; the large inserts are in a shaded box.

enzyme and cyprosin B is very high (96%). The homology to other plant aspartic proteinases is relatively high (61–76% identity). The homology to mammalian aspartic proteinases is somewhat lower (40–55% identity).

Southern analyses of the genomes of *Cynara* and *Centaurea* have shown that small multi-gene families are present in both species. At least two different aspartic proteinase genes are expressed in *C. cardunculus*. The high homology between cyprosin B and the *Centaurea* enzyme may reflect the fact that these enzymes have evolved from the same gene of an ancestral Compositae species and therefore belong to the same sub-family of aspartic proteinase genes. Cyprosin A, with a lower homology to the two other enzymes, would thus represent another sub-family of aspartic proteinase genes in the Compositae family. Based on this assumption it may be suggested that aspartic proteinase gene duplication and divergence have occurred in the Compositae family prior to the emergence of the two species studied.

3. POST-TRANSLATIONAL MODIFICATIONS OF CYPROSINS AND *Centaurea* AP

The aspartic proteinases from *Cynara* and *Centaurea* each contain two putative N-glycosylation sites; one in the specific insert and one in the C- (cyprosin A) or N-terminal (cyprosin B and *Centaurea*) part of the mature protein. All the three purified enzymes contain N-linked glucans of high-mannose type.

N-terminal sequencing of mature cyprosin indicate that the zymogen is not autoactivated since an Arg–Asp bond appears to be hydrolyzed. All three enzymes have an Arg–Asp bond at this position and it is likely that they are all activated by cleavage of this bond (cyprosin A: RGNFR-DSDGE; cyprosin B: GGALR-DSGSD; *Centaurea* AP: GGSLR-DSDSD). However, an activating proteinase has not yet been identified in any of the plants.

The mechanism for conversion of the enzymes to heterodimeric enzymes has not yet been evaluated but it is clear that the larger subunit, containing both catalytic aspartic acids, is derived from the N-terminal part of the mature protein while the smaller subunit originates from the C-terminal part of the protein. A major part of the insert is removed during the posttranslational proteolytic processing of the enzymes. N-terminal sequencing has shown that the small subunits start close to the C-terminal end of the insert. Whether this proteolytic modification is autocatalyzed or not remain to be established.

4. CYPROSIN GENE ORGANIZATION

The genomic organization has been determined for a number of mammalian aspartic proteinases. All genes contain nine exons and eight introns with complete conservation of exon-intron junctions in all cases. Determination of the genomic organisation of the plant aspartic proteinases may give an insight into the evolution of these plant specific regions and may afford an insight into the function of this segment of polypeptide and its transient existence during the biosynthesis and translocation of these aspartic proteinases in plant tissues. To this end, we have initiated studies on the cyprosin gene organisation and regulation.

A genomic fragment (0.9 kB) corresponding to the open reading frame of the plant specific region was obtained by PCR from *Cynara* DNA. It was found that the cyprosin gene contains three small introns (152, 220 and 86 bp) in this region (Figure 2).

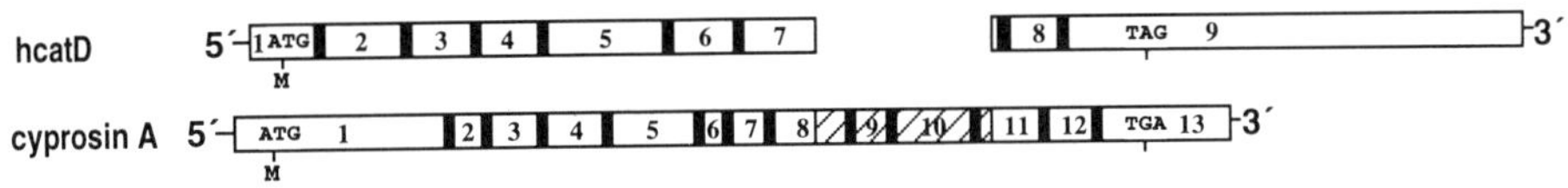

Figure 2. Organization of the cyprosin A gene and human cathepsin D gene. The size of exons (numbered open boxes) is corresponding to the number of bases while the size of introns (black boxes) is independent of size. The plant specific insert is hatched.

Three genomic fragments (1.2, 1.3 and 1.7 kB) corresponding to the entire open reading frame of cyprosin A were recently obtained by PCR. The total size of the gene is around 4.2 kB. Introns corresponding to 2.5 kB are present in the gene. Twelve introns have been localized in the cyprosin gene and. so far, we have sequenced 9 of these introns. The position of these 12 introns is exactly the same as the corresponding introns in the rice (EMBL no. D32165) and *Brassica napus* (EMBL no. U55032) genes even though the sizes of the introns varies. A comparison of the cyprosin gene organization to that of a mammalian gene (human cathepsin D) is shown in Figure 2. The N-terminal end of the plant specific insert is found in exon 8, while the C-terminus is found in exon 11 of the cyprosin gene. The insert is located at a position corresponding exon 7 of the mammalian genes.

5. CONCLUSIONS

Aspartic proteinases are synthesized as preproenzymes and, after removal of the signal peptide, a propart consisting commonly of about 40–50 residues has to be removed, often autocatalytically, in order to generate the mature enzyme form of the aspartic proteinase. The mature enzymes are usually composed of about 340 residues. These features are all faithfully reflected in the plant aspartic proteinases for which sequence data are available and considerable homology exists not only among the plant enzymes but also between the plant aspartic proteinases and their counterparts from other phyla. At present, the complete cDNA sequences of ten different plant aspartic proteinases have been reported. A phylogenetic analysis of 50 aspartic proteinases shows that the plant enzymes are all segregated into one distinct group which is related to the mammalian cathepsin D proteins (Figure 3). The three Compositae enzymes are grouped together and, as expected, cyprosin B and *Centaurea* aspartic proteinase are the two most closely related enzymes.

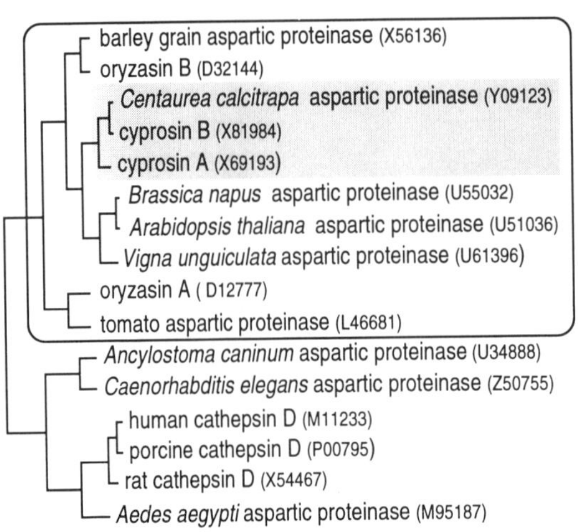

Figure 3. Phylogenetic tree for 16 different aspartic proteinases. The plant aspartic proteinases are within an open box and the Compositae enzymes are in a shaded box. The tree was obtained by using the computer programmes PILEUP, DISTANCES and GROWTREE (Genetics Computer Group, University of Wisconsin). The sequences for the aspartic proteinases were obtained from the EMBL Data Library as indicated in the figure.

The plant aspartic proteinases may be readily distinguished from their counterparts from other phyla. All the plant cDNA sequences elucidated to date predict a large insert of ~100 amino acid residues that is positioned internally in the mature enzyme segment. This plant specific region is not intact in the enzymes as isolated from the plants and must have been affected therefore by post-translational processing *in planta*. Six cysteine residues present in each plant specific region are conserved in all enzymes suggesting that this domain may have a tertiary structure maintained through disulphide bridges. Furthermore, significant similarity (around 40% identity over 35 residue stretches) exists between the plant specific inserts and saposins (due to some rearrangement compared to saposins the plant sequences have been termed swaposins).[6] This suggests a possible functional role for the plant specific region to target the nascent poly peptide to a particular cytomorphological compartment in the plant cell (*e.g.* vacuoles). This may be facilitated further by the presence of a putative glycosylation site that is conserved in the plant specific region in all plant enzymes.

Furthermore, in comparison to aspartic proteinases from other phyla, the sequences flanking the two active site aspartic acid residues are particularly highly conserved except for the substitution of the archetypal hydrophobic–hydrophobic–Asp–Thr–Gly sequence with hydrophobic–hydrophobic–Asp–Ser–Gly for the second of the catalytic Asp residues. This substitution is observed in all plant enzymes.

The high homology between mammalian and plant aspartic proteinases (omitting the plant specific insert) show that all these enzymes have evolved from a common ancestral protein. The genes have, however, evolved in parallel since the number and location of introns are different. In addition to this the plant genes have been modified with a large insert containing three introns.

Like their animal and microbial counterparts, plant aspartic proteinases are inhibited by pepstatin and preferentially cleave peptide bonds between amino acid residues with large hydrophobic side chains.

REFERENCES

1. Tang J, Wong RNS (1987) Evolution in the structure and function of aspartic proteases. *J. Cell. Biochem.* **33:**53–63
2. Guruprasad K, Dhanaraj V, Groves M, Blundell TL. (1995) Aspartic proteinases: The structures and functions of a versatile superfamily of enzymes. *Persp. Drug Dis. Des.* **2:** 329–341
3. Cordeiro MC, Pais MS, Brodelius PE (1997) Milk clotting enzymes from *Cynara cardunculus* subsp. *flavescens* (Cardoon).In: *Biotechnology in Agriculture and Forestry* (Y.P.S. Bajaj, ed.) Springer Verlag, Berlin, in press
4. Heimgartner U, Pietrzak M, Geertsen R, Brodelius P, da Silva Figueiredo AC, Pais MSS (1990) Purification and partial characterization of milk clotting proteases from flowers of *Cynara cardunculus*. *Phytochemistry* **29:**1405–1410
5. Cordeiro M, Xue Z-T, Pietrzak M, Pais MS, Brodelius PE (1994) Isolation and characterization of a cDNA from flowers of *Cynara cardunculus* encoding cyprosin (an aspartic proteinase) and its use to study the organ-specific expression of cyprosin. *Plant Molecular Biology* **24:**733–741
6. Ponting CP, Russell RB (1995) Swaposins: circular permutations within genes encoding saposin homologues. *Trends Biol. Sci.* **20**: 179–180

ASPARTIC PROTEINASES AND INHIBITORS IN PLANT PATHOGENESIS

Jeffrey Hill and Lowri H. Phylip

School of Molecular and Medical Biosciences
University of Wales
Cardiff, CF1 3US, Wales, United Kingdom

Plant disease can be caused by a variety of pathogens including bacteria, viruses and fungi and results in decreased crop yields all over the world. Rice tungro disease,[1] caused by a complex of two viruses, alone causes an annual loss in excess of 1.5×10^9. The ability of pathogens to cause disease in their plant hosts is dependent on the production of virulence factors. In a number of cases, aspartic proteinases have been identified as the factor responsible for invasion and these enzymes are often intimately involved in the host-parasite relationship. As yet, aspartic proteinases have not been implicated in the infection of plants by bacteria but since the existence of aspartic proteinases in bacteria has only recently been demonstrated unequivocally,[2] this situation may soon change.

The role of aspartic proteinases in the infection of plants by fungi has been documented. For example, the chestnut blight fungus, *Cryphonectria parasitica* which has been responsible for devastating the chestnut tree population in the United States, secretes at least three distinct aspartic proteinases, EAPA, EAPB and EAPC.[3] The proteolytic activity of the secretions from this fungus causes tissue damage to the host and permits the parasite to establish itself in a nutrient-rich environment. Other fungi have a more direct economic effect by their impact on cash crops. *Botrytis cinerea* is a ubiquitous fungus which affects fruit and vegetables both before and after harvest and causes extensive crop loss. It too secretes at least one aspartic proteinase[4,5] which is thought to play a key role in pathogenesis. Pepstatin has been shown to prevent the infection of a number of fruit and vegetable types by spores of *Cinerea botrytis*,[6] showing the crucial roles these enzymes play in the infection process.

Viruses cause significant losses to most major crops around the world and they can also prevent the production of specific crops in certain areas.[7] There are many classes of virus which cause disease in plants. The pararetroviruses, one member of the two groups of viruses which use reverse transcription in their replication cycle, contain two plant viral sub-groups:

1. Caulimoviruses. Their type member is cauliflower mosaic virus which is transmitted by aphids and causes extreme damage in *crucifereae*.[8]

Aspartic Proteinases, edited by James.
Plenum Press, New York, 1998.

2. Badna viruses. Their type member is *Commelina* yellow mottle virus which is transmitted by mealybugs and causes disease in a number of commercially important plants including rice, cocao, bananas and sugar cane.[9] The financial damage caused by rice tungro disease is matched by the loss (\$1.5 × 10^9) caused by other members of this viral sub-group. Cocao swollen shoot virus (CSSV) has been responsible for the destruction of 1.9 × 10^8 cocao trees in the last 40 years.

The pararetroviruses share many properties with the well-studied retroviruses that affect vertebrate/mammalian/avian species. For example, the capsid protein gene(s) and the genes for reverse transcriptase and other enzymatic functions are in the same order in the genome as their counterparts in the retrovirus *gag* and *pol* genes.[10] Both retroviruses and pararetroviruses encode polyproteins which are cleaved by virus encoded aspartic proteinases. There is limited sequence similarity between the retroviral aspartic proteinases and those encoded by the plant pararetroviruses (Fig. 1).

Of necessity, the plant viral enzymes contain the ~Hydrophobic–Hydrophobic–Asp–Thr/Ser–Gly~ motif of the active site and the ~Hydrophobic–Hydrophobic–Gly~ sequence which forms the thread of the psi loop.[11] They do differ from their retroviral counterparts in other respects such as in the nature of the residues contributing to the β-hairpin loop or flap.[12] There are distinct differences in the substrate specificity of the proteinases from the two families of viruses. The scissile peptide bond cleaved by retroviral enzymes such as HIV and EIAV proteinases is frequently located between two hydrophobic residues.[13] Alignment of the few cleavage sites that are known for the plant viral enzymes reveals conserved sequences that are unlike those found in the retroviral cleavage junctions. For example, the excision of RT from the polyprotein of the badna virus rice tungro bacilliform virus (RTBV),[14] one of the causative agents of rice tungro disease, requires proteinase cleavage at two sites:

N-terminal site: ~P-I-L-G-Y ↓ K-N-E-I-G~
C-terminal site: ~D-F-I-L-K ↓ C-L-Q-N-P~

The expression of various regions of the polyprotein from cocao swollen shoot virus in *E. coli* has permitted identification of a number of the sites that are cleaved by this pararetroviral proteinase:

HIV PR: P Q I T L W Q R P L V T I R I G G Q L K E
EAIV PR: V T Y N L E K R P T T I V L I N D T P L N
CSSV PR: N M L Y N L D V V H I P E V G R P I K V K

HIV PR: S L L **D T G** A D D T V L E E - - M N - - L
EAIV PR: V L L **D T G** A D T S V L T T A H Y N R L K
CSSV PR: A I L **D T G** A T T C C I N I N S V P Q T A

HIV PR: P G K W K P - - K M I G G I G G F I K V R
EAIV PR: Y R G R K Y Q G T G I I G V G G N V E T F
CSSV PR: I E Q N T F L V Q - F R G I N S T Q S V -

HIV PR: Q Y D Q I P V E I C - - - G H K A I G T V
EAIV PR: S T - P V T I K K K - - - G R H I K T R M
CSSV PR: D - - - - - - K K L - K Y G R M T I S N H

HIV PR: L V G P T P V N - - - - - - - - - - - - -
EAIV PR: L V A D I P V T - - - - - - - - - - - - -
CSSV PR: Q F R I P Y C Y A F P L S L G D G I E M -

HIV PR: I I G R N L L T Q I G C T L N F
EIAV PR: I L G R D I L Q D L G A K L V L
CSSV PR: I L G C N F I R G M Y G G L

Figure 1. Amino acid sequences of the proteinases from HIV, EIAV and cocao swollen shoot virus. The ~Hydrophobic–Hydrophobic–Asp–Thr/Ser–Gly~ motifs are shown in bold type, the ~Hydrophobic–Hydrophobic–Gly~ motifs of the psi loop are underlined with a solid line and the residues contributing to the flap are underlined with a dashed line. Gaps (indicated by a hyphen) are inserted to maximise the similarity.

1. ~G-Q-Q-M-G ↓ R-I-E-F-Y~
2. ~G-I-E-F-Y ↓ K-K-E-A-E~
3. ~E-A-E-E-L ↓ R-L-E-R-D~

The pararetroviral enzymes appear to prefer either a basic or a hydrophobic residue in P1/P1′ and either Glu/Gln or a hydrophobic residue in P3/P3′ with a preponderence of hydrophobic residues occupying the other positions. The genomes of these pararetroviruses encode only a single copy of the ~Hydrophobic–Hydrophobic–Asp–Thr/Ser–Gly~ and ~Hydrophobic–Hydrophobic–Gly~ motifs and so the proteinases must form symmetrical homodimers as in the case of retroviral enzymes. This intrinsic symmetry in the enzymes is thuis reflected in the substrate specificity preferences described above eg a basic residue can be positioned in P1 or P1′ and Glu/Gln are often found in the P3 and/or P3′ positions. Such specificity considerations may facilitate the design of specific inhibitors targetted against these viral enzymes and thus offer an effective means of combatting these costly diseases.

Aspartic proteinases have also been purified from a large number of plant species including monocotyledons[15] and dicotyledons.[16] The hydrolytic specificities of some of these enzymes have been investigated[17,18] and, like the well-characterised mammalian enzymes, the plant aspartic proteinases readily cleave between amino acids with large hydrophobic side chains. Little is known as yet, about the physiological role(s) of these enzymes in plants but it has been suggested that their expression is sometimes upregulated by either injury or infection. As a response to the stress induced by pathogens, some plants such as tomato and tobacco, produce pathogenesis-related proteins (PRPs).[19] It has been demonstrated that the expression of the aspartic proteinases in these plant species coincides with the production of PRPs and thus, it was suggested that the enzymes may be involved in PRP degradation and regulation of their biological function.[20] If then aspartic proteinases regulate the action of PRPs, what in turn regulates the activity of the aspartic proteinases? Potato plants produce a protein (Mr ~22K) which inhibits human cathepsin D but none of the other mammalian aspartic proteinases.[21] This protein shows considerable sequence identity (Fig. 2) to that of a putative aspartic proteinase inhibitor from tomato for which the gene sequence has been published.[21] Expression of the tomato gene in *Pichia pastoris* produces a recombinant protein which is able to inhibit human cathepsin D. It is not known currently whether the potato and tomato inhibitors are produced to regulate endogenous aspartic proteinases from potato/tomato plants and thus serve in a control capacity to limit the effects of proteolysis by aspartic proteinases in the plants themselves, such as has been found for example, in yeast where a low molecular weight protein inhibitor called I(A)3 regulates the vaculolar yeast proteinase A.[23] Alternatively, the inhibitors may exist to prevent the unwanted effects of aspartic proteinases being delivered exogenously, e.g., from an invading pathogen such as a fungus or a virus.

Thus there are crucial roles for aspartic proteinases not only in the pathogenesis of plants but in the responses of the plants to the pathogens. The balance between these conflicting processes may well be influenced by the apparent ability of at least some plants to produce aspartic proteinase inhibitors which may be able to act as regulators of the plant enzymes or provide a defense mechanism against pathogens. Further work is necessary to resolve this issue. However, one solution to the problem of crop loss resulting from fungal and viral attack may lie in the generation of transgenic plants. By engineering naturally-occurring aspartic proteinase inhibitors, their selectivity may be re-targetted against the enzymes of the pathogens. Expression of the genes encoding such adapted inhibitors under the control of cell/tissue specific promoters offers a realistic possibility for the generation of transgenic strains of a variety of commercially important plants that would be resistant to attack by these pathogens which devastate crops and cause widspread economic and human hardships.

```
POT.: M K C L F L L C L C L V P I V V F S S T F T
TOM.: M K C L F L L C L C L F P I V V F S S S F T

POT.: S K N P I N L P S D A T - - - P V L D V A G
TOM.: S Q N P I E L P S A S P K P N P V L D T N G

POT.: K E L D S R L S Y R I I S T F W G A L G G D
TOM.: N E L N P N S S Y R I I S T F W G A L G G D

POT.: V Y L G K S P N S D A P C A N G I F R Y N S
TOM.: V Y L G K S P R S S A P C L D G V F R Y N S

POT.: D V G P S G T P V R F S H F G Q G I F E N E
TOM.: D V G T V G T P V R F I P L S G G I F E D Q

POT.: L L N I Q F A I S T S K L C V S Y T I W K V
TOM.: L M D L Q F N I A T V K L C V S Y T I W K A

POT.: G D Y D A S L G T M L L E T G G T I G Q A D
TOM.: G N L N A Y Y R A M L L E T E G T I G Q V D

POT.: S S W F K I V K S S Q F G Y N L L Y C P V T
TOM.: S S Y F K I V K A S T F G Y N L L Y C P I T

POT.: S T M S C P F S S D D Q F C L K V G V V H Q
TOM.: R P V L C P F C R G D G F C A K V G V I N Q

POT.: N G K R R L A L V K D N P L D V S F K Q V Q
TOM.: D G R R R L A L V N E N P L G V Y F K K V -
```

Figure 2. Amino acid sequences of the aspartic proteinase inhibitors from potato and tomato. Identical residues are in bold type. Gaps (indicated by a hyphen) are inserted to maximise the apparent similarity.

REFERENCES

1. Hay, J., Grieco, F., Arnis, D., Pinner, M., Lee, S.C. and Hull, R. (1994) Virology 205: 430–437.
2. Hill, J. and Phylip, L.H. (1997) FEBS Lett. (m/s in submission).
3. Jara, P., Gilbert, S., Delmas, P., Guillemot, J.C., Kaghad, M., Ferrara, P. and Loison, G. (1996) Mol. and Gen. Genet. 250: 97–105.
4. Staples, R.C. and Mayer, A.M. (1995) FEMS Microbiology Lett. 134: 1–7.
5. Movahedi, S. and Heale, J.B. (1990) Physiol. and Mol. Plant Path. 36: 303–324.
6. Movahedi, S. and Heale, J.B. (1990) Physiol. and Mol. Plant Path. 36: 289–302.
7. Hull, R. (1994) Euphytica 75: 195–205.
8. Harrison, B.D., Finch, J.T., Gibbs, A.J., Hollings, M., Shepherd, R.J., Valenta, V. and Wetter, C. (1971) Virology 45: 356–363.
9. Lockhart, B.E.L. (1990) Phytopathology 80: 127–131.
10. Torruella, M., Gordon, K. and Hohn, T. (1989) EMBO J. 8: 2819–2825.
11. Pearl, L.H. and Taylor, W.R. (1987) Nature 329: 351–354.
12. Gustchina, A., Kervinen, J., Powell, D.J., Zdanov, A., Kay, J. and Wlodawer, A. (1996) Protein Sci. 5: 1453–1465.
13. Powell, D.J., Bur, D., Wlodawer, A., Gustchina, A., Payne, S.L., Dunn, B.M. and Kay, J. (1996) Eur. J. Biochem. 241: 664–674.
14. Laco, G.S., Kent, S.B.H. and Beachy, R.N. (1995) Virology 208: 207–214.
15. Asakura, T., Watanabe,H., Abe, K and Arai, S. (1995) Eur. J. Biochem. 232: 77–83.
16. Voigt, J., Kamaruddin, S., Wrann, D., Senyuk, V. and Biehl, B. (1995). J. Plant Physiol. 145: 299–307.
17. Verissimo, P., Faro, C., Moir, A.J.G., Lin Y., Tang, J., Pires, E. (1996). Eur. J. Biochem. 235: 763–768.
18. Kervinen, J., Sarkkinen, P., Kalkkinen, N., Mikola, L. and Saarma, M. (1993) Phytochemistry 32: 799–803.
19. Rodrigo, I., Vera, P. and Conejero, V. (1989) Eur. J. Biochem. 184: 663–669.
20. Rodrigo, I., Vera, P., Van Loon, L.C. and Conejero, V. (1991) Plant Physiol. 95: 616–622.
21. Strukelj, B., Pungercar, J., Ritonja, A., Krizaj, I., Grubensek, F. Kergar, I. and Turk, V. (1990) Nucl. Acids Res. 18: 4605.
22. Werner, R., Guitton, M.C. and Muhlbach, H.P. (1993) Plant Physiol. 103: 1473.
23. Dreyer, T., Valler, M.J., Kay, J., Charlton, P. and Dunn, B.M. (1985) Biochem. J. 231: 777–779.

61

CRYSTALLISATION, STRUCTURE SOLUTION, AND INITIAL REFINEMENT OF PLANT CARDOSIN-A

I. Bento,[1] R. Coelho,[1] C. Frazão,[1] J. Costa,[1] C. Faro,[2] P. Veríssimo,[2] E. Pires,[2] J. Cooper,[3] Z. Dauter,[3] K. Wilson,[3] and M. A. Carrondo[1,5]

[1]ITQB
Apt. 127, 2780-Oeiras, Portugal
[2]Dep. Bioquímica
Fac. Ciências e Tecnologia
Univ. Coimbra
3000 Coimbra, Portugal
[3]Lab. Molecular Biology
Department of Crystallography
Birkbeck College
Malet Street, London WC1E 7HX, United Kingdom
[4]EMBL
c/o DESY, Notkestrasse 85, 22603 Hamburg, Germany
[5]IST
1000 Lisboa, Portugal

1. INTRODUCTION

In the cheese industry, calf chymosin is universally used as one of the main milk-clotting enzymes. However, in traditional cheese manufacturing, several other coagulating agents have also been used for a long time. A special case in Portugal, involves the use of an enzyme preparation obtained by aqueous extraction of cardoon flowers (*Cynara cardunculus* L.) [1]. Cardosin A, one of its components, is a two chains (31 + 15 KDa) glycosylated aspartic proteinase. Like chymosin, cardosin A cleaves bovine κ-casein between Phe105-Met106 initiating milk-clotting. Similarly to all other known plant aspartic proteinases it is expressed with an extra segment of about 100 residues, which bears no sequence similarity with the aspartic proteinases of mammalian or microbial origins [2]. The total amount of sugar content, present in the mature form of cardosin A, is calculated to correspond to *ca.* 9.6% of its molecular weight [3].

Aspartic Proteinases, edited by James.
Plenum Press, New York, 1998.

2. CRYSTALLISATION

Crystals of glycosylated cardosin A were obtained in PEG 4K 40%, with 0.2M of ammonium acetate and 0.1M of sodium citrate pH 5.5, with a protein concentration of 12mg/ml, using the sitting-drop vapour-diffusion method. They took three months to appear, usually as small needles or as agglomerated thin plates (Fig. 1a). Macroseeding from the plates originated diffracting crystals (Fig. 1b), after about two months.

Crystallisation of complexes with inhibitors has been attempted to study the enzyme specificity with the hope of improving crystal properties. A cardosin A complex with a statine analogue (CP-81,282 [4]) could be crystallised within ten to fifteen days. The macroseeding is again needed to obtain diffracting crystals.

3. DIFFRACTION DATA

Synchrotron diffraction data of native cardosin A to 2.85 Å resolution were collected at the EMBL outstation in Hamburg and were processed with DENZO/SCALEPACK [5]. Table 1 presents their relevant crystallographic and statistics data.

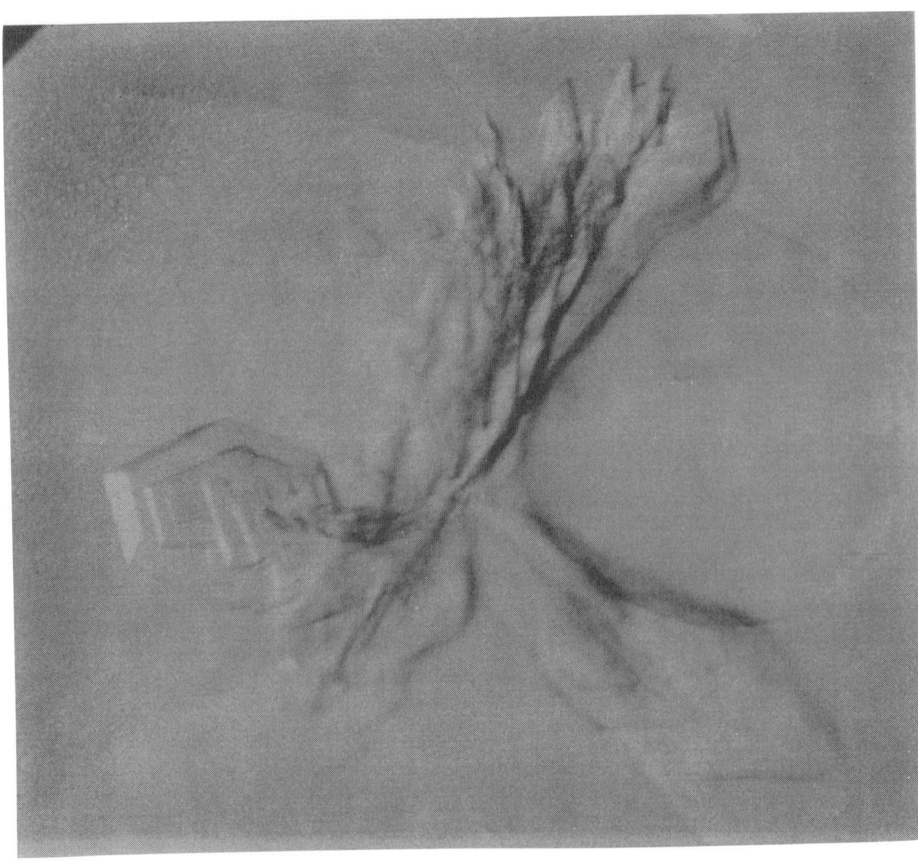

Figure 1a. Cardosin A initial crystals. Fragments of the set of parallel plates, were further used in macro-seeding.

Table 1. Crystallographic data

Space group	C2
Cell dimensions	a=118.9, b=88.1, c=82.3, β=104.4
Resolution (Å)	15.0–2.85
Collected intensities	63522
Unique intensities	18976
Redundancy (%)	3.35
Overall statistics (15.00–2.85 Å) (%)	
R	7.6
I>3sigma(I)	85.1
Signal-noise	14.77
Last shell statistics (2.90–2.85 Å) (%)	
R	23.0
I>3sigma(I)	66.0
Signal-noise	5.75

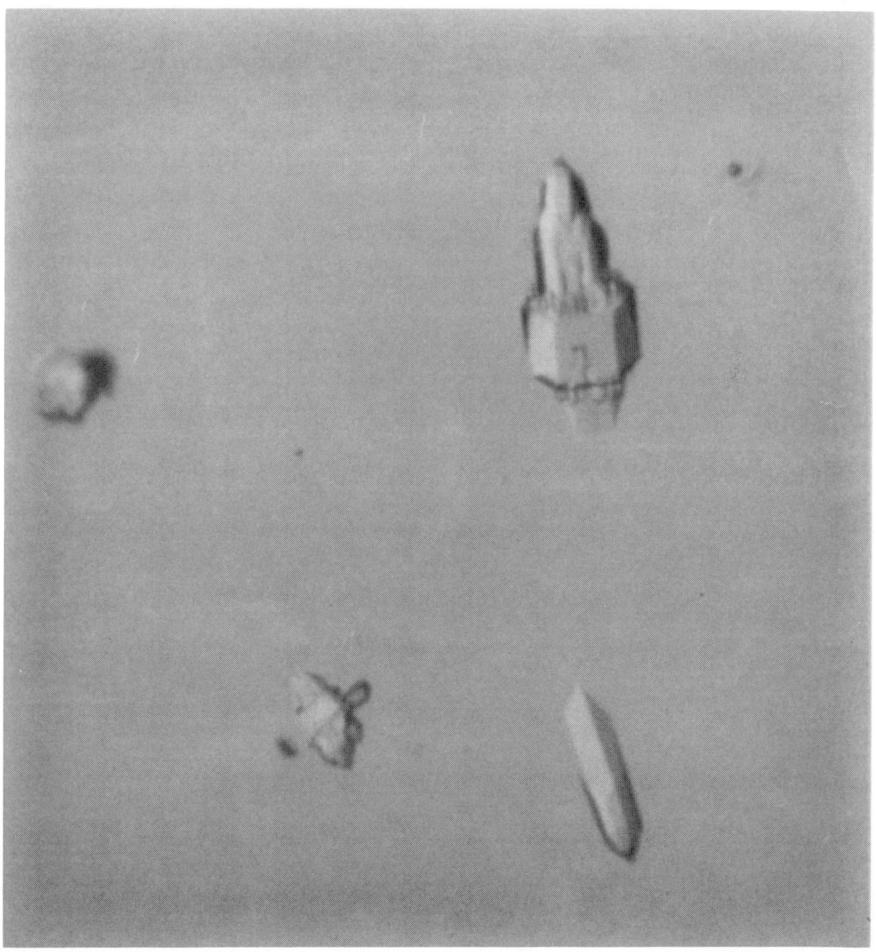

Figure 1b. Cardosin A seeded crystals (biggest dimension reaching *ca.* 1mm).

4. MOLECULAR REPLACEMENT

The molecular replacement problem was solved with AMoRe [7] using the human cathepsin D structure [8] as search model (sequence homology of 44.4% with cardosin A). The solution for the two molecules had a correlation coefficient of 40% (47% above the first noise solution) and was obtained with 15.0–3.5 Å resolution data (Fig. 2).

5. PACKING

The packing of the eight molecules in the unit cell was checked on a graphics workstation using the program O [9]. The packing shows reasonable contacts with no significant clashes and solvents channels are observed in some directions. The non-crystallographic symmetry relates the two molecules in the asymmetric unit with extensive surface contacts (Fig. 3).

6. REFINEMENT

Refinement was carried out with X-PLOR in a first stage, using constrained non-crystallographic symmetry and the simulating annealing protocol [10]. A target sequence

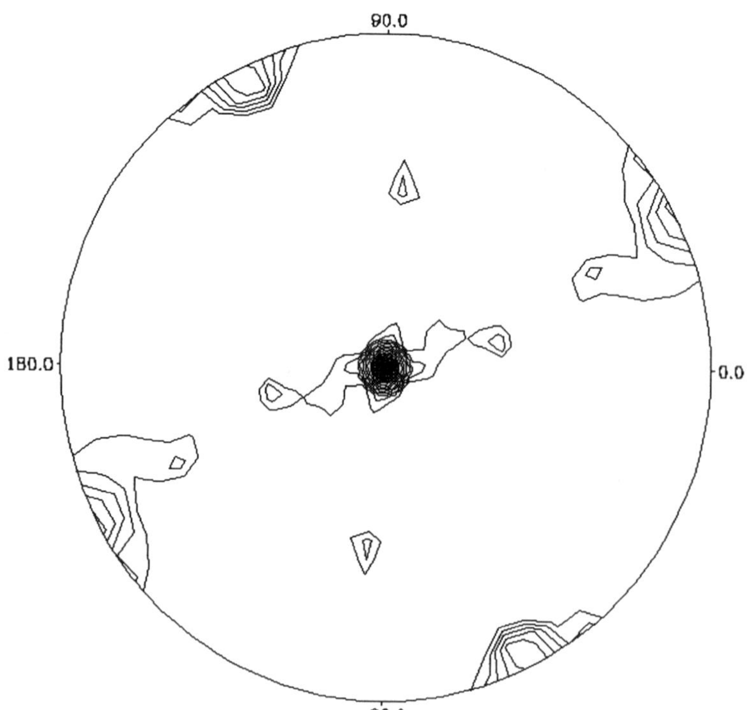

Figure 2. The stereographic projection down the **b** axis of the self rotation function, calculated with program POLARRFN [6], has a peak (5.1σ, ω=92.0, φ=117.5, κ=180.0º) showing a non-crystallographic two-fold rotation axis, deviating *ca.* 13º from axis **a**.

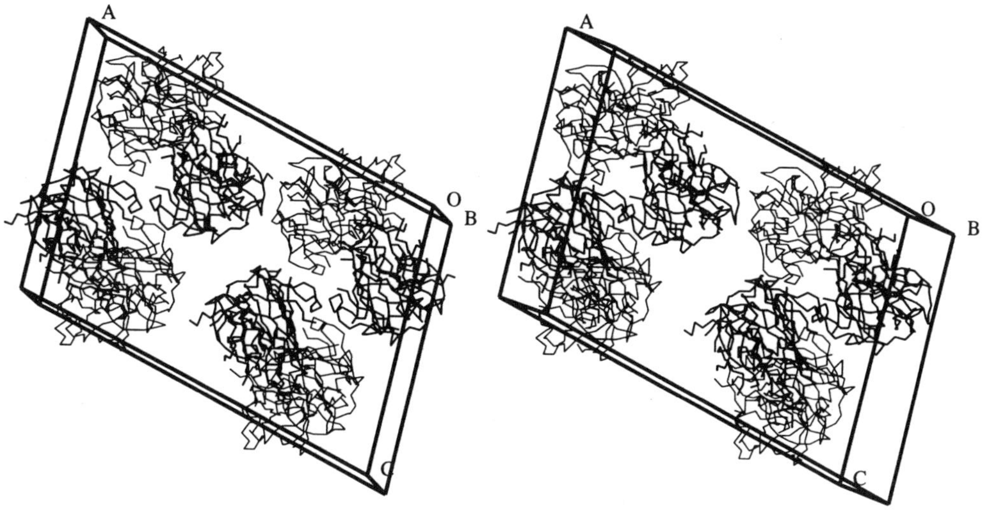

Figure 3. Packing of the non-crystallographic dimers in the unit cell.

composed partially from cardosin A [2] and partially from cyprosin [11] sequences was used in early refinement stages. Initially, the non-identical amino-acids between the target sequence and the human cathepsin D model were mutated into alanines. These were changed to the expected sequence along with the refinement progress. In the latest refinement stage (seventh cycle in Table 2) the complete sequence of cardosin A was used. The refinement was performed with SHELXL96 [12] making use of the restrained non-crystallographic symmetry (restraints for equivalent 1–4 distances) and restraining isotropic temperature factors (restraints between neighbouring atoms and between equivalent atoms of the two independent molecules in the asymmetric unit). Table 2 presents an overview of the refinement parameters.

Electron density maps are generally rather clear. Examination of these maps (Figs. 4 and 5) with O [9] allowed the localisation of the two glycosylation sites. As their primary

Table 2. Refinement progress of cardosin A[a]

Model	Sequence completeness (%)	Refinement with X-Plor CNCS-SA	
		R (%)	R_{free} (%)
First	72.0	46.7 ⇒ 34.5	47.5 ⇒ 41.5
Second	97.6	36.6 ⇒ 31.4	40.2 ⇒ 38.1
Third	95.8	33.6 ⇒ 29.4	37.4 ⇒ 35.0
Fourth	91.2	32.6 ⇒ 29.4	35.9 ⇒ 35.8
Fifth	83.3	32.1 ⇒ 29.6	36.7 ⇒ 35.7
Sixth	89.2	32.7 ⇒ 28.9	36.6 ⇒ 33.7
Seventh*	90.5	32.3 ⇒ 21.4	78.1 ⇒ 29.2

[a]The successive models were obtained after the fit to the electron densities maps. All data (15.00–2.85 Å) were used, B factors were set to 15 Å^2. The last refinement cycle (*) was performed using the program SHELXL96 with restrained non-crystallographic symmetry and restrained temperature factors refinement. No solvent molecules were added so far.

structure was already known [3], several sugar rings were inserted and refined satisfactorily. The plant insertion could not be seen in the density maps. The overall fold is that of a typical aspartic proteinase (Fig. 6). The recently obtained complete sequence of cardosin A and a successful crystal optimisation will enable to extend the resolution, complete the refinement and analyse the structure in more detail.

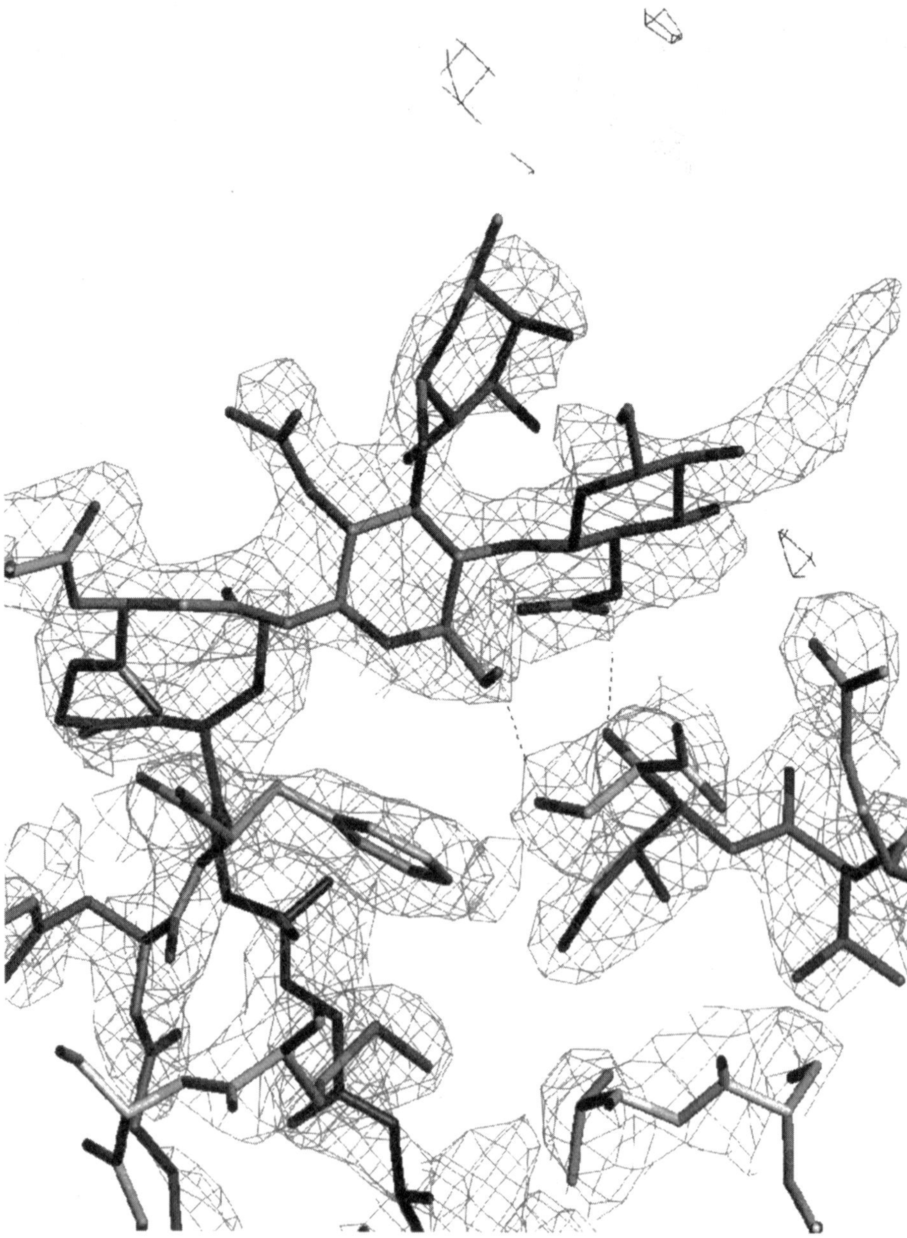

Figure 4. Three NAG rings were modelled at the Asn71 site. Further visible electron density will possibly require the insertion of another sugar ring. The sugar chain spreads over the protein surface, involving many sugar–protein and sugar–sugar interactions. Two hydrogen bonds between the sugar chain and the protein are indicated as dashed lines.

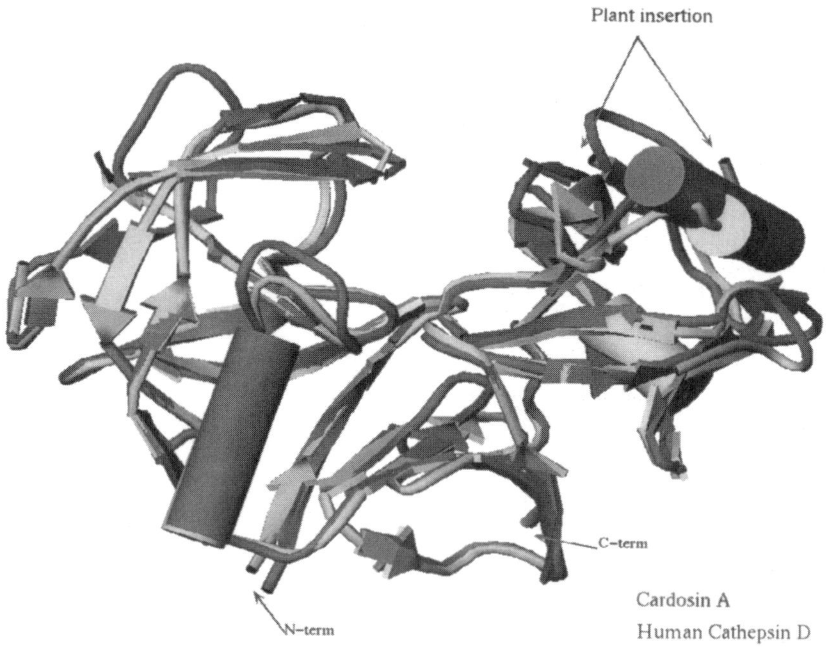

Figure 5. The fit between cardosin (light grey) and human cathepsin D (dark grey) secondary structures reveals that the typical overall fold of aspartic proteinases is observed by the plant member of this family. As expected, differences occur in the plant insertion region (which is absent in the crystallised mature form of cardosin A) and in the shape of some loops.

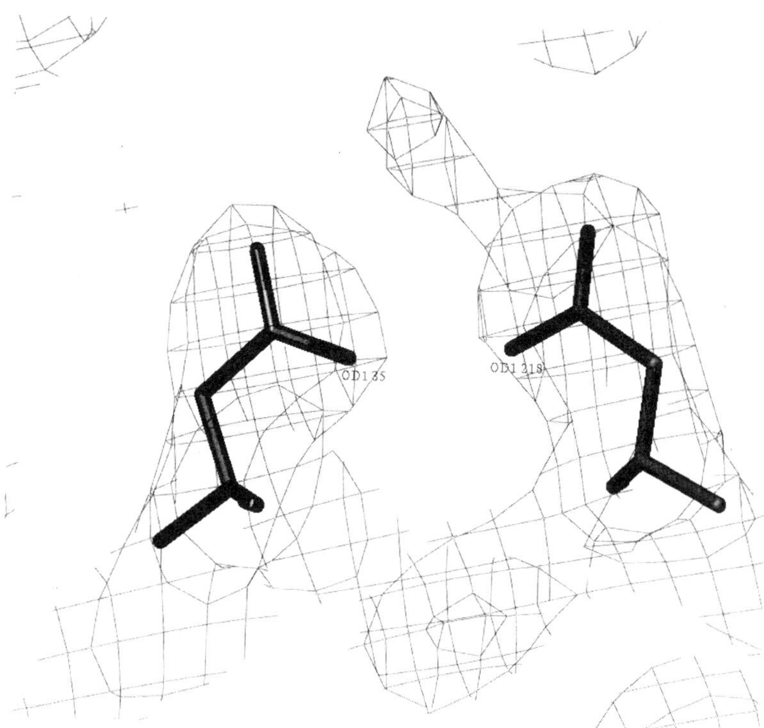

Figure 6. The active-site water, characteristic of crystallographic aspartic proteinase structures, is already visible in the maps of cardosin A, at the present level of refinement (R = 21.4%, averaged map, 3.2σ).

REFERENCES

1. C. Faro and E. Pires et al. (1995) in Aspartic Proteinases: Structure, Function, Biology, and Biomedical Implications. (K. Takahashi ed.), pp. 373–377, Plenum Press.
2. P. Veríssimo and E. Pires *et al.* (1996) *Eur. J. Biochem* **235**, 762–768.
3. J. Costa & M. A. Carrondo *et al.* (1997) *Eur. J. Biochem* **243**, 695–700.
4. D. Bailey and J.Cooper (1994) *Protein Science* **3**, 2129–2143.
5. D. Gewirth, Z. Otwinowski and W. Minor (1995) *The HKL Manual*, Department of Molecular Biophysics and Biochemistry, Yale Yniversity, U.S.A.
6. CCP4 (1979), The SERC (UK) Collaborative Computing Project No.4, a Suite of Programs for Protein Crystallography, distributed from Daresbury Laboratory, Warrington WA4 4AD, UK.
7. J. Navaza (1994) *Acta Cryst., Sect. A*, **50**, 157–163.
8. E.T. Baldwin *et al.* (1993) *Proc. Nat. Acad. Sci. USA* **90**, 6796–6800.
9. [7] A. Jones and M. Kjeldgaard *et al.* (1991) *Acta Cryst. A,* **47**, 110–119.
10. A. T. Brünger (1993). *X-PLOR, Version 3.1.* Yale University, New Haven, CT.
11. M. C. Cordeiro, P. E. Brodelius *et al.* (1994), *Plant. Mol.Biol.* **24**, 733–741.
12. G. M. Sheldrick, T. R. Schneider, *Methods. Enzymol., in press.*

62

ACID PROTEINASE FROM *Nepenthes distillatoria* (BADURA)

Senarath B. P. Athauda,[1,2] Hideshi Inoue,[1] Akihiro Iwamatsu,[3] and Kenji Takahashi[1]

[1]Laboratory of Molecular Biochemistry
Molecular Science Division, School of Life Science
Tokyo University of Pharmacy and Life Science
1432-1, Horinouchi, Hachioji, Tokyo 192-03, Japan
[2]Department of Biochemistry
Faculty of Medicine
University of Peradeniya
Peradeniya, Sri Lanka
[3]Central Laboratories for Key Technology
Kirin Brewery Co. Ltd.
Yokohama, Kanagawa 236, Japan

1. INTRODUCTION

Plant aspartic proteinases have so far received much less attention in contrast to the well characterized mammalian, fungal and viral aspartic proteinases.[1] They are widely distributed in the plant kingdom, and have been detected in seeds, leaves and flowers of different plants as well as in the digestive fluid of some insectivorous species.[1] Aspartic proteinases from barley, rice and cardoon flower have been well characterized and their cDNA-derived primary structures have been reported.[2–4] However, only a few studies were reported on proteinases of insectivorous plants.[5–8] Insectivorous plant Nepenthes distillatoria is available in Sri Lanka in a large quantity and will be a good source of the insectivorous plant proteinases. They are interesting not only from the view point of plant physiology, but also from the view point of structure/function relationship and molecular evolution of aspartic proteinases.

In this study an acid proteinase was purified from the juice of the *Nepenthes* pitchers and its molecular and enzymatic properties were investigated including the NH_2-terminal and internal amino acid sequences and the cleavage specificity toward the B chain of oxidized insulin at pH 3.0.

Aspartic Proteinases, edited by James.
Plenum Press, New York, 1998.

2. MATERIALS AND METHODS

2.1. Materials

Nepenthes juice was collected from the pitchers of *Nepenthes distillatoria* at the forest of Sri Lanka. DEAE-cellulose (DE-52) was a product of Whatman. Sephacryl S-200 and a Mono Q HR5/5 column were purchased from Pharmacia. The B chain of oxidized bovine insulin was obtained from Sigma and reagents for automated amino acid analysis and sequencing were from Applied Biosystems. Other reagents used were of the highest grade available.

2.2. Purification of an Acid Proteinase

Proteolytic activity of the enzyme was determined by the method of Anson[9] with a slight modification[10] using 2% denatured hemoglobin in 0.1M formic acid buffer at pH 3.0 as a substrate.

The collected juice was filtered and dialyzed against 0.02M phosphate buffer, pH 7.5. The proteins in the dialyzed *Nepenthes* juice were adsorbed to DEAE-cellulose gel by batch adsorption and the gel was placed in a column (6 × 70 cm). The column was washed with 0.02M sodium phosphate buffer, pH 7.5, and the protein was eluted with the same buffer containing 0.5M NaCl. The fractions with proteolytic activity were pooled and applied to a second DEAE-cellulose column (3 × 31 cm) equilibrated with 0.02M sodium phosphate buffer, pH 7.5. The column was washed with the same buffer and the protein was eluted with a linear gradient of 0 to 0.5M NaCl in 2 l of the same buffer. The active fractions were pooled and concentrated by using a small DEAE-cellulose column. The concentrated sample was applied to a Sephacryl S-200 column (3.4 × 111 cm) equilibrated and eluted with 0.02M sodium phosphate buffer, 0.2M NaCl, pH 7.5. The active fractions were pooled and dialyzed against 0.02M sodium acetate buffer, pH 4.0, and applied to a pepstatin-Sepharose column (1.3 × 1.6 cm) equilibrated with the same buffer. The column was washed and the protein was eluted with 0.02M Tris/HCl buffer, pH 8.0, containing 1M NaCl. The active fractions were pooled and dialyzed against 0.02M Tris/HCl buffer, pH 7.8, and applied to a Mono Q column. The protein was eluted with a linear gradient of 0 to 1 M NaCl in 0.02M Tris/HCl buffer, pH 7.8.

2.3. Characterization of the Acid Proteinase

The molecular weight of the purified enzyme was determined by SDS-PAGE and gel filtration. The optimum pH and temperature of the enzyme were determined by analyzing the enzyme activity at different pHs or temperatures. To determine the cleavage specificity of the enzyme, the B chain of oxidized insulin was digested with the purified proteinase and the resulting peptides were separated by HPLC using a Hitachi 655A-11/ L-5000 system on a column (0.46 × 25 cm) of TSKgel ODS-120T (Tosoh Co.). The peptides were eluted with a linear gradient of acetonitrile (0–50% in 60 min) in 0.1% trifluoroacetic acid at a flow rate of 0.8ml/min. The effluent was monitored by measuring the absorbance at 215 nm and the peptide peak fractions were lyophilized. An aliquot of each peptide fraction dissolved in water was submitted to amino acid analysis using an Applied Biosystems automated derivatizer analyzer (420A/130–920A) equipped with an on-lined automatic acid hydrolysis apparatus.

2.4. Determination of Partial Amino Acid Sequences of the Acid Proteinase

The NH_2-terminal amino acid sequence was determined by using an Applied Biosystems pulse-liquid protein sequencer model 477A. To determine the partial internal sequence of the enzyme, it was reduced and carboxymethylated and digested with endoproteinases (endopeptidase Asp-N, trypsin and *Staphylococcus aureus* V_8 protease) and the resulting peptides were separated by HPLC in the same manner as described above. An aliquot of each peptide fraction dissolved in water was submitted to automated amino acid sequencing.

3. RESULTS AND DISCUSSION

An acid proteinase was purified from the juice of *Nepenthes distillatoria* by successive chromatograpies on columns of DEAE-cellulose, Sephacryl S-200, pepstatin-Sepharose and Mono Q with a 200-fold purification (Table 1). About 2 mg of the acid proteinase could be purified from 30 l of the *Nepenthes* juice. The purified enzyme was shown to be homogeneous as examined by SDS-PAGE (Figure 1a) and had a specific activity of 872 units/mg of enzyme toward the substrate hemoglobin (Table 1).

The molecular mass of the purified proteinase was estimated to be 45 kDa by SDS-PAGE under reducing conditions (Figure 1a) and 58 kDa by gel filtration. The molecular mass of 45 kDa is similar to those of barley,[2] rice,[3] and cardoon[4] as estimated from SDS-PAGE and/ or cDNA sequence.[1] Upon SDS-PAGE followed by periodic acid-Schiff's base staining, the protein band was stained well, suggesting the presence of carbohydrate (Figure 1b).

The maximum activity of the enzyme toward denatured hemoglobin was observed at pH 2.0–3.0 (Figure 2) and at temperature 50–60°C (Figure 3). The proteinase activity at the acidic pH was completely inhibited by 0.1mM pepstatin, suggesting that it belongs to the aspartic proteinase family. Further the enzyme was stable at neutral pH where pepsin loses its activity, and it was relatively stable at higher temperatures (up to 55°C) (Figure 4).

Cleavage specificity of the proteinase was investigated by using the B chain of oxidized insulin as the substrate. Several peptide bonds, especially Phe_{24}–Phe_{25}, Glu_{13}–Ala_{14}, Leu_6–Cys_7, Leu_{15}–Tyr_{16} and Tyr_{16}–Leu_{17} bonds, were simultaneously cleaved to marked

Table 1. Purification of *Nepenthes* acid proteinase

Step	Total protein (mg)	Total activity (units)	Relative activity (units/mg)	Yield (%)
Crude juice	1,768	7,708	4.4	100
1st DEAE-52	500	5,575	11.2	72.0
2nd DEAE-52				
Major peak*	99	2,595	26.2	34.0
Minor peak	18	1,137	62.5	14.8
Sephacryl S-200	13.6	2,084	152.8	27.0
Pepstatin-Sepharose	2.5	1,809	717.8	24.0
FPLC (Mono Q)	1.8	1,600	872.4	21.0

*The major peak fraction was used for subsequent purification and characterization.

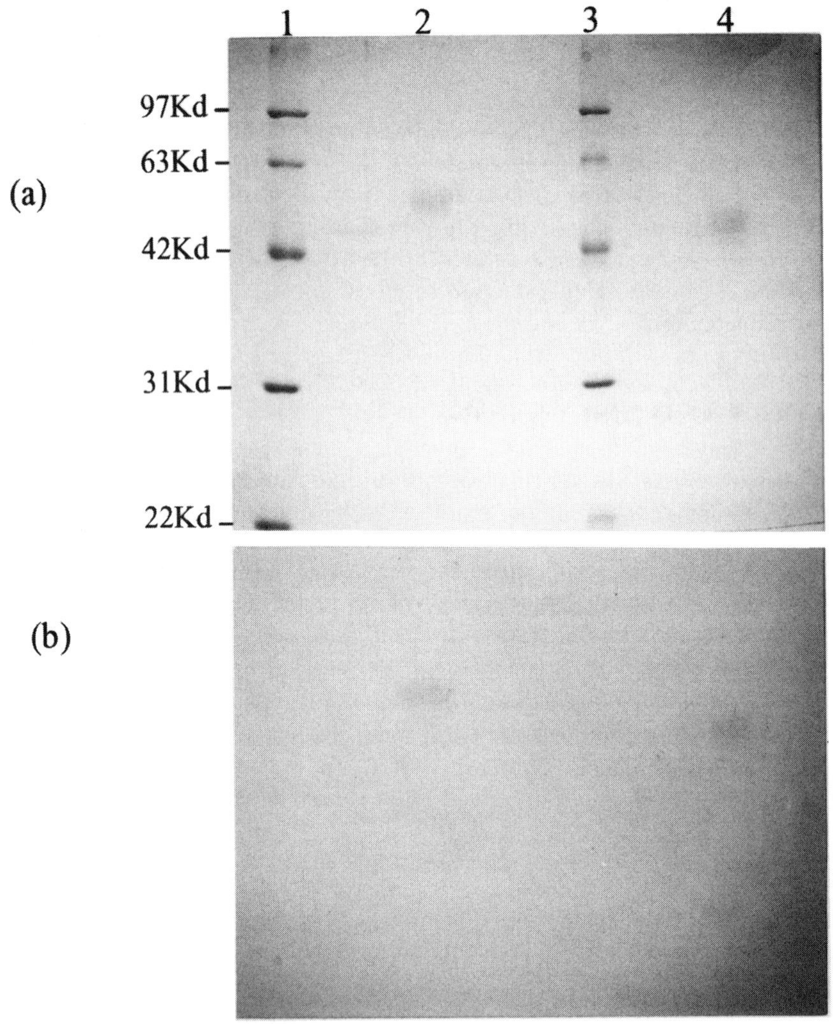

Figure 1. SDS-PAGE of the purified enzyme. (a) Coomassie brilliant blue R-250 staining, (b) periodic acid-Schiff's base staining. Lanes 1 and 3: low molecular weight standard proteins (Pharmacia) under reducing and non-reducing conditions, respectively. Lanes 2 and 4: purified *Nepenthes* acid proteinase under reducing and non-reducing conditions, respectively.

extents by the proteinase. The extents of cleavage of these bonds were estimated to be 80%, 55%, 50%, 38% and 35%, respectively, under the conditions used. Previously rather restricted cleavage specificity of an acid proteinase in Nepenthes was reported by Amagase et al.[6,7] However, the cleavage specificity of the present enzyme toward the B chain of oxidized insulin was rather wide. It was similar to, but not identical with those of human pepsin and cathepsin D. Especially, the relatively rapid hydrolysis of Leu_6–Cys_7 seems to be characteristic of this enzyme since this bond is known to be insensitive to pepsin and most other aspartic proteinases.[11]

The NH_2-terminal amino acid sequence was determined up to the 24th residue as follows: IGPSGVETTVYAGDGEYLMNLSIG. This sequence has little homology with the NH_2-terminal sequences of barley,[2] rice,[3] and cardoon flower.[4]

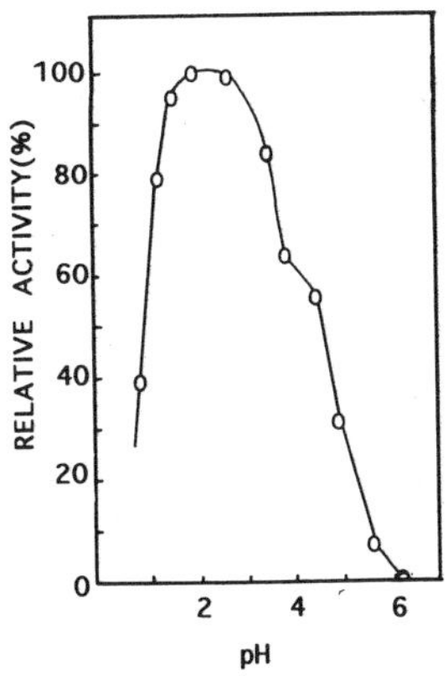

Figure 2. pH dependence of the enzyme activity toward hemoglobin.

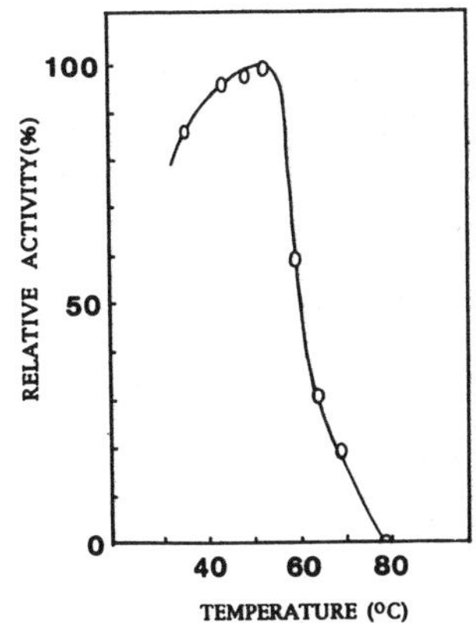

Figure 3. Temperature dependence of the enzyme activity toward hemoglobin.

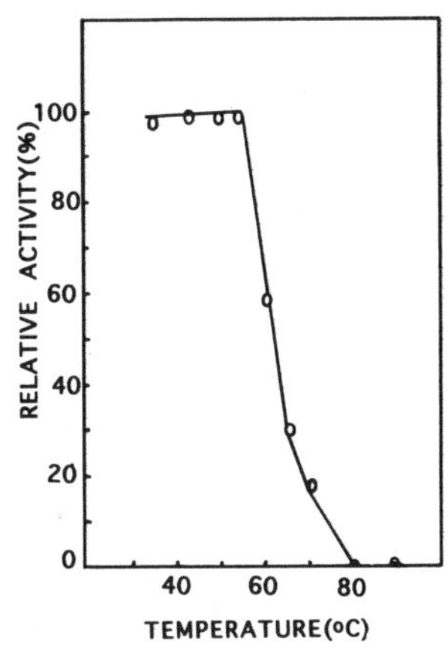

Figure 4. Thermal stability of the enzyme. About 10 μg of enzyme in 0.05M Tris/HCl buffer at pH 8.0 was incubated at different temperature for about 1 h and the remaining activity toward hemoglobin was determined at pH 3.0 and 37°C.

In addition, approximately 30% of the internal amino acid sequence of the proteinase has been determined so far (data not shown) and the two putative active site aspartic acid residues have been identified, indicating that the enzyme belongs to the aspartic proteinase family. It is interesting to note that both aspartic acid residues are present in the sequence Asp–Thr–Gly as in most other aspartic proteinases, whereas in the above three plant as-partic proteinases, one of the catalytic site sequence has been reported to be Asp–Ser–Gly.[2,4] The partial internal amino acid sequence of *Nepenthes* acid proteinase determined so far showed homology with other plant aspartic proteinases of known sequence although the homology was lower than those among the other plant aspartic proteinases. Computer homology search of the partial sequence with the protein data bank also indicated that the enzyme has much less homology with other known aspartic proteinases.

4. CONCLUSION

Nepenthes acid proteinase is a glycoprotein possessing fairly distinguished characteristics, regarding molecular size, optimum temperature, pH and thermal stabilities and cleavage specificity. Its partial amino acid sequence indicates that it belong to the aspartic proteinase family; however, it has less homology with the sequences of other known aspartic proteinases. These properties of *Nepenthes* proteinase suggests that it is a unique proteinase among plant and other aspartic proteinases.

REFERENCES

1. J. Kervinen, K. Törmäkangas, P. Runeberg-Roos, K, Guruprasad, Tom Blundell and T.H. Teeri, Structure and possible function of aspartic proteinases in barley and other plants, in : Aspartic Proteinases. Structure, Function, Biology, and Biomedical Implications (Takahashi, K., ed.) Plenum Press, New York pp 241–255 (1993).
2. P. Runeberg-Roos, K. Törmäkangas and A. Ostman, Primary structure of a barley grain aspartic proteinase—a plant aspartic proteinase resembling mammalian cathepsin D, Eur. J. Biochem. 202, 1021–1027 (1991).
3. H. Hashimoto, R. Nishi, H. Uchimiya and A. Kato, Nucleotide sequence of a cDNA encoding aspartic proteinase in rice, Accession number D 12777, submitted to the DNA database of Japan (August 1, 1992).
4. M. Cordeiro, Z. T. Xue, M. Pietrzak, M.S. Pais and P. Brodelius, Nucleotide sequence of a cDNA encoding cynarase and its tissue specific expression, in : Aspartic Proteinases. Structure, Function, Biology, and Biomedical Implications (Takahashi, K., ed.) Plenum Press, New York pp 367–372 (1993).
5. S. Nakayama and S. Amagase, Acid proteinase in *Nepenthes*. Partial purification and properties of the enzyme, Proc. Japan Acad., 44, 358–362 (1968).
6. S. Amagase, S. Nakayama and A. Tsugita, Acid proteinase in *Nepenthes*, J. Biochem. (Tokyo), 66, 431–439 (1969).
7. S Amagase, Digestive enzymes in insectivorous plants. Acid proteinases in the genus *Nepenthes* and *Drosera peltata*, J. Biochem. (Tokyo), 72, 73–81 (1972).
8. K. Takahashi, W.-J. Chang and J.-S. Ko, Specific inhibition of acid proteinase from brain, kidney, skeletal muscle, and insectivorous plants by diazoacetyl-DL-norleucine methyl ester and by pepstatin. J. Biochem. (Tokyo), 76, 897–899 (1974).
9. M.L. Anson, The estimation of pepsin, trypsin, papain and cathepsin with hemoglobin, J. Gen. Physiol. 22, 77–89 (1938).
10. S.B.P. Athauda, M. Tanji, T. Kageyama and K. Takahashi, A comparative study on the NH_2-terminal amino acid sequences and some other properties of human pepsinogens and pepsins, J. Biochem. (Tokyo), 106, 920–927 (1989).
11. B. Keil, Specificity of proteolysis, Springer, Berlin (1992).

63

A COMPARATIVE STUDY ON THE ASPARTIC PROTEINASES FROM DIFFERENT SPECIES OF CYNARA

Paula Veríssimo, Miguel Ramalho-Santos, Carlos Faro, and Euclides Pires

Departamento de Bioquímica
Universidade de Coimbra
Apartado 3126, 3000 Coimbra, Portugal

1. INTRODUCTION

Aspartic proteinases have been detected in seeds, leaves and in the digestive fluid of some carnivorous plants [1]. The physiological role of these enzymes is not yet as well elucidated as the functions of mammalian or retroviral aspartic proteinases. It has been suggested that they may be involved in hydrolysis of storage and extracellular proteins, but in most cases the evidence is still circumstantial. Recently, aspartic proteinases were also identified in flowers of the cardoon *Cynara cardunculus* L. [2]. Two of them, named cardosin A and cardosin B, are specifically expressed in the female part of the flower (the structure of the pistil is shown in [3]) during the early stages of floral development. Cardosin A is accumulated in protein storage vacuoles of the superficial cell layers of the stigma and style [4] whereas cardosin B is found in the extracellular matrix of the transmitting tissue [3]. A differential localization such as this strongly supports the idea that the two cardosins may have different functions in the flower, although they are probably involved in the same physiological processes. In addition to an involvement of cardosin A in defense against pathogens and/or insects, both cardosins are also thought to participate in plant sexual reproduction. As compared to the proposed functions for other plant aspartic proteinases, this is a unique characteristic of cardosins, raising the question of whether aspartic proteinases may also performed the same function in flowers of other angiosperms. In a first attempt to answer this question we started a screening of aspartic proteinases in flowers of related plants within the Cynarae tribe. Here, we describe the isolation and characterization of cardosin-like enzymes from two species of *Cynara*, namely *C. humilis* and *C. scolymus*. A comparison between these enzymes and cardosins is also presented.

Aspartic Proteinases, edited by James.
Plenum Press, New York, 1998.

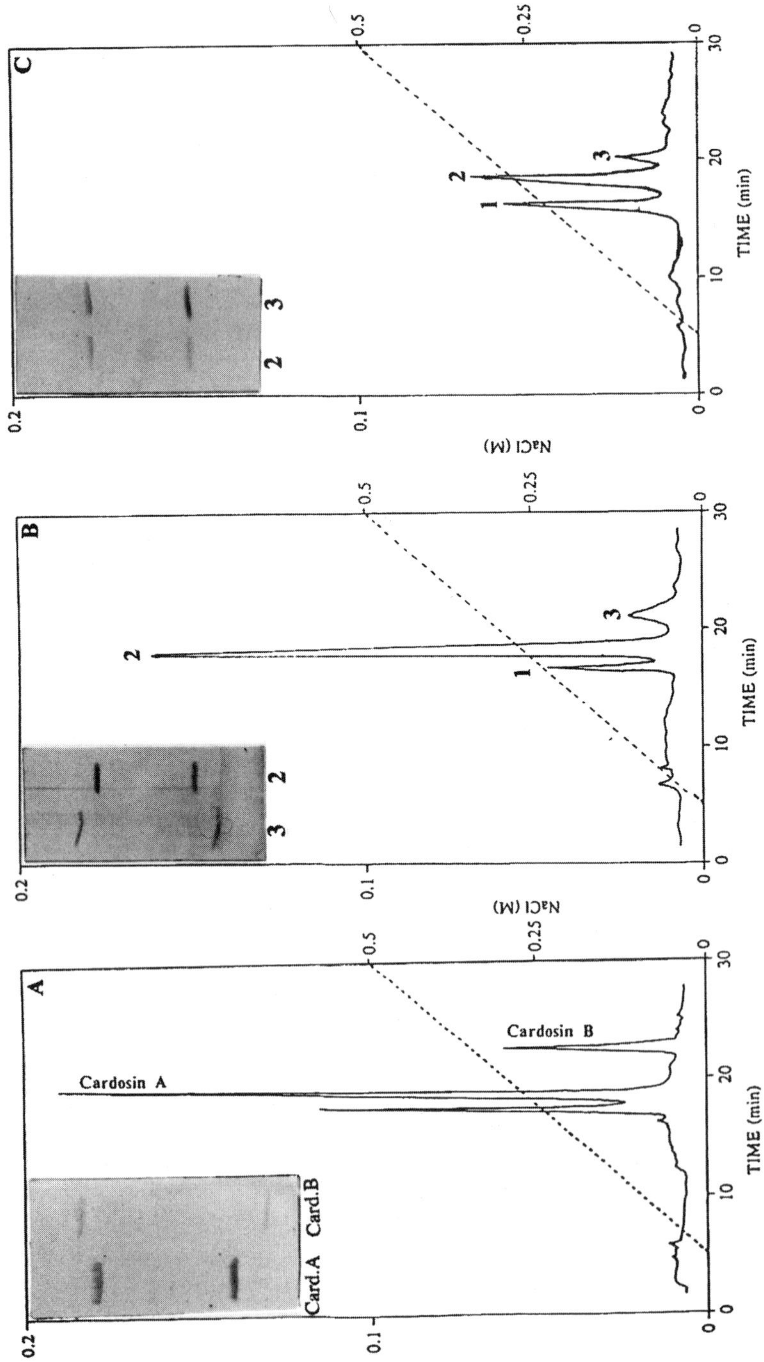

Figure 1. Purification of aspartic proteinases from different species of *Cynara*; mono Q elution profiles and SDS-PAGE patterns of: (A) *Cynara cardunculus*, (B) *Cynara scolymus*, (C) *Cynara humillis*.

2. PURIFICATION OF ASPARTIC PROTEINASES FROM *C. humilis* AND *C. scolymus*

Aspartic proteinases were isolated from pistils of *C. humilis* and *C. scolymus* by a purification procedure similar to the one previously described for cardosins, the aspartic proteinases from *C. cardunculus* L [2]. The proteinases were extracted at pH 3.0 and most protein contaminants were eliminated at this stage of the purification procedure. Similarly to what was found for cardosins, the aspartic proteinases from both species account for the majority of the soluble protein in pistils and for this reason purification was achieved in two chromatographic steps. Non-protein contaminants were eliminated by gel filtration on Superdex 200, from which the proteolytic activity is recovered, in both cases, as a single peak. The proteinase content was finally evaluated by fractionation of the active material on a Mono Q column. The proteolytic activity of all purified proteinases was inhibited by pepstatin A, confirming that they belong to the class of aspartic proteinases. Typical Mono Q elution profiles for each species of *Cynara* are shown in Fig. 1. One active component was isolated from *C. humilis* whereas two components with different specific activities were isolated from *C. scolymus*. Comparison between the chromatographic pattern obtained for *C. cardunculus* and those of the other species of *Cynara* clearly shows that the last component eluted from the column, identified as cardosin B, is absent in *C. humilis* but is present in *C. scolymus*. SDS-PAGE analysis of the active fractions recovered from the Mono Q column confirmed this observation. Indeed, all fractions of *C. humilis* migrated as two bands with apparent molecular masses of 31000 and 15000 daltons. Conversely, the two active fractions isolated from pistils of *C. scolymus* produced different electrophoretic patterns on SDS polyacrylamide gels. The material from the first active fraction migrated as bands of 31000 and 15000 daltons whereas that of the second active peak gave rise to two bands of 34000 and 14000 daltons. These electrophoretic patterns are similar to the ones of cardosin A and cardosin B respectively, suggesting that both types of cardosins occur in *C. scolymus*.

3. HYDROLYSIS OF INSULIN B CHAIN

The cleavage pattern of insulin B chain by cardosin A and cardosin B are clearly different and thus it can be used to distinguish between these two aspartic proteinases. The HPLC peptide maps of insulin B chain digested with each of the proteinases under study are shown in Fig. 2. The proteinase from *C. humilis* cleaved the bonds Leu15–Tyr16, Leu17–Val18 and Phe25–Tyr26 in this polypeptide. The same peptide bonds were cleaved by the first proteinase of *C. scolymus*, whereas the second, more active, enzyme cleaved in

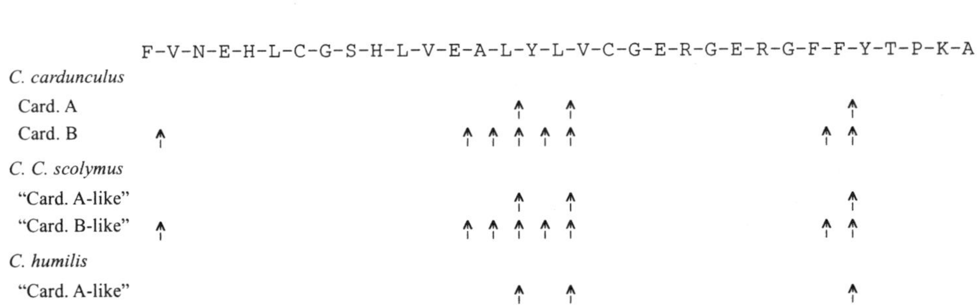

Figure 2. Sites of cleavage by APs from different species of *Cynara* on insulin B chain.

addition to these the bonds Phe1–Val2, Glu13–Ala14, Ala14–Leu15, Tyr16–Leu17 and Phe24–Phe25. The overall specificity of these enzymes seems therefore to be identical to that of the corresponding cardosin suggesting that the enzymic properties may also be very similar.

4. KINETIC STUDIES

In order to investigate the kinetic properties of the aspartic proteinases from *C. humilis* and *C. scolymus*, the synthetic peptide Lys–Pro–Ala–Glu–Phe–Phe(NO_2)–Ala–Leu was used as substrate. The values of the kinetic parameters for the hydrolysis of the bond Phe–Phe(NO_2) in this peptide are shown in Table 1. Interestingly, the Kcat/Km value of the proteinase from *C. humilis* is considerably higher than the values for the proteinases from *C. scolymus*. Further comparison between the proteinases also reveals that the proteinase from *C. humilis* has consistently the lowest Km value for the hydrolysis of this substrate. Taken together these results suggest that the proteinases of *Cynara* probably have slight differences in their active sites that may account for the observed differences in the kinetic parameters.

5. IMMUNOBLOTTING ANALYSIS

In order to gain more insights into the relationship between the *Cynara* proteinases the immunological properties of the proteinases were investigated (Figure 3). Immunoblotting analysis using a polyclonal antibody against cardosins confirmed the nature of the proteinases purified from *C. humilis* and *C. scolymus* as cardosin-like enzymes. Again it is clear that *C. humilis* contains only a cardosin A-like proteinase whereas *C. scolymus* contains both cardosin A and cardosin B-like enzymes. We chose to name the proteinases here described “cardosin-like” at this stage. Cloning and sequence comparison between these enzymes will definitely elucidate their relationships. This research is in progress in our laboratory.

6. CONCLUDING REMARKS

The results described above strongly suggest that while flowers of *C. scolymus* contain cardosin A and cardosin B-like enzymes, the flowers of *C. humilis* contain a protei-

Table 1. Kinetic constants for the hydrolysis of the synthetic peptide Lys–Pro–Ala–Glu–Phe–Phe(NO_2)–Ala–Leu

Enzyme	Km (μM)	Kcat (sec^{-1})	Kcat/Km ($sec^{-1}/\mu M$)
C. cardunculus			
cardosin A	108.02 ± 8.6	55.63 ± 6.12	0.515 ± 0.057
cardosin B	111.2 ± 12	89.36 ± 3.5	0.804 ± 0.032
C. scolymus			
“card. A-like”	136.3 ± 17	18.36 ± 2.52	0.135 ± 0.014
“card. B-like”	106.4 ± 15	33.18 ± 1.09	0.312 ± 0.024
C. humilis			
“card. A-like”	82.1 ± 6	61.59 ± 2.05	0.750 ± 0.034

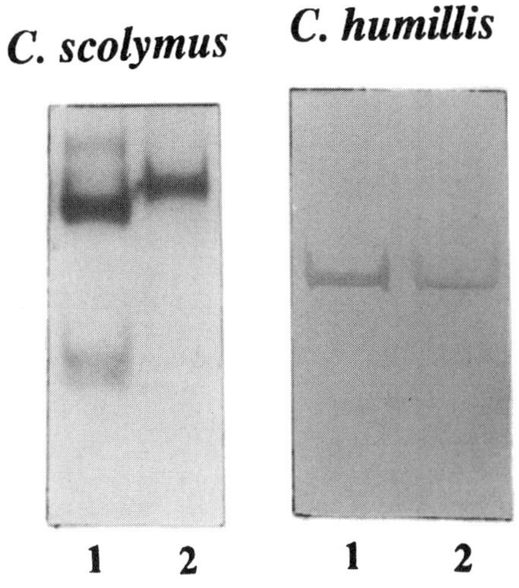

Figure 3. Immunoblotting analysis of aspartic proteinases isolated from different species of *Cynara*; *C. scolymus*, lane 1: cardosin A-like, lane 2: cardosin B-like; and *C. humilis*, lanes 1 and 2: cardosin A-like proteinase.

nase with structural features similar to cardosin A but with unique kinetic properties. It will be worthwhile to further investigate the reasons for the peculiar kinetic properties of the cardosin A-like aspartic proteinase from *C. humilis*.

In view of these results it is possible that cardosin A is the ancestral form of cardosins and that cardosin B may have arisen by gene duplication during the course of evolution. Cardosin B would thus represent a specialization within cardosins required for a particular role in the pistil. The involvement of aspartic proteinases in plant sexual reproduction may therefore be a common theme in the *Cynarae* tribe.

REFERENCES

1. Kervinen, J., Tormakangas, K., Runeberg-Roos, P., Guruprasad, K., Blundell, T. and Teeri, T.H. (1995) Structure and possible function of aspartic proteinases in barley and other plants In Aspartic proteinases: structure, function, biology and biomedical implications (Takahashi, K.ed.) New York: Plenum Press, pp. 241–254.
2. Veríssimo, P., Faro, C., Moir, A., Lin, Y., Tang, J. and Pires, E. (1996) Purification, characterization and partial amino acid sequencing of two novel aspartic proteinases from the flowers of *Cynara cardunculus* L. *Eur. J. Biochem.* **235**, 762–768.
3. Faro, C.,Ramalho-Santos, M., Veríssimo, P., Pissarra, J., Frazão, C., Costa, J., Lin, X., Tang, J., Pires, E., (1997) Structural and functional aspects of cardosins in structure and function of aspartic proteinases: retroviral and cellular enzymes.(M.N.G.James.ed.)New York:Plenum Press.
4. Ramalho-Santos, M., Pissarra, J., Verissimo, P., Pereira, S., Salema, R., Pires, E. and Faro, C.(1997) Cardosin A accumulates in protein storage vacuoles of the stigmatic pappilae of *Cynara cardunculus* L. Planta (accepted for publication).

64

AN ASPARTIC PROTEINASE FROM FLOWERS OF *Centaurea calcitrapa*

Purification, Characterization, Molecular Cloning, and Modelling of Its Three-Dimensional Structure

A. Domingos,[1] Z.-T. Xue,[2] K. Guruprasad,[3] A. Clemente,[1] T. Blundell,[3] M. S. Pais,[4] and P. E. Brodelius[2]

[1]Departamento de Biotecnologia, INETI
Estrada das Palmeiras, 2745 Queluz, Portugal
[2]Department of Plant Biochemistry, Lund University
P.O. Box 117, S-22100 Lund, Sweden
[3]Department of Crystallography, Birkbeck College
Malet Street, London WC1E 7HX, United Kingdom
[4]Departamento Biologia Vegetal, FCL
Bloco C2, Campo Grande, 1700 Lisboa, Portugal

1. INTRODUCTION

Enzyme technology developments associated with cheese manufacturing have been concerned with replacement of expensive and limited rennet supplies. Several substitutes have been evaluated, including microbial enzymes, animal pepsins and plant proteinases. Only a few have proved to be useful, showing similar catalytic properties to chymosin. *Centaurea calcitrapa*, a plant widely distributed in Portugal accumulates in flowers and leaves proteolytic enzymes with milk clotting activity.[1,2] This thistle is a wild species of Portuguese vegetation and is used, in some regions of Portugal, to make artisanal cheeses in a similar way as *Cynara* species.[3] Taxonomically the plant is included in Compositae family and in genus *Centaurea*.[1]

2. PURIFICATION OF *Centaurea* ASPARTIC PROTEINASES

Aspartic proteinases have been isolated from fresh and dried flowers of *C. calcitrapa*. After ammonium sulphate fractionation, the enzymes were further purified by three chromatographic steps (DEAE-Sepharose t-Butyl and MonoQ FPLC columns). The purified

Aspartic Proteinases, edited by James.
Plenum Press, New York, 1998.

enzymes are strongly inhibited by pepstatin A, but not by other group-specific protease inhibitors, indicating that the enzymes belong to the aspartic proteinase family. The enzymes are N–glycosylated (binding to ConA-Sepharose and endoglucosidase F treatment) and express maximum activity at pH 4 to 5. The enzymes isolated from fresh flowers appear to be a single peptide chain (around 50 kD) while the enzymes isolated from dried flowers are heterodimeric enzymes with one large (around 30 kD) and one small (around 16 kD) subunit. The heterodimeric form of the enzyme is very similar to that of the cyprosins.[3] Preliminary experiments have shown that the purified single chain enzyme can be converted to the dimeric form by incubating the protein at an acidic pH. During this apparently autocatalyzed proteolytic processing of the enzyme additional bands can be seen on a SDS-PAGE (43 and 21 kD). The *Centaurea* enzyme may, thus, serve as an experimental system to establish a mechanism for the removal of the plant specific insert in plant aspartic proteinases.

3. ISOLATION OF A *Centaurea* ASPARTIC PROTEINASE cDNA CLONE

3.1. Construction and Screening of the cDNA Library

A λZapII-cDNA library was constructed from poly(A)$^+$ RNA isolated from young flowers of *C. calcitrapa* according to the protocols supplied by Stratagene. This library was screened for aspartic proteinase cDNA clones using the 1.6 kb insert of *cypro1s* from *Cynara cardunculus* as probe.[4] This probe was ^{32}P-labelled by random oligonucleotide-primed synthesis (Pharmacia, Uppsala, Sweden). The phagemids (pBluescript SK-) carrying aspartic proteinase cDNA were excised from λZap clones by the helper phage R408 and identified by DNA blot hybridization using the above-mentioned *Cynara* probe.

Eighteen putative *Centaurea* aspartic proteinase clones were identified by a two round screening procedure followed by Southern hybridization of *in vivo* excised DNAs. The sizes of these cDNA clones were up to 1.7 kb. Among them, six 1.7 kb cDNA clones were selected for further study on primary structure.

3.2. DNA Sequencing

DNA sequencing of the putative aspartic proteinase cDNAs was performed according to the dideoxy-chain termination method.[5] Purification of single-stranded template DNA from pBluescript SK- cDNA clone and sequencing were carried out as in the protocol for the Sequenace 2.0 kit (United States Biochemical Corp.). Initial DNA sequencing was carried out using reverse primer or M13–40 primer. In addition, oligonucleotides (15-mers) were synthesized according to sequence information obtained and used directly as primers for further sequencing.

The six selected cDNA clones were in the first sequencing reaction shown to be identical. One clone, *cent8*, was selected for further sequencing in both directions. The nucleotide sequence and deduced amino acid sequence of this clone are shown in Figure 1.

4. COMPARISON OF DEDUCED AMINO ACID SEQUENCE OF *Centaurea* ASPARTIC PROTEINASE TO OTHER ASPARTIC PROTEINASES

The deduced amino acid sequence of *Centaurea* aspartic proteinase has been compared to the sequences of nine other plant aspartic proteinases and five mammalian aspar-

Table 1. Comparison of deduced amino acid sequence of mature *Centaurea* aspartic proteinase to nine plant and five mammalian mature aspartic proteinases

Aspartic proteinase	AC number	Identity (%)	Similarity (%)
Centaurea AP	Y09123	100.0	100.0
Cyprosin A	X69163	85.3	92.5
Cyprosin B	X81984	96.2	98.4
Brassica AP	U55032	75.5	87.5
Arabidopsis AP	U51036	76.2	86.8
Vigna AP	U61396	73.7	86.6
Tomato AP	L46681	69.8	82.3
Barley AP	X56136	74.0	86.7
Rice AP A	D32144	63.8	79.8
Rice AP B	D32165	74.4	86.6
Human pepsin C	J04443	43.0	59.9
Human cathepsin D	M11233	52.1	74.3
Human cathepsin E	J05036	49.3	68.1
Human pepsin A	M26032	45.8	63.6
Bovine chymosin b	J00003	40.2	62.2

tic proteinases. The identity and similarity between mature *Centaurea* aspartic proteinase and these aspartic proteinases are shown in Table 1. The size of the mature aspartic proteinases of plant origin is somewhat larger (around 440 amino acids) than that of the mammalian enzymes (320–350 amino acids). This difference in size is due to an internal additional sequence of around 100 amino acids in the plant enzymes. This insert is removed by a proteolytic processing of the enzymes resulting in heterodimeric enzymes. The *Centaurea* enzyme exhibits varying homology to the plant aspartic proteinases (64–96% identity). The identity to the mammalian enzymes (omitting the plant specific region) is somewhat lower (40–52%).

5. EVOLUTIONARY RELATIONSHIP BETWEEN ASPARTIC PROTEINASE SEQUENCES

The deduced amino acid sequence of 47 aspartic proteinases (10 plant enzymes and 37 randomly picked enzymes from various phyla) have been used to establish a phylogenetic tree (Figure 2). The plant enzymes are grouped together and are most closely related to a mosquito enzyme and to the cathepsin D enzymes. It is interesting to note that there is no clear segregation of the plant enzymes into monocotyledons and dicotyledons. However, enzymes from plants of the same family are grouped together (e.g. *Centaurea* and *Cynara*) and as expected from a comparison of the deduced amino acid sequences the *Centaurea* aspartic proteinase is most closely related to cyprosin B (identity 96.2%).

6. GENOMIC ORGANIZATION OF THE *Centaurea* ASPARTIC PROTEINASE GENES

Genomic DNA from young flower tissue of *C. calcitrapa* was digested with restriction endonucleases and analyzed by Southern hybridization. The *Bam* HI, *Eco*RI and *Hin-*

```
  1                                    M  G  T  A  I  K  A  S  L  L  A  L  F  L  F  V  L  L  S  P  T  A     22
  1 CGCATGTTTGCTTCCATGAAAATCATCGTGAAACATGGGTACCGCAATCAAAGCAAGCCTGCTTGCCTTGTTTTTGTTTGTTCTTCTATCACCTACTGCT  100

 23 F  S  A  S  N  G  G  L  L  R  V  G  L  K  K  R  K  V  D  Q  I  N  Q  L  R  N  H  G  A  S  M  E  G  K  56
101 TTTTCGGCCTCAAATGGTGGATTACTTAGAGTTGGACTAAAAAAGAGGAAGGTTGACCAAATTAACCAACTTAGAAACCATGGTGCGTCTATGGAAGGAA  200
                                         ¤
 57  A  R  K  D  F  G  F  G  G  S  L  R  D  S  D  S  D  I  I  E  L  K  N  Y  M  D  A  Q  Y  Y  G  E  I    89
201 AGGCTAGAAAAGATTTTGGCTTCGGTGGTAGCCTGAGGGACTCGGACAGCGACATTATTGAACTGAAGAACTACATGGATGCTCAGTATTATGGTGAGAT  300
                                                  *  *  *
 90 G  I  G  S  P  A  Q  K  F  T  V  I  F  D  T  G  S  S  N  L  W  V  P  S  A  K  C  Y  F  S  V  A  C    122
301 TGGTATTGGATCTCCAGCTCAAAAGTTCACCGTGATTTTTGACACCGGAAGTTCTAATCTATGGGTGCCTTCTGCAAAGTGCTACTTTTCAGTAGCTTGC  400
                                                           +  +  +
123 L  F  H  S  K  Y  K  S  S  H  S  S  T  Y  K  K  N  G  T  S  A  A  I  Q  Y  G  T  G  S  I  S  G  F  V  156
401 CTTTTTCACTCAAAGTATAAGTCAAGCCATTCAAGCACCTACAAGAAAAACGGGACATCTGCCGCTATCCAATATGGAACTGGATCAATCTCTGGTTTTG  500

157  S  Q  D  S  V  K  L  G  D  L  V  V  K  E  Q  D  F  I  E  A  T  K  E  P  G  V  T  F  L  A  A  K  F   189
501 TTAGTCAAGACTCTGTCAAACTCGGCGATCTTGTTGTTAAAGAGCAGGATTTTATAGAGGCAACCAAAGAGCCTGGCGTCACTTTCTTGGCAGCCAAGTT  600

190 D  G  I  L  G  L  G  F  Q  E  I  S  V  G  K  S  V  P  V  W  Y  N  M  V  N  Q  G  L  V  Q  E  P  V    222
601 TGATGGTATACTTGGCCTTGGATTTCAGGAGATCTCTGTTGGGAAATCTGTTCCTGTCTGGTACAACATGGTTAATCAAGGTCTTGTTCAAGAACCCGTG  700

223 F  S  F  W  F  N  R  N  A  D  E  E  E  G  G  E  L  V  F  G  G  V  D  P  N  H  F  K  G  K  H  T  Y  V  256
701 TTTTCCTTTTGGTTTAATCGCAATGCTGACGAGGAAGAAGGGGGTGAACTCGTGTTTGGTGGAGTTGACCCTAATCATTTTAAGGGGAAGCATACATACG  800

257  P  V  T  Q  K  G  Y  W  Q  F  N  M  G  D  V  L  I  E  D  K  T  T  G  F  C  A  D  G  C  A  A  I  A   289
801 TCCCCGTGACTCAAAAGGGATATTGGCAGTTTAATATGGGCGACGTTCTTATTGAAGATAAAACTACGGGATTTTGCGCTGATGGTTGTGCAGCAATTGC  900
```

```
          *  *  *
290  D  S  G  T  S  L  L  A  G  P  T  A  I  I  T  Q  I  N  H  A  I  G  A  K  G  V  M  S  Q  Q  C  K  T     322
901  CGACTCTGGAACCTCTTTGTTGGCAGGTCCAACGGCTATTATTACTCAAATCAATCATGCAATTGGTGCCAAGGGGGTCATGAGCCAGCAATGCAAAACA 1000

323  L  V  D  Q  Y  G  K  T  I  I  E  M  L  L  S  E  A  Q  P  D  K  I  C  S  Q  M  K  L  C  T  F  D  G  A  356
1001 TTGGTTGATCAGTATGGAAAGACTATCATTGAGATGCTCCTGTCTGAGGCGCAACCTGATAAAATCTGTTCGCAAATGAAGTTATGCACTTTTGATGGTG 1100

357   R  D  V  S  S  I  I  E  S  V  V  D  K  N  N  G  K  S  S  G  G  V  H  D  E  M  C  T  F  C  E  M  A    389
1101 CTCGCGATGTTAGTTCAATAATCGAGAGTGTGGTTGATAAGAATAACGGCAAGTCTTCTGGTGGTGTACACGATGAGATGTGTACCTTCTGTGAGATGGC 1200
                                        +  +  +
390   V  V  W  M  Q  N  Q  I  K  R  N  Q  T  E  D  N  I  I  N  Y  V  N  E  L  C  D  R  L  P  S  P  M  G    422
1201 AGTCGTTTGGATGCAAAACCAAATCAAACGAAACCAGACTGAAGATAACATAATCAACTATGTCAATGAGCTGTGTGATCGCTTGCCCAGTCCAATGGGA 1300

423  E  S  A  V  D  C  N  D  L  S  S  M  P  N  I  A  F  T  I  G  G  K  V  F  E  L  C  P  E  Q  Y  I  L  K  456
1301 GAATCAGCAGTAGACTGCAACGATCTTTCCTCCATGCCCAATATTGCCTTCACTATTGGTGGCAAAGTATTTGAGCTCTGCCCAGAACAGTACATCCTGA 1400

457   I  G  E  G  E  A  A  Q  C  I  S  G  F  T  A  M  D  V  A  P  P  R  G  P  L  W  I  L  G  D  V  F  M    489
1401 AAATTGGTGAGGGAGAAGCCGCACAATGCATCAGTGGATTCACAGCTATGGATGTCGCCCCTCCTCGTGGACCTCTATGGATCTTGGGAGATGTTTTTAT 1500

490   G  Q  Y  H  T  V  F  D  Y  G  K  L  R  V  G  F  A  E  A  A  *                                        509
1501 GGGTCAATACCATACCGTGTTCGATTATGGCAAGTTGCGAGTTGGATTTGCTGAAGCAGCTTGAGAGAGAGATGTTAATCGGTGTTCTGTGAATATCCAT 1600

1601 CCTACCGAATGAATGTCCAATAAATTTATGAGTTTTAGTACCCTAAATTTGTTGGAATATGCTAGAAACGTATGTTGGTTTGCTTTTACTTTATGTGTGA 1700

1701 TGCAACTTTCGTTTCAAAAAA                                                                                 1721
```

Figure 1. Nucleotide and deduced amino acid sequences for Centaurea calcitrapa aspartic proteinase. The nycleotide sequence is from cDNA clone cent8. (¤) Putative start of mature protein; (*) active site Asp–Thr/Ser–Gly; (+) putative N-glucosylation sites; the plant specific insert is in italic. The start codon (position 35–37), stop codon (position 1562–1564) and the polyadenylation signal (position 1619–1624) are underlined in the nucleic acid sequence.

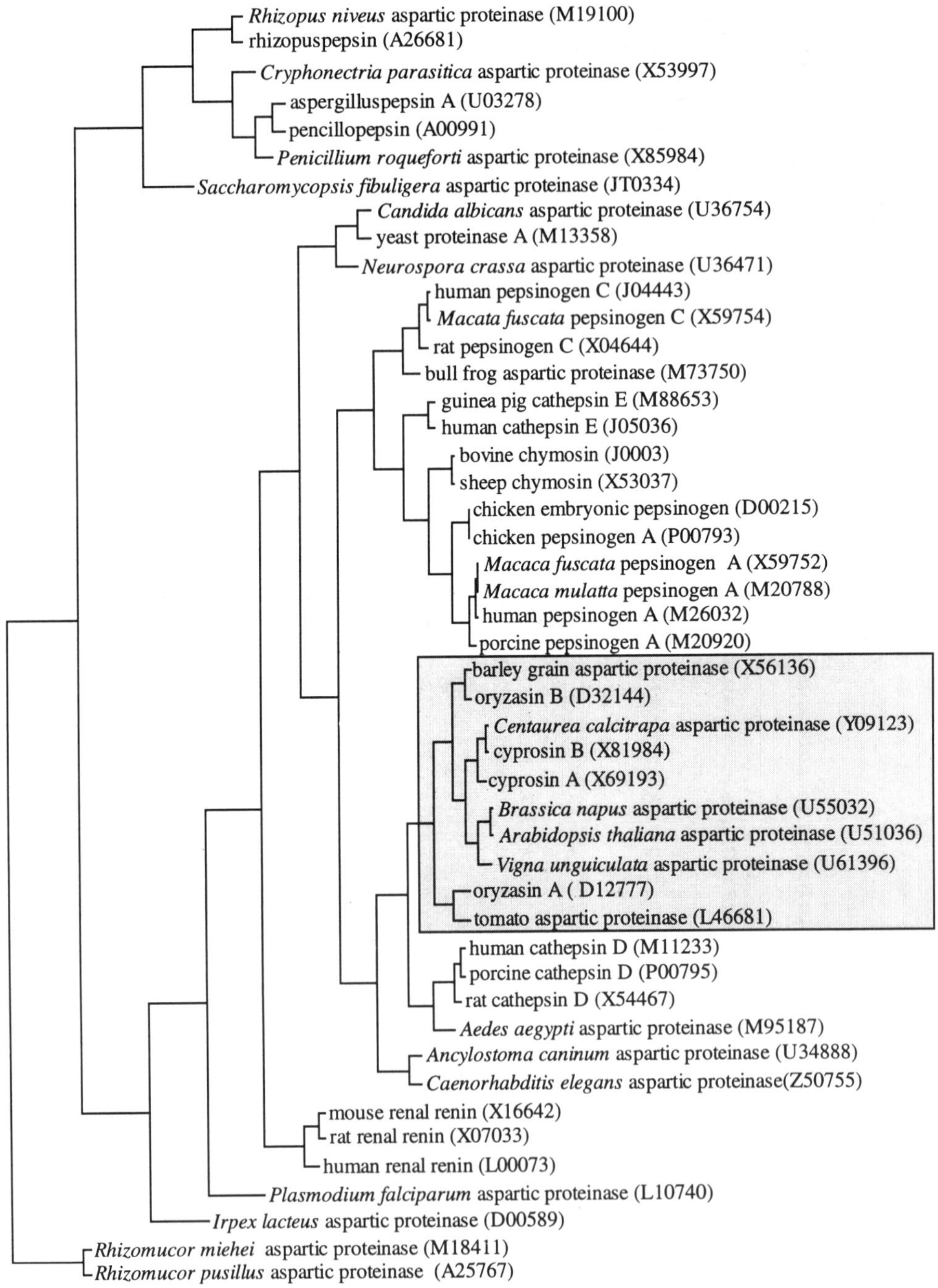

Figure 2. Phylogenetic tree for 46 different aspartic proteinases. The plant aspartic proteinases are within a shaded box. The tree was obtained by using the computer programmes PILEUP, DISTANCES and GROWTREE (Genetics Computer Group, University of Wisconsin). The sequences for the aspartic proteinases were obtained from the EMBL Data Library as indicated in the figure.

*d*III digests each showed several hybridizing bands when the entire 1.7 kb insert of *cent8* was used as probe. From these hybridization patterns it may be suggested that the *Centaurea* aspartic proteinase genes are organized as a small multi-gene family. However, only one of these genes appear to be expressed in flowers of *C. calcitrapa*, since only one type of cDNA clones could be isolated from the flower cDNA library.

7. MODELLING

7.1. *Centaurea* Aspartic Proteinase Model

A three-dimensional model of the *Centaurea* aspartic proteinase has been constructed using the deduced amino acid sequence of *cent8* cDNA. This model was constructed using COMPOSER—a rule-based comparative modelling approach as in SYBYL software.[6–8] The basic steps involved in COMPOSER are: identification and selection of homologous protein structures, construction of the 3-D framework for the model, alignment of the model with the structurally conserved regions (SCRs), modelling the main chains and side chains for the SCRs and finally construction of the regions that connect the fragments of the conserved core—structurally variable regions (SVRs). The crystal structures of four highly homologous aspartic proteinases (human cathepsin D (1LYB) 52.1, bovine chymosin (4CMS) 40.2, mouse renin (1SMR) 42.6, porcine pepsin (5PEP) 42.6% identity) were used in the modelling procedure. Model refinement was carried out in two stages. First the COMPOSER built model was refined using options in FRODO. Next the model was energy minimised using SYBYL (a) by fixing the backbone and allowing the side-chains to move then (b) allowing the whole enzyme to relax into an energy minima by assigning the disulphide constraints and the distance constraints of 3 C-alpha atom that lie at ends of the polypeptide chain on either side of the plant specific region.

The SCRs selected to model *Centaurea* aspartic proteinase are shown in Table 2. SVR no. 10 was modelled using the 'collar' approach as in the barley-grain aspartic proteinase.[9]

Table 2. Selection of fragments for structurally conserved regions (SCRs) in the model

SCR no.	SCR	PDB code of protein used in 'framework'
1	Ser 4 – Tyr 49	1SMR
2	Ser 50 – Asn 70	1LYB
3	Thr 72 – Glu 77	1LYB
4	Gly 83 – Leu 96	1SMR
5	Asp 98 – Val 101	1SMR
6	Gln 104 – Gln 113	1LYB
7	Thr 117 – Gly 137	1LYB
8	Ser 139 – Arg 163	1LYB
9	Gly 171 – Ile 207	1LYB
10	Gly 204 – Gly 239	5PEP
11	Ser 253 – Lys 285	1LYB
12	Cys 294 – Met 301	1LYB
13	Pro 309 – Ala 338	1LYB

7.2. Comparison of the Model of *Centaurea* Aspartic Proteinase to Models of Other Plant Aspartic Proteinases

In addition to the 3-D structure of *Centaurea* aspartic proteinase, two other plant aspartic proteinases, *i.e.* barley-grain aspartic proteinase[9] and cyprosin[10] have been modelled. The overall folding of these three enzymes is very similar. A comparison of the active site of these three models shows that a major difference is the residue at position 295 (pepsin numbering) which may affect S2′ binding. This residue is Arg in barley AP and *Centaurea* aspartic proteinase and His in cyprosin. In addition, at position 291 there is Val in *Centaurea* aspartic proteinase and cyprosin but Ile in barley and this may affect S1′ binding. Furthermore, at position 292 there is Ala in *Centaurea* aspartic proteinase and cyprosin whereas there is Pro in the barley enzyme. The Pro at 292 is similar to the Pro in renins and cathepsin D and is likely to orient the loop in the C-flap slightly different compared with *Centaurea* aspartic proteinase and cyprosin and therefore affect substrate binding at the prime site. Pro292 is near the turn on the C-flap loop.

REFERENCES

1. Franco JA (1984) Nova Flora de Portugal (Continente e Acores), Vol II, Lisboa
2. Domingos A, Clemente A, Pais MS (1993) Studies on proteases with milk-clotting activity from *Centaurea calcitrapa*. *Proceedings of the 6th Forum for Applied Biotechnology*, pp. 1909–1911
3. Heimgartner U, Pietrzak M, Geertsen R, Brodelius P, da Silva Figueiredo AC, Pais MSS (1990) Purification and partial characterization of milk clotting proteases from flowers of *Cynara cardunculus*. *Phytochemistry* **29:**1405–1410
4. Cordeiro M, Xue Z-T, Pietrzak M, Pais MS, Brodelius PE (1994) Isolation and characterization of a cDNA from flowers of Cynara cardunculus encoding cyprosin (an aspartic proteinase) and its use to study the organ-specific expression of cyprosin. *Plant Molecular Biology* **24**:733–741.
5. Sanger F, Nicklen S and Coulson AB (1977) DNA sequencing with chain-terminating inhibitors. *Proc. Natl. Acad. Sci. USA* **74:**5463–5467
6. Blundell TL, Sibanda BL, Sternberg M, and Thornton J M (1987) Knowledge-based prediction of protein structures and the design of novel molecules. *Nature* **326**:347–352.
7. Sutclife MJ, Haneef I, Carney D, and Blundell TL (1987) Knowledge-based modelling of homologous proteins. Part 1: Three-dimensional frameworks derived from simultaneous superposition of multiple structures. *Prot. Eng.* **1**:377–384.
8. Sutcliffe MJ, Hayes FRF, and Blundell TL (1987) Knowledge-based modelling of homologous proteins. Part 2: Rules for the conformations of substituted side-chains. *Prot. Eng.* **1**:385–392.
9. Guruprasad K, Tormakangas K, Kervinen J and Blundell T (1994) Comparative modelling of barley-grain aspartic proteinase: A structural rationale for observed hydrolytic specificity. *FEBS Lett.* **352**: 131–136
10. Cordeiro M, Lowther T, Dunn BM, Guruprasad K,. Blundell T, Pais MS and Brodelius PE (1997) Substrate specificity and molecular modelling of aspartic proteinases (cyprosins) from flowers of *Cynara cardunculus* subsp. *flavescens* cv. cardoon. In: *Aspartic proteinases* (M. James, ed.) Plenum Press, New York, submitted

SUBSTRATE SPECIFICITY AND MOLECULAR MODELLING OF ASPARTIC PROTEINASES (CYPROSINS) FROM FLOWERS OF *Cynara cardunculus* SUBSP. *flavescens* CV. CARDOON

M. Cordeiro,[1] T. Lowther,[2] B. M. Dunn,[2] K. Guruprasad,[3] T. Blundell,[3] M. S. Pais,[4] and P. E. Brodelius[1]

[1]Department of Plant Biochemistry
Lund University
P.O. Box 117, 22100 Lund, Sweden
[2]Department of Biochemistry and Molecular Biology
University of Florida, College of Medicine
Gainesville, Florida 32610-0245
[3]Department of Crystallography
Birkbeck College
Malet Street, London WC1E 7HX, United Kingdom
[4]Centro Biotech Vegetal, FCL
Bloco C2, Campo Grande, 1700 Lisboa, Portugal

1. INTRODUCTION

Water extracts of dried flowers of *Cynara cardunculus* subsp. *flavescens* cv. cardoon have been traditionally used in Portugal to produce artisanal cheeses (e.g. Serpa and Serra cheeses) with ewe's milk. The clotting activity is due to three heterodimeric aspartic proteinases (cyprosin 1, cyprosin 2 and cyprosin 3) which have been purified from flowers.[1] A native M_r of around 49000 was determined for the three cyprosins by SDS-PAGE. The subunit sizes are 32.5 + 16.5, 33.5 + 16.5, and 35.5 + 13.5 kD for cyprosin 1, 2 and 3, respectively. Gel filtration chromatography indicated somewhat lower native M_rs (41–45000). The enzymes are N–glycosylated (high mannose type) and express maximum activity around pH 4.1 using a synthetic peptide as substrate.

The proteolytic and milk clotting activities of the cyprosins were compared to those of chymosin.[2] Cyprosin 3 showed the highest specific activity which is comparable to that of chymosin. Slight differences in the specificity towards milk caseins were observed for the two enzymes.

Aspartic Proteinases, edited by James.
Plenum Press, New York, 1998.

Studies on the organ–specific expression of the enzyme were carried out at different stages of flower development and in leaves and seeds.[3] Immunostained Western blots revealed the presence of cyprosin in very young flowers. The amount of enzyme increased towards later stages of development. The enzyme could not be detected in leaves or seeds. Immunogold labelling and silver enhancement of transversal cuts from fixed flowers using polyclonal anti cyprosin 3 antibodies have shown tissue-specific expression of cyprosin in the epidermal cell layer of styles.

Two cDNA clones, *cyproIs* and *cyproII*, encoding cyprosins have been sequenced.[4,5] The sequences contain 1515/1527 bp open reading frames coding for 505/509 amino acids including putative full length mature proteins (440/441 amino acids). Plant specific regions of 103/104 amino acids are well identified in these cDNAs.

Our studies on cyprosins have now been extended with more extensive kinetic measurements and modelling of the 3-D structure.

2. SUBSTRATE SPECIFICITY OF CYPROSIN

Three isozymes of cyprosin 3 (cyprosin 3α: pI 3.85; cyprosin 3β: pI 4.00; cyprosin 3γ: pI 4.15) were isolated by preparative isoelectric focusing from a crude flower extract. Each isozyme was further purified by MonoQ FPLC. The specific clotting activity of the three isoforms was distinctly different.

Two series of related chromophoric octa-peptide substrates with systematic variation in the amino acid residues have been used to study the proteolytic activity of the three cyprosin 3 isozymes. The rate of hydrolysis was determined to investigate the effects of various substitutions in the P_5–P_1 and the P_2'–P_3' positions on enzyme activity. Enzyme concentrations were determined by titration with the competitive inhibitor pepstatin A as described by Henderson[6]. The K_i-values were < 0.1 nM for all three isozymes.

The rates of hydrolysis obtained for substrates of the form Lys(P_5)–Pro(P_4)–Ala(P_3)–Lys(P_2)–Phe(P_1)*Nph(P_1')–Arg(P_2')–Leu(P_3') with systematic substitutions in all the positions flanking the scissile peptide bond (Phe*Nph) were obtained for 31 different peptides. All the peptides were cleaved with measurable rates by the three isozymes under the assay conditions used (0.1 M Na-formate buffer, pH 4.0; 50 μM substrate), with the exception of peptides containing Ala or Leu in the P_1 position or Asp in the P_2' position.

To elucidate the S_2 subsite preferences of the enzymes further, another series of substrates based on the parent peptide, Lys–Pro–Ile–Glu–Phe*Nph–Arg–Leu, with systematic substitutions in the P_2 position was used (15 different peptides). The replacement of Lys by Glu in this series leads to a significant improvement in the overall cleavage efficiencies. This is clearly seen by comparison of Lys–Pro–Ile–Lys–Phe*Nph–Arg–Leu (Å4 M/s/[E]) and Lys–Pro–Ile–Glu–Phe*Nph–Arg–Leu (Å38–56 M/s/[E]). Furthermore, the substrate Lys–Pro–Ile–Glu–Leu*Nph–Arg–Leu (Å22–38 M/s/[E]) is hydrolyzed at a much higher rate than the substrate Lys–Pro–Ala–Lys–Leu*Nph–Arg–Leu (>1 M/s/[E]).

Four peptides were selected for detailed kinetic measurements and the results are summarized in Table 1. The kinetic constants of two mammalian aspartic proteinases have been included for comparison.[7,8]

The k_{cat}-values are in the range 13–94 s^{-1} for all five enzymes showing that the rate of hydrolysis of cyprosins are comparable to those of mammalian aspartic proteinases. Even though the K_m-values for the substrate Lys–Ser–Ala–Lys–Phe*Nph–Arg–Leu are considerably higher (> 200 μM) than those for the other substrates (< 100 μM) no significant difference in the rate of hydrolysis is observed. The K_m-values observed for the cyprosins are comparable to those of cathepsin E and pepsin for the substrates used. The k_{cat}/K_m-values are somewhat higher for human cathepsin E (2–5 $\mu M^{-1}s^{-1}$) than for the other four enzymes (0.4–2.5 $\mu M^{-1}s^{-1}$). Also here the cyprosins exhibit similar values as the mammalian enzymes.

It is evident that cyprosins, like other aspartic proteinases, preferentially cleave peptide bonds between hydrophobic amino acids and that the sequence P_3–P_2' is of particular importance for binding of the substrate. The three cyprosin 3 isozymes show some small but distinct differences in kinetic properties. We assume that these differences are important for the function of these enzymes in the *Cynara* plant.

3. MODELLING

In this work a three-dimensional model of cyprosin has been constructed using the deduced amino acid sequence of *cypro1s* cDNA. This model was constructed using COMPOSER—a rule-based comparative modelling approach as in SYBYL software.[9–11] The basic steps involved in COMPOSER are: identification and selection of homologous protein structures, construction of the 3-D framework for the model, alignment of the model with the structurally conserved regions (SCRs), modelling the main chains and side chains for the SCRs and finally construction of the regions that connect the fragments of the conserved core—θ structurally variable regions (SVRs). The crystal structures of six highly

```
          1                                            46 47
  1  DSDGELIALKNYMDAQYFGEIGIGTPPQKFTVIFDTGSSNLWVPSSKCYFSVACLFHSKYRSTDSTTYKK   70
        : |:||||:||:||||||||||:|:|:|||||||:|||||||  |:||::||||  ::|::||
        TSSVILTNYMDTQYYGEIGIGTPPQTFKVVFDTGSSNVWVPSSKCS SLACFLHSKYDHEASSSYKA
        1LYB                              ***          ypra

     67 69  74   78            95  98      107    111                131 133
 71  NGKSAAIQYGTGSISGFFSQDSVKLGDLLVKEQDFIEATKEPGITFLAAKFDGILGLGFQEISVGDAVPV  140
     | |: :|:   |::|||:||| :::|:::  :||| |||:|   ||:||||||||||:::: |||: ::||
     N KPLSIH   GTVSGFLSQDIITVGGIV  KQDFAEATSE   TFIAAKFDGILGMAYPRISVN VLPV
       4CMS     1BBS                ypra         1LYB                   1LYB

                         157       163                        187 189     199
141  WYTMLNQGLVQEPVFSFWLNRNADEQEGGELVFGGVDPNHFKGEHTYVPVTQKGYWQFEMGDVLIGDKTT  210
     : ::::| ||:: :|||:|:|       ||:|:|| ||:|::|:  ||::::: ||||::::| |
     FDNLMQQKLVDQNIFSFYLSR       GEVVLGGSDPQHYQGDFHYVSLSKT YWQFTVDSVTI
                                 1SMR                      4CMS

           210                           239                  245250 252
211  GFCASGCAAIADSGTSLLAGTTTIVTQINQAIGAAGVM.. 248  352 ..GESAVDCSSLSSMPNIAF  369
           |:||:|:||||||:|:|: |::|: :|||:                   ::||: ::::|:|:|
           CQAIVDTGTSLLTGPTSAIANIQSDIGASE                     YTLDCN VPTLPDISF
           5PEP ***                                           ypra   1SMR

                    276         282  289       298                       326
370  TVGGKTFNLSPEQYVLKVGEGATAQCISGFYAMDVAPPHGPLWILGDVFMGQYHTVFDYGNLRVGFAEAA  439
     ::||::::||:::|||         |:||| :|        |||||||:|:|:|||| :| ||||||||
     NLGGRAYTLSSTDYVL         CLSGFMGM        LWILGDVFIGRYYTVFDRDNNRVGFAEAA

                              1LYB            1LYB
```

Figure 1. Alignment of the various SCRs to the deduced amino acid sequence of cyprosin. Pepsin numbering is over the sequence and cyprosin numbering is shown left and right of the sequence. Note that the plant specific insert has been removed (amino acids 249–351). Active site aspartic acids are marked with *.

homologous aspartic proteinases (human cathepsin D (1LYB), bovine chymosin (4CMS), yeast proteinase A (ypra), human renin (1BBS), mouse renin (1SMR), porcine pepsin (5PEP)) were used in the modelling procedure. Model refinement was carried out in two stages. First the COMPOSER built model was refined using options in FRODO. Next the model was energy minimised using SYBYL (a) by fixing the backbone and allowing the side-chains to move then (b) allowing the whole enzyme to relax into an energy minima by assigning the disulphide constraints and the distance constraints of 3 C-alpha atom that lie at ends of the polypeptide chain on either side of the plant specific region.

The SCRs selected to model cyprosin are aligned to the deduced amino acid sequence of cyprosin in Figure 1. A schematic representation of the model showing the overall fold and the probable site of the plant specific region is presented in Figure 2.

3.1. Stereochemical Quality of the Model

The PROCHECK suite of programs was used to evaluate the stereochemical quality of the model.[13] A Ramachandran plot for the model of cyprosin enzyme modelled with a

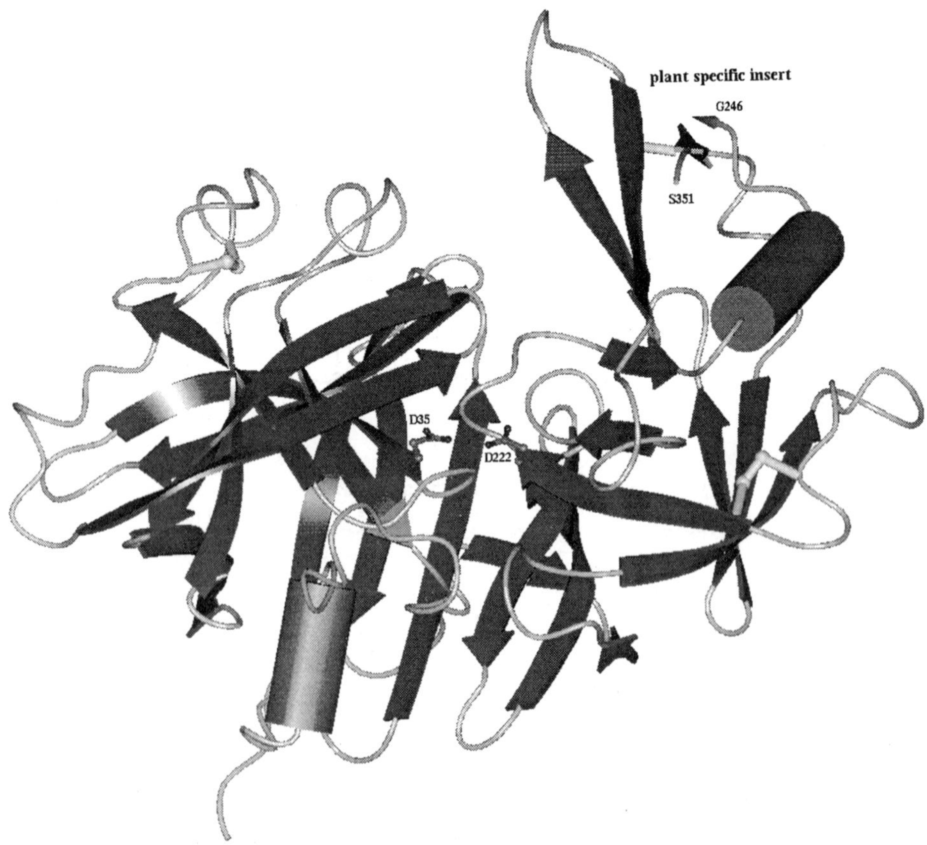

Figure 2. Schematic representation showing the model of cyprosin. Helices are shown as cylinders and sheets as flat arrows. The remaining polypeptide chain is shown as a rope which represents the turn/coil regions in the molecule. The location of the plant specific insert between G246 and S351 is indicated. The catalytic aspartate side chains (D35 and D222) and SS-bonds (C48-C54, C213-C217, C360-C397) are also shown. (This figure was made with the assistance of Dr. S. Al-Karadaghi).

Table 1. Kinetic constants for the cyprosin 3 isozymes, human cathepsin E and porcine pepsin

Peptide	k_{cat} (s^{-1})					K_m (μM)					k_{cat}/K_m ($\mu M^{-1}s^{-1}$)				
	3α	3β	3γ	Cat E	Pepsin	3α	3β	3γ	Cat E	Pepsin	3α	3β	3γ	Cat E	Pepsin
KSAKF*XRL	16 ± 3	15 ± 3	37 ± 7	—	—	221 ± 54	267 ± 68	279 ± 67	—	—	0.1 ± 0.02	0.1 ± 0.02	0.1 ± 0.04	—	—
KPIQF*XRL	21 ± 2	13 ± 1	70 ± 7	75 ± 16	24 ± 8	38 ± 3	30 ± 3	38 ± 4	22 ± 2	72 ± 28	0.5 ± 0.07	0.4 ± 0.06	1.8 ± 0.3	3.0 ± 0.8	0.3 ± 0.1
KPIVF*XRL	18 ± 2	15 ± 2	64 ± 7	94 ± 19	17 ± 5	15 ± 1	19 ± 3	26 ± 3	20 ± 3	31 ± 8	1.2 ± 0.2	0.8 ± 0.15	2.5 ± 0.4	5.0 ± 1.0	0.6 ± 0.2
KPILF*XRL	35 ± 4	54 ± 6	86 ± 11	79 ± 17	20 ± 6	98 ± 15	58 ± 8	66 ± 11	36 ± 6	14 ± 5	0.4 ± 0.07	0.9 ± 0.16	1.3 ± 0.3	2.0 ± 0.6	1.4 ± 0.6

Table 2. Specificity pockets in models of cyprosin with ligands in the active site[a]

Pockets	Substrate[b] ($P_4 \rightarrow P_3'$)	CH-66[c] ($P_6 \rightarrow P_4'$)	Chymosin with substrate[b] ($P_4 \rightarrow P_3'$)
S_6	—	M10, A12, L276, V278	—
S_5	—	A12	—
S_4	—	S219	S12, S219
S_3	I30, T77, F117, G217, T218, S219	Q13, T111, A115, F117, G217, T218, S219	S12, Q13, Y75, F117, G217, T218, S219
S_2	Y75, G76, T77, T218, A222, T287, M289	Y75, G76, T77, G217, T218, A222, T287, M289, I300	G217, T218, V222, Q287
S_1	D32, Y75, G76, T77, F117, D215, G217, T218	D32, G34, Y75, G76, I120, D215, G217, T218	L30, D32, Y75, G76, F117, D215, G217, T218
S_1'	Y75, G76, Y189, I213, D215, V291, H295, I300	Y75, G76, Y189, I213, D215, H295	G34, Y189, I213, D215, T218, E293, !300
S_2'	G34, Q74, I128, Y189, H295	G34, S35, I128, Y189, H295	G34, L128, Y189
S_3'	Q74, V291, H295	Q74, Y189, H295	—
S_4'	—	—	—

[a]A distance cut-off of 4.0 Å was used to define the specificity pockets; residues are numbered according to porcine pepsin (Protein Data Bank code: 5PEP)
[b]Substrate: Lys-Pro-Ile-Glu-Phe*Nph-Arg-Leu.
[c]CH-66: trimethylacetyl-His-Pro-Phe-His-Leu ($CH(OH)CH_2$)–Leu–Tyr–Tyr–Ser–NH_2.

good substrate into the active site showed that about 87% of the residues (in a total 333) fall in the most favoured regions of this plot suggesting that this model is a good representation of the crystal structure.

3.2. Modelling the Substrate/Inhibitor

In order to correlate observations from the kinetic measurements with structural features in the enzyme, we constructed models of cyprosin complexed with ligands in the active site to mimic the transition-state analogue. The 3-D structure of endothiapepsin complexed with a pepsin inhibitor H256 was used to model the substrate Lys–Pro–Ile–Glu–Phe*Nph–Arg–Leu in the active site of cyprosin, and the mouse renin structure complexed with a decapeptide inhibitor CH-66 was used to model this inhibitor in the active site of cyprosin. The specificity pockets showing the residues interactions in the substrate model, inhibitor model and a chymosin substrate model are given in Table 2.

Most of the pockets in cyprosin are very similar to those in the barley-grain aspartic proteinase[12]; the only difference in near the S2′ and S3′ pockets. This difference is due to His295 in cyprosin compared with Arg295 in barley enzyme.

3.3. Comparison of the Model of Cyprosin with Crystal Structure of Chymosin

Cyprosin and chymosin are both milk clotting enzymes used in cheese making. The proteolytic and clotting activities of these enzymes have been compared.[2] Cyprosin and chymosin have similar clotting activities with slight differences in specificity towards the caseins in milk. Both enzymes cleave the phe_{105}-met_{106} bond of K-casein, but cyprosin also hydrolyzes other proteins in milk, such as α_{s1}-casein, γ-casein and β-casein, at a lower rate.

In order to look for similarities, we carried out a comparison of the active sites of chymosin and cyprosin. This comparison showed a number of significant differences. Alignment of the two enzymes with other aspartic proteinases has shown that His295, involved in specificity of the S2′ and S3′ region of cyprosin (Table 2) is present only in cyprosin and chymosin. Therefore, we expected that the residue His295 in chymosin would interact with the substrate in the S2′ and S3′ pocket in an analogous way. However, in the chymosin structure, His295 is distant from the active site and thus may not play any role in catalysis (Table 2). The S2 pocket in these two enzymes are also different due to residue at position 287.

Another difference observed is the interaction of Tyr75 in the chymosin structure, which makes steric clashes with P1 (Phe in the K-casein molecule). Such steric clashes do not occur in the cyprosin model.

In conclusion, no specific similarities of the active sites of cyprosin and chymosin could be observed that would determine the capacity for milk clotting activity in these enzymes. The 3-D model of cyprosin is more similar to human cathepsin D, an enzyme that does not clot milk, than to chymosin.

REFERENCES

1. Heimgartner U, Pietrzak M, Geertsen R, Brodelius P, da Silva Figueiredo AC, Pais MSS (1990) Purification and partial characterization of milk clotting proteases from flowers of *Cynara cardunculus*. *Phytochemistry* **29:** 1405–1410

2. Cordeiro M, Jakob E, Puhan Z, Pais MS, Brodelius PE (1992) Milk clotting and proteolytic activities of purified cynarases from *Cynara cardunculus*; A comparison to chymosin. *Milchwissenschaft* **47:** 683–687.
3. Cordeiro M, Pais MS, Brodelius PE (1994) Tissue-specific expression of multiple forms of cyprosin (aspartic proteinase) in flowers of*Cynara cardunculus*. *Physiol. Plant.* **92:** 645–653
4. Cordeiro M, Xue Z-T, Pietrzak M, Pais MS, Brodelius PE (1994) Isolation and characterization of a cDNA from flowers of Cynara cardunculus encoding cyprosin (an aspartic proteinase) and its use to study the organ-specific expression of cyprosin. *Plant Mol. Biol.* **24:** 733–741.
5. Pietrzak M, Brodelius PE and Pais MS (1994) EMBL Data Entry no X81984
6. Henderson PJF (1972) A linear equation that describes the steady-state kinetics of enzymes and subcellular particles interacting with tightly bound inhibitors. *Biochem. J.* **127:** 321–333
7. Rao NC, Guruprasad K, Batley B, Rapundalo S, Hill J, Blundell T, Kay J and Dunn BM (1995) Exploring the binding preferences-specificity in the active site of human cathepsin E. *Prot. Struc. Func. Gen.* **22:** 168–181
8. Rao CM (1994) A detailed investigation into the enzymatic specificity of porcine pepsin. *PhD dissertation, University of Florida, Gainesville, FL.*
9. Blundell TL, Sibanda BL, Sternberg M, and Thornton J M (1987) Knowledge-based prediction of protein structures and the design of novel molecules. *Nature* **326:** 347–352.
10. Sutclife MJ, Haneef I, Carney D, and Blundell TL (1987) Knowledge-based modelling of homologous proteins. Part 1: Three-dimensional frameworks derived from simultaneous superposition of multiple structures. *Prot. Eng.* **1:** 377–384.
11. Sutcliffe MJ, Hayes FRF, and Blundell TL (1987) Knowledge-based modelling of homologous proteins. Part 2: Rules for the conformations of substituted side-chains. *Prot. Eng.* **1:** 385–392.
12. Guruprasad K, Tormakangas K, Kervinen J and Blundell T (1994) Comparative modelling of barley-grain aspartic proteinase: A structural rationale for observed hydrolytic specificity. *FEBS Lett.* **352:** 131–136
13. Laskowski RA, MacArthur MW, Moss DS, and Thornton J. (1993) PROCHECK- A program to check the stereochemical quality of protein structures. *J. Appl. Cryst.* **26:** 283–291.

ABSTRACTS

STRUCTURAL PROPERTIES OF BARLEY ASPARTIC PROTEINASE

Jukka Kervinen

Macromolecular Structure Laboratory, NCI-Frederick Cancer Research and Development Center, ABL-Basic Research Program
Frederick, Maryland 21702

Aspartic proteinases (APs) are widely dispersed in the plant kingdom, and are found in monocotyledonous and dicotyledonous species, as well as gymnosperms (Törmäkangas *et al.*, 1994, Kervinen *et al.*, 1995). The cDNA sequences are now available for plant APs from barley, rice, cardoon, and tomato. In addition, partial cDNA clones have been sequenced from cauliflower and *Arabidopsis*. In plants, the similarity of APs with other species is split into two regions separated by an internal sequence of approximately 100 residues, called the plant specific insert. Although this region carries no similarity to animal or microbial APs, the plant specific insert has recently been noticed to be highly homologous to saposins, sphingolipid-activating proteins from mammalian cells. This similarity includes six conserved cysteines in saposins, glycosylation site, hydrophobicity pattern, and an invariant Tyr-55 in saposins. It has been proposed that the saposin-like region could have a role in the vacuolar (lysosomal) targeting of plant APs (Guruprasad *et al.*, 1994, Ponting & Jones, 1995, Paris *et al.*, 1996). These and other recent findings on the physiological properties of barley and other plant APs are discussed.

Research sponsored by the National Cancer Institute, DHHS, under contract with ABL, and by a grant from the Academy of Finland.

REFERENCES

Guruprasad, K., Törmäkangas , K., Kervinen, J. & Blundell, T. 1994: Comparative modelling of barley-grain aspartic proteinase: a structural rationale for observed hydrolytic specificity. FEBS Lett. 352, 131–136

Kervinen, J., Törmäkangas, K., Runeberg-Roos, P., Guruprasad, K., Blundell, T. & Teeri, T. 1995: Structure and possible function of aspartic proteinase in barley and other plants. In: "Aspartic Proteinases: Structure, Function, Biology, and Biomedical Implications, ed. K. Takahashi, Plenum Press, New York, 1995, p. 241–254

Paris, N., Stanley, C.M., Jones, R.L. & Rogers, J.C. 1996: Plant cells contain two functionally distinct vacuolar compartments. Cell 85, 563–572

Ponting, C.P. & Russell, R.B. 1995: Swaposins: circular permutations within genes encoding saposin homologues. TIBS 20, 179–180

Törmäkangas, K., Kervinen, J., Östman, A. & Teeri, T. 1994: Tissue-specific localization of aspartic proteinase in developing and germinating barley grains. Planta: 195, 116–125

PLANT ASPARTIC PROTEINASES: RECOMBINANT FORMS OF AN ENZYME FROM *Cynara cardunculus*

P. C. White,[1] J. Hill,[1] P. E. Brodelius,[2] and J. Kay

[1]School of Molecular and Medical Biosciences, University of Wales, College of Cardiff, Cardiff, CF1 3US, Wales, United Kingdom
[2]Department of Plant Biochemistry, University of Lund, S-22100 Lund, Sweden

The aspartic proteinases that have been identified in plant species have some features in common and others that are different. All of the genes sequenced to date for plant aspartic proteinases predict the second active site motif to be –Asp215–Ser216–Gly217– (pepsin numbering) and contain an insert coding for 100 amino acid residues, always positioned at the same location in the C-terminal domain. This plant-specific insert or loop is not present in an intact form in mature aspartic proteinases purified from plant tissues so that these enzymes are commonly heterodimers. They are present in dissimilar locations within different plants, e.g. in the seeds of barley, but within the flower buds of the cardoon, *Cynara cardunculus*, where a number of isoenzymes with milk-clotting activity have been identified. In order to characterise such plant aspartic proteinases in detail, a recombinant approach is necessary since only relatively small amounts of each isoform can be isolated from specific plants and tissues.

Using flower buds from *Cynara cardunculus* as a source of mRNA, a full length clone called procyprosin11 (EMBL: X81984) was identified and sequenced. This contained a 1530 bp open reading frame coding for 510 amino acids, consisting of 68 residues in the signal peptide and propart together, followed by 442 residues of the mature proteinase (including 104 amino acids in the putative plant specific loop). Within the mature enzyme region, the predicted cyprosin11 sequence was 85% identical to that reported previously for a different clone (EMBL: X69193) but greater diversity was observed between the propart regions encoded by the two clones. The cDNA for procyprosin11 (minus the 54 bp encoding the putative signal peptide) was subcloned into the expression plasmid, pET-3a. Upon induction of the recombinant plasmid with IPTG, high levels of recombinant protein were readily detected to accumulate in the *E. coli*, at a molecular weight (55,000) consistent with that predicted for the full length protein (i.e. consisting of propart and mature enzyme containing the plant-specific loop). The protein was insoluble in the *E. coli* lysates and so a wide variety of solubilisation and refolding conditions were evaluated in attempts to generate correctly folded precursor that would be capable of autoactivation. No conversion to mature cyprosin11 was observed under any of the conditions employed.

In attempts to determine the refolding/autoactivation conditions necessary for generation of mature cyprosin11 in an active form, mutagenesis by overlapping PCR was used to engineer a number of procyprosin11 constructs e.g. introduction of a potential autoactivation site (–Phe–Ala–Ala–Phe*Leu–Arg–Asp–) in place of the authentic procyprosin11 sequence (–Phe–Gly–Gly–Ala–Leu–Arg*Asp–) immediately adjacent to the putative N-terminal residue (Asp) of mature cyprosin11; and deletion of the 104 residues of the plant-specific loop and replacement with a shorter loop consisting of the 5 residues from the corresponding region of Human Cathepsin D. The expression of these constructs in *E. coli* together with characterisation of the resultant recombinant proteins will be described.

CONTACTS

Abad-Zapatero, Cele
D-46Y, AP-10, L-07
Laboratory of Protein Crystallography
Abbott Laboratories
Abbott Park, Illinois 60064, USA
Tel: 847-937-0294
Fax: 847-937-2625
abad@mozart.pprd.abbott.com

Andreeva, Natalia
W. Engelhardt Institute of Molecular Biology
Vavilov Str. 32
117984 Moscow B-334, Russia
Tel: 7095-135-0237
Fax: 7095-135-1405
andreeva@imb.imb.ac.ru

Berry, Colin
Department of Biochemistry
University of Wales
P.O. Box 911, Museum Avenue
Cardiff, Wales CF1 3US, United Kingdom
Tel: 44-1222-874508
Fax: 44-1222-874116
Berry@Cardiff.AC.UK

Blundell, Tom
Department of Biochemistry
University of Cambridge
Tennis Court Road
Cambridge CB2 1QW, United Kingdom
Tel: 44-1223-333600
Fax: 44-1223-333637
t.blundell@mail.cryst.bbk.ac.uk

Brodelius, Peter
Department of Plant Biochemistry
Lund University
P.O. Box 117
S-22100 Lund, Sweden
Tel: 46-46-2228115; Fax: 46-46-2224116
peter.brodelius@plantbio.lu.se

Carrondo, Maria Armenia
ITQB, Apartado 127
2780 Oeiras, Portugal P-2780
Tel: 28351-1-4418407 X11
Fax: 28351-1-4433644

Chirgwin, John
Medicine/Endocrinology
The University of Texas
U.T. Health Sci. Ctr., 7703 Floyd Curl Drive
San Antonio, Texas, USA
Tel: 210-567-4900; Fax: 210-567-6693
chirgwin@uthscsa.edu

Craik, Charly
Dept. of Pharmaceutical Chemistry
University of California, San Francisco
San Francisco, CA 94143-0446, USA
Tel: 415-476-8146; Fax: 415-476-0688
craik@cgl.ucsf.edu

Danielson, U. Helena
Department of Biochemistry
Uppsala University
Box 576 BMC, Uppsala, Sweden
Tel: 46-18-174545; Fax: 46-18-550733
helena@bmc.uu.se

DiIanni Carroll, Carolyn
Department of Biology
Pharmacopeia
101 College Road East
Princeton, New Jersey 08540, USA
Tel: 609-452-3762
Fax: 609-655-4187
diianni@pharmacop.com

Dunn, Ben
Department of Biochemistry and Molecular Biology
University of Florida, College of Medicine
Health Science Center, P.O. Box 100245
Gainesville, FL 32610-0245, USA
Tel: 904-392-3362
Fax: 904-846-0412
yagi@nervm.nerdc.ufl.edu

Erickson, Ann
Department of Biochemistry and Biophysics
The University of North Carolina
CB #7260, Faculty Laboratory Office Building
Chapel Hill, North Carolina 27599-7260, USA
Tel: 919-966-4694
Fax: 919-966-2852
aerickso.biochem@mhs.unc.edu

Erickson, John
Structural Biochemistry Program
Frederick Biomedical Supercomputing Center
NCI-Frederick Cancer Research & Development Center
P.O. Box B
Frederick, Maryland 21702-1201
Tel: 301-846-1979
Fax: 301-846-6066
erickson@ncifcrf.gov

Faro, Carlos
Departamento de Bioquimica
Universidade de Coimbra
Apartado 3126
3000 Coimbra, Portugal
Tel: 351-39-480110
Fax: 351-39-480117
cfaro@imagem.ibili.uc.pt

Gómez, Javier
Department of Biology
The Johns Hopkins University
3400 N. Charles Street
106 Mudd Hall
Baltimore, Maryland 21218, USA
Tel: 410-516-8433; Fax: 410-516-6469
javier@biocal2.bio.jhu.edu

Harboe, Marianne
Chr. Hansen
Boege Alle 10-12
2970 Hoersholm, Denmark
Tel: 45-45-767676; Fax: 45-45-765455

Hill, Jeffrey
School of Molecular and Medical Biosciences
University of Wales, College of Cardiff
P.O. Box 911
Cardiff CF1 3US, Wales, United Kingdom
Tel: 44-1222-874000 Ext. 6800
Fax: 44-1222-874116
hillj@cardiff.ac.uk

Hube, Bernhard
Universitat Hamburg
Institut fur Allgemeine Botanik
AMP III, Ohnhorst Str 18
D-22609 Hamburg, Germany
Tel: 40-822-82-396; Fax: 40-822-82-513

James, Michael
Department of Biochemistry
University of Alberta
Edmonton, Alberta, Canada T6G 2H7
Tel: 403-492-4550; Fax: 403-492-0886
michael.james@ualberta.ca

Kashparov, Ilya
Laboratory of Chemistry of Proteolytic Enzymes
Shemyakin-Ovchinnikov Inst. of Bioorganic Chem.
Russian Academy of Sciences
Miklukho-Maklaya Street 16
Moscow 117871, Russia
Tel: 7-29-330-6374
Fax: 7-29-335-7103
ilko@enzyme.siobc.ras.ru

Kay, John
Department of Biochemistry
University of Wales
P.O. Box 903, Museum Avenue
Cardiff, Wales CF1 1ST, United Kingdom
Tel: 44-1222-874124; Fax: 44-1222-874116
kayj@cardiff.ac.uk

Korant, Bruce
DuPont Merck Company
Experimental Station
Bldg. 336, Room 31
Wilmington, Delaware 19880-0336, USA
Tel: 302-695-9493; Fax: 302-695-4920
korantbd@a1.lldmpc.umc.dupont.com

Loh, Peng
Section on Cellular Neurobiology
Laboratory of Developmental Neurobiology
National Inst. of Child Health and Human Development
Bldg. 49, Rm. 5A38, MSC 4480
Bethesda, Maryland 20892, USA
Tel: 301-496-3239; Fax: 301-496-9938
ypl@codon.nih.gov

Nakanishi, Hiroshi
Department of Pharmacology
Faculty of Dentistry
Kyushu University
Higashi-ku, Fukuoka 812, Japan
Tel: 81-92-641-1151 (4262)
Fax: 81-92-632-1226
nakandeg@mbox.nc.kyushu-u.ac.jp

Oda, Kohei
Faculty of Textile Science
Kyoto Institute of Technology
Matsugasaki, Sakyo-ku
Kyoto 606, Japan
Tel: 81-75-724-7763; Fax: 81-75-724-7710
bika@ipc.kit.ac.jp

Pazhanisamy, Sam
Vertex Pharmaceuticals Inc.
130 Waverly Street
Cambridge, Massachusetts 02139-4242, USA
Tel: 617-577-6484; Fax: 617-577-6437
pazhanisamy@vpharm.com

Phylip, Lowri
School of Molecular and Medical Biosciences
University of Wales, College of Cardiff
P.O. Box 911
Cardiff CF1 3US, Wales, United Kingdom
Tel: 44-1222-874000 Ext. 6800
Fax: 44-1222-874116
phylip@cardiff.ac.uk

Pichová, Iva
Department of Biochemistry
Inst. of Organic Chemistry & Biochemistry AS CR
Flemingovo nám. 2
166 10 Praha 6, Czech Republic
Tel: 422-3312250
Fax: 422-24310090
iva.pichova@uochb.cas.cz

Popov, Michael
Laboratory of Chemistry of Proteolytic Enzymes
Shemyakin-Ovchinnikov Inst. of Bioorganic Chem.
Russian Academy of Sciences
Miklukho-Maklaya Street 16
Moscow 117871, Russia
Tel: 7-29-335-4222
Fax: 7-29-335-7103
michael@enzyme.siobc.ras.ru

Quail, Wilson
Department of Chemistry
University of Saskatchewan
Saskatoon, Saskatchewan S7N 0W0, Canada
Tel: 306-966-4663
Fax: 306-966-4730
quail@sask.usask.ca

Reudelhuber, Timothy
Laboratory of Molecular Biochemistry of Hypertension
Clinical Research Institute of Montreal
110, avenue des Pins Quest
Montreal, Quebec H2W 1R7, Canada
Tel: 514-987-5716
Fax: 514-987-5717
reudelt@ircm.umontreal.ca

Ridley, Robert
Pharma Division
F. Hoffmann-La Roche Ltd.
Basel 4002, Switzerland
Tel: 41-61-688-2575; Fax: 41-61-688-2729
ROBERT_G.RIDLEY@roche.com

Roberts, R.M.
Department of Animal Sciences
University of Missouri
158 Animal Science Research Center
Columbia, Missouri 65211, USA
Tel: 573-882-0908
Fax: 573-882-6827
C570938@mizzou1.missouri.edu

Silva, Abelardo M.
National Cancer Institute
Building 322
Frederick, Maryland 21702, USA
Tel: 301-846-1977
Fax: 301-846-6066
abelardo@ncifcrf.gov

Sussman, Fredy
Research and Education Computing
Oklahoma University H.S.C.
800 N.E. 15th
Oklahoma, Oklahoma 73104, USA
Tel: 405-271-2202
Fax: 405-236-3923
fredy@rev.uokhsc.edu

Swanstrom, Ronald
Department of Biochemistry
University of North Carolina
Lineberger Comprehensive Cancer Center
Chapel Hill, North Carolina 27599-7295, USA
Tel: 919-966-5710
Fax: 919-966-3015
risunc@med.unc.edu

Takahashi, Kenji
Laboratory of Molecular Biochemistry
School of Life Science
Tokyo University of Pharmacy and Life Science
1432-1 Horinouchi, Hachioji
Tokyo 192-03, Japan
Tel: 81-426-7146; Fax: 81-426-7149
kenjitak@ls.toyaku.ac.jp

Tang, Jordan
Protein Studies Program
Oklahoma Medical Research Foundation
825 Northeast 13th Street
Oklahoma City, Oklahoma 73104, USA
Tel: 405-271-7291; Fax: 405-271-7249
jordan-tang@omrf.uokhsc.edu

Tanokura, Masaru
Laboratory of Structural Biology
Biotechnology Research Center

University of Tokyo
Yayoi, Tokyo 113, Japan
Tel: 81-3-3812-2111 Ext. 3074
Fax: 81-3-5689-7225
utanok@hongo.ecc.u-tokyo.ac.jp

Tarasova, Nadya I.
ABL-BRP, NCI-FCRDC
P.O. Box B, Building 538
Frederick, Maryland 21702, USA
Tel: 301-846-1216; Fax: 301-846-6231

Wormser, Uri
Department of Pharmacology
Faculty of Medicine
The Hebrew University
P.O. Box 12065, 91120 Jerusalem, Israel
Tel: 972-2-757515
Fax: 972-2-431094
wormser@yam-suff.cc.huji.ac.il

AUTHOR INDEX

SUBJECT INDEX